Other Titles in This Series

(Continued in the back of this publication)

Moonshine, the Monster,
and Related Topics

CONTEMPORARY MATHEMATICS

193

Moonshine, the Monster, and Related Topics

Joint Research Conference on
Moonshine, the Monster, and Related Topics
June 18–23, 1994
Mount Holyoke College,
South Hadley, Massachusetts

Chongying Dong
Geoffrey Mason
Editors

American Mathematical Society
Providence, Rhode Island

The AMS-IMS-SIAM Joint Summer Research Conference in the Mathematical Sciences on Moonshine, the Monster, and Related Topics was held at Mount Holyoke College, South Hadley, Massachusetts, June 18–23, 1994, with support from the National Science Foundation, Grant DMS-9221892-001 and the National Security Agency, Grant MDA904-94-H-2045.

1991 *Mathematics Subject Classification.* Primary 20–XX, 81–XX, 11F22; Secondary 20D08, 20H10, 20G45, 80R10, 81T40.

Library of Congress Cataloging-in-Publication Data
AMS-IMS-SIAM Summer Research Conference on Moonshine, the Monster, and Related Topics (1994: Mount Holyoke College)
    Moonshine, the monster, and related topics : AMS-IMS-SIAM Summer Research Conference on Moonshine, the Monster, and Related Topics, June 18–23, 1994, Mount Holyoke College, S. Hadley, MA / Chongying Dong, Geoffrey Mason, editors.
    p. cm. — (Contemporary mathematics, ISSN 0271-4132; v. 193)
    Includes bibliographical references.
    ISBN 0-8218-0385-9 (alk. paper)
    1. Quantum field theory—Congresses. 2. Vertex operator algebras— Congresses. 3. Mathematical physics—Congresses. I. Dong, Chongying, 1958– . II. Mason, Geoffrey, 1948– . III. Title. IV. Series: Contemporary mathematics (American Mathematical Society); v. 193.
QC174.45.A1A15   1994
530.1′43—dc20   95-40534
                                                           CIP

Contents

Preface

One of the great legacies of the classification of the finite simple groups is the existence of the Monster. It was the study of this group that first suggested that there might be interesting relations between finite groups and certain elliptic modular functions, and it was this possibility—fuelled by the Conway-Norton conjectures—that led to what one might call the first version of "Moonshine", that is, the study of class functions on groups with values in a ring of modular functions.

Work of Borcherds and Frenkel-Lepowsky-Meurman led to the notion of a vertex (operator) algebra, which was seen to be the same as the chiral algebras used by physicists in conformal field theory. Nowadays one considers the Monster as the group of automorphisms of a certain vertex operator algebra—the so-called Moonshine Module—and "Moonshine" may be construed as the representation theory of certain vertex operator algebras, namely so-called orbifolds.

The connections with physics have proven to be invaluable, and it seems likely that another branch of mathematics whose origins are eerily similar to those of moonshine—that is, elliptic cohomology—will turn out to be very relevant too.

Most of the talks at the Moonshine Conference were devoted to one or more of these subjects, as are the accompanying papers in this volume.

Way back in 1980 in the Proceedings of the Santa Cruz conference on finite groups, when Moonshine was just emerging, Andy Ogg wrote, "...(we) should rejoice at the emergence of a new subject...rich and deep, with all the theorems yet to be proved." His remarks are as true today as they were then.

Contemporary Mathematics
Volume **193**, 1996

HIGHER GENUS MOONSHINE

P. BÁNTAY

1. Introduction

Since their discovery, string theory [1] and the closely related Conformal Field Theory [2] have shed light on interesting new connections beween mathematics and physics. The vertex operator realization of affine Kac-Moody algebras [3] and the quantum field theoretic approach to knot and link invariants [4] are famous examples.

A most interesting example is the relation between Moonshine and the so-called orbifold CFT. It was Dixon, Ginsparg and Harvey [5] who first realized, that Norton's 'generalized Moonshine' postulates [6] - apart from the famous 'genus zero' postulate - are essentially equivalent to the existence of a suitable orbifold model, whose twist group is the Fischer-Griess Monster. That such a model does indeed exist follows from the work of Frenkel, Lepowsky and Meurmann [7] who constructed its untwisted Hilbert-space, the celebrated Moonshine module.

The above connection does not only allow for a reinterpretation of Moonshine in physics terms, but it also opens the way for interesting generalizations and extensions. An obvious one is wether Moonshine is a specific feature of the Monster, or there exists some analogous phenomena for other finite groups. This question has been studied for some time [8], and was answered in the affirmative by the construction by Dong and Mason of a 'moonshine module' for the Mathieu group M_{24} [9]. Physical intuition suggests that there should exist moonshine-like phenomena for a large number of finite groups - perhaps for all of them?

Another possible extension follows from the observation, that the connection discovered by Dixon, Ginsparg and Harvey relates the Thompson-McKay series entering the formulation of the Moonshine conjectures with physical quantities characterizing the relevant orbifold model defined over a complex torus, i.e. a Riemann surface of genus one. But a CFT may be defined over Riemann surfaces of arbitrary topology, and this allows one to define some kind of higher genus analogues of the ordinary Thompson-McKay series. The title 'higher genus Moonshine' refers to the study of these quantities, and my goal is to sketch the relevant ideas leading to their definition.

For a start, let me recall you the 'generalized' Moonshine postulates of Norton [6]: If G is the Fischer-Griess Monster, then for each pair $x, y \in G$ of commuting elements, there exists a Thompson-McKay series $\mathcal{Z}(x, y \,|\, \tau)$ meromorphic on $\mathcal{T}_1 = \{\tau \,|\, Im(\tau) > 0\}$, such that

(1) $\mathcal{Z}(x, y \,|\, \frac{a\tau+b}{c\tau+d}) = \mathcal{Z}(x^a y^b, x^c y^d \,|\, \tau)$ for $\left(\begin{smallmatrix} a & b \\ c & d \end{smallmatrix}\right) \in SL(2, \mathbb{Z})$.

1991 *Mathematics Subject Classification.* Primary 81R50, 16W30.
Partially supported by OTKA grant No. F-4010.

(2) $\mathcal{Z}(x^z, y^z \,|\, \tau) = \mathcal{Z}(x, y \,|\, \tau)$ for all $z \in G$.

(3) For any $x \in G$ and any rational number r, the coefficient of q^r in the expansion of $\mathcal{Z}(x, y \,|\, \tau)$ ($q = e^{2\pi i \tau}$) is, as a function of y, a character of $C_G(x)$.

(4) Unless $\mathcal{Z}(x, y \,|\, \tau)$ is a constant, its invariance group is a genus zero subgroup of $SL(2, \mathbb{R})$ commensurable with $SL(2, \mathbb{Z})$, whose Hauptmoduln is $\mathcal{Z}(x, y \,|\, \tau)$.

The interpretation of the last 'genus zero' condition in physics terms is still not clear, although there has been some progress in this direction [10]. It seems that it is related to some specific property of the Monster and/or the Moonshine module of FLM. Consequently, I shall drop this last postulate from the definition of a Thompson-McKay series. As for the remaining postulates, we'll see later that they are automatically satisfied by the twisted genus 1 character of a holomorphic orbifold model [5]. From this observation follows the natural expectation that the twisted higher genus characters form some kind of higher genus generalization of the above series, and as we shall see they do indeed have many analogous properties. To understand what these higher genus characters are, let's take a look at CFT and orbifold models.

2. Concepts from CFT

To define a quantum field theory, the first basic ingredient is a Riemannian manifold Σ, the 'space-time' of the theory. That Σ is Riemannian and not pseudo-Riemannian means that we are considering the Euclidian version of our field theory instead of the usual Minkowskian version.

The next ingredient we need is some fibre bundle ξ over Σ. Usually ξ is topologically trivial, but we shall need the more general definition later when studying orbifold models. The smooth sections ϕ of ξ are the fields of the theory.

Finally, we need an action functional $S[\phi]$, which is a smooth real-valued function on the space $\Gamma[\xi]$ of smooth sections of ξ. The physics of such a theory is encoded in functional integrals of the form

$$(1) \qquad\qquad Z[\chi] = \int_{\phi_{|\partial\Sigma} = \chi} \mathcal{D}\phi \exp\left(-S[\phi]\right),$$

where the argument χ of Z is a smooth section of the restriction of ξ to the boundary $\partial\Sigma$ of space-time - it prescribes the boundary conditions -, and one should integrate over those ϕ-s whose restriction to $\partial\Sigma$ equals χ. Of course, the above functional integral is generally ill-defined, and some work is needed to make it meaningful, but we shall bypass this problem as we only need the functional integrals to illustrate the basic notions of CFT.

If space-time has no boundary, then the functional integral - which is then called the partition function - depends only on the metric of Σ. If the partition function is invariant under conformal rescalings of the metric, i.e. diffeomorphisms which leave invariant the ratio of the matrix elements of the metric tensor, then the corresponding quantum field theory is called a CFT.

Conformal field theories are especially interesting when $\dim \Sigma = 2$. There are several reasons for this, but only one is relevant for us : in two dimension, the conformal equivalence classes of metrics are in one-to-one correspondence with the equivalence classes of complex structures that one can put on the manifold. In other words, a 2-dimensional CFT assigns a number to each inequivalent complex structure, i.e. to closed Riemann surfaces.

The classification of closed Riemann surfaces is well-known : the different topological equivalence classes are characterized by a non-negative integer - the genus g - related to the Euler-characteristic. The different complex structures on a surface of genus g form a complex space, the moduli space $\mathcal{M}_g$, which is the quotient $\mathcal{T}_g/\Gamma_g$ of a contractible complex manifold $\mathcal{T}_g$ - the Teichmüller-space of genus g surfaces - by the proper discontinuous action of a finitely presented group Γ_g - the mapping class group of genus g. Note that the action of Γ_g on $\mathcal{T}_g$ has fixed points corresponding to Riemann surfaces with non-trivial automorphisms, consequently the moduli space $\mathcal{M}_g$ is not a manifold.

If $g = 0$, i.e. we are on the sphere, there is just one complex structure by the Riemann Uniformization Theorem, that is Teichmüller-space has only one point. For $g = 1$ the situation is more interesting : the different complex structures may be characterized by one complex parameter τ lying in the upper half-plane, so $\mathcal{T}_1 = \{\tau \,|\, Im(\tau) > 0\}$, and the relevant mapping class group is the classical modular group $\Gamma_1 = SL(2, \mathbb{Z})$. For $g > 1$ Teichmüller-space is a Kähler manifold of complex dimension $3g - 3$, and the mapping class group Γ_g is generated by Dehn-twists around a suitable system of simple non-contractible curves (c.f. [11]). If one introduces complex coordinates $\tau_1, \dots, \tau_{3g-3}$ on $\mathcal{T}_g$, then the genus g partition function will be a function $Z_g(\tau_i, \bar{\tau}_i)$ invariant under the action of Γ_g.

There is a class of conformal theories of special interest, the so-called rational CFT, characterized by the fact that the Vertex Operator Algebra associated to the theory has only a finite number of irreducible modules. This property implies that at each genus the partition function is the sum of the modulus squared of a finite number of holomorphic functions of the Teichmüller coordinates :

$$(2) \qquad Z_g(\tau_i, \bar{\tau}_i) = \sum_{p=1}^{N_g} |\,\Psi_p^{(g)}(\tau_i)\,|^2$$

The holomorphic functions $\Psi_p^{(g)}(\tau_i)$ appearing in this decomposition are called the genus g characters of the theory. They span a finite dimensional Γ_g module, which is an important characteristic of the theory.

For example, at genus one there is just one Teichmüller coordinate $\tau \in \mathcal{T}_1$, and the mapping class group $SL(2, \mathbb{Z})$ is generated by the transformations $T : \tau \mapsto \tau + 1$ and $S : \tau \mapsto -1/\tau$. It may be shown that the action of T on the genus 1 characters is diagonal, i.e.

$$(3) \qquad \Psi_p^{(1)}(\tau + 1) = \lambda_p \Psi_p^{(1)}(\tau)$$

for some complex phases λ_p. As for the other generator S,

(4)
$$\Psi_p^{(1)}\left(-1/\tau\right) = \sum_{q=1}^{N_1} S_{pq}\Psi_q^{(1)}\left(\tau\right),$$

and one has $S_{pq} = S_{qp}$. The matrix elements λ_q and S_{qp} contains much information about the CFT, e.g. one has the famous formula of Verlinde [12] determining the number of genus g characters :

(5)
$$N_g = \sum_{p=1}^{N_1} S_{0p}^{2-2g},$$

where N_1 is the number of genus 1 characters - which equals the number of irreps of the VOA associated to the theory -, and 0 labels the trivial character.

There is an even more special class of CFT of interest, the so-called holomorphic (or self-dual) theories, for which

(6)
$$Z_g\left(\tau_i, \bar\tau_i\right) = \mid \Psi^{(g)}\left(\tau_i\right) \mid^2,$$

i.e. there is only one character for each g. Famous examples of such theories are the Moonshine module of FLM and the theories constructed from even unimodular lattices. They are characterized algebraically by the fact that the associated VOA has only one irrep.

3. Orbifold models

Suppose now that we have a 2 dimensional CFT. Let G be a finite group of symmetries, i.e. a group of transformations acting on the fibers of ξ whose induced action on the sections leaves the action functional invariant. In such a situation one can construct a new CFT, called a G-orbifold of the original one, by the procedure below [13].

To each homomorphism $\gamma \in Hom\left(\pi_1(\Sigma), G\right)$, there corresponds a principal G-bundle over Σ. Because the group G acts on the fibers of ξ, one can construct the associated bundle ξ_γ to the above principal bundle. This allows one to define the following functional integral :

(7)
$$Z[\gamma] = \int_{\phi \in \Gamma(\xi_\gamma)} \mathcal{D}\phi \exp\left(-S[\phi]\right),$$

where one integrates over the smooth sections of the associated bundle ξ_γ. This expression will depend - besides the complex structure of Σ - on the homomorphism γ, and it is called the twisted partition function. Note that homomorphisms whose images are conjugate in G correpond to the same principal bundle and thus determine the same associated bundle, in other words the twisted partiton function is invariant under conjugation by an element of G :

(8)
$$Z[\gamma^z] = Z[\gamma] \qquad \forall z \in G.$$

The partition function of the orbifold model is then given by

$$(9) \qquad Z_{orb} = \sum_{\gamma \in Hom(\pi_1(\Sigma), G)} Z[\gamma].$$

If the theory we started with was holomorphic, i.e. the untwisted partition function correponding to the trivial homomorphism γ_0 sending each element of $\pi_1(\Sigma)$ to the identity of G was the modulus squared of a holomorphic function of the Teichmüller coordinates, then the same will be true for the twisted partition function, that is

$$(10) \qquad Z_g[\gamma] = |\Psi^{(g)}[\gamma]|^2$$

for some holomorphic function $\Psi^{(g)}[\gamma]$ of the Teichmüller coordinates. To simplify the notation, here and in the sequel the dependence on the Teichmüller coordinates is supressed. The functions $\Psi^{(g)}[\gamma]$ are called the twisted characters of the theory.

The mapping class group Γ_g, besides acting on the Teichmüller coordinates, also acts via (outer) automorphisms on the fundamental group $\pi_1(\Sigma)$, and this induces an action on $Hom(\pi_1(\Sigma), G)$. It is obvious from the functional integral representation of the partition function that it should be equivariant under Γ_g.

Our primary aim is to understand the group theoretic properties of the twisted characters. To achieve this goal, it is convenient to introduce a presentation of the fundamental group $\pi_1(\Sigma)$. The standard presentation of $\pi_1(\Sigma)$ is in terms of $2g$ generators $a_1, b_1, \ldots, a_g, b_g$ for which the product of the commutators $[a_j, b_j] = a_j^{-1} b_j^{-1} a_j b_j$ is the identity :

$$(11) \qquad \pi_1(\Sigma) = \langle a_1, b_1, \ldots, a_g, b_g : \prod_{j=1}^{g} [a_j, b_j] = 1 \rangle.$$

Using this presentation, an element of $Hom(\pi_1(\Sigma), G)$ is specified by a $2g$-tuple of elements of G satisfying the relation $\prod_{j=1}^{g} [a_j, b_j] = 1$. In the obvious notation, the twisted genus g characters satisfy

(1) $\Psi^{(g)} \begin{pmatrix} a_1 & \cdots & a_g \\ b_1 & \cdots & b_g \end{pmatrix} = 0 \qquad$ if $\quad \prod_{j=1}^{g} [a_j, b_j] \neq 1.$

(2) $\Psi^{(g)} \begin{pmatrix} a_1^z & \cdots & a_g^z \\ b_1^z & \cdots & b_g^z \end{pmatrix} = \omega_z \begin{pmatrix} a_1 & \cdots & a_g \\ b_1 & \cdots & b_g \end{pmatrix} \Psi^{(g)} \begin{pmatrix} a_1 & \cdots & a_g \\ b_1 & \cdots & b_g \end{pmatrix} \qquad$ for all $z \in G.$

(3) $\Psi^{(g)} \begin{pmatrix} a_1 & \cdots & a_g \\ b_1 & \cdots & b_g \end{pmatrix} \qquad$ is equivariant under Γ_g up to some complex phases.

Note the appearance in (2) and (3) of the complex phases, which are independent of the Teichmüller coordinates. At first sight it may seem that these phases are quite arbitrary, but it turns out that there are only a finite number of possibilities, classified by the third cohomology group $H^3(G, \mathbb{T})$ of G with coefficients in the complex unit circle $\mathbb{T} = \{z \in \mathbb{C} \mid z\bar{z} = 1\}$ [14]. The corresponding cohomology class is an important characteristic of the orbifold model, and it is determined by the action of the twist group G on the original theory. If ϕ is some representative 3-cocycle of the cohomology class, then

(12) $\omega_z \begin{pmatrix} a_1 \dots a_g \\ b_1 \dots b_g \end{pmatrix} = \prod_{j=1}^{g} \dfrac{\eta_z(A_{j-1}, [a_j, b_j])\eta_z(a_j \subset, a_j^{b_j})\eta_z(a_j, b_j)}{\eta_z(a_j \subset, a_j)\eta_z(b_j, a_j^{b_j})},$

where we have used the shorthands $A_j = \prod_{k \leq j} [a_k, b_k]$ and

$$\eta_z(x, y) = \frac{\phi(x, z, y^z)}{\phi(x, y, z)\phi(z, x^z, y^z)}.$$

The phases appearing in the transformation formula under mapping class group transformations may also be computed explicitly (c.f. [15]).

Let's take a closer look at the first non-trivial case, that of $g = 1$. The twisted characters satisfy [16]:

(1) $\Psi^{(1)}(x, y \,|\, \tau) = 0$ if $xy \neq yx$.

(2) $\Psi^{(1)}(x^z, y^z \,|\, \tau) = \frac{\eta_z(y, x)}{\eta_z(x, y)} \Psi^{(1)}(x, y \,|\, \tau)$ for all $z \in G$.

(3) $\Psi^{(1)}\left(x^a y^b, x^c y^d \,|\, \tau\right) = \alpha(x, y)\Psi^{(1)}\left(x, y \,\middle|\, \frac{a\tau+b}{c\tau+d}\right)$ for $\begin{pmatrix} a & b \\ c & d \end{pmatrix} \in SL(2, \mathbb{Z})$,

where the phases $\alpha(x, y)$ are complicated expressions involving the 3-cocycle ϕ. For example, for the generators S and T of $SL(2, \mathbb{Z})$ one has

(13) $\Psi^{(1)}\left(y, x^{-1} \,|\, \tau\right) = \dfrac{\phi(x, y, x^{-1})}{\phi(y, x, x^{-1})\phi(x, x^{-1}, y)} \Psi^{(1)}(x, y \,|\, -1/\tau)$

and

(14) $\Psi^{(1)}(x, xy \,|\, \tau) = \phi(x, y, x)\Psi^{(1)}(x, y \,|\, \tau + 1).$

We see from the above that these twisted genus 1 characters fall short of being Thompson-McKay series. Namely, if the cocycle ϕ is trivial, then the phase factors disappear, and we recover the postulates (1) and (2) of Norton. As for the remaining postulate (3), its truth follows from the representation theoretic interpretation of the twisted genus 1 characters, which I'll explain briefly.

The point is, that besides their representation in terms of functional integrals, the genus 1 characters may be interpreted as the trace of a suitable linear operator acting on the irreps of the VOA associated to the CFT (this is why their number equals the number of irreps of the VOA). The operator in question is of the form $\exp(2\pi i \tau(L_0 - c/24))$, where τ is the usual Teichmüller coordinate, L_0 is a distinguished element of the VOA, and c - the central charge - is a rational number, an important characteristic of the theory. Because the eigenvalues of L_0 are integrally spaced, the characters are just a kind of graded trace in the variable $q = e^{2\pi i \tau}$.

The related expression for the twisted characters is a little bit more involved. It turns out that the Hilbert-space of an orbifold model affords a representation of the so-called double of the group G [17], which is a finite-dimensional semisimple algebra. This algebra is generated by operators $P(x)$ and $Q(x)$ for $x \in G$, subject to the relations

(1) $P(x)P(y) = \delta_{x,y}P(x),$
(2) $Q(x)Q(y) = Q(xy),$

(3) $Q^{-1}(x)P(y)Q(x) = P(y^x)$,

(4) $\sum_{x \in G} P(x) = Q(1)$.

With the aid of the above operators, the twisted genus 1 characters may be written as

$$\Psi^{(1)}(x, y \mid \tau) = Tr\left(P(x)Q(y)q^{(L_0 - c/24)}\right),$$

and the representation theoretic properties of the double of G imply that the twisted characters do indeed satisfy the remaining Norton postulate, so they are indeed Thompson-McKay series for G.

Unfortunately, such simple expressions for the characters are not available for $g > 1$. It is still possible to write the higher genus characters as traces of suitable operators, but these are complicated expressions which depend on the VOA and whose representation theoretic meaning is obscure. So, although the twisted higher genus characters have a lot of similar features with the genus 1 characters, the lack of a representation theoretic interpretation is a severe limitation. Nevertheless, I think that these quantities deserve further study.

4. Conclusions

We have seen, by exploiting the close relationship between Moonshine and orbifold models, namely that Thompson-McKay series for a group G are nothing but the twisted genus 1 characters of a holomorphic G-orbifold, that it is possible to define higher genus analogues of the Thompson-McKay series, the twisted higher genus characters of the orbifold model. These quantities posses properties analogous to their genus 1 counterparts, e.g. their holomorphicity in the Teichmüller coordinates and their behavior under mapping class group transformations. The basic open question is to give them a representation theoretic interpretation, which may have also some interest for pure group theory. Another interesting open problem is to find the generalization of the 'genus zero' postulate, if there exists any. While it is not yet clear what is the significance of these higher genus phenomena, one may hope that they could shed new light on Moonshine.

REFERENCES

1. M. B. Green, J. H. Schwarz and E. Witten, Superstring Theory, Cambridge monographs on mathematical physics, Cambridge University Press, 1987.
2. A. A. Belavin, A. M. Polyakov and A. B. Zamolodchikov, Infinite conformal symmetry in two-dimensional quantum field theory, Nuclear Physics **B241** (1984), 333.
3. I. B. Frenkel and V. G. Kac, Basic representations of affine Lie algebras and dual resonance models, Invent. Math., 62 (1980), 62; G. Segal, Unitary representations of some infinite dimensional groups, Commun. Math. Phys., 81 (1981), 301.
4. E. Witten, Quantum field theory and the Jones polynomial, Commun. Math. Phys., 121 (1989), 351.
5. L. Dixon, P. Ginsparg and J. Harvey, Beauty and the Beast : superconformal symmetry in a Monster module, Commun. Math. Phys., 119 (1988), 221.
6. S. Norton, Generalized Moonshine (appendix to G. Mason's paper), Proc. Symp. Pure Math., 47 (1987), 208.
7. I. Frenkel, J. Lepowsky and A. Meurman, Vertex operator algebras and the Monster, Pure and Applied Math., Vol. 134, Academic Press, New York, 1988.
8. G. Mason, Finite groups and modular functions, Proc. Symp. Pure Math., 47 (1987), 181.
9. C. Dong and G. Mason, An orbifold theory of genus zero associated to the sporadic group M_{24}, Commun. Math. Phys., 164 (1994), 87.

10. M. P. Tuite, Monstrous Moonshine and the uniqueness of the Moonshine module, preprint DIAS-STP-92-29.
11. J. S. Birman, Mapping class groups of surfaces, Contemp. Math., Vol. 78 (1988), 13.
12. E. Verlinde, Modular invariance and fusion rules, Nuclear Physics **B300** (1988), 360.
13. L. Dixon, J. Harvey, C. Vafa and E. Witten, Strings on orbifolds, Nuclear Physics **B274** (1986), 286.
14. R. Dijkgraaf and E. Witten, Topological field theories and group cohomology, Commun. Math. Phys., 129 (1990), 393.
15. P. Bantay, Algebraic aspects of orbifold models, Int. J. of Modern Physics **A9** (1994), 1443.
16. R. Dijkgraaf, C. Vafa, E. Verlinde and H. Verlinde, The operator algebra of orbifold models, Commun. Math. Phys., 123 (1989), 485.
17. P. Bantay, Orbifolds, Hopf algebras and the Moonshine, Lett. Math. Phys., 22 (1991), 187.

INSTITUTE FOR THEORETICAL PHYSICS, ROLLAND EÖTVÖS UNIVERSITY, BUDAPEST

E-mail: bantay@ludens.elte.hu

Contemporary Mathematics
Volume **193**, 1996

Superstring Twisted Conformal Field Theory

L. DOLAN

ABSTRACT. We discuss four-dimensional superstring theory and the vertex
operators corresponding to ground states in the twisted sector. In particular
we show how such states may carry non-trivial charge under a gauge group
as they do in the twisted bosonic conformal field theory construction of
twisted affine E_8 and the Monster group.

1. Introduction

Many interesting applications of conformal field theory involve twisted fields.
In superstring models, the entire set of space-time fermionic fields is an example
of twisted conformal fields, and an analysis of these fermion emission operators
gives information about the symmetries described by the theory in analogy with
that provided earlier for bosonic 'spin' fields.

In the introduction, we shall briefly review the construction[**11,12,6-8**] of the
Z_2-twisted bosonic conformal field theory associated with the reflection twist
applied to a suitable d-dimensional momentum lattice, Λ, in the sense of establishing explicit expressions for a complete system of mutually local vertex
operators $V(\psi, z)$ or conformal fields which are in one-to-one correspondence
with a basis of states ψ for the theory. Our approach applies whenever both
Λ and $\sqrt{2}\Lambda^*$ are even and d is a multiple of 8. Of particular interest are the
cases of the theory associated with the Leech lattice (for which $d = 24$) [**12**], and
with the root lattice of E_8 (for which $d = 8$)[**11**]. The former theory provides
the natural module for the Monster group[**16,5**]. In sections 2-5 analogous superstring twisted fields are described, with an emphasis on the covariant BRST
ghost formalism. An analysis of the symmetry carried by the ground states is
given.

In the bosonic, meromorphic, chiral Z_2-twisted conformal field theory, we
note two special states, the vacuum $|0\rangle$ and the "conformal state" ψ_L associated

1991 *Mathematics Subject Classification.* Primary 81R10, 20C35; Secondary 83E30, 20C34.
The author is supported in part by DOE Grant #DE-FG 05-85ER40219/Task A.

9

with the vertex operator $V(\psi_L, z) = L(z) = \sum_n L_n z^{-n-2}$, the Virasoro current. Construction of the conformal field theory corresponds to proving 1) a defining property:

$$V(\psi, z)|0\rangle = e^{zL_{-1}}\psi \tag{1.1}$$

and 2) locality:

$$V(\psi, z)V(\phi, \zeta) = V(\phi, \zeta)V(\psi, z) \tag{1.2}$$

where the equality in (1.2) is in the sense of analytic continuation for the left side defined for $|z| > |\zeta|$ and the right side defined for $|\zeta| > |z|$. From (1.1,2) it then follows that we have
1) s-t duality:

$$V(\psi, z)V(\phi, \zeta) = V(V(\psi, z - \zeta)\phi, \zeta) \tag{1.3}$$

since

$$\begin{aligned}
V(\psi, z)V(\phi, \zeta)|0\rangle &= V(\psi, z)e^{\zeta L_{-1}}\phi \\
&= e^{\zeta L_{-1}}V(\psi, z - \zeta)\phi \\
&= V(V(\psi, z - \zeta)\phi, \zeta)|0\rangle ;
\end{aligned} \tag{1.4}$$

2) the operator product expansion:

$$V(\psi, z)V(\phi, \zeta) = \sum_{n \geq 0}(z - \zeta)^{n - h_\psi - h_\phi}V(\chi_n, \zeta) \tag{1.5}$$

where $\chi_n = V_{-n+h_\psi}(\psi)\phi$ and the moments of vertex operators are defined as $V(\psi, z) = \sum V_n(\psi)z^{-n-h_\psi}$; and
3) all weight one states ($L_0\psi^a = \psi^a$) in a conformal field theory close to form an affine algebra:

$$V(\psi^a, z) = T^a(z) = \sum_n T_n^a z^{-n-1} \tag{1.6}$$

where we let $\langle\psi^a|\psi^b\rangle = k\delta^{ab}$; $T_0^a\psi^b = if_{abc}\psi^c$; so that

$$[T_n^a, T_m^b] = if_{abc}T_{n+m}^c + k\delta^{ab}n\delta_{n,-m} . \tag{1.7}$$

T_0^a are elements of the finite-dimensional Lie algebra g which generates a continuous symmetry group of the automorphisms of the conformal field theory. In superstring theory, this symmetry is the gauge group.

The twisted conformal field theory $\tilde{\mathcal{H}}(\Lambda)$ associated with a lattice Λ of dimension d is defined for a Z_2 reflection twist by keeping the $\theta = 1$ subset of the states created by integrally-moded bosonic operators a_m^j, $1 \leq j \leq d$, $m \in Z$ from momentum states $|\lambda\rangle, \lambda \in \Lambda$. Here $\theta|\lambda\rangle = |-\lambda\rangle$ and $\theta a_m^j \theta^{-1} = -a_m^j$. To this we add in the $\theta = 1$ subspace of the space $\mathcal{H}_T(\Lambda)$ generated from an irreducible representation space $\mathcal{X}(\Lambda)$ for the gamma matrix algebra $\{\gamma_\lambda : \lambda \in \Lambda\}$ associated with Λ, by half-integrally moded oscillators c_r^j, $1 \leq j \leq d$, $r \in Z + \frac{1}{2}$. In this case, the involution θ is defined by $\theta c_r^j \theta^{-1} = -c_r^j$. The oscillators satisfy the commutation relations $[a_m^i, a_n^j] = m\delta_{m,-n}\delta^{ij}$, $[c_r^i, c_s^j] = r\delta_{r,-s}\delta^{ij}$, and $[a_m^i, c_r^i] = 0$.

In the twisted CFT, the untwisted and twisted sectors of $\tilde{\mathcal{H}}(\Lambda)$ are the subspaces $\mathcal{H}^+(\Lambda)$ and $\mathcal{H}_T^+(\Lambda)$ on which $\theta = 1$. If $\psi \in \mathcal{H}^+(\Lambda)$, $\mathcal{V}(\psi, z)$ maps $\mathcal{H}^+(\Lambda) \to \mathcal{H}^+(\Lambda)$ and $\mathcal{H}_T^+(\Lambda) \to \mathcal{H}_T^+(\Lambda)$ whereas $\mathcal{V}(\chi, z)$ maps $\mathcal{H}^+(\Lambda) \to \mathcal{H}_T^+(\Lambda)$ and $\mathcal{H}_T^+(\Lambda) \to \mathcal{H}^+(\Lambda)$ if $\chi \in \mathcal{H}_T^+(\Lambda)$. Thus we can write these vertex operators in matrix form

$$\mathcal{V}(\psi, z) = \begin{pmatrix} V(\psi, z) & 0 \\ 0 & V_T(\psi, z) \end{pmatrix} ; \qquad \mathcal{V}(\chi, z) = \begin{pmatrix} 0 & \overline{W}(\chi, z) \\ W(\chi, z) & 0 \end{pmatrix} . \quad (1.8)$$

In this notation, the vertex operators of the twisted CFT $\tilde{\mathcal{H}}(\Lambda)$ are given by, for the untwisted states: $\psi = \left(\prod_{a=1}^M \alpha_{-m_a}^{j_a} \right) |\lambda\rangle$,

$$V(\psi, z) = \sum_{\lambda' \in \Lambda} \langle \lambda' | : e^{F(-z)} : |\psi\rangle \sigma_{\lambda'}$$

$$= : \left(\prod_{a=1}^M \frac{i}{(m_a - 1)!} \frac{d^{m_a} X^{j_a}(z)}{dz^{m_a}} \right) \exp\{i\lambda \cdot X(z)\} \sigma_\lambda : \quad (1.9)$$

and

$$V_T(\psi, z) = V_T^0(e^{\Delta(z)}\psi, z)$$

$$= \sum_{\lambda' \in \Lambda} \gamma_{\lambda'} \langle \lambda' | : e^{B(-z)} : e^{A(-z)} |\psi\rangle , \quad (1.10)$$

where

$$V_T^0(\psi, z) = \sum_{\lambda' \in \Lambda} (4z)^{-\frac{1}{2}\lambda'^2} \gamma_{\lambda'} \langle \lambda' | : e^{B(-z)} : |\psi\rangle$$

$$= : \left(\prod_{a=1}^M \frac{i}{(m_a - 1)!} \frac{d^{m_a} R^{j_a}(z)}{dz^{m_a}} \right) \exp\{i\lambda \cdot R(z)\} :$$

$$\cdot (4z)^{-\frac{1}{2}\lambda^2} \gamma_\lambda \quad (1.11)$$

and

$$X^j(z) = q^j - ip^j \log z + i \sum_{n \neq 0} \frac{a_n^j}{n} z^{-n} , \quad (1.12)$$

$$R(z) = i \sum_{r=-\infty}^{\infty} \frac{c_r}{r} z^{-r} ; \quad (1.13)$$

and for the twisted states: $\chi = \left(\prod_{a=1}^M c_{-m_a}^{j_a} \right) \mathcal{X}$, the analogue of the fermion emission operator is

$$W(\chi, z) = e^{z L_{-1}^c} \tilde{W}(\psi, z) , \quad (1.14)$$

where

$$\tilde{W}(\chi, z) = \sum_{\lambda \in \Lambda} \gamma_\lambda \langle \lambda | : e^{B(z)} : e^{A(z)} |\chi\rangle , \quad (1.15)$$

and

$$\overline{W}(\chi, z) = z^{-2h_\chi} W(e^{z^* L_1^c} \overline{\chi}, 1/z^*)^\dagger . \quad (1.16)$$

In the above expressions we define

$$\Delta(z) = \frac{1}{2} \sum_{\substack{m,n \geq 0 \\ (m,n) \neq (0,0)}} \begin{pmatrix} -\frac{1}{2} \\ m \end{pmatrix} \begin{pmatrix} -\frac{1}{2} \\ n \end{pmatrix} \frac{z^{-m-n}}{m+n} a_m \cdot a_n . \qquad (1.17)$$

Expressions for A(z), B(z), and F(z) are also written as bilinears in oscillators and are given in Ref.[6], together with the cocyle operators σ_λ and γ_λ. Note that the special state ψ_L is given in these CFT's by $\frac{1}{2} a_{-1} \cdot a_{-1} |0\rangle$ and that its vertex operator is $L_n = \frac{1}{2} \sum_{m=-\infty}^{\infty} : a_m \cdot a_{n-m} :$ and $L_n^c = \frac{1}{2} \sum_{m=-\infty}^{\infty} : c_m \cdot c_{n-m} : + \frac{d}{16}$ represented on the untwisted and twisted sectors respectively.

The space of states for the Z_2-twisted fermionic theory, $\tilde{\mathcal{H}}$, is obtained by starting with the states of the untwisted Neveu-Schwarz theory, $\mathcal{H}$, adding in a twisted Ramond sector, $\mathcal{H}_T$, and keeping only the subspace of each defined by $\theta = 1$, with $\theta^2 = 1$. The states of the untwisted theory are generated by the action of d infinite sets of half-integrally moded oscillators, b_s^j, $1 \leq j \leq d$, on the vacuum state, Ψ_0. The twisted sector is obtained from the action of d infinite sets of integrally moded oscillators, d_n^j, on the twisted ground states which form a $2^{d/2}$ irreducible representation, $\mathcal{X}$, of the gamma matrix Clifford algebra, $\{\gamma^j\}$. The involution θ is defined on the untwisted space $\mathcal{H}$ by $\theta|0\rangle = (-1)^{d/8}|0\rangle$, $\theta b_s^i \theta^{-1} = -b_s^i$, and on the twisted space, $\mathcal{H}_T$, by $\theta|0\rangle_R^{\pm} = \pm|0\rangle_R$, $\theta d_n^i \theta^{-1} = -d_n^i$, where $\mathcal{X} = |0\rangle_R^+ + |0\rangle_R^-$ and we are assuming d is a multiple of 8, (which is necessary for the spectrum of L_0^d to contain half-integral values).

This theory consists of fermionic and bosonic fields. As in the bosonic case, the conformal field theories discussed here are defined on the complex plane, or rather the Riemann sphere, and are *chiral, i.e.* holomorphic. In this case, the intertwining relation (1.2) is generally defined by

$$\mathcal{V}(\psi, z)\mathcal{V}(\phi, \zeta) = \epsilon_{\psi\phi}\mathcal{V}(\phi, \zeta)\mathcal{V}(\psi, z) \qquad (1.18)$$

in the sense of analytic continuation, where $\epsilon_{\psi\phi} = 1$ if either of the states ψ or ϕ are bosons, and $\epsilon_{\psi\phi} = -1$ if both of them are fermions. We will construct (in the F_1- picture, i.e. the $q = -1$ superconformal ghost picture) the vertex operators $\mathcal{V}(\psi, z)$; these are the conformal fields which are in one-to-one correspondence with a basis of states for the theory:

$$\mathcal{V}(\psi, z)|0\rangle = e^{zL_{-1}}\psi . \qquad (1.19)$$

In the language of superconformal field theory, these vertex operators are the lower components of the superfields. Here $|0\rangle \equiv \Psi_0$ is the vacuum and L_{-1} one of the moments of the special vertex operator $V(\psi_L, z) = \sum_n L_n z^{-n-2}$, which satisfy the Virasoro algebra: $[L_m, L_n] = (m-n)L_{m+n} + \frac{d}{24}m(m^2-1)\delta_{m,-n}$, where m, n run over the integers, $L_n^\dagger = L_{-n}$, and $L_n|0\rangle = 0$ for $n \geq -1$.

The oscillators satisfy the anti-commutation relations $\{b_r^i, b_s^j\} = \delta^{ij}\delta_{r,-s}$, $\{d_m^i, d_n^j\} = \delta^{ij}\delta_{m,-n}$, and $\{b_r^i, d_n^j\} = 0$, where $b_s^{i\dagger} = b_{-s}^i$, $b_s^j|0\rangle = 0$, $s > 0$, $d_n^{j\dagger} = d_{-n}^j$, $d_n^j|0\rangle_R = 0$, $n > 0$. In these theories, the special state ψ_L is given by

$\frac{1}{2}b_{-\frac{3}{2}} \cdot b_{-\frac{1}{2}}|0\rangle$ and that its vertex operator is defined by $L_n = \frac{1}{2}\sum_{s=-\infty}^{\infty}(\frac{1}{2}n - s):$ $b_s \cdot b_{n-s}:$ in the untwisted sector, and by $L_n^d = \frac{1}{2}\sum_{m=-\infty}^{\infty}(\frac{1}{2}n - m):d_m \cdot d_{n-m}:$ $+\frac{d}{16}\delta_{n0}$ in the twisted sector.

The states in the Neveu-Schwarz sector are given by $\psi = \left(\prod_{a=1}^{M}b_{-s_a}^{j_a}\right)|0\rangle$, where each s_a is a positive half-odd integer, and the product is understood to be written down in a definite order, e.g. left to right, in order to avoid a sign ambiguity,, and each oscillator occurs at most once. The vertex operators for these states with $(m_a = s_a - \frac{1}{2})$ are given by

$$V(\psi, z) = :\left(\prod_{a=1}^{M}\frac{1}{m_a!}\frac{d^{m_a}b^{j_a}(z)}{dz^{m_a}}\right):$$
$$= \langle 0'|:e^{F(-z)}:|\psi\rangle, \qquad (1.20)$$

where we have introduced the Neveu-Schwarz fermion conformal fields $b^j(z) = \sum_{s=-\infty}^{\infty}b_s^j z^{-s-\frac{1}{2}}$ and

$$V_T(\psi, z) = V_T^0(e^{\Delta(z)}\psi, z))$$
$$= \langle 0|:e^{B(-z)}:e^{A(-z)}|\psi\rangle, \qquad (1.21)$$

where

$$V_T^0(\psi, z) =: \left(\prod_{a=1}^{M}\frac{1}{(m_a)!}\frac{d^{m_a}d^{j_a}(z)}{dz^{m_a}}\right):$$
$$= \langle 0|:e^{B(-z)}:|\psi\rangle, \qquad (1.22)$$

with the Ramond fermion fields defined as

$$d^j(z) = \sum_{n=-\infty}^{\infty}d_n^j z^{-n-\frac{1}{2}}. \qquad (1.23)$$

For the twisted states $\chi = \left(\prod_{a=1}^{M}d_{-m_a}^{j_a}\right)|0\rangle_R^{\pm}$, the fermion emission operator is

$$W(\chi, z) = e^{zL_{-1}^d}\tilde{W}(\psi, z), \qquad (1.24a)$$

where

$$\tilde{W}(\chi, z) = \langle 0|:e^{B(z)}:e^{A(z)}|\chi\rangle. \qquad (1.24b)$$

In the above expressions, define $A(z) = \Delta(-z)$ where

$$\Delta(z) = \frac{1}{4}\sum_{r,s>0}\binom{-\frac{1}{2}}{r-\frac{1}{2}}\binom{-\frac{1}{2}}{s-\frac{1}{2}}\frac{r-s}{r+s}z^{-r-s}b_r \cdot b_s. \qquad (1.25)$$

Similar expressions for B(z) and F(z) are also written as bilinears in oscillators. In general these "lower component" vertex operators are not meromorphic, for *eg.* $V_T(b^j_{-\frac{1}{2}}|0\rangle, z) = d^j(z) = \sum_n d^j_n z^{-n-\frac{1}{2}}$. Therefore although the intertwining relation is satisfied: $V_T(b^j_{-\frac{1}{2}}|0\rangle, z)W(|0\rangle^+_R, \zeta) = W(|0\rangle^+_R, \zeta)V(b^j_{-\frac{1}{2}}|0\rangle, z)$, the operator product expansion $\tilde{W}(|0\rangle^+_R, \zeta)V(b^j_{-\frac{1}{2}}|0\rangle, z)$ is double valued.

2. Ramond states

We consider the spin field vertex operator described in (1.24) directly in four dimensions[10].

$$V_{-\frac{3}{2}}(k, z) = v^{\dot{\alpha}}(k)S_{\dot{\alpha}}(z)e^{ik\cdot X(z)}\Sigma(z)e^{-\frac{3}{2}\phi(z)} \qquad (2.1)$$

A suitable choice of supercurrent is given by:

$$F(z) = a_\mu(z)h^\mu(z) + \bar{F}(z) \qquad (2.2)$$

where $0 \le \mu \le 3$ and $\bar{F}(z)$ corresponds to internal degrees of freedom. The vertex operator for the Ramond states in the canonical $q = -\frac{1}{2}$ superconformal ghost picture is given for $k\cdot\frac{1}{\sqrt{2}}\gamma v \sim u$ by

$$\begin{aligned}
V_{-\frac{1}{2}}(k, \zeta) &= \lim_{z\to\zeta} e^{\phi(z)}F(z)V_{-\frac{3}{2}}(k, \zeta) \\
&= [u^\alpha(k)S_\alpha(\zeta)e^{ik\cdot X(\zeta)}\Sigma(\zeta) \\
&\quad + \lim_{z\to\zeta}(z-\zeta)^{\frac{3}{2}}\bar{F}(z)v^{\dot{\alpha}}(k)S_{\dot{\alpha}}(\zeta)e^{ik\cdot X(\zeta)}\Sigma(\zeta)]e^{-\frac{1}{2}\phi(\zeta)}. \quad (2.3)
\end{aligned}$$

BRST invariance of a vertex operator requires its commutator with the BRST charge Q to be a total divergence; for a vertex operator in the $q = -\frac{3}{2}$ superconformal ghost picture such as (2.1), this invariance is assured whenever its operator product with the supercurrent has at most a $(z-\zeta)^{-\frac{3}{2}}$ singularity. We see this as follows:

The ghost superfields[13,14,17] are $B(z) = \beta(z) + \theta b(z)$ and $C(z) = c(z) + \theta\gamma(z)$ with conformal spin $h_\beta = \frac{3}{2}$, $h_c = -1$. Then $h_b = 2$ and $h_\gamma = -\frac{1}{2}$. The modings on the Ramond sector and the commutation relations are $\{b_n, c_m\} = \delta_{n,-m}$, $[\beta_n, \gamma_m] = \delta_{n,-m}$. Normal ordering is defined by putting the annihilation operators b_n for $n \ge -1$, c_n for $n \ge 2$ to the right of the creation operators b_n for $n \le -2$, c_n for $n \le 1$, then

$$b(z)c(\zeta) = {}^{\times}_{\times}b(z)c(\zeta){}^{\times}_{\times} + \frac{1}{z-\zeta} \quad ; \quad c(z)b(\zeta) = {}^{\times}_{\times}c(z)b(\zeta){}^{\times}_{\times} + \frac{1}{z-\zeta}. \quad (2.4)$$

This is a natural definition for normal ordering as the "vacuum" expectation value of this normal ordered product including its finite part is zero. For the superconformal ghosts, normal ordering is defined analogously so that

$$\beta(z)\gamma(\zeta) = {}^{\times}_{\times}\beta(z)\gamma(\zeta){}^{\times}_{\times} - \frac{1}{z-\zeta} \quad ; \quad \gamma(z)\beta(\zeta) = {}^{\times}_{\times}\gamma(z)\beta(\zeta){}^{\times}_{\times} + \frac{1}{z-\zeta}. \quad (2.5)$$

In order to make contact with the field $\phi(z)$ appearing in (2.1), we can then "bosonize" this system, i.e. write it as a theory where operators are associated with vectors q in the weight lattice of some algebra. For the bosonic β, γ conformal field theory, we define the boson fields

$$\phi(z)\phi(\zeta) =: \phi(z)\phi(\zeta) : -\ln(z - \zeta) \tag{2.6}$$

so that

$$: e^{\phi(z)} :: e^{\phi(\zeta)} := : e^{\phi(z)} e^{\phi(\zeta)} : (z - \zeta)^{-1}. \tag{2.7}$$

Because $: e^{\phi(z)} :$ is a fermion field and $\beta(z), \gamma(z)$ are bosons, an additional bosonic field $\chi(z)$ is introduced and

$$\gamma(z) =: e^{\phi(z)} :: e^{-\chi(z)} : \quad ; \qquad J(z) = -{}^{\times}_{\times}\beta\gamma{}^{\times}_{\times} = \partial\phi$$
$$\beta(z) =: e^{-\phi(z)} :: \partial e^{\chi(z)} : \tag{2.8}$$

Here

$$\chi(z)\chi(\zeta) =: \chi(z)\chi(\zeta) : +\ln(z - \zeta) \tag{2.9}$$

Because the $\beta\gamma$ spectrum is unbounded from below, it is useful to define an infinite number of $\beta\gamma$ 'vacua' $|q\rangle_{\beta\gamma}$ where $\beta_n|q\rangle = 0, n > -q - \frac{3}{2}$, $\gamma_n|q\rangle = 0, n \geq q + \frac{3}{2}$, where $|q\rangle_{\beta\gamma} = e^{q\phi(0)}|0\rangle_{\beta\gamma}$ and $L_0^{\beta\gamma}|q\rangle = -\frac{1}{2}q(q + 2)|q\rangle$. The bosonic β, γ ghost system has two sectors: one is Neveu-Schwarz where $q \in \mathcal{Z}$, and the fields $\beta(z)$, $\gamma(z)$, and $: e^{\phi(z)} :$ are periodic, i.e. half-integrally moded; the other sector is Ramond, where $q \in \mathcal{Z} + \frac{1}{2}$, and the fields $\beta(z)$, $\gamma(z)$, and $: e^{\phi(z)} :$ are anti-periodic, i.e. integrally moded. We note that the conformal fields $: e^{q\phi(z)} :$ for q odd have the same periodicity and statistics as the supercurrent. Their conformal dimensions given by $L_0^{\beta\gamma}|q\rangle = -\frac{1}{2}q(q + 2)|q\rangle$ are $\frac{1}{2}$ for $q = -1$; $-\frac{3}{2}$ for $q = 1, -3$; and $-\frac{15}{2}$ for $q = 3, -5$; etc.

The superVirasoro ghost representation has central charge $c = -15$:

$$L(z) = -2{}^{\times}_{\times}b\partial c{}^{\times}_{\times} - {}^{\times}_{\times}(\partial b)c{}^{\times}_{\times} - \frac{3}{2}{}^{\times}_{\times}\beta\partial\gamma{}^{\times}_{\times} - \frac{1}{2}{}^{\times}_{\times}(\partial\beta)\gamma{}^{\times}_{\times} \tag{2.10a}$$
$$F(z) = b\gamma - 3\beta\partial c - 2(\partial\beta)c \tag{2.10b}$$

For the $N = 1$ worldsheet supersymmetry system, the BRST charge $Q \equiv \frac{1}{2\pi i}\oint dzQ(z)$ is given from the general form

$$Q(z) \sim c(L^{matter} + \frac{1}{2}L^{ghost}) - \gamma\frac{1}{2}(F^{matter} + \frac{1}{2}F^{ghost}) \tag{2.11a}$$

by the BRST current

$$Q(z) = Q_0(z) + Q_1(z) + Q_2(z)$$
$$Q_0(z) = cL^{X,\psi} - {}^{\times}_{\times}cb\partial c{}^{\times}_{\times} + \frac{3}{2}\partial^2 c + cL^{\beta\gamma} + \partial(\frac{3}{4}{}^{\times}_{\times}\gamma\beta{}^{\times}_{\times})$$
$$Q_1(z) = -\gamma\frac{1}{2}F^{X,\psi}$$
$$Q_2(z) = -\frac{1}{4}\gamma b\gamma \tag{2.11b}$$

Here the matter fields $L^{X,\psi}(z)$ and $F^{X,\psi}(z)$ close the superconformal algebra with $c = 15$; and L^{ghost} and F^{ghost} denoted in (2.11a) are given in (2.10). From the operator product expansion of the BRST current with itself it follows that

$$Q^2 = \tfrac{1}{2}\{Q, Q\} = 0. \tag{2.12}$$

The commutator which must vanish for BRST invariance is

$$[Q, V_{-\frac{3}{2}}(k, z)] = [Q_0, V_{-\frac{3}{2}}(k, z)] + [Q_1, V_{-\frac{3}{2}}(k, z)] + [Q_2, V_{-\frac{3}{2}}(k, z)]. \tag{2.13a}$$

By inspection, we find

$$[Q_0, V_{-\frac{3}{2}}(k, z)] = \tfrac{d}{dz}[c(z)V_{-\frac{3}{2}}(k, z)]; \qquad [Q_2, V_{-\frac{3}{2}}(k, z)] = 0 \tag{2.13b}$$

and

$$\begin{aligned}
Q_1(z)V_{-\frac{3}{2}}(k, \zeta) &= -\tfrac{1}{2} : e^{-\chi(z)} : e^{\phi(z)} F(z) V_{-\frac{3}{2}}(k, \zeta) \\
&= -\tfrac{1}{2} : e^{-\chi(z)} : (z - \zeta)^{\frac{3}{2}} e^{-\frac{1}{2}\phi(z)} F(z) v^{\dot\alpha}(k) S_{\dot\alpha}(z) e^{ik \cdot X(z)} \mathcal{S}(z) \\
&= regular\ terms
\end{aligned} \tag{2.13c}$$

so that

$$[Q_1, V_{\frac{3}{2}}(k, z)] = 0. \tag{2.13d}$$

3. Bispinor notation

In a Weyl representation, we can represent the four-dimensional γ matrix Clifford algebra given by $\{\gamma^\mu, \gamma^\nu\} = 2\eta^{\mu\nu}$ with $\eta^{\mu\nu} = diag(-1, 1, 1, 1)$ as

$$(\gamma^\mu)^A_{\ B} = \begin{pmatrix} 0 & (\bar\sigma^\mu)^\alpha_{\ \dot\beta} \\ (\sigma^\mu)^{\dot\alpha}_{\ \beta} & 0 \end{pmatrix}; \qquad C^{AB} = \begin{pmatrix} (i\sigma^2)^{\alpha\beta} & 0 \\ 0 & (i\sigma^2)^{\dot\alpha\dot\beta} \end{pmatrix} \tag{3.1}$$

where $\sigma^\mu = (\sigma^0, \sigma^i)$ and $\bar\sigma^\mu = (-\sigma^0, \sigma^i)$ are given by $\sigma^0 \equiv \begin{pmatrix} 1 & 0 \\ 0 & 1 \end{pmatrix}$ and the Pauli matrices σ^i. We define $\gamma^5 = (i\gamma^0\gamma^1\gamma^2\gamma^3)^A_{\ B} = \begin{pmatrix} (\sigma^0)^\alpha_{\ \beta} & 0 \\ 0 & -(\sigma^0)^{\dot\alpha}_{\ \dot\beta} \end{pmatrix}$. The charge conjugation matrices C^{AB} and $(C^{-1})_{AB}$ are tensors used to raise and lower indices: $C^{-1}_{AD}(\gamma^\mu)^D_{\ B} \equiv (\gamma^\mu)_{AB}$ and $C^{BD}(\gamma^\mu)^A_{\ D} \equiv (\gamma^\mu)^{AB}$. The transpose relation which defines C^{AB} is $C^{-1}_{AB}(\gamma^\mu)^B_{\ C}C^{CD} = -(\gamma^{\mu T})^D_{\ A}$ and it implies the matrices $(\gamma^\mu)^{AB}$ and $(\gamma^\mu)_{AB}$ are symmetric.

Since we are in the Weyl representation, we can use van der Waerden notation[1,19] for spinor indices $1 \leq \alpha, \dot\alpha \leq 2$. We have the following symmetry properties

$$\begin{aligned}
\sigma^\mu\bar\sigma^\nu &= -\sigma^\nu\bar\sigma^\mu + 2\eta^{\mu\nu} & \bar\sigma^\mu\sigma^\nu &= -\bar\sigma^\nu\sigma^\mu + 2\eta^{\mu\nu} \\
\sigma^\mu\bar\sigma^\nu &= \eta^{\mu\nu} - \tfrac{i}{2}\epsilon^{\mu\nu\rho\sigma}\sigma_\rho\bar\sigma_\sigma & \bar\sigma^\mu\sigma^\nu &= \eta^{\mu\nu} + \tfrac{i}{2}\epsilon^{\mu\nu\rho\sigma}\bar\sigma_\rho\sigma_\sigma
\end{aligned} \tag{3.2a}$$

$$\begin{aligned}
\sigma^\mu\bar\sigma^\nu\sigma^\rho + \sigma^\rho\bar\sigma^\nu\sigma^\mu &= 2(\eta^{\mu\nu}\sigma^\rho - \eta^{\mu\rho}\sigma^\nu + \eta^{\nu\rho}\sigma^\mu) \\
\bar\sigma^\mu\sigma^\nu\bar\sigma^\rho + \bar\sigma^\rho\sigma^\nu\bar\sigma^\mu &= 2(\eta^{\mu\nu}\bar\sigma^\rho - \eta^{\mu\rho}\bar\sigma^\nu + \eta^{\nu\rho}\bar\sigma^\mu)
\end{aligned} \tag{3.2b}$$

and the trace properties $tr\sigma^\mu\bar\sigma^\nu = 2\eta^{\mu\nu}$,

$$tr\,\bar\sigma^\rho\sigma^\mu\bar\sigma^\lambda\sigma^\kappa = 2\eta^{\rho\mu}\eta^{\lambda\kappa} - 2\eta^{\rho\lambda}\eta^{\mu\kappa} + 2\eta^{\rho\kappa}\eta^{\mu\lambda} - 2i\epsilon^{\rho\mu\lambda\kappa}$$
$$tr\,\sigma^\rho\bar\sigma^\mu\sigma^\lambda\bar\sigma^\kappa = 2\eta^{\rho\mu}\eta^{\lambda\kappa} - 2\eta^{\rho\lambda}\eta^{\mu\kappa} + 2\eta^{\rho\kappa}\eta^{\mu\lambda} + 2i\epsilon^{\rho\mu\lambda\kappa}\,. \qquad (3.2c)$$

The two linearly independent solutions to the massless Dirac equation $k\cdot\gamma u^\ell(k) = 0$ are now given by solutions of the Weyl equations

$$k_\mu\sigma^{\mu\dot\alpha}{}_\beta u^{1\beta} = 0 \qquad\qquad k_\mu\bar\sigma^{\mu\alpha}{}_{\dot\beta} u^{2\dot\beta} = 0$$

as

$$u^{1\beta}(k) = \begin{pmatrix} k^0 + k^3 \\ k^1 + ik^2 \end{pmatrix}(k^0 + k^3)^{-\frac{1}{2}}\,; \qquad u^{2\dot\beta}(k) = \begin{pmatrix} -k^1 + ik^2 \\ k^0 + k^3 \end{pmatrix}(k^0 + k^3)^{-\frac{1}{2}}\,.$$
$$(3.3)$$

We define two additional spinors $v^\ell(k)$ by $k\cdot\gamma v^\ell \sim u^\ell$ i.e.

$$\frac{1}{\sqrt{2}}k_\mu\bar\sigma^{\mu\alpha}{}_{\dot\beta} v^{1\dot\beta} = u^{1\alpha} \qquad\qquad \frac{1}{\sqrt{2}}k_\mu\sigma^{\mu\dot\alpha}{}_\beta v^{2\beta} = -u^{2\dot\alpha}$$

as

$$v^{1\dot\beta}(k) = \begin{pmatrix} k^0 + k^3 \\ k^1 + ik^2 \end{pmatrix}(2(k^0)^2(k^0 + k^3))^{-\frac{1}{2}}\,;$$
$$v^{2\beta}(k) = \begin{pmatrix} -k^1 + ik^2 \\ k^0 + k^3 \end{pmatrix}(2(k^0)^2(k^0 + k^3))^{-\frac{1}{2}}\,. \qquad (3.4)$$

Note that formally $v^{1\dot\beta} = \frac{1}{\sqrt{2}k^0}u^{1\beta}$ and $v^{2\beta} = \frac{1}{\sqrt{2}k^0}u^{2\dot\beta}$. In (3.3,4) we have $k_\mu k^\mu = 0$. The spin decomposition of the Weyl bispinors into the two helicity states of the massless vector is as follows:

$$u^{1\alpha}u^{1\beta} = -\epsilon^+_\lambda k_\kappa(\bar\sigma^\lambda\sigma^\kappa\sigma^2)^{\alpha\beta}$$
$$u^{2\delta}u^{2\dot\gamma} = \epsilon^-_\lambda k_\kappa(\sigma^\lambda\bar\sigma^\kappa\sigma^2)^{\dot\delta\dot\gamma}\,. \qquad (3.5)$$

We also find that

$$u^{1\alpha}v^{1\dot\beta} = \epsilon^+_\lambda(\bar\sigma^\lambda\sigma^2)^{\alpha\dot\beta}\sqrt{2}\,; \qquad u^{2\delta}v^{2\gamma} = \epsilon^-_\lambda(\sigma^\lambda\sigma^2)^{\dot\delta\gamma}\sqrt{2}$$
$$v^{1\dot\alpha}u^{1\beta} = \epsilon^+_\lambda(\sigma^\lambda\sigma^2)^{\dot\alpha\beta}\sqrt{2}\,; \qquad v^{2\delta}u^{2\dot\gamma} = \epsilon^-_\lambda(\bar\sigma^\lambda\sigma^2)^{\delta\dot\gamma}\sqrt{2} \qquad (3.6)$$

where the expressions in (3.6) are defined only up to a gauge transformation. In the Lorentz gauges defined by $k \cdot \epsilon^\pm(k) = 0$, we have that

$$i\epsilon^{\mu\nu\lambda\rho}\epsilon^\pm_\lambda k_\rho = \pm(\epsilon^{\mu\pm}k^\nu - \epsilon^{\nu\pm}k^\mu) \qquad (3.7)$$

holds generally for the polarization vectors[9]. We also note here the spin decomposition of the spin-vector state $\psi^A_\mu = \epsilon^+_\mu u^A$ separated into its spin-$\frac{3}{2}$ and spin-$\frac{1}{2}$ content. This is simple in van der Waerden notation and eliminates the need for introducing[15,18] a noncovariant momentum vector $\bar k$ where $k \cdot \bar k = 1$. We find the spin-$\frac{3}{2}$ part to be given by

$$\epsilon^+_\mu u^{1\alpha} \qquad helicity = \tfrac{3}{2}\,; \qquad \epsilon^-_\mu u^{2\dot\beta} \qquad helicity = -\tfrac{3}{2} \qquad (3.8)$$

since the spin-vectors in (3.8) satisfy the on-shell $k^2 = 0$ Rarita-Schwinger equation

$$k \cdot \psi = 0 \,; \qquad \gamma \cdot k\psi_\mu = 0 \,; \qquad \gamma \cdot \psi = 0 \,. \tag{3.9}$$

For example, for the spin-$\frac{3}{2}$ helicity,

$$\gamma \cdot \psi = \sigma^{\mu\dot\alpha}_{\ \ \alpha} \epsilon^+_\mu u^{1\alpha} = 0 \,. \tag{3.10}$$

To prove (3.10), consider

$$\begin{aligned}
\sigma^{\mu\dot\alpha}_{\ \ \alpha} \epsilon^+_\mu u^{1\alpha} v^{1\dot\beta} &= \epsilon^+_\mu \sigma^{\mu\dot\alpha}_{\ \ \alpha} \epsilon^+_\lambda (\bar\sigma^\lambda \sigma^2)^{\dot\alpha\dot\beta} \sqrt{2} \\
&= \sqrt{2}\epsilon^+ \cdot \epsilon^+ \sigma^2 = 0 \,.
\end{aligned} \tag{3.11}$$

The spin-$\frac{1}{2}$ part is

$$\epsilon^+_\mu u^{2\dot\beta} \qquad helicity = \tfrac{1}{2} \,; \qquad \epsilon^-_\mu u^{1\alpha} \qquad helicity = -\tfrac{1}{2} \tag{3.12}$$

since for the spin vectors in (3.12) we have

$$k \cdot \psi = 0 \,; \qquad \gamma \cdot k\psi_\mu = 0 \,; \qquad \gamma \cdot \psi \neq 0 \,. \tag{3.13}$$

4. BRST invariance and picture changing

We consider the following vertex operator in four dimensions.

$$\begin{aligned}
V^{(1)}_{-\frac{3}{2}}(k, z) &= v^{1\alpha}(k) S_{\dot\alpha}(z) e^{ik \cdot X(z)} \mathcal{S}(z) V(z) \, e^{-\frac{3}{2}\phi(z)} \\
V^{(2)}_{-\frac{3}{2}}(k, z) &= -v^{2\alpha}(k) S_\alpha(z) e^{ik \cdot X(z)} \tilde{\mathcal{S}}(z) V(z) \, e^{-\frac{3}{2}\phi(z)} \,.
\end{aligned} \tag{4.1}$$

For the supercurrent given by $F(z) = a_\mu(z) h^\mu(z) + \bar F(z)$ we choose

$$\bar F(z) = \tilde\gamma^5 \otimes \widehat F(z) + \tilde\gamma^5 \otimes \tilde\Gamma^7 \otimes \tilde F(z) \,. \tag{4.2}$$

Here $\tilde\gamma^5 \equiv \gamma^5 (-1)^{\sum_{n>0} \psi^\mu_{-n}\psi^\mu_n}$ and $\tilde\Gamma^7 \equiv (i\Gamma^1\Gamma^2\Gamma^3\Gamma^4\Gamma^5\Gamma^6)(-1)^{\sum_{n>0} \psi^a_{-n}\psi^a_n}$ where $\Gamma^7 = i\Gamma^1\Gamma^2\Gamma^3\Gamma^4\Gamma^5\Gamma^6 = \begin{pmatrix} (I_4)^\ell_{\ \dot m} & 0 \\ 0 & -(I_4)^{\dot\ell}_{\ \dot m} \end{pmatrix}$. Also $1 \le a \le 6$ and we use a Weyl representation for the internal gamma matrices given for the $c = 3$ system for $1 \le a \le 3$, $1 \le \ell, \dot\ell \le 4$ by

$$\begin{aligned}
\Gamma^a &= \begin{pmatrix} 0 & (\alpha^a)^\ell_{\ \dot m} \\ -(\alpha^a)^{\dot\ell}_{\ m} & 0 \end{pmatrix} \,; \qquad \Gamma^{a+3} = i \begin{pmatrix} 0 & (\beta^a)^\ell_{\ \dot m} \\ (\beta^a)^{\dot\ell}_{\ m} & 0 \end{pmatrix} \,; \\
C &= \begin{pmatrix} 0 & (I_4)^{\ell\dot m} \\ -(I_4)^{\dot\ell m} & 0 \end{pmatrix} \,.
\end{aligned} \tag{4.3}$$

For $\widehat{F}(z) \equiv (-\frac{i}{6})\frac{1}{\sqrt{\frac{c_\psi}{2}}} f_{abc}\psi^a(z)\psi^b(z)\psi^c(z)$, and f_{abc} given by the structure constants of $SU(2) \otimes SU(2)$, we have

$$\widehat{F}(z)\,\mathcal{S}(\zeta) = (z-\zeta)^{-\frac{3}{2}}\,\tilde{\mathcal{S}}(\zeta) + \ldots$$
$$\widehat{F}(z)\,\tilde{\mathcal{S}}(\zeta) = (z-\zeta)^{-\frac{3}{2}}\tfrac{1}{4}\mathcal{S}(\zeta) + \ldots \qquad (4.4)$$

Here $\tilde{F}(z)$ corresponds to the remaining internal degrees of freedom with $c = 6$. We assume there exists a pair of states in the Ramond sector of this piece of the internal system corresponding to weight zero conformal spin fields $V(z)$ and $U(z)$ such that

$$\tilde{F}(z)V(\zeta) = (z-\zeta)^{-\frac{3}{2}}U(\zeta)$$
$$\tilde{F}(z)U(\zeta) = (z-\zeta)^{-\frac{3}{2}}(-\tfrac{c}{24})V(\zeta).$$

$$(4.5a)$$

This corresponds to a non-hermitian choice for the operator $\tilde{F}_0$, since the norm of a state $\tilde{F}_0|\Psi\rangle$ is given by $||\tilde{F}_0|\Psi\rangle|| = \langle\Psi|\tilde{F}_0^\dagger\tilde{F}_0|\Psi\rangle$, and therefore non-hermitian $\tilde{F}_0$ does not imply $\tilde{F}_0|\Psi\rangle$ is a negative norm state, even though $\langle\Psi|\tilde{F}_0^2|\Psi\rangle < 0$. The operator product expansion leading order of the spin fields is

$$U(z)V(\zeta) = O(z-\zeta)^{-\frac{1}{2}}; \qquad U(z)U(\zeta) = O(z-\zeta)^0$$
$$V(z)U(\zeta) = O(z-\zeta)^{-\frac{1}{2}}; \qquad V(z)V(\zeta) = O(z-\zeta)^0. \qquad (4.5b)$$

The spin fields are nonlocal with respect to the supercurrent because they make states in the Ramond sector of the theory, i.e. in order to generate the appropriate Fourier series expansion of Ramond fermions[14,2,3,17]. Spin fields change the boundary condition on the fermion superconformal fields between periodic and anti-periodic. In the Ramond sector, the spin fields have a $(z-\zeta)^{-\frac{3}{2}}$ singularity in their OPE with the supercurrent except for the "ground" states for which $h = \frac{c}{24}$; these have only a $(z-\zeta)^{-\frac{1}{2}}$ singularity. This implies all excited states come in pairs that are mapped into each other by F_0.

Since F_0 and L_0 commute these spin fields come in pairs $S^\pm(z)$ such that

$$|h^+\rangle = S^+(0)|0\rangle$$
$$|h^-\rangle = S^-(0)|0\rangle = F_0|h^+\rangle$$
$$(h - \tfrac{c}{24})|h^+\rangle = F_0|h^-\rangle$$
$$L_0|h^\pm\rangle = h|h^\pm\rangle \qquad (4.6a)$$

or as an operator product:

$$F(z)S^+(\zeta) = (z-\zeta)^{-\frac{3}{2}}S^-(\zeta)$$
$$F(z)S^-(\zeta) = [h - \tfrac{c}{24}](z-\zeta)^{-\frac{3}{2}}S^+(\zeta). \qquad (4.6b)$$

If $h = \frac{c}{24}$, global worldsheet supersymmetry is unbroken in the Ramond sector. In this case the states need not be paired, i.e. only $|h^-\rangle$ survives since if F_0 is hermitian, then for $h = \frac{c}{24}$ we have $\langle h^-|h^-\rangle = \langle h^+|F_0^2|h^+\rangle = \langle h^+|L_0 - \frac{c}{24}|h^+\rangle = 0$, i.e. $|h^-\rangle$ is null, i.e. has zero norm.

If F_0 is not hermitian, as for example the spacetime part $F(z) = a_\mu(z)\psi^\mu(z)$, then $h = \frac{c}{24}$ does not imply $|h^-\rangle$ is null:

$$F(z)S_\alpha(\zeta)\,e^{ik\cdot X(\zeta)} = (z - \zeta)^{-\frac{3}{2}}\,k_\mu\frac{1}{\sqrt{2}}\Gamma^{\mu\dot\beta}_\alpha S_{\dot\beta}(\zeta)\,e^{ik\cdot X(\zeta)}$$

$$F(z)k_\mu\frac{1}{\sqrt{2}}\Gamma^{\mu\dot\beta}_\alpha S_{\dot\beta}(\zeta)\,e^{ik\cdot X(\zeta)} \sim [\tfrac{1}{4} - \tfrac{6}{24}] = 0 \qquad (4.7)$$

but the norm of $|h^-\rangle = k_\mu\frac{1}{\sqrt{2}}\Gamma^{\mu\dot\beta}_\alpha S_{\dot\beta}(0)\,e^{ik\cdot X(0)}|0\rangle$ is not zero:

$$\langle h^-|h^-\rangle = \langle h^+|F_0^\dagger F_0|h^+\rangle \neq \langle h^+|L_0 - \tfrac{c}{24}|h^+\rangle = 0\,. \qquad (4.8)$$

The vertex operator for the Ramond states in the canonical $q = -\frac{1}{2}$ super-conformal ghost picture is given for $k\cdot\frac{1}{\sqrt{2}}\gamma v = u$ and $\Sigma^{(1)}(z) \equiv \mathcal{S}(z)V(z)$ and $\tilde\Sigma^{(1)}(z) \equiv \tilde{\mathcal{S}}(z)V(z) + \mathcal{S}(z)U(z)$, by

$$V^{(1)}_{-\frac{1}{2}}(k,\zeta) = [u^{1\alpha}(k)S_\alpha(\zeta)\Sigma^{(1)}(\zeta) - v^{1\dot\alpha}(k)S_{\dot\alpha}(\zeta)\tilde\Sigma^{(1)}(\zeta)]\,e^{ik\cdot X(\zeta)}\,e^{-\frac{1}{2}\phi(\zeta)}\,.$$

$$(4.9a)$$

For $\Sigma^{(2)}(z) \equiv \tilde{\mathcal{S}}(z)V(z)$ and $\tilde\Sigma^{(2)}(z) \equiv \frac{1}{4}\mathcal{S}(z)V(z) - \tilde{\mathcal{S}}(z)U(z)$, we have

$$V^{(2)}_{-\frac{1}{2}}(k,\zeta) = [u^{2\dot\alpha}(k)S_{\dot\alpha}(\zeta)\Sigma^{(2)}(\zeta) - v^{2\alpha}(k)S_\alpha(\zeta)\tilde\Sigma^{(2)}]\,e^{ik\cdot X(\zeta)}\,e^{-\frac{1}{2}\phi(\zeta)}\,.$$

$$(4.9b)$$

We note that inside the fermion scattering amplitudes, the modified fermion emission vertex given above effectively normalizes the Ramond-Ramond bosons to 1 not k^0, i.e. the same normalization as the conventional gauge bosons found in the NS-NS sector, a required feature when both kinds of bosons are used together to form the adjoint representation of an enhanced (larger) gauge symmetry group[9].

BRST invariance of the vertex operators in the $q = -\frac{1}{2}$ picture (4.9) is assured from BRST invariance of the picture changed $q = -\frac{3}{2}$ operators and from (2.12). It can also be checked directly.

The vertex operators for the massless Neveu-Schwarz states in the canonical $q = -1$ superconformal ghost picture are

$$V_{-1}^a(k, z) = \tilde{\gamma}^5 \otimes \psi^a(z) e^{ik \cdot X(z)} e^{-\phi(z)} \tag{4.10a}$$

$$V_{-1}(k, z, \epsilon) = \epsilon \cdot \psi(z) e^{ik \cdot X(z)} e^{-\phi(z)}. \tag{4.10b}$$

States in the Neveu-Schwarz matter system (i.e. without ghosts) form superconformal fields $V(z, \theta) = V_q(z) + \theta V_{q+1}(z)$ with upper and lower components related by

$$G(z) V_q(\zeta) = (z - \zeta)^{-1} V_{q+1}(\zeta)$$

$$G(z) V_{q+1}(\zeta) = (z - \zeta)^{-2} 2 h_q V_q(\zeta) + (z - \zeta)^{-1} \partial V_q(\zeta). \tag{4.11}$$

BRST invariance holds for both the vertices (4.10) since

$$[Q_0, V_{-1}(k, z)] = \tfrac{d}{dz}[c(z) V_{-1}(k, z)]; \qquad [Q_2, V_{-1}(k, z)] = 0 \tag{4.12}$$

and

$$
\begin{aligned}
Q_1(z) V_{-1}(k, \zeta) &= -\tfrac{1}{2} : e^{-\chi(z)} : e^{\phi(z)} G(z) V_{-1}(k, \zeta) \\
&= -\tfrac{1}{2} : e^{-\chi(z)} : (z - \zeta)^1 G(z) V_{-1}^{matter}(k, \zeta) \\
&= -\tfrac{1}{2} : e^{-\chi(z)} : V_0^{matter}(k, \zeta) \\
&= regular\ terms
\end{aligned} \tag{4.13a}
$$

so that

$$[Q_1, V_{-1}k, z)] = 0. \tag{4.13b}$$

In the $q = 0$ superconformal ghost picture, (4.10) is

$$
\begin{aligned}
V_0^a(k, z) &= \lim_{z \to \zeta} e^{\phi(z)} G(z) V_{-1}^a(k, \zeta) \\
&= [k \cdot \psi(\zeta) \tilde{\gamma}^5 \otimes \psi^a(\zeta) - \tfrac{i}{2} f_{abc} \psi^b(\zeta) \psi^c(\zeta)] e^{ik \cdot X(\zeta)} \tag{4.14a}
\end{aligned}
$$

$$
\begin{aligned}
V_0(k, z, \epsilon) &= \lim_{z \to \zeta} e^{\phi(z)} G(z) V_{-1}(k, \zeta \epsilon) \\
&= [k \cdot \psi(\zeta) \epsilon \cdot \psi(\zeta) + \epsilon \cdot a(\zeta)] e^{ik \cdot X(\zeta)} \tag{4.14b))}
\end{aligned}
$$

The tree correlation functions of BRST invariant vertex operators are independent of the distribution of ghost charges given that $\sum_i q_i = -2$. As an example, we compute the three point amplitude for the left-movers of the closed superstring as

$$\langle 0 | V_{-\frac{1}{2}}^{(1)}(k_1, z_1) \, V_{-1}(k_2, z_2, \epsilon_2^-) \, V_{-\frac{1}{2}}^{(1)}(k_3, z_3) c(z_1) c(z_2) c(z_3) | 0 \rangle$$

$$= \langle 0 | V_{-\frac{3}{2}}^{(1)}(k_1, z_1) \, V_0(k_2, z_2, \epsilon_2^-) \, V_{-\frac{1}{2}}^{(1)}(k_3, z_3) c(z_1) c(z_2) c(z_3) | 0$$

$$= -v^{1\dot{\gamma}}(k_1)[k_{2\kappa} \epsilon_{2\mu} \tfrac{1}{2} (\gamma^\kappa \gamma^\mu)_{\dot{\gamma}\dot{\alpha}} + k_3^\mu \epsilon_{2\mu}^- C_{\dot{\gamma}\dot{\alpha}}^{-1}] v^{1\dot{\alpha}}(k_3)(-\tfrac{i}{\sqrt{2}} f^{\dot{m}} f^\ell C_{\dot{m}\ell}^{-1}) \cdot 4$$

$$= [u^{1\gamma}(k_1) \epsilon_{2\mu}^- (i\sigma^2 \bar{\sigma}^\mu)_{\gamma\dot{\alpha}} v^{1\dot{\alpha}}(k_3) + v^{1\dot{\gamma}}(k_1) \epsilon_{2\mu}^- (i\sigma^2 \sigma^\mu)_{\dot{\gamma}\alpha} u^{1\alpha}(k_3)](-i f^{\dot{m}} f^\ell C_{\dot{m}\ell}^{-1}) \cdot 2 \tag{4.15}$$

If we use

$$V_{-\frac{1}{2}}^{(1)}(k,\zeta)$$
$$= [u^{1\alpha}(k)S_{\alpha}(\zeta)e^{ik\cdot X(\zeta)}(f^{\ell}\Sigma_{\ell}(\zeta)+if^{\dot{\ell}}\Sigma_{\dot{\ell}}(\zeta))\,[U(\zeta)+2V(\zeta)]e^{-\frac{1}{2}\phi(z)} \quad (4.16)$$

for the right-movers of the closed superstring, we find for

$$\langle 0|V_{-\frac{1}{2}}^{(1)}(k_1,\bar{z}_1)\,V_{-1}^{a}(k_2,\bar{z}_2)\,V_{-\frac{1}{2}}^{(1)}(k_3,\bar{z}_3)c(\bar{z}_1)c(\bar{z}_2)c(\bar{z}_3)|0\rangle$$
$$= \langle 0|V_{-\frac{3}{2}}^{(1)}(k_1,\bar{z}_1)\,V_0^{a}(k_2,\bar{z}_2)\,V_{-\frac{1}{2}}^{(1)}(k_3,\bar{z}_3)c(\bar{z}_1)c(\bar{z}_2)c(\bar{z}_3)|0\rangle$$
$$= u^{1\delta}(k_1)k_{2\kappa}\frac{1}{\sqrt{2}}(\gamma^{\kappa}\gamma^5)_{\delta\dot{\beta}}v^{1\dot{\beta}}(k_3)f^n f^k(-\sqrt{2})\alpha_{nk}^{a}$$
$$= u^{1\delta}(k_1)(-i\sigma^2)_{\delta\beta}u^{1\beta}(k_3)((-\sqrt{2})f^n f^k\alpha_{nk}^{a}\,.$$
$$\tag{4.17}$$

Combining the left and right pieces we have

$$[u^{1\delta}v^{1\dot{\gamma}}\epsilon_2^{-}(\sigma^2\sigma^{\mu})_{\dot{\gamma}\alpha}u^{1\alpha}u^{1\beta}+u^{1\delta}u^{1\gamma}\epsilon_2^{-}(\sigma^2\bar{\sigma}^{\mu})_{\gamma\dot{\alpha}}v^{1\dot{\alpha}}u^{1\beta}]$$
$$\cdot f^{\dot{m}}f^{\ell}f^n f^k(2\sqrt{2})(C_{\dot{m}\ell}^{-1}\alpha_{nk}^{a})$$
$$= if_{IaJ}(\epsilon_2^{a-}\cdot\epsilon_3^{J+}\epsilon_1^{I+}\cdot k_2+\epsilon_1^{I+}\cdot\epsilon_2^{a-}\epsilon_3^{J+}\cdot k_1) \quad (4.18)$$

which is the on-shell three gluon tree coupling for these polarizations. Here the structure constants f_{IaJ} are defined from α_{nk}^{a}, and together with the symmetric subgroup structure constants f_{abc} of $SU(2)^2$ they form $SO(8)$, see Ref.[9].

5. Modular invariance

The cosmological constant Λ in a string theory is called the partition function. It is related to the function which counts the number of states at each mass level. To describe consistent interacting strings, one must in general check that all the scattering amplitudes are modular invariant, finite, and unitary. A guide to this program is the calculation of the one-loop partition function which can be checked for modular invariance, albeit a quantity equal to zero. For closed strings, the one-loop cosmological constant is defined by

$$\Lambda \equiv \tfrac{1}{2}tr\,\ln\Delta^{-1}$$

where

$$\Delta^{-1}=\alpha'(p^2+m^2)\,;\qquad \tfrac{1}{2}\alpha'm^2=\alpha'm_L^2+\alpha'm_R^2$$

so in D space-time dimensions, for $\omega=e^{2\pi i\tau}$,

$$\Lambda = -\tfrac{1}{2}(2\pi)^{-1}(\alpha')^{-\frac{D}{2}}\int_F d^2\tau(Im\tau)^{-2-\frac{1}{2}(D-2)}$$
$$\cdot \sum_{all\ sectors} tr[\bar{\omega}^{\alpha'm_L^2}\omega^{\alpha'm_R^2}possible\,projections] \quad (5.1)$$

F is a fundamental region of the modular group: $\frac{1}{2}\le Re\tau\le\frac{1}{2}$; $|\tau|>1$. For the $D=4$, $N=8$ model considered above, the partition function computed for the degrees of freedom donoted in (4.2) by a_μ, ψ^μ and ψ^a with $\alpha' m_L^2 = L_0 - \frac{c}{24}$, etc. is

$$\Lambda = -(4\pi\alpha'^2)^{-1} \int_F d^2\tau (Im\tau)^{-3} |f(\omega)|^{-12} |\omega|^{-\frac{1}{2}}$$
$$\cdot \frac{1}{4}[\theta_3^4 - \theta_4^4 - \theta_2^4][\bar{\theta}_3^4 - \bar{\theta}_4^4 - \bar{\theta}_2^4] \tag{5.2}$$

We see this is already modular invariant using the transformation properties under $SL(2, I)$ given by

$$\tau \to \tau + 1 : \quad \theta_3 \to \theta_4; \, \theta_4 \to \theta_3; \, \theta_2 \to e^{i\frac{\pi}{4}}; \, Im\tau \to Im\tau$$
$$\tau \to -\frac{1}{\tau} : \quad \theta_2 \to (-i\tau)^{\frac{1}{2}}\theta_4; \, \theta_4 \to (-i\tau)^{\frac{1}{2}}\theta_2; \, \theta_3 \to (-i\tau)^{\frac{1}{2}}\theta_3; \, Im\tau \to |\tau|^{-2}Im\tau;$$
$$\omega^{\frac{1}{24}} f(\omega) \to (-i\tau)^{\frac{1}{2}} \omega^{\frac{1}{24}} f(\omega) \tag{5.3}$$

where $f(\omega) = \prod_{n=1}(1 - \omega)$ is related to the Dedekind eta function $\eta(\omega) = \omega^{\frac{1}{24}} f(\omega)$, $\omega = e^{2\pi i\tau}$ and $\theta_i(0|\tau)$ are the Jacobi theta functions. We suggest that 1) the inclusion of the remaining internal conformal field theory will leave the modular invariance of (5.2) unchanged, by multiplying the integrand by another modular invariant function, and 2) the physical states[4] may be in one to one correspondence with the states of the partition function given in (5.2).

6. Conclusions

We have considered properties of bosonic and fermionic twisted conformal field theories. In the fermionic superstring case, the superconformal fields are discussed as the generators of an affine Kac-Moody Lie algebra whose zero modes describe the gauge symmetry of the particle spectrum. In addition to this conventional origin of gauge symmetry, we have analyzed the role twisted emission vertex operators or spin fields in connection with symmetry enhancement.

REFERENCES

1. J. Bagger and J. Wess, *Supersymmetry and Supergravity*, Princeton University Press, Princeton, 1992.
2. T. Banks, J. Dixon, D. Friedan, and E. Martinec, *Phenomenology and conformal field theory*, Nucl. Phys. **B299** (1988), 613-626.
3. T. Banks and J. Dixon, *Constraints on string vacua with spacetime supersymmetry*, Nucl. Phys. **B307** (1988), 93-108.
4. N. Berkovits and C. Vafa, *On the uniqueness of string theory*, hep-th/9310170 preprint (October 1993), 12 pages.
5. J. H. Conway and N.J.A. Sloane, *Sphere packings, lattices and groups*, Springer, Berlin, 1988.
6. L. Dolan, P. Goddard and P. Montague, *Conformal field theory of twisted vertex operators*, Nucl. Phys. **B338** (1990), 529-601.
7. L. Dolan, P. Goddard and P. Montague, *Conformal field theory, triality and the monster group*, Phys. Lett. **B236** (1990), 165-172.
8. L. Dolan, P. Goddard and P. Montague, *Conformal field theories, representations and lattice constructions*, hep-th/9410029 preprint (October 1994); 65 pages.
9. L. Dolan and S. Horvath, *Gauge bosons in superstring theory*, Nucl. Phys. **B416** (1994), 87–118.

10. L. Dolan and S. Horvath, *BRST properties of spin fields*, IFP/499/UNC (November 1994); 25 pages.

11. I. Frenkel, J. Lepowsky, and A. Meurman, *A Moonshine module for the monster*, Vertex Operators in Mathematics and Physics, Proc. 1983 MSRI Conf., eds. J. Lepowsky et al., Springer, New York, 1985, pp. 231-276.

12. I. Frenkel, J. Lepowsky, and A. Meurman, *Vertex operator algebras and the monster*, Academic Press, New York, 1988.

13. D. Friedan, *Notes on string theory and two-dimensional conformal field theory*, Workshop on Unified String Theories, World Scientific, Singapore, 1986, pp. 162-213.

14. D. Friedan, E. Martinec, and S. Shenker, *Conformal invariance, supersymmetry, and string theory*, Nucl. Phys. **B271** (1986), 93–165.

15. F. Gliozzi, J. Scherk and D. Olive, *Supersymmetry, supergravity theories and the dual spinor model*, Nucl. Phys. **B122** (1977), 253-290.

16. B. Griess, *The Friendly giant*, Invent. Math. **69** (1982), 1–102.

17. D. Lust and S. Theisen, *Lectures on String Theory*, Springer, New York, 1989.

18. J. Scherk and J. Schwarz, *Dual models for non-hadrons*, Nucl. Phys. **B81** (1974), 118-144.

19. B. L. van der Waerden, *Group theory and quantum mechanics*, Springer, New York, 1974.

DEPARTMENT OF PHYSICS, UNIVERSITY OF NORTH CAROLINA, CHAPEL HILL, NORTH CAROLINA 27599-3255

E-mail address: dolan@physics.unc.edu

Contemporary Mathematics
Volume **193**, 1996

SOME TWISTED SECTORS FOR THE MOONSHINE MODULE

CHONGYING DONG, HAISHENG LI AND GEOFFREY MASON

ABSTRACT. The construction of twisted sectors, or g-twisted modules, for a vertex operator algebra V and automorphism g, is a fundamental problem in algebraic conformal field theory and the theory of orbifold models. For the moonshine module $V^\natural$, whose automorphism is the Monster $\mathbb{M}$, Tuite has shown that this problem is intimately related to the generalized moonshine conjecture which relates hauptmoduln to twisted sectors.

In this paper we show how to give a uniform existence proof for irreducible g-twisted $V^\natural$-modules for elements of type $2A$, $2B$ and $4A$ in $\mathbb{M}$. The most interesting of these is the twisted sector $V^\natural(2A)$, whose automorphism group is essentially the centralizer of $2A$ in $\mathbb{M}$. This is a 2-fold central extension of the Baby Monster, the second largest sporadic simple group.

We also establish uniqueness of the twisted sectors and the hauptmodul property for the graded traces of automorphisms of odd order, as predicted by conformal field theory.

1. INTRODUCTION

The purpose of this paper is to give a uniform existence proof for irreducible g-twisted $V^\natural$-modules for elements g in $\mathbb{M}$ of type $2A, 2B$ and $4A$. Here $V^\natural$ is the moonshine module [FLM] with automorphism group the Monster $\mathbb{M}$, and the notation for elements in $\mathbb{M}$ follows the conventions of [Cal]. We assume the reader to be familiar with the definition of *vertex operator algebra* (VOA) (see [B1] and [FLM]) which we do not repeat here, noting only that $V^\natural$ is a particularly famous example of a VOA.

1991 *Mathematics Subject Classification*. Primary 17B69; Secondary 17B68, 81T40.

This paper is in final form and no version of it will be submitted for publication elsewhere.

C.D. and G.M. are partially supported by NSF grants and a research grant from the Committee on Research, US Santa Cruz.

If V is a VOA and g an automorphism of finite order, there is the notion of a (*weak*) *g-twisted V-module*. Briefly, a weak g-twisted V-module is a pair (M, Y_M) consisting of a $\mathbb{C}$-graded linear space $M = \oplus_{n \in \mathbb{C}} M_n$ locally truncated below in the sense that $M_{n+q} = 0$ for fixed $n \in \mathbb{C}$ and sufficiently small $q \in \mathbb{Q}$, together with a linear map

$$V \to (\text{End } M)[[z^{1/T}, z^{-1/T}]]$$
$$v \mapsto Y_M(v, z) = \sum_{n \in \frac{1}{T}\mathbb{Z}} v_n z^{-n-1} \quad (v_n \in \text{End } M),$$

where T is the order of g. One requires a certain Jacobi identity to be satisfied by these maps, as well as an appropriate action of the Virasoro algebra. See [DM1] for the complete definition. A *g-twisted V-module* is a weak g-twisted V-module such that all of the homogeneous spaces M_n are of finite dimension. We should point out that the definition in [DM1] differs slightly from that used here. Namely, we allow M to be $\mathbb{C}$-graded in the present paper (in order to be able to apply results in [DM2]) and we admit the possibility of infinite-dimensional homogeneous spaces.

It can be easily shown that any weak g-twisted module is a direct sum of submodules of type

$$N = \oplus_{n=0}^{\infty} N_{c+\frac{n}{T}} \tag{1.1}$$

where $c \in \mathbb{C}$ such that $N_c \neq 0$. The subspace N_c is called the *top level* of N. Note that the grading of a simple weak g-twisted module always has the form (1.1).

Theorem 1. *Let $g \in \mathbb{M}$ be of type $2A, 2B$ or $4A$. The following hold:*

(i) Every simple weak g-twisted $V^\natural$-module is a g-twisted $V^\natural$-module.

(ii) Up to isomorphism there is exactly one simple g-twisted $V^\natural$-module.

(iii) If g has type $2A$ then every g-twisted $V^\natural$-module is completely reducible.

Twisted sectors for $V^\natural$ of type $2B$ have been constructed by [Hu] and also by two of us [DM2], but until now that has been the extent of our knowledge of the existence of twisted sectors for $V^\natural$.

Continuing with earlier notation, we define an *extended automorphism* of (M, Y_M) to be a pair $(x, \alpha(x))$ where $x : M \to M$ and $\alpha(x) : V \to V$ are invertible linear maps satisfying

$$xY_M(v, z)x^{-1} = Y_M(\alpha(x)v, z)$$
$$\alpha(x)g = g\alpha(x), \alpha(x)1 = 1, \alpha(x)\omega = \omega \tag{1.2}$$

for $v \in V$ where $1, \omega$ are the vacuum and conformal element respectively. This definition is a slight modification of [DM4], where it is explained that if V and M are both simple then $x \mapsto \alpha(x)$ is a group homomorphism from the group $Aut^e(M)$ of extended automorphisms of M (which we identify with the group of linear maps x) into $Aut(V)$. Moreover, the kernel is a central subgroup of $Aut^e(M)$ consisting of the scalar operators.

We will show that if $V^\natural(2A)$ is the simple $2A$-twisted $V^\natural$-module whose existence is guaranteed by Theorem 1, then $Aut^e(V^\natural(2A))$ is isomorphic to $\mathbb{C}^* \times 2Baby$ where $2Baby$ denotes the 2-fold central extension of the Baby Monster which is known to be isomorphic to the centralizer of $2A$ in $\mathbb{M}$ (cf. [Cal] and references therein). So in this case, the map $x \mapsto \alpha(x)$ splits, though this is not always the case. (E.g., for $g = 2B$ it does not. See Section 6 for details.) We identify $2Baby \subset Aut^e(V^\natural(2A))$ with the corresponding centralizer in $\mathbb{M}$. If $V^\natural$ has grading

$$V^\natural = \oplus_{n=0}^\infty V_n^\natural \tag{1.3}$$

then for $h \in \mathbb{M}$ we define

$$Z(1, h, \tau) = q^{-1} \sum_{n=0}^\infty tr(h|V_n^\natural)q^n, \tag{1.4}$$

sometimes called the *McKay-Thompson* series of h. Here $q = e^{2\pi i \tau}$ as usual. It is known (the Conway-Norton conjecture = Borcherds' theorem [B2]) that each $Z(1, h, \tau)$ is a so-called hauptmodul. For example, $Z(1, 1, \tau) = J(q) = q^{-1} + 196884q + \cdots$ is the absolute modular invariant with constant term 0.

It transpires that if $M = V^\natural(2A)$ then the constant c in (1.1) is equal to $1/2$. Then for $g \in \mathbb{M}$ of type of $2A$ and $h \in C_\mathbb{M}(g) \simeq 2Baby$ we define

$$Z(g, h, \tau) = q^{-1} \sum_{n=1}^\infty tr(h|V^\natural(2A)_{n/2})q^{n/2}. \tag{1.5}$$

Remark that from (1.2) with v equal to the conformal element ω, it follows that $Aut^e(V^\natural(2A))$ preserves the $\mathbb{Q}$-grading on $V^\natural(2A)$, so that (1.5) makes sense. We prove

Theorem 2. *The following hold:*
(i) If h has odd order then

$$Z(g, h, 2\tau) = Z(1, gh, \tau). \tag{1.6}$$

(ii) If h either has odd order or satisfies $g \in \langle h \rangle$, then $Z(g, h, \tau)$ is a hauptmodul.

Remark 3. *1. These results amount to establishing the generalized moonshine conjecture due to Norton (cf. the appendix to [M]) for the commuting pairs (g, h) such that $\langle g, h \rangle$ is cyclic and $g = 2A$.*

2. If $g \in \langle h \rangle$ there are formulas which are analogous to (1.6), though more complicated.

3. Similar results also hold for the twisted sectors of type $2B$ and $4A$.

4. We can compute $Z(g, h, \tau)$ in other cases too. For example if $\langle g, h \rangle \simeq \mathbb{Z}_2 \times \mathbb{Z}_2$ has all 3 involutions of type of $2A$ then we will show in Section 5 that $Z(g, h, \tau)$ is precisely the hauptmodul $t_{2/2} = \sqrt{J(q) - 984}$ as predicted by Conway-Norton [CN].

The proofs of the two theorems depend heavily on the papers [DM2] and [DMZ]. In [DMZ] it is explained that $V^\natural$ contains a sub VOA which is isomorphic to $L = L(\frac{1}{2}, 0)^{\otimes 48}$, the tensor product of 48 vertex operator algebras $L(\frac{1}{2}, 0)$ associated with the highest weight unitary representation of the Virasoro algebra with central charge 1/2. Moreover, L is a rational VOA.

There are many sub VOAs of $V^\natural$ isomorphic to L; they may be constructed from 48-dimensional associative subalgebras A of the Griess algebra (cf. [MN] and [Mi] for example) which themselves are related to what we have called marked Golay codes [CM]. One can show easily that if $g \in \mathbb{M}$ has type $2A, 2B$ or $4A$ then g fixes a suitable A (element-wise), so that we may take L to lie in the sub VOA $(V^\natural)^{\langle g \rangle}$ of g-invariants. Then we are in a position to invoke several results from [DM2] to conclude Theorem 1.

Theorem 1 is purely an existence theorem. To get further information one needs to know the representation of the extended automorphism group on the top level M_c of the corresponding twisted sector. This can vary greatly: for $M = V^\natural(2B)$ the top level is of dimension 24, whereas for $M = V^\natural(2A)$ it is 1. In each case the extended automorphism group acts irreducibly. To establish Theorem 2 we argue first that the top level of $V^\natural(2A)$ is a *trivial* module for $2Baby$, using the structure of the Leech lattice as well as the fusion rules for $L(\frac{1}{2}, 0)$ and its modules established in [DMZ]. Then we can prove that the dimension is 1 by combining some monstrous calculation of Norton [N] together with further properties of the Virasoro algebra.

At this point we are in a position to combine the modular-invariance properties of [DM2], which relate graded traces of elements on $V^\natural(2A)$ to those on $V^\natural$, with Borcherds' proof of the original moonshine conjecture. This leads to the proof of Theorem 2.

We thank the referee for useful comments.

2. TRANSPOSITIONS AND VIRASORO ALGEBRAS

With the conformal grading (1.3) of the moonshine module one knows that $V_0^\natural = \mathbb{C}1$ is spanned by the vacuum and $V_1^\natural = 0$. $V_2^\natural$ carries the structure of a commutative non-associative algebra which we denote by B. As Monster module, B is the direct sum of two irreducible modules $\mathbb{C}\omega \oplus B_0$. We refer to both B and B_0 as 'the' *Griess algebra*. Note that $\frac{1}{2}\omega$ is the identity of B.

A *transposition* of $\mathbb{M}$ is an involution of type $2A$. If x is a transposition then $C_{\mathbb{M}}(x) \simeq 2Baby$. We fix such an x and let $H = C_{\mathbb{M}}(x)$. The space B^H of H-invariants on B is 2-dimensional and spanned by ω together with the so-called *transposition axis* t_x (cf. [C] or [MN]). Moreover H is the subgroup of $\mathbb{M}$ leaving t_x invariant. From this we easily conclude

Lemma 2.1. *Let $x_1, ..., x_k$ be transpositions, let $A = \langle t_{x_1}, ..., t_{x_k}, \omega \rangle$ be the subalgebra of B spanned by w and the corresponding transpositions axes, and let $E = \langle x_1, ..., x_k \rangle$ be the subgroup of $\mathbb{M}$ generated by the x_i. Then the subgroup $C_{\mathbb{M}}(E)$ of elements of $\mathbb{M}$ commuting with E coincides with the subgroup $C_{\mathbb{M}}(A)$ of $\mathbb{M}$ fixing A (elementwise).* $\square$

The subalgebra A of Lemma 2.1 is associative if, and only if, each product $x_i x_j$ $(i \neq j)$ is an involution of type $2B$ (Theorem 5, Corollary 1 of [MN]). In particular, E is elementary abelian in this case.

It was shown in [MN] that we may choose an associative algebra A of dimension 48 satisfying the conclusions of Lemma 2.1. It is sufficient to give 24 mutually orthogonal vectors v_i in the Leech lattice Λ such that v_i has squared length 4, in which case the 48 elements $\pm v_i$ correspond to 48 mutually orthogonal transposition axes of B (cf. [DMZ] and [Mi]).

One has some latitude in choosing the v_i. We may take, for example, $v_1 = (4, 4, 0^{22})$, $v_2 = (4, -4, 0^{22})$, $v_3 = (0^2, 4, 4, 0^{20})$, $v_4 = (0^2, 4, -4, 0^{20})$,. . . where the squared length of v_i is $\frac{1}{8}v_i \cdot v_i$. Similarly, we could choose any twelve pairs of coordinates and take the corresponding vectors. Such choices are not all equivalent, and in combination with a choice of Golay Code give rise to the notion of a "marked" Golay Code [CM].

Lemma 2.2. *We may choose the v_i so that the corresponding elementary abelian group E satisfies $|E| \leq 2^9$.*

Proof. We may take $E \leq Q = O_2(C) \simeq 2_+^{1+24}$ where C is the centralizer of an involution of type $2B$ in M. We identify $Q/Z(Q)$ with $\Lambda/2\Lambda$, so that each $\pm v_i$ becomes an involution (of type $2A$) in Q. Clearly $E = \langle \pm v_i \rangle = \langle -1, v_i \rangle$. We have $v_1 + v_2 = (8, 0^{23})$ and for $i \geq 1$, $v_{2i+1} + v_{2i+2} = (0^{2i}, 8, 0^{23-2i}) = v_1 + v_2 + 2(-4, 0^{2i-1}, 4, 0^{23-2i}) \equiv v_1 + v_2 \pmod{2\Lambda}$. So $E = \langle -1, v_2, v_{2i-1}, 1 \leq i \leq 12 \rangle$.

Suppose now that we mark our Golay Code so that the six 4-element sets $\{1, 2, 3, 4\}$, ..., $\{21, 22, 23, 24\}$ constitute a *sextet* i.e., the union of any two of them is an *octad* (=block in the Witt design). Then $v_1 + v_3 + v_5 + v_7 = (4^8, 0^{16}) = 2(2^8, 0^{16}) \in 2\Lambda$ and similarly $v_1 + v_3 + v_{4i+1} + v_{4i+3} \in 2\Lambda$ for $1 \leq i \leq 5$. Hence $E = \langle -1, v_2, v_1, v_3, v_5, v_9, v_{13}, v_{17}, v_{21} \rangle$. $\square$

Corollary 2.3. *With the choice made in Lemma 2.2, the corresponding associative algebra A of dimension 48 is fixed (pointwise) by elements in $\mathbb{M}$ of types $2A$, $2B$ and $4A$.*

Proof. These are the three types of Monster elements (apart from 1) contained in Q. Since Q is extra-special of order 2^{25} and $E \leq Q$ has order less than or equal to 2^9, then certainly $C_Q(E)$ contains elements of these three types. Now apply Lemma 2.1. $\square$

It was shown in [DMZ] that corresponding to the 48 transpositions $x_1, ..., x_{48}$ with transposition axes spanning a 48-dimensional associative algebra we obtain a certain Virasoro algebra of central charge 24. To explain this, first recall that the Virasoro algebra Vir is spanned by L_n, $n \in \mathbb{Z}$ and a central element c satisfying the relation

$$[L_m, L_n] = (m - n)L_{m+n} + \frac{m^3 - m}{12}\delta_{m+n,0}c. \qquad (2.1)$$

Lemma 2.4. *Let ω_1 and ω_2 be two orthogonal idempotents such that the components of the vertex operator $Y(\omega_i, z)$ generate a copy Virasoro algebra with central charge $1/2$. Then the actions of these two Virasoro algebras are commutative on $V^\natural$.*

Proof. Set $Y(\omega_i, z) = \sum_{n \in \mathbb{Z}} L^i(n)z^{-n-2}$ for $i = 1, 2$. Then the inner product of ω_1 with ω_2 is given by $L^1(2)\omega_2$, which is 0 by assumption. By Norton's inequality (cf [C] or [MN]) $0 = (\omega_1, \omega_2) = (\omega_1^2, \omega_2^2) \geq (\omega_1\omega_2, \omega_1\omega_2) \geq 0$, the product $\omega_1\omega_2$ (which is $L^1(0)\omega_2$) in the Griess algebra is also 0. Thus $L^1(n)\omega_2 = 0$ for all nonnegative integrals n

and ω_2 is a highest weight vector with highest weight 0 for the Virasoro algebra Vir_1 generated by $L^1(m)$. The submodule of $V^\natural$ for Vir_1 generated by ω_2 is necessarily isomorphic to $L(1/2,0)$. Here $L(k,h)$ is the simple highest weight representation of Vir with central change k and highest weight h. From the module structure of $L(1/2,0)$ we see immediately that $L^1(-1)\omega_2 = 0$. Now use the commutator formula $[L^1(m), L^2(n)] = \sum_{i=-1}^{\infty} \binom{m+1}{i+1}(L^1(i)\omega_2)_{m+n+1-i} = 0$ to complete the proof. $\square$

Some multiple ω_i of the transposition axis t_{x_i} is an idempotent such that the components of the vertex operator $Y(\omega_i, z)$ generate a copy Virasoro algebra with central charge $1/2$. As the ω_i are orthogonal the algebras mutually commute as operators on $V^\natural$ by Lemma 2.4, yielding an action of the sub VOA

$$L = L(\frac{1}{2}, 0)^{\otimes 48} \tag{2.2}$$

on $V^\natural$.

Note that if $g \in \mathbb{M}$ fixes the associative algebra A spanned by the t_{x_i} then g fixes each ω_i. Hence we get

Lemma 2.5. *If g is of type 2A, 2B or 4A then the fixed subalgebra $(V^\natural)^{\langle g \rangle}$ contains L.*

The sum of all the ω_i is the conformal vector ω. If $H = 2Baby$ fixes ω_1, say, then it also fixes ω_0 where $\omega = \omega_1 + \omega_0$. Now the component operators of $Y(\omega_0, z)$ generate a copy of the Virasoro algebra of central charge $47/2$, and in any case we get a subspace of $(V^\natural)^H$ corresponding to $Y(\omega_1, z)$ and $Y(\omega_0, z)$, that is to say the subspace $L(\frac{1}{2},0) \otimes L(\frac{47}{2},0)$ which is again a vertex operator algebra [FHL].

Now one knows (cf. [KR] for example) that the q-characters of these two algebras are as follows:

$$ch_q L(\frac{1}{2}, 0) = \frac{1}{2}\left(\prod_{n=1}^{\infty}(1 + q^{n-1/2}) + \prod_{n=1}^{\infty}(1 - q^{n-1/2})\right)$$
$$ch_q L(\frac{47}{2}, 0) = \prod_{n=2}^{\infty}(1 - q^n)^{-1}. \tag{2.3}$$

We thus calculate that the subspace $L(\frac{1}{2},0) \otimes L(\frac{47}{2},0) \subset (V^\natural)^H$ has q-character

$$1 + 2q^2 + 2q^3 + 5q^4 + 6q^5 + 12q^6 + \cdots. \tag{2.4}$$

Lemma 2.6. *Up to weight six, all elements of* $(V^\natural)^H$ *lie in* $L(\frac{1}{2},0) \otimes L(\frac{47}{2},0)$.

Proof. Norton has calculated (Table 5 of [N]) the decomposition of the permutation representation of $\mathbb{M}$ on the conjugacy class of transpositions into simple characters for $\mathbb{M}$. The representation is multiplicity free and the characters occurring are precisely $\chi_1, \chi_2, \chi_4, \chi_5, \chi_9, \cdots$ where we order the simple characters according to their degrees as in [Cal].

By elementary character theory, if V_i is the $\mathbb{M}$-module affording χ_i then the dimension of the space of H $(=2Baby)$ invariants V_i^H is precisely the multiplicity of χ_i above. Hence it is either 0 or 1 and is 1 precisely if $i = 1, 2, 4, 5, 9, \cdots$.

Now the decomposition of the first few homogeneous spaces $V_n^\natural$ into simple Monster module is known (e.g. [MS]). We have

$$V_0^\natural = V_1, \quad V_1^\natural = 0$$

$$V_2^\natural = V_1 \oplus V_2, \quad V_3^\natural = V_1 \oplus V_2 \oplus V_3$$

$$V_4^\natural = 2V_1 \oplus 2V_2 \oplus V_3 \oplus V_4 \qquad (2.5)$$

$$V_5^\natural = 2V_1 \oplus 3V_2 \oplus 2V_3 \oplus V_4 \oplus V_6$$

$$V_6^\natural = 4V_1 \oplus 5V_2 \oplus 3V_3 \oplus 2V_4 \oplus V_5 \oplus V_6 \oplus V_7.$$

So to compute $\sum_{n=0}^6 \dim(V_n^\natural)^H q^n$, from the foregoing it is equivalent to counting the number of occurrences of $\chi_1, \chi_2, \chi_4, \chi_5$ in (2.5). We find that the multiplicities are precisely those of (2.4). $\qquad\square$

3. Proof of Theorem 1

We need to quote some results from [DM2]. First, following [Z] and [DM2], we say that the VOA V satisfies the *Virasoro condition* if V is a sum of highest weight modules for the Virasoro algebra generated by the components of $Y(\omega, z)$; we say that V satisfies the C_2 *condition* if $V/C_2(V)$ is finite-dimensional, where $C_2(V)$ is spanned by $u_{-2}v$ for $u, v \in V$. What is important for us is that because $V^\natural$ contains L as a rational subalgebra, $V^\natural$ satisfies the C_2 condition as well as the Virasoro condition. See [Z] for more information on this point.

Theorem 3.1. *[DM2] Suppose that V is a holomorphic VOA satisfying both Virasoro and C_2 conditions. Let $g \in Aut(V)$ have finite order. Then the following hold:*

(i) V has at most one simple g-twisted module.

(ii) V has at least one weak simple g-twisted module.

So to prove parts (i) and (ii) of Theorem 1, it is enough to prove that a weak simple g-twisted $V^\natural$-module for g of type $2A, 2B$ or $4A$ has finite-dimensional homogeneous spaces. Let M be such a module.

Let $W = (V^\natural)^{\langle g \rangle}$. Then M is an ordinary weak W-module, in fact it is the sum of a finite number of simple weak W-modules (see [DM3], for example). So it suffices to prove that a weak simple W-module N, say, has finite-dimensional homogeneous spaces.

It is shown in [DMZ] that $L(\frac{1}{2}, 0)$ is a rational VOA with three irreducible modules $L(\frac{1}{2}, h_i)$, $h_i = 0, 1/2, 1/16$ which are exactly the highest weight unitary representations of Vir with central charge $1/2$. Moreover all weak modules are ordinary modules. The characters of these modules are as follows: as well as $L(\frac{1}{2}, 0)$ given in (2.3) we have

$$ch_q L(\frac{1}{2}, \frac{1}{2}) = \frac{1}{2}\left(\prod_{n=1}^{\infty}(1 + q^{n-1/2}) - \prod_{n=1}^{\infty}(1 - q^{n-1/2})\right)$$
$$ch_q L(\frac{1}{2}, \frac{1}{16}) = q^{1/16}\prod_{n=1}^{\infty}(1 + q^n). \tag{3.1}$$

The fusion rules are as follows:

$$L(\frac{1}{2}, \frac{1}{2}) \times L(\frac{1}{2}, \frac{1}{2}) = L(\frac{1}{2}, 0)$$

$$L(\frac{1}{2}, \frac{1}{2}) \times L(\frac{1}{2}, \frac{1}{16}) = L(\frac{1}{2}, \frac{1}{16}) \tag{3.2}$$

$$L(\frac{1}{2}, \frac{1}{16}) \times L(\frac{1}{2}, \frac{1}{16}) = L(\frac{1}{2}, 0) + L(\frac{1}{2}, \frac{1}{2})$$

where the "product" $\times$ is commutative and $L(\frac{1}{2}, 0)$ is the identity.

From these facts we conclude (cf. [FHL]) that L has just 3^{48} simple modules, namely

$$L(h_1, ..., h_{48}) = L(\frac{1}{2}, h_1) \otimes \cdots \otimes L(\frac{1}{2}, h_{48}) \tag{3.3}$$

where $h_i \in \{0, 1/2, 1/16\}$. The corresponding fusion rules (see Proposition 2.10 of [DMZ]) are

$$L(h_1, ..., h_{48}) \times L(h_1', ..., h_{48}')$$
$$= (L(\frac{1}{2}, h_1) \times L(\frac{1}{2}, h_1')) \otimes \cdots \otimes (L(\frac{1}{2}, h_{48}) \times L(\frac{1}{2}, h_{48}')) \tag{3.4}$$

The fusion rules can be interpreted in terms of a tensor product $\boxtimes$ of modules for vertex operator algebras (see [HL] and [L]). Then the product $\times$ in (3.2) and (3.4) can be replaced by the tensor product $\boxtimes$ to get the tensor product decomposition. We refer the reader to [L] for more details.

Now if $0 \neq n \in N$ is a highest weight vector with highest weight $(h_1, ..., h_{48})$ for the Virasoro algebra generated by the components of $Y(\omega_i, z)$, $i = 1, ..., 48$, then the L-module generated by n is isomorphic to $L(h_1, ..., h_{48})$. Let $N_{h'_1, ..., h'_{48}}$ be the multiplicity of $L(h'_1, ..., h'_{48})$ in W. Then as an L-module W has decomposition

$$W = \oplus_{h'_i \in \{0, 1/2, 1/16\}} N_{h'_1, ..., h'_{48}} L(h'_1, ..., h'_{48}).$$

Then we have the tensor product of L-modules

$$W \boxtimes L(h_1, ..., h_{48}) = \oplus_{h'_i} N_{h'_1, ..., h'_{48}} L(h'_1, ..., h'_{48}) \boxtimes L(h_1, ..., h_{48})$$

which is an ordinary L-module with finite-dimensional homogeneous subspaces. From Proposition 4.1 of [DM3] we see that N is spanned by $w_m n$ for $w \in W$ and $m \in \mathbb{Z}$. Now using the universal property of tensor product, we conclude that N is a submodule of $W \boxtimes L(h_1, ..., h_{48})$ as L-modules. Thus each homogeneous subspace of N is finite-dimensional. This completes the proof of parts (i) and (ii) of Theorem 1.

If $H = 2Baby$ is the centralizer of $g = 2A$ in $\mathbb{M}$ then, as explained in [DM1] and [DM4], the *uniqueness* of $V^{\natural}(2A)$ yields a projective representation of H on $V^{\natural}(2A)$. This must be an ordinary and faithful representation since $H^2(Baby, \mathbb{C}^*) \simeq \mathbb{Z}_2$. Thus the group of extended automorphisms of $V^{\natural}(2A)$ is precisely $\mathbb{C}^* \times H$, as claimed in the introduction.

4. TOP LEVEL OF $V^{\natural}(2A)$

First we review some further results from [DM2]. Under the assumptions of the Virasoro condition and the C_2 condition, which we know hold for $V^{\natural}$, together with the complete reducibility of $V^{\natural}$-modules [D], the modular-invariance properties established in [DM2] may be stated as follows:

$$Z(g, h, \gamma \tau) = \sigma(\gamma^{-1}, g, h) Z((g, h)\gamma, \tau). \tag{4.1}$$

Here (g, h) is a pair of elements which generate a cyclic group, $Z(g, h, \tau)$ is a function in the so-called (g, h)-conformal block, $\gamma \in SL(2, \mathbb{Z})$ and $\sigma(\gamma^{-1}, g, h)$ is a nonzero constant. The notation $(g, h)\gamma$ is the action of $SL(2, \mathbb{Z})$ on pairs of commuting elements:

$$(g, h) \begin{pmatrix} a & b \\ c & d \end{pmatrix} = (g^a h^c, g^b h^d). \tag{4.2}$$

What is important is that if g is either $1A$ or one of $2A, 2B$ or $4A$, so that Theorem 1 applies, then $qZ(g, h, \tau)$ is precisely the graded trace of h on the g-twisted sector $V^\natural(g)$. In particular, taking $g = 1$, $h = 2A$ and $\gamma = \begin{pmatrix} 0 & -1 \\ 1 & 0 \end{pmatrix} = S$ in (4.1) yields

$$Z(1, 2A, S\tau) = \sigma(S^{-1}, 1, 2A)Z(2A, 1, \tau). \tag{4.3}$$

By Borcherds' theorem [B2], $Z(1, 2A, \tau)$ is a hauptmodul, specifically the one denoted by $2+$ in [CN]. This means that $Z(1, 2A, \tau)$ is invariant under the Fricke involution $W_2 = \begin{pmatrix} 0 & -1 \\ 2 & 0 \end{pmatrix} = S \begin{pmatrix} 2 & 0 \\ 0 & 1 \end{pmatrix}$, so that (4.3) yields

$$\begin{aligned} Z(1, 2A, \tau/2) &= Z(1, 2A, W_2(\tau/2)) \\ &= Z(1, 2A, S\tau) \\ &= \sigma(S^{-1}, 1, 2A)Z(2A, 1, \tau). \end{aligned}$$

Now $Z(1, 2A, \tau/2) = q^{-1/2} + O(q^{1/2})$, and $qZ(2A, 1, \tau)$ is the q-character of $V^\natural(2A)$. We conclude that

Lemma 4.1. *The q-character of $V^\natural(2A)$ has the form*

$$k(q^{1/2} + O(q^{3/2}))$$

for some constant $k \neq 0$.

We will establish

Theorem 4.2. *The constant k is equal to 1.*

In the following we let $M = V^\natural(2A)$ with $M_{1/2}$ the top level of M. Now $M_{1/2}$ is spanned by highest weight vectors corresponding to simple L-modules $L(h'_1, ..., h'_{48}) \subset M$ satisfying $\sum_i h'_i = 1/2$. Then $M^1 = \sum_{n \in \mathbb{Z}} M_{n+\frac{1}{2}}$ and $M^0 = \sum_{n \in \mathbb{Z}} M_n$ are irreducible W-modules where $W = (V^\natural)^{\langle 2A \rangle}$ (see Theorem 6.1 of [DM4]). Since we are mainly concerned with the top level $M_{1/2}$ we will only pay attention to the W-module M^1. We have already explained that W is the sum of simple L-modules $W = W_1 + \cdots + W_t$ and M^1 is spanned by $w_l m$ for $w \in W$, $l \in \mathbb{Z}$, and fixed $0 \neq m \in M_{1/2}$. So if we choose m to be a highest weight vector in the simple L-module $N = L(h'_1, ..., h'_{48}) \subset M^1$ we get information about M^1 by considering the fusion rules $W_i \times N$ or the tensor product $W_i \boxtimes N$.

In particular, we can bound $\dim M_{1/2}$ by estimating how many indices i are such that $W_i \boxtimes N$ contains a simple L-module of highest weight $1/2$. Let $W_i = L(h_1, ..., h_{48})$ and assume that $W_i \boxtimes N \supset L(k_1, ..., k_{48})$ with $\sum_i k_i = 1/2$.

Lemma 4.3. *If* $N = L(\frac{1}{2}, 0^{47})$ *with '1/2' in any position then Theorem 4.2 holds.*

Proof. Without loss take '1/2' in the first position. Then we have $k_j = h_j$, $2 \le j \le 48$, and $k_1 = 1/2$ if $h_1 = 0$; $k_1 = 0$ if $h_1 = 1/2$; $k_1 = 1/16$ if $h_1 = 1/16$. This follows from the fusion rules (3.2).

Since $\sum_j h_j$ is a nonnegative integer not equal to 1 (since $V_n^\natural = 0$ for $n < 0$ or $n = 1$), the condition $\sum_i k_i = 1/2$ forces $h_1 = \cdots h_{48} = 0$, that is $W_i = L$. Since L has multiplicity 1 in $V^\natural$ and $L \boxtimes N = N$, the lemma follows. $\square$

From now on we assume that all highest weight vectors in $M_{1/2}$ generate L-modules of type $L((\frac{1}{16})^8, 0^{40})$ for some distribution of the 1/16's. For now we take $N = L((\frac{1}{16})^8 | 0^{40})$, meaning that the 1/16's are in the first 8 coordinate positions (this is purely for notational convenience).

Lemma 4.4. *If* W_i *is such that* $W_i \times N \supset N' = L(k_1, ..., k_{48})$ *with* $\sum_i k_i = 1/2$ *then one of the following holds:*
    (a) $W_i = L(0, (\frac{1}{2})^3, (\frac{1}{16})^4 | (\frac{1}{16})^4, 0^{36})$ *and* $N' = L((\frac{1}{16})^4, 0^4 | (\frac{1}{16})^4, 0^{36})$;
    (b) $W_i = L((\frac{1}{2})^8 | 0^{40})$ *and* $N' = N$
    (c) $W_i = L((\frac{1}{2})^6, 0^2 | 0^{40})$ *and* $N' = N$;
    (d) $W_i = L((\frac{1}{2})^4, 0^4 | 0^{40})$ *and* $N' = N$;
    (e) $W_i = L$ *and* $N' = N$.
(No specific ordering of the first 8 coordinates or the last 40 entries is implied.)

Proof. Let $W_i = (0^a, (\frac{1}{2})^b, (\frac{1}{16})^c | h_9, ..., h_{48})$. Using the fusion rules (3.2), we see that $N' = ((\frac{1}{16})^{a+b}, 0^{c-d}, (\frac{1}{2})^d | h_9, ..., h_{48})$ for some $0 \le d \le c$.
We have, setting $s = \sum_{i=9}^{48} h_i$, that

$$\frac{a+b}{16} + \frac{d}{2} + s = \frac{1}{2}$$

$$\frac{b}{2} + \frac{c}{16} + s = 0 \ \ or \ \ge 2 \ , lies \ in \ \mathbb{Z}$$

$$a + b + c = 8.$$

It follows that $c \equiv 0 \pmod 4$ and $d = 0$. If $c = 8$ then $a + b = 0$, so $s = 1/2$ and $c/16 + s = 1$, contradiction. So $c = 0$ or 4. If $c = 4$ then $a + b = 4$, $s = 1/4$, and then $b = 3$. If $c = 0$ then $a + b = 8, s = 0, b = 4, 6$ or 8. The lemma follows. $\square$

Lemma 4.5. *In the notation of Lemma 4.4, if $W_i \times N = N' \subset M^1$ then $w_{wt\, w-1}m \neq 0$ where w is a nonzero highest weight vector of W_i. Moreover $v_{wtv-1}m$ is a scalar multiple of $w_{wtw-1}m$ for any homogeneous $v \in W_i$.*

Proof. We need to use results on the Zhu algebra $A(V)$ and its bimodule $A(W_i)$ (which is a quotient of W_i by a subspace) to prove this result. We refer the reader to [Z] and [FZ] for the definitions. Let m' be a nonzero highest weight vector of N'. By Theorem 1.5.3 of [FZ] and Theorem 4.2.4 of [L], $A(W^i) \otimes_{A(L)} \mathbb{C}m$ is isomorphic to $\mathbb{C}m'$ as $A(L)$-modules under the map $\bar{v} \otimes m \to v_{wtv-1}m$ where $\bar{v}$ is the image of $v \in W_i$ in $A(W_i)$. By Lemma 2.8 and Proposition 3.3 of [DMZ], $A(L)$ is isomorphic to the associative commutative algebra $\mathbb{C}[t_j | j = 1, ..., 48]/I$ where I is the ideal generated by $t_j(t_j - \frac{1}{2})(t_j - \frac{1}{16})$. By Lemma 2.9, Propositions 3.1 and 3.4 of [DMZ], $A(W_i)$ is isomorphic to $\mathbb{C}[x_j, y_j | j = 1, ..., 48]/I_i$ where I_i is a certain ideal of $\mathbb{C}[x_j, y_j | j = 1, ..., 48]$ and the left and right actions of $t_j + I$ on $A(W_i)$ are multiplications by x_j and y_j respectively. Moreover, under this identification, $\bar{w}$ is mapped to $1 + I_i$. Since $A(W_i)$ is generated by $\bar{w}$ as $A(L)$-bimodule, the lemma follows immediately. $\square$

Lemma 4.6. *$M_{1/2}$ is a trivial module for $H = 2Baby$.*

Proof. The minimal nontrivial representation of H has degree 4371 [Cal] so it suffices to show that $\dim M_{1/2}$ is less than this. Now the multiplicity of $L(h_1, ..., h_{48})$ in $V^\natural$ is less than or equal to 1 if all $h_i \in \{0, 1/2\}$ by Proposition 5.1 of [DMZ]. So by Lemma 4.4 we conclude that the multiplicity of $N = L((\frac{1}{16})^8 | 0^{40})$ in M^1 is at most $\binom{8}{0} + \binom{8}{4} + \binom{8}{6} + \binom{8}{8} = 100$. All other simple L-submodules of M^1 of weight $1/2$ arise from the action of $L(0, (\frac{1}{2})^3, (\frac{1}{16})^4 | (\frac{1}{16})^4, 0^{36}) \subset V_2^\natural$, and the multiplicity[1] μ of *all* modules of this type in $V^\natural$ is given in Theorem 6.5 [DMZ]. Since $\mu = 24 \cdot 2^6$, the lemma follows. $\square$

Lemma 4.7. *$M_{1/2}$ is contained in the span of $v_{wtv-1}m$ for $v \in V_n^\natural$ with $n \leq 4$.*

---

[1]Some of the multiplicities stated in Theorem 6.5 of [DMZ] are inaccurate. See [Ho] for corrections.

Proof. First we observe from Lemma 4.4 that the highest weight vectors in W_i have weights 0,2,3 or 4. By Corollary 4.2 of [DM3], $M_{1/2}$ is spanned by $v_{wt\,v-1}m$ for $v \in W_i$ with W_i occurring in Lemma 4.4. So it is enough to show that the span of $v_{wt\,v-1}m$ for $v \in W_i$ is contained in $w_{wt\,w-1}m$ where w is a nonzero highest weight vector of W_i. This follows from Lemma 4.5. □

Now we can complete the proof of Theorem 4.2. We may represent the conclusion of Lemma 4.7 by the containment $(\oplus_{n=0}^4 V_n^\natural)m \supset M_{1/2}$. Since $M_{1/2}$ is a trivial H-module by Lemma 4.6, it follows that $M_{1/2} \subset (\oplus_{n=0}^4 (V_n^\natural)^H)m$.

Now use Lemma 2.6 to see that $M_{1/2}$ lies in the space spanned by $u_{wt\,u-1}m$ for $u \in L(\frac{1}{2},0) \otimes L(\frac{47}{2},0)$ homogeneous. But m is an eigenvector for such operators. So we get $M_{1/2} = \mathbb{C}m$, as required. □

We are now ready for the proof of Theorem 1 (iii). Let M be an irreducible g-twisted $V^\natural$-module so that the top level of M is $\mathbb{C}m$. Let N be the L-submodule of M generated by m. Then by Lemmas 4.3 and 4.4, N is either $L(\frac{1}{2},0^{47})$ or $L((\frac{1}{16})^8|0^{40})$. Then $W_i \times N = N$ if, and only, if in the former case $W_i = L$, and in the latter case W_i appears in (b)-(e) of Lemma 4.4. In either case, since M is simple we see that the subspace of N spanned by $v_l m$ for $v \in W_i$ and $l \in \mathbb{Z}$ is N (by Proposition 11.9 of [DL]). Again by Lemma 4.5, $w_{wt\,w-1}m \neq 0$.

Let X be a g-twisted $V^\natural$-module. Then X has a finite composition series (see [DLM]). Using induction on the number of composition-factors of X, we only need to prove that X is completely reducible if X has two factors. It is shown in [DLM] that X is a completely reducible g-twisted $V^\natural$-module if, and only, if $X_{\frac{1}{2}}$ is a semisimple $A_g(V^\natural)$-module via the action $v_{wt\,v-1}$ for homogeneous $v \in V^\natural$. If $N = L(\frac{1}{2},0^{47})$ it is clear that $X_{\frac{1}{2}}$ is a semisimple $A_g(V^\natural)$-module from the discussion above. Here $A_g(V^\natural)$ is the twisted Zhu algebra as defined and used in [DLM].

Now we assume that $N = L((\frac{1}{16})^8|0^{40})$. Let $X_{\frac{1}{2}} = \mathbb{C}x_1 + \mathbb{C}x_2$ such that x_1 generates an irreducible g-twisted $V^\natural$-module X^1 (which is necessarily isomorphic to M). Using the associativity of vertex operators (see the proof of Proposition 4.1 of [DM3], for example) we see that $w_{wt\,w-1}w_{wt\,w-1}x_2$ is a nonzero multiple of x_2. Thus $w_{wt\,w-1}$ acts semisimply on $X_{\frac{1}{2}}$, hence acts as a scalar. Thus for all homogeneous $v \in W_i$ the action of $v_{wt\,v-1}$ on $X_{\frac{1}{2}}$ is semisimple. Note that the image of $A_g(V^\natural)$

in $End(X_{\frac{1}{2}})$ is a subalgebra of dimension less than or equal to 2. So we conclude that $X_{\frac{1}{2}}$ is indeed a semisimple $A_g(V^\natural)$-module.

5. PROOF OF THEOREM 2

Let $g \in \mathbb{M}$ be of type $2A$ and let h be an element of $C_{\mathbb{M}}(g)$ of *odd* order N. We use the notation of (4.1)-(4.2)

Lemma 5.1. *Let $F \leq SL(2, \mathbb{R})$ be the fixing group of $Z(1, gh, \tau)$. Then F contains the Atkin-Lehner involution W_2.*

Proof. By Borcherds' theorem [B2], each $Z(1, x, \tau)$ for $x \in \mathbb{M}$ is a hauptmodul on a discrete group $F = F_x \leq SL(2, \mathbb{R})$. Then F is precisely the group conjectured in [CN], Table 2. This has been established by the work of several authors; see [CN] and [F] for further references.

It is a fact that if $x = gh$ then W_2 always lies in F. In fact, the group F is of the form $2N+$, or $2N + 2$ in all but one case. The exception is the element $30F$, where the group is $30+2$, $15,30$. In any case, the lemma follows. $\square$

Now we may take $W_2 = \begin{pmatrix} a & b \\ cN & 2d \end{pmatrix} \begin{pmatrix} 2 & 0 \\ 0 & 1 \end{pmatrix}$ where $\gamma = \begin{pmatrix} a & b \\ cN & 2d \end{pmatrix} \in$ $SL(2, \mathbb{Z})$. Then we have from (4.1) that

$$\begin{aligned} Z(1, gh, \tau/2) &= Z(1, gh, W_2(\tau/2)) \\ &= \sigma Z((1, gh)\gamma, \tau) \\ &= \sigma Z((gh)^{cN}, (gh)^{2d}, \tau) \\ &= \sigma Z(g, h', \tau). \end{aligned}$$

Here, σ is a constant and $h' = h^{a'}$ where $aa' \equiv 1 \ (mod \ N)$. But $Z(1, gh, \tau/2) = q^{-1/2} + \cdots$ and $Z(g, h', \tau) = q^{-1/2} + \cdots$ by Theorem 4.2. So $\sigma = 1$. Now Theorem 2, part (i) follows immediately.

As for part (ii) of Theorem 2, if we now take $h \in \mathbb{M}$ such that $g \in \langle h \rangle$ then we can find $\gamma \in SL(2, \mathbb{Z})$ such that $(g, h) = (1, h)\gamma$. Then (4.1) yields

$$Z(g, h, \tau) = \sigma Z(1, h, \gamma\tau) \tag{5.1}$$

for some constant σ. As $Z(1, h, \tau)$ is a hauptmodul by Borcherds' theorem then so is $Z(g, h, \tau)$ by (5.1).

Now let us consider $Z(g, h, \tau)$ where g, h and gh are all of type $2A$. From [DM2] we know that $Z(g, h, \tau)$ is holomorphic on the upper half-plane and meromorphic at the cusps. Furthermore (4.1) still holds because of Theorem 1 (iii). We see that if $\gamma \in SL(2, \mathbb{Z})$ then

$$Z(g, h, \gamma\tau) = \sigma(\gamma)Z(g, h, \tau)$$

for some constant $\sigma(\gamma)$. So σ is a character of $SL(2, \mathbb{Z})$. As $T = \begin{pmatrix} 1 & 1 \\ 0 & 1 \end{pmatrix}$ covers the abelianization of $SL(2, \mathbb{Z})$ then the kernel K of σ is the subgroup of $SL(2, \mathbb{Z})$ of index 2. As $i\infty$ is the unique cusp for K we see easily that K is indeed of genus zero with hauptmodul $Z(g, h, \tau)$. It is in fact the function denoted $t_{2/2} = \sqrt{J(q) - 984} = q^{-1/2} - 492q^{1/2} - 22590q^{3/2} + \cdots$ in [CN]. To see this, note that from Theorem 2 (i) with $h = 1$, combined with tables in [CN] and the results of [B2], we see that $Z(g, 1, \tau) = q^{-1/2} + 4372q^{1/2} + \cdots$. This tells us that the weight $\frac{3}{2}$ subspace of $V^\natural(2A)$ is the module $1 \oplus 4371$ for $2Baby$ (see [Cal]), on which h has trace -492 (ibid). Thus $Z(g, h, \tau) = q^{-1/2} - 492q^{1/2} + \cdots$ as claimed.

6. Final comments

We have rather ignored the twisted sectors $V^\natural(2B)$ and $V^\natural(4A)$. As we have said, there is a construction of $V^\natural(2B)$ in [Hu], and its existence also follows from [DM2] without the necessity of the effort we needed to understand $V^\natural(2A)$. Huang also constructs the $2B$-orbifold, i.e., puts an abelian intertwining algebra structure [DL] on $V^\natural \oplus V^\natural(2B)$. This is closely related to the construction of $V^\natural$ in [FLM].

Concerning the extended automorphism group it is known [G] that the centralizer C of $2B$ in $\mathbb{M}$ is a non-split extension of $\cdot 1$ (largest simple Conway group) by the extra-special group 2_+^{1+24}. Furthermore (loc.cit.) $H^2(C, \mathbb{C}^*) \simeq \mathbb{Z}_2$. In fact if $\cdot 0$ is the 2-fold cover of $\cdot 1$ (=automorphism group of the Leech lattice) there is a diagram

$$
\begin{array}{ccccccccc}
1 & \to & 2_+^{1+24} & \to & C & \to & \cdot 1 & \to & 1 \\
& & \| & & \uparrow & & \uparrow & & \\
1 & \to & 2_+^{1+24} & \to & \hat{C} & \to & \cdot 0 & \to & 1
\end{array}
$$

and $\hat{C}$ is the universal central extension of C.

Griess also shows that $\hat{C}$ has a simple module of degree 2^{12}, whereas C has no such representation. The smallest faithful irreducible representation for C has dimension $24 \cdot 2^{12}$.

Now the weight $3/2$ subspace of $V^\natural(2B)$ has dimension 2^{12} (see [Hu] and [DM2]) and there is a projective representation of C on $V^\natural(2B)$ by [DM1] and [DM4]. The conclusion is thus

Lemma 6.1. *The extended automorphism group of $V^\natural(2B)$ is a non-split extension*

$$1 \to \mathbb{C}^* \to Aut^e(V^\natural(2B)) \to C \to 1.$$

It has commutator subgroup isomorphic to $\hat{C}$.

Theorem 6.2. *Let $h \in \mathbb{M}$ be such that $\langle h \rangle$ contains $g = 2B$ or $4A$. Then $Z(g, h, \tau)$ is a hauptmodul.*

Proof. Same as the proof of Theorem 2 (ii).

Finally, we consider the q-characters of $V^\natural(2B)$ and $V^\natural(4A)$. Note by [CN] or [FLM] that we have for $g \in \mathbb{M}$ of type $2B$ that

$$Z(1, g, \tau) = 24 + q^{-1} \prod_{n \, odd} (1 - q^n)^{24} = 24 + \frac{\eta(\tau)^{24}}{\eta(2\tau)^{24}}$$

where $\eta(\tau)$ is the Dedekind eta-function. By (4.1) and the transformation law for $\eta(\tau)$ [K] we get the q-character of $V^\natural(2B)$ equal to

$$Z(g, 1, \tau) = \sigma Z(1, g, S\tau) = \sigma \left\{ 24 + \frac{2^{12}\eta(\tau)^{24}}{\eta(\tau/2)^{24}} \right\}.$$

In fact $\sigma = 1$, as we know from [Hu] and [DM2].

Similarly for t of type $4A$ we have

$$Z(1, t, \tau) = -24 + \frac{\eta(2\tau)^{48}}{\eta(\tau)^{24}\eta(4\tau)^{24}}$$

which leads to

$$Z(t, 1, \tau) = \sigma \left\{ -24 + \frac{\eta(\tau/2)^{48}}{\eta(\tau)^{24}\eta(\tau/4)^{24}} \right\}.$$

Presumably $\sigma = 1$ in this case too, but a proof would be more difficult than that for $2A$ given above.

REFERENCES

[B1]   R. E. Borcherds, Vertex algebras, Kac-Moody algebras, and the Monster, *Proc. Natl. Acad. Sci. USA* **83** (1986), 3068-3071.

[B2]   R. E. Borcherds, Monstrous moonshine and monstrous Lie superalgebras, *Invent. Math.* **109** (1992), 405-444.

[CM]   M. Conder and J. McKay, Solution of the marking problem for M_{24}, preprint.

[C]   J. H. Conway, A simple construction for the Fischer-Griess Monster group, *Invent. Math.* **79** (1984), 513-540.

[Cal]   J. H. Conway et al, Altas of Finite Groups, Clarendon Press, Oxford, 1985.

[CN]   J. H. Conway and S. P. Norton, Monstrous Moonshine, *Bull. London. Math. Soc.* **12** (1979), 308-339.

[D]   C. Dong, Representations of the moonshine module vertex operator algebra, *Contemporary Math.* **175** (1994), 27-36.

[DL]   C. Dong and J. Lepowsky, Generalized Vertex Algebras and Relative Vertex Operators, *Progress in Math.* Vol. 112, Birkhäuser, Boston, 1993.

[DLM]   C. Dong, H. Li and G. Mason, Twisted representations of vertex operator algebras, preprint.

[DM1]   C. Dong and G. Mason, Nonabelian orbifolds and the boson-fermion correspondence, *Commu. Math. Phys.* **163** (1994), 523-559

[DM2]   C. Dong, G. Mason, Elliptic functions and orbifold theory, preprint, USCS 1994.

[DM3]   C. Dong, G. Mason, On quantum Galois theory, preprint UCSC 1994, hep-th/9412039.

[DM4]   C. Dong, G. Mason, On the operator content of nilpotent orbifold models, preprint, UCSC 1994, hep-th/9412109.

[DMZ]   C.Dong, G. Mason and Y. Zhu, Discrete series of the Virasoro algebra and the moonshine module, *Proc. Symp. Pure. Math., American Math. Soc.* **56** II (1994), 295-316.

[F]   C. Ferenbaugh, The genus-zero problem for $n|h$-type groups, *Duke. Math. J.* **72** (1993), 31-63.

[FHL]   I. B. Frenkel, Y.-Z. Huang and J. Lepowsky, On axiomatic approaches to vertex operator algebras and modules, *Memoirs American Math. Soc.* **104**, 1993.

[FLM]   I. B. Frenkel, J. Lepowsky and A. Meurman, Vertex Operator Algebras and the Monster, *Pure and Applied Math.,* Vol. **134**, Academic Press, 1988.

[FZ]   I. B. Frenkel and Y. Zhu, Vertex operator algebras associated to representations of affine and Virasoro algebras, *Duke Math. J.* **66** (1992), 123-168.

[G]   R. Griess, The Friendly Giant, *Invent. Math.* **69** (1982), 1-102.

[Ho]   G. Höhn, Ph.D Dissertation, University of Bonn, 1995.

[Hu]   Y.-Z. Huang, A non-meromorphic extension of the moonshine module vertex operator algebra, *Contemporary Math.,* to appear.

[HL]   Y.-Z. Huang and J. Lepowsky, Toward a theory of tensor products for representations of a vertex operator algebra, in: *Proc. 20th Intl. Conference on Differential Geometric Methods in Theoretical Physics, New York, 1991,* ed. S. Catto and A. Rocha, World Scientific, Singapore, 1992, Vol. 1, 344-354.

[KR]   V. G. Kac and A. K. Raina, Highest Weight Representations of Infinite Dimensional Lie Algebras, *Adv. Series Math. Phy.* Vol. **2**, World Scientific, 1987.

[K]   N. Koblitz, Introduction to Elliptic Curves and Modular Forms, Springer-Verlay, New York, 1984.

[L]   H. Li, Representation theory and tensor product theory for vertex operator algebras, Ph.D. Dissertation, Rutgers University, 1994.

[M]   G. Mason, Finite groups and modular functions, *Proc. Symp. Pure. Math., American Math. Soc.* **47** (1987).

[MS]   J. McKay and H. Strauss, The q-series of monstrous moonshine and the decomposition of the head characters, *Commu. Alg.* **18** (1990), 253-278.

[MN]   W. Meyer and W. Neutsch, Associative subalgebras of the Griess algebra, *J. Algebra* **158** (1993),1-17.

[Mi]   M. Miyamoto, Griess algebras and Virasoro elements in vertex operator algebras, preprint.

[N]   S. Norton, The uniqueness of the Fischer-Griess Monster, *Contemporary Math.* **45** (1985).

[Z]   Y. Zhu, Vertex operator algebras, elliptic functions and modular forms, Ph.D. dissertation, Yale University (1990).

DEPARTMENT OF MATHEMATICS, UNIVERSITY OF CALIFORNIA, SANTA CRUZ
E-mail address: dong@cats.ucsc.edu, hli@cats.ucsc.edu, gem@cats.ucsc.edu

Contemporary Mathematics
Volume **193**, 1996

Spinor Constructior. of the
$c = 1/2$ Minimal Model

ALEX J. FEINGOLD, JOHN F. X. RIES, MICHAEL D. WEINER

1. Introduction

The representation theories of affine Kac-Moody Lie algebras and the Virasoro algebra have been found to have interesting connections with many other parts of mathematics and have played a vital role in the development of conformal field theory in theoretical physics.(See [**BPZ**], [**GO**], [**BMP**], [**MS**], [**TK**], the Introduction and references in [**FLM**].) An important contribution to Kac-Moody representation theory was the rigorous mathematical construction of representations using vertex operators, discovered independently by mathematicians and physicists, but known earlier by physicists. Once the connection was discovered, an exciting dialog began which has enriched both sides. The precise axiomatic definition of vertex operator algebras (VOA's) by mathematicians [**FLM**] gave a rigorous foundation to the algebraic aspects of conformal field theory called chiral algebras by physicists. This theory includes, unifies and vastly extends the representation theories of affine Kac-Moody algebras and the Virasoro algebra.

In the theory of VOA's one has the notion of a module and of intertwining operators going between modules [**FHL**], [**F**]. The main axiom for a VOA is an identity called the Jacobi-Cauchy identity because it combines the usual Jacobi identity for Lie algebras with the Cauchy residue formula for rational functions whose possible poles are limited to three points, 0, 1 and ∞. A slight modification is needed for the appropriate definition of a module, and intertwining operators are defined by a similar axiom relating them to vertex operators. We believe that

1991 *Mathematics Subject Classification.* Primary 17B65, 17B67, 81R10; Secondary 17B68, 17A70, 81T40.

A.J.F. was partially supported by the National Security Agency under Grant number MDA904-94-H-2019. The United States Government is authorized to reproduce and distribute reprints notwithstanding any copyright notation heron.

J.F.X.R. died July 6, 1993. This paper is dedicated to his memory.

This paper is in final form and no version of it will be submitted for publication elsewhere.

an important next step is the understanding of a new kind of "matrix" Jacobi-Cauchy identity relating any two intertwining operators. This would lead to a larger unifying structure, incorporating a VOA, its modules and its intertwining operators.

There are several new features which appear when trying to understand intertwining operators and the new kind of Jacobi-Cauchy identity they obey. First one has to deal with fusion rules,

$$\mathcal{N}(M_1, M_2, M_3) = \dim(\mathcal{I}(M_1, M_2, M_3)),$$

which give the dimension of the space of intertwining operators determined by a triple of modules. It is a basic principle of VOA's that there is a one-to-one correspondence between vectors v in a simple VOA V and vertex operators $Y_M(v, z)$ acting on an irreducible V-module M. One can think of this as a map

$$Y_M(\cdot, z) : V \to End(M)[[z, z^{-1}]]$$

which obeys the various axioms defining a V-module. The fusion rule in this case is always $\mathcal{N}(V, M, M) = 1$ and one axiom normalizes Y_M so it is uniquely determined. Given three V-modules, M_1, M_2, M_3, one can think of an intertwining operator as a map

$$Y(\cdot, z) : M_1 \to Hom(M_2, M_3)\{\{z, z^{-1}\}\}$$

which obeys the axioms for intertwining operators. (The notation $\{\{z, z^{-1}\}\}$ indicates rational powers of z.) It is quite possible that the fusion rule is $\mathcal{N}(M_1, M_2, M_3) = n > 1$, and in that case one does not have a one-to-one correspondence between vectors w in module M_1 and operators $Y(w, z)$ whose components send M_2 to M_3. It would seem that this is a kind of labeling problem, there not being enough "copies" of the vectors in M_1 to distinguish the n linearly independent intertwiners which could be taken as a basis for the space of all intertwiners. It is also possible to have four modules $M_1, \cdots, M_4$ with $M_3 \neq M_4$, with fusion rules $\mathcal{N}(M_1, M_2, M_3) \geq 1$ and $\mathcal{N}(M_1, M_2, M_4) \geq 1$. This also indicates a labeling problem, showing the inadequacy of the notation $Y(w_1, z)w_2$, where knowing that $w_i \in M_i$ still does not determine which module contains the outcome.

Another new feature which appears is the nature of the correlation functions,

$$(Y(w_1, z_1)Y(w_2, z_2)w_3, w_4) \quad \text{and} \quad (Y(Y(w_1, z)w_2, z_2)w_3, w_4),$$

made from two intertwiners. These are series which converge in certain domains to functions which, after factoring out some rational powers of z_1 and z_2, can be expressed as power series in z_2/z_1 and z/z_2 satisfying certain differential equations. One way of thinking of the usual Jacobi-Cauchy identity for a VOA V is as follows. The three series

$$(Y(v_1, z_1)Y(v_2, z_2)v_3, v_4), \qquad (Y(v_2, z_2)Y(v_1, z_1)v_3, v_4)$$

and
$$(Y(Y(v_1, z_1 - z_2)v_2, z_2)v_3, v_4)$$
converge in their respective domains, $|z_1| > |z_2|$, $|z_2| > |z_1|$, and $|z_2| > |z_1 - z_2|$, to the same rational function $f(z_1, z_2)$ in the ring
$$\mathcal{R} = \mathbb{C}[z_1, z_1^{-1}, z_2, z_2^{-1}, (z_1 - z_2)^{-1}].$$

For fixed $z_2 \neq 0$, these are functions of z_1 with possible poles only at $z_1 = 0$, $z_1 = \infty$ and $z_1 = z_2$. The Jacobi-Cauchy identity is just the statement of the Cauchy residue theorem applied to $f(z_1, z_2)g(z_1, z_2)$, where $g(z_1, z_2) \in \mathcal{R}$ is arbitrary, and the residues at each of the three points are computed from the three series giving f and appropriate series expansions of g. But the kinds of correlation functions obtained from intertwiners generally involve hypergeometric functions to which the treatment just described does not apply.

The purpose of this paper is to show how both of these new features can be handled in a simple but nontrivial example which may have important applications. In the monograph [FFR] we constructed certain vertex operator superalgebras (VOSA's) and their twisted modules from Clifford algebras and their spinor representations. These "fermionic" constructions extend the corresponding constructions of the orthogonal affine Kac-Moody algebras of type $D_n^{(1)}$ in [FF]. (See [Fr1], [Fr2].) In [W] vertex operator para-algebras were constructed from the bosonic constructions of level $-\frac{1}{2}$ representations of symplectic affine Kac-Moody algebras of type $C_n^{(1)}$. It is well known to physicists, and follows immediately from the work in [FFR], that a spinor construction from one fermion gives a VOSA and a twisted module for it. These are known in the physics literature as the Neveu-Schwarz and Ramond sectors, respectively, and they each decompose into two irreducible modules for the Virasoro algebra with central charge $c = \frac{1}{2}$. Using the usual notation for labeling Virasoro modules, the two components into which the Neveu-Schwarz sector decomposes are labeled by $h = 0$ and $h = \frac{1}{2}$, and the two coming from the Ramond sector are both $h = \frac{1}{16}$. It is very interesting and important that in this construction one naturally has two copies of the $h = \frac{1}{16}$ Virasoro module. That enables us to define unique intertwining operators for each vector in the Ramond sector, such that the usual Ising fusion rules for just three modules are replaced by fusion rules given by the group $\mathbb{Z}_4$. This behavior is just like the behavior of vertex operator para-algebras defined in [FFR]. (See also [DL].) It means that the VOA V and its modules are indexed by a finite abelian group Γ such that $V = V_0$ ($0 \in \Gamma$ is the identity element) and the fusion rules are $\mathcal{N}(V_a, V_b, V_c) = 1$ if $a + b = c$ in Γ, zero otherwise. The important question is whether this is a rare special situation, or if there are other natural constructions of VOA's where multiple copies of the modules allow unique labeling of intertwining operators and where fusion rules are replaced by a group law.

In the example studied here we find that the hypergeometric functions to which the correlation function converge, come in pairs, as bases for the two dimensional spaces of solutions of certain differential equations. In order to relate them to each other, we must use Kummer's quadratic transformation formula. This involves certain substitutions which lift the correlation functions of $t = z_2/z_1$ to functions of x on a four-sheeted covering of the t-sphere, branched at $t = 0, 1, \infty$ with possible poles only at $x = \alpha \in \{0, \infty, 1, -1, \mathbf{i}, -\mathbf{i}\}$. We may then find the 2×2 matrices B_α which relate the 2×4 matrices of transformed correlation functions at $x = 0$ to those at $x = \alpha$. The matrix-valued functions have possible poles only at those six points, so after multiplying the matrix-valued functions by any function $g(x)$ in the ring

$$\mathcal{R}_x = \mathbb{C}[x, x^{-1}(x^4 - 1)^{-1}]$$

the Cauchy residue theorem gives that the sum of their residues at the six points adds up to zero. Expressing the residues in terms of the series and expanding the function $g(x)$ as an appropriate series, we get a "matrix" Jacobi-Cauchy identity. It is, of course, actually a generating function of identities, equivalent to infinitely many identities for the components of the intertwining operators. It may take some time to sort out which ones are the most important, but we can already point out some interesting ones.

A very interesting aspect of this example is the structure of the matrices B_α. In order to find those matrices one must choose some linear fractional transformations relating the variable $x = x_0$ to the variables x_α which are local variables at the poles. There are several sign choices which can be made, and these correspond to choices of braiding. It is not so surprising that these matrices provide a representation of a braid group. The details of this aspect of the example have not been completely worked out yet, and will be studied later. There should be an interesting connection with the work in [**MaS**].

As a final motivation for the detailed study of this special example, we would like to mention the possible application to a spinor construction of the moonshine module for the monster group. It has been noted by Dong, Mason and Zhu [**DMZ**] that there are 48 commuting $c = \frac{1}{2}$ Virasoro algebras in the moonshine module $V^\natural$ considered as a VOA. It means that $V^\natural$ decomposes into a sum of tensor products of 48 Virasoro modules, each of which is one of the $h = 0$, $h = \frac{1}{2}$ or $h = \frac{1}{16}$ modules. Although some information is known about this decomposition, the complete picture is not clear. But since there is a spinor construction of each of these modules, there is some hope that a spinor construction of $V^\natural$ is possible. In fact, we hope that the new light we have shed on the $c = \frac{1}{2}$ minimal model will be of help in achieving that goal.

2. Construction of Vertex Operator Superalgebra and Module

Let $E = \mathbb{C}e$ with $\langle e, e \rangle = 2$ and let $Z = \mathbb{Z}$ or $Z = \mathbb{Z} + \frac{1}{2}$. Let $E(Z)$ be the vector space with basis $\{e(m) \mid m \in Z\}$ and the symmetric form

$$\langle e(m), e(n) \rangle = \langle e, e \rangle \delta_{m,-n} = 2\delta_{m,-n}.$$

Let $\mathbf{Cliff}(Z)$ be the Clifford algebra generated by $E(Z)$ and that form. Let $E(Z) = E(Z)^+ \oplus E(Z)^-$ be the polarization where $E(Z)^+$ is spanned by $\{e(m) \mid 0 < m \in Z\}$ and $E(Z)^-$ is spanned by $\{e(m) \mid 0 \geq m \in Z\}$. Define $\mathfrak{I}(Z)$ to be the left ideal in $\mathbf{Cliff}(Z)$ generated by $E(Z)^+$, so that

$$\mathbf{CM}(Z) = \mathbf{Cliff}(Z)/\mathfrak{I}(Z)$$

is a left $\mathbf{Cliff}(Z)$-module. One has the parity decomposition

$$\mathbf{CM}(Z) = \mathbf{CM}(Z)^0 \oplus \mathbf{CM}(Z)^1$$

where $\mathbf{CM}(Z)^i$, $i = 0, 1$, is the subspace with basis

$$\{e(-m_1) \ldots e(-m_r)\mathbf{vac}(Z) \mid m_1 > \ldots > m_r \geq 0, \ m_1, \ldots, m_r \in Z, \ r \equiv i \bmod 2\}$$

and $\mathbf{vac}(Z) = 1 + \mathfrak{I}(Z)$. Define the vacuum space

$$\mathbf{VAC}(Z) = \{v \in \mathbf{CM}(Z) \mid E(Z)^+ \cdot v = 0\}.$$

Then $\mathbf{VAC}(\mathbb{Z} + \frac{1}{2})$ is one-dimensional, spanned by

$$\mathbf{vac} = \mathbf{vac}(\mathbb{Z} + \tfrac{1}{2}) = 1 + \mathfrak{I}(\mathbb{Z} + \tfrac{1}{2}),$$

and $\mathbf{VAC}(\mathbb{Z})$ is two-dimensional, spanned by

$$\mathbf{vac}' = \mathbf{vac}(\mathbb{Z}) = 1 + \mathfrak{I}(\mathbb{Z}) \qquad \text{and} \qquad e(0)\mathbf{vac}'.$$

It is easy to see that $\mathbf{CM}(\mathbb{Z} + \frac{1}{2})$ is an irreducible $\mathbf{Cliff}(\mathbb{Z} + \frac{1}{2})$-module, but $\mathbf{CM}(\mathbb{Z})$ decomposes into two irreducible $\mathbf{Cliff}(\mathbb{Z})$-modules. Note that

$$\mathbf{vac}'_+ = \mathbf{vac}' + e(0)\mathbf{vac}' \qquad \text{and} \qquad \mathbf{vac}'_- = \mathbf{vac}' - e(0)\mathbf{vac}'$$

are eigenvectors for $e(0)$ with eigenvalues $+1$ and -1, respectively, because the relations in $\mathbf{Cliff}(\mathbb{Z})$ give $e(0)^2 = 1$. We then have the alternative decomposition into two irreducible $\mathbf{Cliff}(\mathbb{Z})$-modules,

$$\mathbf{CM}(\mathbb{Z}) = \mathbf{CM}(\mathbb{Z})^+ \oplus \mathbf{CM}(\mathbb{Z})^-$$

where $\mathbf{CM}(\mathbb{Z})^\pm$ is the subspace with basis

$$\{e(-m_1) \ldots e(-m_r)\mathbf{vac}'_\pm \mid m_1 > \ldots > m_r > 0, \ m_1, \ldots, m_r \in \mathbb{Z}\}.$$

Later we will need the vector space isomorphism

$$\theta : \mathbf{CM}(\mathbb{Z}) \to \mathbf{CM}(\mathbb{Z})$$

defined by

$$\theta(\mathbf{vac}') = e(0)\mathbf{vac}' \qquad \text{and} \qquad \theta(e(m)v) = e(m)\theta(v)$$

for any $m \in \mathbb{Z}$ and $v \in \mathbf{CM}(\mathbb{Z})$. It is clear that θ is an involution switching $\mathbf{CM}(\mathbb{Z})^0$ with $\mathbf{CM}(\mathbb{Z})^1$, and that $\mathbf{CM}(\mathbb{Z})^+$ and $\mathbf{CM}(\mathbb{Z})^-$ are the $+1$ and -1 eigenspaces, respectively, for θ.

Define the **Vir** operators

$$L(k) = -\tfrac{1}{4} \sum_{n \in Z} (n + \tfrac{1}{2}) \; {}^{\circ}_{\circ} e(n)e(k-n) {}^{\circ}_{\circ} \quad \text{for} \;\; k \neq 0,$$

$$L(0) = \frac{1 + \iota}{32} - \tfrac{1}{4} \sum_{n \in Z} (n + \tfrac{1}{2}) \; {}^{\circ}_{\circ} e(n)e(-n) {}^{\circ}_{\circ}$$

where $\iota = 1$ if $Z = \mathbb{Z}$ and $\iota = -1$ if $Z = \mathbb{Z} + \tfrac{1}{2}$.

THEOREM 1. *The operators $L(k)$, $k \in \mathbb{Z}$, and the identity operator, represent a $c = \tfrac{1}{2}$ Virasoro algebra* **Vir** *on* $\mathbf{CM}(Z)$. *In particular, for $k, n \in \mathbb{Z}$, $m \in Z$, we have*

$$[L(k), e(m)] = -(m + \tfrac{1}{2}k) \; e(m+k),$$

$$[L(k), L(n)] = (k-n)L(k+n) + \tfrac{1}{24}(k^3 - k)\delta_{k,-n}\mathbf{1}.$$

The parity decomposition of $\mathbf{CM}(Z)$ *is a decomposition into two irreducible* **Vir**-*modules. The highest weight vectors in these modules are* **vac**, $e(-\tfrac{1}{2})$**vac**, **vac**′ *and $e(0)$**vac**′, whose weights are 0, $\tfrac{1}{2}$, $\tfrac{1}{16}$ and $\tfrac{1}{16}$, respectively. The decomposition of* $\mathbf{CM}(\mathbb{Z})$ *into two irreducible* **Cliff**$(\mathbb{Z})$-*modules is also a decomposition into two irreducible* **Vir**-*modules. The highest weight vectors in these modules are* **vac**′$_+$ *and* **vac**′$_-$, *whose weights are both $\tfrac{1}{16}$. The operators $L(k)$ commute with θ on* $\mathbf{CM}(\mathbb{Z})$.

As usual we can define a positive Hermitian form $(\, , \,)$ on $\mathbf{CM}(Z)$ such that $e(m)^* = e(-m)$ and $L(k)^* = L(-k)$, where $*$ denotes adjoint. The eigenspaces of $L(0)$ provide $\mathbf{CM}(Z)$ with a grading. If u is an eigenvector for $L(0)$ write $wt(u)$ for the eigenvalue and write $(\mathbf{CM}(Z))_n$ for the n-eigenspace. For $u = e(-m_1) \ldots e(-m_r)\mathbf{vac}(Z) \in \mathbf{CM}(Z)$ we have

$$L(0)u = \left(m_1 + \ldots + m_r + \frac{1 + \iota}{32} \right) u.$$

Let $\mathbf{W}$ be any subspace of $\mathbf{CM}(Z)$ which is a direct sum of $L(0)$ eigenspaces, $(\mathbf{W})_n$. The homogeneous character of $\mathbf{W}$ is defined to be

$$ch(\mathbf{W}) = \sum_n dim(\mathbf{W})_n q^n,$$

a formal series in q. The homogeneous characters of the Clifford modules are then

$$ch(\mathbf{CM}(\mathbb{Z} + \tfrac{1}{2})) = \prod_{0 \leq n \in \mathbb{Z}} (1 + q^{n + \frac{1}{2}})$$

and

$$ch(\mathbf{CM}(\mathbb{Z})) = 2q^{1/16} \prod_{1 \leq n \in \mathbb{Z}} (1 + q^n).$$

As in the spinor construction of $D_n^{(1)}$ we construct a vertex operator superalgebra on $\mathbf{CM}(\mathbb{Z} + \tfrac{1}{2})$ and a (twisted) representation on $\mathbf{CM}(\mathbb{Z})$. This is done by defining a vertex operator $Y(v, \zeta)$ on $\mathbf{CM}(\mathbb{Z} + \frac{1}{2})$ and on $\mathbf{CM}(\mathbb{Z})$ for any $v \in \mathbf{CM}(\mathbb{Z} + \frac{1}{2})$. Actually $Y(v, \zeta)$ is a generating function of operators

$$Y(v, \zeta) = \sum_{n \in \frac{1}{2}\mathbb{Z}} Y_{n+1-wt(v)}(v)\zeta^{-n-1} = \sum_{n \in \frac{1}{2}\mathbb{Z}} \{v\}_n \zeta^{-n-1}$$

where

$$wt(Y_m(v)w) = wt(w) - m.$$

The definition of $Y(v, \zeta)$ on $\mathbf{CM}(\mathbb{Z})$ is more complicated than it is on $\mathbf{CM}(\mathbb{Z} + \frac{1}{2})$, but there is a common part which we denote by $\bar{Y}(v, \zeta)$.

Recall that $\mathbf{vac} = \mathbf{vac}(\mathbb{Z} + \frac{1}{2})$ and $\mathbf{vac}' = \mathbf{vac}(\mathbb{Z})$. On $\mathbf{CM}(Z)$ define $\bar{Y}(\mathbf{vac}, \zeta) = 1$ (the identity operator), and for $0 \leq n \in \mathbb{Z}$ let

$$\bar{Y}(e(-n - \tfrac{1}{2})\mathbf{vac}, \zeta) = n!^{-1}(d/d\zeta)^n \sum_{m \in Z} e(m)\zeta^{-m-\frac{1}{2}}.$$

Using the fermionic normal ordering ${}^{\circ}_{\circ}e(n_1) \cdots e(n_r){}^{\circ}_{\circ}$ of a product of Clifford generators, for vectors of the form

$$v = e(-n_1 - \tfrac{1}{2}) \cdots e(-n_r - \tfrac{1}{2})\mathbf{vac} \in \mathbf{CM}(\mathbb{Z} + \tfrac{1}{2})$$

we define

$$\bar{Y}(v, \zeta) = {}^{\circ}_{\circ}\bar{Y}(e(-n_1 - \tfrac{1}{2})\mathbf{vac}, \zeta) \cdots \bar{Y}(e(-n_r - \tfrac{1}{2})\mathbf{vac}, \zeta){}^{\circ}_{\circ}$$

and extend the definition to all $v \in \mathbf{CM}(\mathbb{Z} + \frac{1}{2})$ by linearity. Then

$$\bar{Y}_m(e(-\tfrac{1}{2})\mathbf{vac}) = e(m)$$

is a Clifford generator. Furthermore, with

$$\omega = L(-2)\mathbf{vac} = \tfrac{1}{4}e(-\tfrac{3}{2})e(-\tfrac{1}{2})\mathbf{vac},$$

we have

$$\bar{Y}(\omega, \zeta) = \begin{cases} L(\zeta) & \text{on } \mathbf{CM}(\mathbb{Z} + \tfrac{1}{2}) \\[2ex] L(\zeta) - \tfrac{1}{16}\zeta^{-2} & \text{on } \mathbf{CM}(\mathbb{Z}) \end{cases}$$

where $L(\zeta) = \sum_{k \in \mathbb{Z}} L(k)\zeta^{-k-2}$ is the generating function of the Virasoro operators.

In order to define the operators $Y(v, \zeta)$ we need the additional quadratic operator

$$\Delta(\zeta) = \frac{1}{4} \sum_{0 \leq m, n \in \mathbb{Z}} C_{mn} e(m + \tfrac{1}{2})e(n + \tfrac{1}{2})\zeta^{-m-n-1}$$

whose definition involves the combinatorial coefficients

$$C_{mn} = \frac{1}{2} \frac{m - n}{m + n + 1} \binom{-\frac{1}{2}}{m}\binom{-\frac{1}{2}}{n}.$$

For $v \in \mathbf{CM}(\mathbb{Z} + \frac{1}{2})$ we define

$$
Y(v, \zeta) = \begin{cases} \bar{Y}(v, \zeta) & \text{on } \mathbf{CM}(\mathbb{Z} + \frac{1}{2}) \\[2mm] \bar{Y}(\exp(\Delta(\zeta))v, \zeta) & \text{on } \mathbf{CM}(\mathbb{Z}). \end{cases}
$$

Note that $\exp(\Delta(\zeta))v$ is a finite sum of vectors whose weights are between 0 and $wt(v)$. In particular,

$$
\exp(\Delta(\zeta))\omega = \omega + \tfrac{1}{16}\zeta^{-2}\mathbf{vac},
$$

so $Y(\omega, \zeta) = L(\zeta)$ on both $\mathbf{CM}(\mathbb{Z} + \frac{1}{2})$ and $\mathbf{CM}(\mathbb{Z})$.

THEOREM 2. *We have* $(\mathbf{CM}(\mathbb{Z}+\frac{1}{2}), Y(\ , z), \mathbf{vac}, \omega)$ *is a vertex operator super-algebra and* $(\mathbf{CM}(\mathbb{Z}), Y(\ , z))$ *is a (twisted) vertex operator superalgebra module.*

COROLLARY 3. *Let* $v_i \in \mathbf{CM}(\mathbb{Z} + \frac{1}{2})^{\alpha_i}$ *for* $i = 1, 2$. *Then for any* $r \in \mathbb{Z}$, *for* $m, n \in \mathbb{Z}$ *on* $\mathbf{CM}(\mathbb{Z} + \frac{1}{2})$, *and for* $m \in \mathbb{Z} + \frac{1}{2}\alpha_1$, $n \in \mathbb{Z} + \frac{1}{2}\alpha_2$ *on* $\mathbf{CM}(\mathbb{Z})$, *we have*

$$
\sum_{0 \le i \in \mathbb{Z}} (-1)^i \binom{r}{i} (\{v_1\}_{m+r-i}\{v_2\}_{n+i} - (-1)^{\alpha_1 \alpha_2 + r}\{v_2\}_{n+r-i}\{v_1\}_{m+i})
$$

$$
= \sum_{0 \le k \in \mathbb{Z}} \binom{m}{k} \{\{v_1\}_{r+k} v_2\}_{m+n-k} \ .
$$

The sum over k *is finite, and we may take* $0 \le k \le wt(v_1) + wt(v_2) - r - 1$.

COROLLARY 4. *For* $v \in \mathbf{CM}(\mathbb{Z} + \frac{1}{2})$, $m \in \mathbb{Z}$, *on* $\mathbf{CM}(Z)$ *we have*

$$
[L(m), Y(v, z)] = \sum_{0 \le k \in \mathbb{Z}} \binom{m+1}{k} z^{m+1-k} Y(L(k-1)v, z)
$$

and the sum is finite. If $L(n)v = 0$ *for all* $n > 0$, *then we have*

$$
[L(m), Y(v, z)] = z^{m+1}(d/dz)Y(v, z) + (m+1)z^m Y(L(0)v, z)
$$

which means that

$$
[L(m), Y_{n-m}(v)] = (-n + m \ wt(v))Y_n(v).
$$

COROLLARY 5. *For* $v \in \mathbf{CM}(\mathbb{Z} + \frac{1}{2})$, $m \in \mathbb{Z}$, *on* $\mathbf{CM}(Z)$ *we have*

$$
Y(L(-m-1)v, z) =
$$

$$
\sum_{0 \le i \in \mathbb{Z}} \binom{m+i-1}{i} [z^i L(-m-i-1)Y(v, z) - (-1)^m z^{-m-i} Y(v, z)L(i-1)].
$$

Let

$$
\mathbf{V}_0 = \mathbf{CM}(\mathbb{Z} + \tfrac{1}{2})^0, \quad \mathbf{V}_2 = \mathbf{CM}(\mathbb{Z} + \tfrac{1}{2})^1,
$$

$$
\mathbf{V}_1 = \mathbf{CM}(\mathbb{Z})^0, \quad \mathbf{V}_3 = \mathbf{CM}(\mathbb{Z})^1,
$$

$$
\mathcal{V} = \mathbf{V}_0 \oplus \mathbf{V}_1 \oplus \mathbf{V}_2 \oplus \mathbf{V}_3,
$$

$$
\mathbf{V} = \mathbf{V}_0 \oplus \mathbf{V}_2,
$$

and extend the Hermitian form (,) to $\mathcal{V}$ so that $\mathbf{CM}(\mathbb{Z} + \frac{1}{2})$ and $\mathbf{CM}(\mathbb{Z})$ are orthogonal. We give the set of subscripts $\{0, 1, 2, 3\}$ used to index these **Vir**-modules (sectors) the group structure of $\mathbb{Z}_4$. In the following table we give an orthonormal basis for the space $(\mathbf{V}_i)_{\Delta_i}$ of vectors of minimal weight Δ_i in sector $\mathbf{V}_i$. These are vacuum vectors for **Vir**.

$\mathbf{V}_0$	$u_0 = \mathbf{vac}$	$\Delta_0 = 0$
$\mathbf{V}_1$	$u_1 = \mathbf{vac}'$	$\Delta_1 = \frac{1}{16}$
$\mathbf{V}_2$	$u_2 = \frac{1}{\sqrt{2}}e(-\frac{1}{2})\mathbf{vac}$	$\Delta_2 = \frac{1}{2}$
$\mathbf{V}_3$	$u_3 = e(0)\mathbf{vac}'$	$\Delta_3 = \frac{1}{16}$

For $m, n \in \mathbb{Z}_4$, define

$$\Delta(m, n) = \Delta_m + \Delta_n - \Delta_{m+n}.$$

Then we have $\Delta(0, n) = 0$ and

$$\Delta(1, 1) = -\tfrac{3}{8}, \qquad \Delta(1, 2) = \tfrac{1}{2}, \qquad \Delta(1, 3) = \tfrac{1}{8},$$
$$\Delta(2, 2) = 1, \qquad \Delta(2, 3) = \tfrac{1}{2}, \qquad \Delta(3, 3) = -\tfrac{3}{8}.$$

For $m, n, p \in \mathbb{Z}_4$, define the totally symmetric function

$$\Delta(m, n, p) = \Delta(m, n) + \Delta(m, p) - \Delta(m, n + p)$$
$$= \Delta_m + \Delta_n + \Delta_p - \Delta_{m+n} - \Delta_{m+p} - \Delta_{n+p} + \Delta_{m+n+p}.$$

Then we have $\Delta(0, n, p) = 0$ and

$$\Delta(1, 1, 1) = -\tfrac{5}{4}, \quad \Delta(1, 1, 2) = 0, \quad \Delta(1, 1, 3) = -\tfrac{1}{4},$$
$$\Delta(1, 2, 2) = 1, \quad \Delta(1, 2, 3) = 1, \quad \Delta(1, 3, 3) = -\tfrac{1}{4},$$
$$\Delta(2, 2, 2) = 2, \quad \Delta(2, 2, 3) = 1, \quad \Delta(2, 3, 3) = 0, \quad \Delta(3, 3, 3) = -\tfrac{5}{4}.$$

3. Intertwining Operators

We wish to extend the definition of vertex operators so that $Y(v_1, z)v_2$ is defined for all $v_1, v_2 \in \mathcal{V}$. The new operators we need to define, when $v_1 \in \mathbf{V}_1 \oplus \mathbf{V}_3$ are called "intertwining operators". If $v_i \in \mathbf{V}_{n_i}$ then our basic assumptions are that the components of $Y(v_1, z)v_2$ are in $\mathbf{V}_{n_1+n_2}$, that

$$[L(-1), Y(v_1, z)] = \frac{d}{dz}Y(v_1, z)$$

and that Corollaries 4 and 5 are valid for all $v \in \mathcal{V}$. Let us see to what extent these assumptions determine the intertwiners. Since the form $(\cdot, \cdot)$ on $\mathcal{V}$ is non-degenerate, $Y(v_1, z)v_2$ is determined if the one-point function $(Y(v_1, z)v_2, v_3)$ is known for all $v_3 \in \mathcal{V}$. Suppose $v_i \in \mathbf{V}_{n_i}$ for $1 \leq i \leq 3$, $L(0)v_i = \lambda_i v_i$ and

$n_1 + n_2 = n_3$ (otherwise the one-point function is zero). For any $0 < m \in \mathbb{Z}$ we have

$$(Y(v_1, z)v_2, L(-m)v_3) = (L(m)Y(v_1, z)v_2, v_3)$$

$$= (Y(v_1, z)L(m)v_2, v_3) + ([L(m), Y(v_1, z)]v_2, v_3)$$

$$= (Y(v_1, z)L(m)v_2, v_3) + \sum_{0 \leq k \in \mathbb{Z}} \binom{m+1}{k} z^{m+1-k}(Y(L(k-1)v_1, z)v_2, v_3)$$

$$= (Y(v_1, z)L(m)v_2, v_3) + z^{m+1}\frac{d}{dz}(Y(v_1, z)v_2, v_3)$$

$$+ (m+1)z^m \lambda_1 (Y(v_1, z)v_2, v_3)$$

$$+ \sum_{1 < k \in \mathbb{Z}} \binom{m+1}{k} z^{m+1-k}(Y(L(k-1)v_1, z)v_2, v_3)$$

where we have used

$$Y(L(-1)v_1, z) = [L(-1), Y(v_1, z)] = \frac{d}{dz}Y(v_1, z)$$

which comes from Corollary 5 with $m = 0$, and our assumptions. This shows that $(Y(v_1, z)v_2, L(-m)v_3)$ is determined by the one-point functions $(Y(v_1', z)v_2', v_3)$ where $wt(v_1') \leq wt(v_1)$ and $wt(v_2') \leq wt(v_2)$. This reduces the problem to the case when v_3 has minimal weight in its sector, that is, when v_3 is a vacuum vector for **Vir**.

Assuming that $L(k)v_3 = 0$ for all $0 < k \in \mathbb{Z}$, for $0 \leq m \in \mathbb{Z}$ we have

$$(Y(L(-m-1)v_1, z)v_2, v_3) =$$

$$\sum_{0 \leq i \in \mathbb{Z}} \binom{m+i-1}{i} [z^i (L(-m-i-1)Y(v_1, z)v_2, v_3)$$

$$- (-1)^m z^{-m-i}(Y(v_1, z)L(i-1)v_2, v_3)]$$

$$= \sum_{0 \leq i \in \mathbb{Z}} \binom{m+i-1}{i} [z^i (Y(v_1, z)v_2, L(m+i+1)v_3)$$

$$- (-1)^m z^{-m-i}(Y(v_1, z)L(i-1)v_2, v_3)].$$

But since $m \geq 0$ and $i \geq 0$, $L(m+i+1)v_3 = 0$ by assumption, so the above equals

$$- (-1)^m \sum_{1 < i \in \mathbb{Z}} \binom{m+i-1}{i} z^{-m-i}(Y(v_1, z)L(i-1)v_2, v_3)$$

$$- (-1)^m z^{-m}(Y(v_1, z)L(-1)v_2, v_3) - (-1)^m z^{-m-1}\lambda_2(Y(v_1, z)v_2, v_3).$$

Now we use the fact that

$$(Y(v_1, z)L(-1)v_2, v_3) = (L(-1)Y(v_1, z)v_2, v_3) - ([L(-1), Y(v_1, z)]v_2, v_3)$$

$$= (Y(v_1, z)v_2, L(1)v_3) - \frac{d}{dz}(Y(v_1, z)v_2, v_3)$$

$$= -\frac{d}{dz}(Y(v_1, z)v_2, v_3).$$

So $(Y(L(-m-1)v_1, z)v_2, v_3)$ for $m \geq 0$ and v_3 a vacuum vector is determined by the one-point functions $(Y(v_1, z)v_2', v_3)$ for $wt(v_2') \leq wt(v_2)$. Thus we are reduced to the case when v_1 and v_3 are vacuum vectors.

With that assumption, for $0 < m \in \mathbb{Z}$, we have

$$(Y(v_1, z)L(-m)v_2, v_3) = (L(-m)Y(v_1, z)v_2, v_3) - ([L(-m), Y(v_1, z)]v_2, v_3)$$

$$(Y(v_1, z)v_2, L(m)v_3) - \sum_{0 \leq k \in \mathbb{Z}} \binom{-m+1}{k} z^{-m+1-k}(Y(L(k-1)v_1, z)v_2, v_3)$$

$$= -z^{-m+1}(Y(L(-1)v_1, z)v_2, v_3) - (-m+1)z^{-m}(Y(L(0)v_1, z)v_2, v_3)$$

$$= -z^{-m+1}\frac{d}{dz}(Y(v_1, z)v_2, v_3) + (m-1)z^{-m}\lambda_1(Y(v_1, z)v_2, v_3)$$

showing that $(Y(v_1, z)L(-m)v_2, v_3)$ is determined by $(Y(v_1, z)v_2, v_3)$. Thus, we have reduced the general one-point function to the special case when v_1, v_2 and v_3 are vacuum vectors. These complex numbers are called the structure constants, and we will see if they can be consistently determined by conditions we want for the two-point functions.

4. Two-Point Functions and Hypergeometric Differential Equations

For $1 \leq i \leq 4$ let $v_i \in \mathbf{V}_{n_i}$ with $wt(v_i) = |v_i| = N_i + \Delta_{n_i}$ for $0 \leq N_i \in \mathbb{Z}$ and suppose $n_1 + n_2 + n_3 = n_4$. Let

$$G(v_1, v_2, v_3, v_4; z_1, z_2) = (Y(v_1, z_1)Y(v_2, z_2)v_3, v_4),$$

$$H(v_1, v_2, v_3, v_4; z, z_2) = (Y(Y(v_1, z)v_2, z_2)v_3, v_4).$$

Then we have

$$G(v_1, v_2, v_3, v_4; z_1, z_2) =$$

$$z_1^{N_4-N_1-\Delta(n_1, n_2+n_3)} z_2^{-N_2-N_3-\Delta(n_2, n_3)} \sum_{0 \leq k \in \mathbb{Z}} \left(\frac{z_2}{z_1}\right)^k \Phi_k(v_1, v_2, v_3, v_4)$$

where

$$\Phi_k = \Phi_k(v_1, v_2, v_3, v_4) = (Y_{k+\Delta_{n_2+n_3}-|v_4|}(v_1) Y_{-k-\Delta_{n_2+n_3}+|v_3|}(v_2)v_3, v_4)$$

and

$$H(v_1, v_2, v_3, v_4; z, z_2) =$$

$$z^{-N_1-N_2-\Delta(n_1,n_2)} z_2^{N_4-N_3-\Delta(n_1+n_2,n_3)} \sum_{0 \leq k \in \mathbb{Z}} \left(\frac{z}{z_2}\right)^k \Psi_k(v_1, v_2, v_3, v_4)$$

where

$$\Psi_k = \Psi_k(v_1, v_2, v_3, v_4) = (Y_{|v_3|-|v_4|}(Y_{-k-\Delta_{n_1+n_2}+|v_2|}(v_1)v_2)v_3, v_4).$$

DEFINITION 6. For $m, n, p \in \mathbb{Z}_4$ let

$$A_{mn} = (Y_{\Delta_n - \Delta_{m+n}}(u_m)u_n, u_{m+n}),$$

$$K_{mnp} = \Phi_0(u_m, u_n, u_p, u_{m+n+p}) = A_{np}A_{m,n+p},$$

$$M_{mnp} = \Psi_0(u_m, u_n, u_p, u_{m+n+p}) = A_{mn}A_{m+n,p}.$$

LEMMA 7. Let $v \in \mathbf{V}_i$, $0 \leq i \leq 3$, be of weight Δ_i. Then we have $L(-1)^2 v = \gamma L(-2)v$ where γ is given by the following table:

i	0	1	2	3
γ	0	$\frac{3}{4}$	$\frac{4}{3}$	$\frac{3}{4}$

THEOREM 8. For $1 \leq i \leq 4$ let $v_i \in \mathbf{V}_{n_i}$ be vacuum vectors for **Vir**. For $i = 1, 2$ let $Y_i = Y(v_i, z_i)$, suppose that $L(-1)^2 v_3 = \gamma L(-2)v_3$ and let

$$G = G(v_1, v_2, v_3, v_4; z_1, z_2) = (Y_1 Y_2 v_3, v_4).$$

Then $\dot{G}$ satisfies the partial differential equation

$$(\partial_1 + \partial_2)^2 G + \gamma(z_1^{-1}\partial_1 + z_2^{-1}\partial_2)G - \gamma(z_1^{-2}\Delta_{n_1} + z_2^{-2}\Delta_{n_2})G = 0.$$

PROOF. We have

$$(Y_1 Y_2 L(-1)^2 v_3, v_4) = (Y_1 L(-1)Y_2 L(-1)v_3, v_4) - (Y_1[L(-1), Y_2]L(-1)v_3, v_4)$$

$$= (L(-1)Y_1 Y_2 L(-1)v_3, v_4) - ([L(-1), Y_1]Y_2 L(-1)v_3, v_4)$$

$$- \partial_2(Y_1 Y_2 L(-1)v_3, v_4)$$

$$= -(\partial_1 + \partial_2)(Y_1 Y_2 L(-1)v_3, v_4)$$

$$= (\partial_1 + \partial_2)^2(Y_1 Y_2 v_3, v_4)$$

and

$$(Y_1 Y_2 L(-2)v_3, v_4) = (Y_1 L(-2)Y_2 v_3, v_4) - (Y_1[L(-2), Y_2]v_3, v_4)$$

$$= -([L(-2), Y_1]Y_2 v_3, v_4) - (Y_1[L(-2), Y_2]v_3, v_4)$$

$$= -(z_1^{-1}\partial_1 - z_1^{-2}\Delta_{n_1})(Y_1 Y_2 v_3, v_4)$$

$$- (z_2^{-1}\partial_2 - z_2^{-2}\Delta_{n_2})(Y_1 Y_2 v_3, v_4)$$

so the relation

$$(Y_1 Y_2 L(-1)^2 v_3, v_4) = \gamma(Y_1 Y_2 L(-2)v_3, v_4)$$

gives the result.

Let $v_i = u_{n_i}$ for $1 \leq i \leq 4$, and suppose $n_4 = n_1 + n_2 + n_3$. If we write

$$G_{n_1 n_2 n_3}(z_1, z_2) = K_{n_1 n_2 n_3} z_1^{-A} z_2^{-B} \left(1 - \frac{z_2}{z_1}\right)^{-C} F(z_2/z_1)$$

for

$$A = \Delta(n_1, n_2 + n_3), \qquad B = \Delta(n_2, n_3), \qquad C = \Delta(n_1, n_2)$$

then the differential equation for G becomes a differential equation for F. Letting $x = z_2/z_1$ and using the fact that

$$\Delta(n, n_3)(\Delta(n, n_3) + 1) = \gamma(\Delta(n, n_3) + \Delta_n)$$

for any $n = 0, 1, 2, 3$, we get the ordinary differential equation

$$x(1 - x)F'' + [(\gamma - 2B) - (2A + 2 - 2C - \gamma)x]F' + [2AB - 2BC + \gamma C]F = 0$$

for F. The hypergeometric function

$$ {}_2F_1(a, b, c; z) = \sum_{n=0}^{\infty} \frac{(a)_n (b)_n}{n! (c)_n} z^n$$

where $(a)_n = a(a+1)\ldots(a+n-1)$, is a solution to the hypergeometric differential equation

$$x(1 - x)w'' + (c - (a + b + 1)x)w' - abw = 0.$$

Therefore, we find that

$$G_{n_1 n_2 n_3}(z_1, z_2) = K_{n_1 n_2 n_3} z_1^{-A} z_2^{-B} \left(1 - \frac{z_2}{z_1}\right)^{-C} {}_2F_1(a, b, c; z_2/z_1)$$

where

$$c = \gamma - 2B, \qquad ab = (2B - \gamma)C - 2AB, \qquad a + b = 2A - 2C - \gamma + 1.$$

In fact, we find that $a = -\Delta(n_1, n_2, n_3)$.

THEOREM 9. *For $1 \leq i \leq 4$ let $v_i \in \mathbf{V}_{n_i}$ be vacuum vectors for* **Vir**. *Suppose that $L(-1)^2 v_3 = \gamma L(-2)v_3$ and let*

$$H = H(v_1, v_2, v_3, v_4; z, z_2) = (Y(Y(v_1, z)v_2, z_2)v_3, v_4).$$

Then H satisfies the partial differential equation

$$\partial_2^2 H + \gamma(z_2^{-1}\partial_2 - (z_2 + z)^{-1} z_2^{-1} z\partial)H - \gamma((z_2 + z)^{-2}\Delta_{n_1} + z_2^{-2}\Delta_{n_2})H = 0.$$

Let $v_i = u_{n_i}$ for $1 \leq i \leq 4$, and suppose $n_4 = n_1 + n_2 + n_3$. If we write

$$H_{n_1 n_2 n_3}(z, z_2) = M_{n_1 n_2 n_3} z_2^{-A'} z^{-B'} \left(1 + \frac{z}{z_2}\right)^{-C'} F(z/z_2)$$

for

$$A' = \Delta(n_1 + n_2, n_3), \qquad B' = \Delta(n_1, n_2), \qquad C' = \Delta(n_1, n_3)$$

then the differential equation for H becomes a differential equation for F. Letting $x = z/z_2$ and using the facts that

$$\Delta(n, n_3)(\Delta(n, n_3) + 1) = \gamma(\Delta(n, n_3) + \Delta_n)$$

for any $n = 0, 1, 2, 3$, and

$$(A' - C')(A' + C' + 1 - \gamma) = \gamma(\Delta_{n_2} - B')$$

we get the ordinary differential equation

$$x(1+x)F'' + [(2A'+2-2\gamma) + (2A'+2-2C'-\gamma)x]F' - [2(A'-C')C' + \gamma B']F = 0$$

for F. Therefore, we find that

$$H_{n_1 n_2 n_3}(z, z_2) = M_{n_1 n_2 n_3} z_2^{-A'} z^{-B'} \left(1 + \frac{z}{z_2}\right)^{-C'} {}_2F_1(a', b', c'; -z/z_2)$$

where

$$c' = 2(A'+1-\gamma), \qquad a'b' = 2(C'-A')C' - \gamma B', \qquad a'+b' = 2A'-2C'-\gamma+1.$$

In fact, we find that $a' = a = -\Delta(n_1, n_2, n_3)$.

For $n_1, n_2 \in \{1, 3\}$ we give the values for $a, b, c, a', b', c', A, B, C, A', B', C'$ as $n_3 = 0, 1, 2, 3$ in tables at the end of the paper.

5. Transformation Formulas and Rationalization

The following Lemmas are well-known results in the theory of hypergeometric functions. The formula in Lemma 10 can easily be obtained from formula (2) on page 92 of [**L**], and the formulas in Lemma 11 are the formulas numbered (1.4.5) and (1.4.13) on page 14 of [**S**].

LEMMA 10. *(Kummer's Quadratic Transformation Formula) For $|4z| < |1 - z|^2$ we have*

$${}_2F_1(a, b, 1 + a - b; z) = (1 - z)^{-a} {}_2F_1(\tfrac{1}{2}a, \tfrac{1}{2} + \tfrac{1}{2}a - b, 1 + a - b; -4z/(1 - z)^2).$$

LEMMA 11. *(Gauss Recurrence Relations for Contiguous Functions) For a and b nonzero, we have*

$$(c - a - 1) {}_2F_1(a, b, c; z) + a {}_2F_1(a + 1, b, c; z) = (c - 1) {}_2F_1(a, b, c - 1; z),$$

$$c(1 - z) {}_2F_1(a, b, c; z) + (c - a)z {}_2F_1(a, b, c + 1; z) = c {}_2F_1(a, b - 1, c; z).$$

COROLLARY 12. *For $|4z| < |1 - z|^2$ we have*

$$_2F_1(\tfrac{1}{4}, \tfrac{3}{4}, \tfrac{3}{2}; -4z/(1 - z)^2) = (1 - z)^{\frac{1}{2}},$$

$$_2F_1(-\tfrac{1}{4}, \tfrac{1}{4}, \tfrac{1}{2}; -4z/(1 - z)^2) = (1 - z)^{-\frac{1}{2}},$$

$$_2F_1(\tfrac{5}{4}, \tfrac{3}{4}, \tfrac{3}{2}; -4z/(1 - z)^2) = \frac{(1 - z)^{\frac{3}{2}}}{1 + z},$$

$$_2F_1(\tfrac{1}{4}, \tfrac{3}{4}, \tfrac{1}{2}; -4z/(1 - z)^2) = \frac{(1 - z)^{\frac{1}{2}}}{1 + z}.$$

PROOF. The first two equations follow from Kummer's quadratic transformation formula by taking $a = \tfrac{1}{2}$ or $a = -\tfrac{1}{2}$, $b = 0$. The last two equations follow from the Gauss recurrence relations by taking $a = \tfrac{1}{4}$, $b = \tfrac{3}{4}$, $c = \tfrac{3}{2}$ or $c = \tfrac{1}{2}$.

We wish to rationalize the correlation functions so that we can apply the usual contour integration techniques to obtain algebraic relations for the vertex operators. To do this we will use substitutions which give four sheeted coverings of the t-plane, where $t = z_2/z_1$. The functions G and H have possible poles at $t = 0$, $t = 1$ and $t = \infty$, as well as various cuts, but the four sheeted coverings are branched at these points, and each of these three points has only two points lying above it. First, in $G_{n_1 n_2 n_3}(z_1, z_2)$, we use the substitution

$$t^{1/2} = \left(\frac{z_2}{z_1}\right)^{\frac{1}{2}} = \frac{2x_0}{1 + x_0^2} = \frac{2x_\infty}{1 + x_\infty^2}.$$

In $G_{n_2 n_1 n_3}(z_2, z_1)$ we use the substitution

$$t^{-1/2} = \left(\frac{z_1}{z_2}\right)^{\frac{1}{2}} = \frac{2x_\mathbf{i}}{1 + x_\mathbf{i}^2} = \frac{2x_{-\mathbf{i}}}{1 + x_{-\mathbf{i}}^2}.$$

In $H_{n_1 n_2 n_3}(z, z_2)$ we use the substitution

$$(t^{-1} - 1)^{1/2} = \left(\frac{z_1 - z_2}{z_2}\right)^{\frac{1}{2}} = \frac{2x_1}{1 - x_1^2} = \frac{2x_{-1}}{1 - x_{-1}^2}.$$

In this notation, for $\alpha \in \{0, \infty, \mathbf{i}, -\mathbf{i}, 1, -1\}$, the variable x_α is local at the point α on the four sheeted cover. These points are all related to each other by linear fractional transformations, and we let x be a global variable on the genus zero covering. Below we will discuss the regions of absolute convergence after the substitutions have been made. Let

$$\mathcal{R}_x = \mathbb{C}[x, x^{-1}(x^4 - 1)^{-1}]$$

be the ring of rational functions in x with possible poles at those six points. Note that if $f(x) \in \mathcal{R}_x$ and $\mu(x)$ is any of the linear fractional transformations $I(x) = x$,

$$A(x) = \frac{1}{x}, \quad B(x) = \frac{x + \mathbf{i}}{\mathbf{i}x + 1}, \quad C(x) = \frac{\mathbf{i}x + 1}{x + \mathbf{i}}, \quad D(x) = \frac{x - 1}{x + 1} \text{ or } E(x) = \frac{1 + x}{1 - x}$$

then $f(\mu(x)) \in \mathcal{R}_x$. Also note that $\mathcal{R}_x$ is preserved by the operator d/dx.

Note that

$$1 - \frac{z_2}{z_1} = 1 - t = \frac{(1 - x_0^2)^2}{(1 + x_0^2)^2} = \frac{(1 - x_0^2)^4}{(1 - x_0^4)^2}$$

and

$$1 + \frac{z}{z_2} = t^{-1} = \frac{(1 + x_1^2)^2}{(1 - x_1^2)^2} = \frac{(1 + x_1^2)^4}{(1 - x_1^4)^2}.$$

The S_3 permutation group acting on the three points $t = 0, 1, \infty$, lifts to an S_4 permutation group acting on the six points $0, \infty, \mathbf{i}, -\mathbf{i}, 1, -1$. For $a \in \{1, -1, \mathbf{i}, -\mathbf{i}\}$ let $F_a(x) = ax$. The group of 24 linear fractional transformations giving these permutations is generated by $B(x)$ and $D(x)$, each of which has order 4. It consists of the compositions of the four functions F_a, with the six functions I, A, B, C, D and E. One has the relations $BDB^{-1} = F_{\mathbf{i}}$, $DBD^{-1} = F_{-\mathbf{i}}$, $F_{-1}DF_{-1} = D^{-1}$ and $F_{-\mathbf{i}}DF_{\mathbf{i}} = B$. It is easy to check that the correspondence $B \leftrightarrow (1,2,3,4)$, $D \leftrightarrow (1,3,2,4)$ and $F_{\mathbf{i}} \leftrightarrow (1,2,4,3)$ determines an isomorphism between this group of 24 transformations and the permutation group S_4. In this paper we will not use all 24 transformations, but we will just choose enough to relate local variables at each of the six points. However, we believe that it will be necessary to use all 24 transformations in order to fully understand the algebraic structure of the 2×2 B matrices defined just before Theorem 15. That will be a subject for future investigation.

We can choose local variables x_α such that

$$x_\infty = \frac{1}{x_0}, \qquad x_{\mathbf{i}} = \frac{1}{x_{-\mathbf{i}}}, \qquad x_1 = \frac{-1}{x_{-1}}.$$

The relations between x_0 and x_α are determined, up to some sign choices, by their relationship to t. We make the choices

$$x_0 = \frac{x_{\mathbf{i}} + \mathbf{i}}{\mathbf{i}x_{\mathbf{i}} + 1} \quad \text{so} \quad x_0 = \frac{-x_{-\mathbf{i}} + \mathbf{i}}{\mathbf{i}x_{-\mathbf{i}} - 1}, \qquad x_0 = \frac{x_1 + 1}{-x_1 + 1} \quad \text{so} \quad x_0 = \frac{x_{-1} - 1}{x_{-1} + 1},$$

so that

$$x_{\mathbf{i}} = \frac{-x_0 + \mathbf{i}}{\mathbf{i}x_0 - 1} \quad \text{and} \quad x_1 = \frac{x_0 - 1}{x_0 + 1}.$$

Until further notice we assume that $n_1, n_2, n_3 \in \{1, 3\}$.

THEOREM 13. *For $\alpha = 0$ or $\alpha = \infty$, after the substitution $(z_2/z_1)^{1/2} = 2x_\alpha/(1 + x_\alpha^2)$, the series $z_2^{\frac{1}{8}} G_{n_1 n_2 n_3}(z_1, z_2)$ converges absolutely to a function of x_α in the domain $|x_\alpha| < \sqrt{3 - 2\sqrt{2}}$ and in that domain we have*

$$(1 - x_\alpha^4)^{\frac{1}{4}} z_2^{\frac{1}{8}} G_{n_1 n_2 n_3}(z_1, z_2) = \begin{cases} 2x_\alpha K_{n_1 n_2 n_3} & \text{if } n_2 = n_3 \\[2ex] K_{n_1 n_2 n_3} & \text{if } n_2 \neq n_3 \end{cases}$$

For $\alpha = \mathbf{i}$ or $\alpha = -\mathbf{i}$ the above assertions with z_1 and z_2 switched and with n_1 and n_2 switched, are true.

PROOF. In the expression for $G_{n_1 n_2 n_3}(z_1, z_2)$ obtained after Theorem 8 we see it as a series in $t = z_2/z_1$ which converges absolutely for $|t| < 1$. But in order to use Corollary 12 with $z = -x_\alpha^2$ we need $\frac{|2x_\alpha|}{|1+x_\alpha^2|} < 1$. Using polar coordinates $x_\alpha = re^{i\theta}$ this condition is equivalent to $4r^2 < 1 + 2r^2 \cos(2\theta) + r^4$. This is certainly true when $4r^2 < 1 - 2r^2 + r^4$, that is, when $0 < r^4 - 6r^2 + 1$. The parabola $y = x^2 - 6x + 1$ is positive for $x < 3 - 2\sqrt{2}$ so with $0 < r < \sqrt{3 - 2\sqrt{2}}$ we are guaranteed to have the desired condition.

Using the values of A, B, C, a, b, c in the tables, and Corollary 12, after some algebra one gets the explicit formula stated in the theorem. ∎

THEOREM 14. *For $\alpha = 1$ or $\alpha = -1$, after the substitution $(z/z_1)^{1/2} = 2x_\alpha/(1 - x_\alpha^2)$, the series $z^{\frac{1}{8}} H_{n_1 n_2 n_3}(z, z_2)$ converges absolutely to a function of x_α in the domain $|x_\alpha| < \sqrt{3 - 2\sqrt{2}}$ and in that domain we have*

$$(1 - x_\alpha^4)^{\frac{1}{4}} z^{\frac{1}{8}} H_{n_1 n_2 n_3}(z, z_2) = \begin{cases} 2x_\alpha M_{n_1 n_2 n_3} & \text{if } n_1 = n_2 \\ \\ M_{n_1 n_2 n_3} & \text{if } n_1 \neq n_2 \end{cases}$$

PROOF. In the expression for $H_{n_1 n_2 n_3}(z, z_2)$ obtained after Theorem 9 we see it as a series in $t^{-1} - 1 = z/z_2$ which converges absolutely for $|z/z_2| < 1$. But in order to use Corollary 12 with $z = x_\alpha^2$ we need $\frac{|2x_\alpha|}{|1-x_\alpha^2|} < 1$. Using polar coordinates $x_\alpha = re^{i\theta}$ this condition is equivalent to $4r^2 < 1 - 2r^2 \cos(2\theta) + r^4$. This is certainly true when $4r^2 < 1 - 2r^2 + r^4$. So the analysis proceeds just as in Theorem 13.

Using the values of A', B', C', a', b', c' in the tables, and Corollary 12, after some algebra one gets the explicit formula stated in the theorem. ∎

In order to get the most general form of the Jacobi identity, we will eventually apply the Cauchy residue theorem to matrix valued differential forms on the four sheeted covering. Let

$$[G(z_1, z_2)] = \begin{bmatrix} G_{111}(z_1, z_2) & G_{311}(z_1, z_2) & G_{333}(z_1, z_2) & G_{133}(z_1, z_2) \\ G_{331}(z_1, z_2) & G_{131}(z_1, z_2) & G_{113}(z_1, z_2) & G_{313}(z_1, z_2) \end{bmatrix},$$

$$[G(z_2, z_1)] = \begin{bmatrix} G_{111}(z_2, z_1) & G_{311}(z_2, z_1) & G_{333}(z_2, z_1) & G_{133}(z_2, z_1) \\ G_{331}(z_2, z_1) & G_{131}(z_2, z_1) & G_{113}(z_2, z_1) & G_{313}(z_2, z_1) \end{bmatrix}$$

and

$$[H(z, z_2)] = \begin{bmatrix} H_{111}(z, z_2) & H_{331}(z, z_2) & H_{113}(z, z_2) & H_{333}(z, z_2) \\ H_{311}(z, z_2) & H_{131}(z, z_2) & H_{313}(z, z_2) & H_{133}(z, z_2) \end{bmatrix}.$$

Note that in the matrix $[G(z_1, z_2)] = [G_{n_1 n_2 n_3}(z_1, z_2)]$ the first row has $n_2 = n_3$, the second row has $n_2 \neq n_3$, while each column has the same n_3 but different values of n_1. The $[G(z_2, z_1)]$ matrix has the same pattern of subscripts as the $[G(z_1, z_2)]$ matrix, but the variables z_1 and z_2 are switched. In the matrix

$[H(z, z_2)]$ the first row has $n_1 = n_2$, the second row has $n_1 \neq n_2$, while each column has the same n_2 and the same n_3. Then, for $\alpha = 0$ or $\alpha = \infty$, we have

$$(1 - x_\alpha^4)^{\frac{1}{4}} z_2^{\frac{1}{8}} \, [G(z_1, z_2)]_\alpha = \begin{bmatrix} 2K_{111}x_\alpha & 2K_{311}x_\alpha & 2K_{333}x_\alpha & 2K_{133}x_\alpha \\ K_{331} & K_{131} & K_{113} & K_{313} \end{bmatrix},$$

for $\alpha = \mathbf{i}$ or $\alpha = -\mathbf{i}$, we have

$$(1 - x_\alpha^4)^{\frac{1}{4}} z_1^{\frac{1}{8}} \, [G(z_2, z_1)]_\alpha = \begin{bmatrix} 2K_{111}x_\alpha & 2K_{311}x_\alpha & 2K_{333}x_\alpha & 2K_{133}x_\alpha \\ K_{331} & K_{131} & K_{113} & K_{313} \end{bmatrix},$$

and for $\alpha = 1$ or $\alpha = -1$, we have

$$(1 - x_\alpha^4)^{\frac{1}{4}} z^{\frac{1}{8}} \, [H(z, z_2)]_\alpha = \begin{bmatrix} 2M_{111}x_\alpha & 2M_{331}x_\alpha & 2M_{113}x_\alpha & 2M_{333}x_\alpha \\ M_{311} & M_{131} & M_{313} & M_{133} \end{bmatrix}.$$

These equalities mean that the left sides are series which, after an appropriate substitution, converge absolutely in a small disk around the appropriate x_α to the globally defined matrix valued functions given on the right sides.

For any $f(x) \in \mathcal{R}_x$ we can use linear fractional transformations to express the globally defined matrix valued differential form

$$\begin{bmatrix} 2K_{111}x & 2K_{311}x & 2K_{333}x & 2K_{133}x \\ K_{331} & K_{131} & K_{113} & K_{313} \end{bmatrix} f(x)dx$$

in terms of the appropriate local coordinate variable x_α at each of the six possible poles. The residue at each pole is then easily found. By the residue theorem, the sum of all the residues is zero, giving us a relation among the correlation functions. That relation is the generalization of the Jacobi Identity which we seek. The residue at $x = 0$ can be found immediately from the global expression. To find the residue at $x = \infty$, use $x_\infty = x_0^{-1}$ and find the residue at $x_\infty = 0$. Leaving off the function $f(x)$ and the differential dx for the moment, we have

$$\begin{bmatrix} 2K_{111}x_0 & 2K_{311}x_0 & 2K_{333}x_0 & 2K_{133}x_0 \\ K_{331} & K_{131} & K_{113} & K_{313} \end{bmatrix}$$

$$= \begin{bmatrix} 2K_{111}x_\infty^{-1} & 2K_{311}x_\infty^{-1} & 2K_{333}x_\infty^{-1} & 2K_{133}x_\infty^{-1} \\ K_{331} & K_{131} & K_{113} & K_{313} \end{bmatrix}$$

$$= \frac{1}{x_\infty} \begin{bmatrix} 2K_{111} & 2K_{311} & 2K_{333} & 2K_{133} \\ K_{331}x_\infty & K_{131}x_\infty & K_{113}x_\infty & K_{313}x_\infty \end{bmatrix}$$

$$= \frac{1}{x_\infty} \begin{bmatrix} 0 & \frac{2K_{111}}{K_{331}} \\ \frac{K_{331}}{2K_{111}} & 0 \end{bmatrix} \begin{bmatrix} 2K_{111}x_\infty & 2K_{311}x_\infty & 2K_{333}x_\infty & 2K_{133}x_\infty \\ K_{331} & K_{131} & K_{113} & K_{313} \end{bmatrix}$$

$$= (1 - x_\infty^4)^{\frac{1}{4}} z_2^{\frac{1}{8}} \frac{1}{x_\infty} \begin{bmatrix} 0 & \frac{2K_{111}}{K_{331}} \\ \frac{K_{331}}{2K_{111}} & 0 \end{bmatrix} [G(z_1, z_2)]_\infty$$

if and only if the following conditions are consistent:

$$K_{111}K_{131} = K_{331}K_{311}, \qquad K_{111}K_{113} = K_{331}K_{333}, \qquad K_{111}K_{313} = K_{331}K_{133}.$$

To find the residue at $x = \mathbf{i}$ use $x_0 = (x_{\mathbf{i}} + \mathbf{i})/(\mathbf{i}x_{\mathbf{i}} + 1)$ and find the residue at $x_{\mathbf{i}} = 0$. Without imposing any further conditions, we have

$$
\begin{bmatrix} 2K_{111}x_0 & 2K_{311}x_0 & 2K_{333}x_0 & 2K_{133}x_0 \\ K_{331} & K_{131} & K_{113} & K_{313} \end{bmatrix}
$$
$$
= \begin{bmatrix} 2K_{111}\frac{x_{\mathbf{i}}+\mathbf{i}}{\mathbf{i}x_{\mathbf{i}}+1} & 2K_{311}\frac{x_{\mathbf{i}}+\mathbf{i}}{\mathbf{i}x_{\mathbf{i}}+1} & 2K_{333}\frac{x_{\mathbf{i}}+\mathbf{i}}{\mathbf{i}x_{\mathbf{i}}+1} & 2K_{133}\frac{x_{\mathbf{i}}+\mathbf{i}}{\mathbf{i}x_{\mathbf{i}}+1} \\ K_{331} & K_{131} & K_{113} & K_{313} \end{bmatrix}
$$
$$
= \frac{1}{\mathbf{i}x_{\mathbf{i}}+1} \begin{bmatrix} 2K_{111}(x_{\mathbf{i}}+\mathbf{i}) & 2K_{311}(x_{\mathbf{i}}+\mathbf{i}) & 2K_{333}(x_{\mathbf{i}}+\mathbf{i}) & 2K_{133}(x_{\mathbf{i}}+\mathbf{i}) \\ K_{331}(\mathbf{i}x_{\mathbf{i}}+1) & K_{131}(\mathbf{i}x_{\mathbf{i}}+1) & K_{113}(\mathbf{i}x_{\mathbf{i}}+1) & K_{313}(\mathbf{i}x_{\mathbf{i}}+1) \end{bmatrix}
$$
$$
= \frac{1}{\mathbf{i}x_{\mathbf{i}}+1} \begin{bmatrix} 1 & \frac{2\mathbf{i}K_{111}}{K_{331}} \\ \frac{\mathbf{i}K_{331}}{2K_{111}} & 1 \end{bmatrix} \begin{bmatrix} 2K_{111}x_{\mathbf{i}} & 2K_{311}x_{\mathbf{i}} & 2K_{333}x_{\mathbf{i}} & 2K_{133}x_{\mathbf{i}} \\ K_{331} & K_{131} & K_{113} & K_{313} \end{bmatrix}
$$
$$
= (1 - x_{\mathbf{i}}^4)^{\frac{1}{4}} z_1^{\frac{1}{8}} \frac{1}{\mathbf{i}x_{\mathbf{i}}+1} \begin{bmatrix} 1 & \frac{2\mathbf{i}K_{111}}{K_{331}} \\ \frac{\mathbf{i}K_{331}}{2K_{111}} & 1 \end{bmatrix} [G(z_2, z_1)]_{\mathbf{i}}.
$$

To find the residue at $x = -\mathbf{i}$ use $x_0 = (-x_{-\mathbf{i}} + \mathbf{i})/(\mathbf{i}x_{-\mathbf{i}} - 1)$ and find the residue at $x_{-\mathbf{i}} = 0$. Without imposing any new conditions, we have

$$
\begin{bmatrix} 2K_{111}x_0 & 2K_{311}x_0 & 2K_{333}x_0 & 2K_{133}x_0 \\ K_{331} & K_{131} & K_{113} & K_{313} \end{bmatrix}
$$
$$
= \frac{-1}{\mathbf{i}x_{-\mathbf{i}}-1} \begin{bmatrix} 2K_{111}(x_{-\mathbf{i}}-\mathbf{i}) & 2K_{311}(x_{-\mathbf{i}}-\mathbf{i}) & 2K_{333}(x_{-\mathbf{i}}-\mathbf{i}) & 2K_{133}(x_{-\mathbf{i}}-\mathbf{i}) \\ K_{331}(-\mathbf{i}x_{-\mathbf{i}}+1) & K_{131}(-\mathbf{i}x_{-\mathbf{i}}+1) & K_{113}(-\mathbf{i}x_{-\mathbf{i}}+1) & K_{313}(-\mathbf{i}x_{-\mathbf{i}}+1) \end{bmatrix}
$$
$$
= \frac{1}{\mathbf{i}x_{-\mathbf{i}}-1} \begin{bmatrix} -1 & \frac{2\mathbf{i}K_{111}}{K_{331}} \\ \frac{\mathbf{i}K_{331}}{2K_{111}} & -1 \end{bmatrix} \begin{bmatrix} 2K_{111}x_{-\mathbf{i}} & 2K_{311}x_{-\mathbf{i}} & 2K_{333}x_{-\mathbf{i}} & 2K_{133}x_{-\mathbf{i}} \\ K_{331} & K_{131} & K_{113} & K_{313} \end{bmatrix}
$$
$$
= (1 - x_{-\mathbf{i}}^4)^{\frac{1}{4}} z_1^{\frac{1}{8}} \frac{1}{\mathbf{i}x_{-\mathbf{i}}-1} \begin{bmatrix} -1 & \frac{2\mathbf{i}K_{111}}{K_{331}} \\ \frac{\mathbf{i}K_{331}}{2K_{111}} & -1 \end{bmatrix} [G(z_2, z_1)]_{-\mathbf{i}}.
$$

To find the residue at $x = -1$ use $x_0 = (x_{-1} - 1)/(x_{-1} + 1)$ and find the residue at $x_{-1} = 0$. Assuming the previous consistency conditions are true, we have

$$
\begin{bmatrix} 2K_{111}x_0 & 2K_{311}x_0 & 2K_{333}x_0 & 2K_{133}x_0 \\ K_{331} & K_{131} & K_{113} & K_{313} \end{bmatrix}
$$
$$
= \frac{1}{x_{-1}+1} \begin{bmatrix} 2K_{111}(x_{-1}-1) & 2K_{311}(x_{-1}-1) & 2K_{333}(x_{-1}-1) & 2K_{133}(x_{-1}-1) \\ K_{331}(x_{-1}+1) & K_{131}(x_{-1}+1) & K_{113}(x_{-1}+1) & K_{313}(x_{-1}+1) \end{bmatrix}
$$
$$
= \frac{1}{x_{-1}+1} \begin{bmatrix} \frac{K_{111}}{M_{111}} & \frac{-2K_{111}}{M_{311}} \\ \frac{K_{331}}{2M_{111}} & \frac{K_{331}}{M_{311}} \end{bmatrix} \begin{bmatrix} 2M_{111}x_{-1} & 2M_{331}x_{-1} & 2M_{113}x_{-1} & 2M_{333}x_{-1} \\ M_{311} & M_{131} & M_{313} & M_{133} \end{bmatrix}
$$
$$
= (1 - x_{-1}^4)^{\frac{1}{4}} z^{\frac{1}{8}} \frac{1}{x_{-1}+1} \begin{bmatrix} \frac{K_{111}}{M_{111}} & \frac{-2K_{111}}{M_{311}} \\ \frac{K_{331}}{2M_{111}} & \frac{K_{331}}{M_{311}} \end{bmatrix} [H(z, z_2)]_{-1}
$$

if and only if the following conditions hold:

$$
\frac{K_{111}}{K_{133}} = \frac{M_{111}}{M_{333}} = \frac{M_{311}}{M_{133}}, \quad \frac{K_{111}}{K_{311}} = \frac{M_{111}}{M_{331}} = \frac{M_{311}}{M_{131}}, \quad \frac{K_{111}}{K_{333}} = \frac{M_{111}}{M_{113}} = \frac{M_{311}}{M_{313}}.
$$

To find the residue at $x = 1$ use $x_0 = (x_1 + 1)/(-x_1 + 1)$ and find the residue at $x_1 = 0$. Without imposing any new conditions, we have

$$\begin{bmatrix} 2K_{111}x_0 & 2K_{311}x_0 & 2K_{333}x_0 & 2K_{133}x_0 \\ K_{331} & K_{131} & K_{113} & K_{313} \end{bmatrix}$$

$$= \frac{1}{-x_1 + 1} \begin{bmatrix} 2K_{111}(x_1+1) & 2K_{311}(x_1+1) & 2K_{333}(x_1+1) & 2K_{133}(x_1+1) \\ K_{331}(-x_1+1) & K_{131}(-x_1+1) & K_{113}(-x_1+1) & K_{313}(-x_1+1) \end{bmatrix}$$

$$= \frac{1}{-x_1 + 1} \begin{bmatrix} \frac{K_{111}}{M_{111}} & \frac{2K_{111}}{M_{311}} \\ \frac{-K_{331}}{2M_{111}} & \frac{K_{331}}{M_{311}} \end{bmatrix} \begin{bmatrix} 2M_{111}x_1 & 2M_{331}x_1 & 2M_{113}x_1 & 2M_{333}x_1 \\ M_{311} & M_{131} & M_{313} & M_{133} \end{bmatrix}$$

$$= (1 - x_1^4)^{\frac{1}{4}} z^{\frac{1}{8}} \frac{1}{-x_1 + 1} \begin{bmatrix} \frac{K_{111}}{M_{111}} & \frac{2K_{111}}{M_{311}} \\ \frac{-K_{331}}{2M_{111}} & \frac{K_{331}}{M_{311}} \end{bmatrix} [H(z, z_2)]_1.$$

Using Definition 6, the consistency conditions we have found translate into the following conditions on the structure constants A_{mn}:

$$\frac{A_{30}}{A_{10}} = \frac{A_{12}}{A_{32}} = \frac{A_{11}}{A_{33}} = \frac{A_{31}}{A_{13}}, \quad \frac{A_{30}A_{31}}{A_{10}A_{13}} = 1, \quad A_{21} = A_{23}, \quad A_{01} = A_{03}.$$

In fact, we know quite a bit more about the structure constants A_{mn}. Since $Y(u_0, z) = Y(\mathbf{vac}, z) = 1$, we have $A_{0n} = 1$ for any n. We also want

$$Y(v, z)\mathbf{vac} = e^{zL(-1)}v \in \mathcal{V}[z]$$

which implies the "creation property"

$$lim_{z \to 0} Y(v, z)\mathbf{vac} = Y_{-wt(v)}(v)\mathbf{vac} = v.$$

These give $A_{m0} = 1$ for any m. We also have

$$A_{21} = (Y_0(u_2)u_1, u_3) = \frac{1}{\sqrt{2}}(e(0)\mathbf{vac}', e(0)\mathbf{vac}') = \frac{1}{\sqrt{2}},$$

$$A_{22} = (Y_{1/2}(u_2)u_2, u_0) = \tfrac{1}{2}(e(\tfrac{1}{2})e(-\tfrac{1}{2})\mathbf{vac}, \mathbf{vac}) = \tfrac{1}{2}\langle e, e\rangle(\mathbf{vac}, \mathbf{vac}) = 1,$$

$$A_{23} = (Y_0(u_2)u_3, u_1) = \frac{1}{\sqrt{2}}(e(0)e(0)\mathbf{vac}', \mathbf{vac}') = \frac{1}{\sqrt{2}}.$$

We also want the "symmetry" condition, generalizing what is called "skew-symmetry" in [**FHL**],

$$z^{\Delta(n_1, n_2)}Y(v_1, z)v_2 = (-z)^{\Delta(n_1, n_2)}e^{zL(-1)}Y(v_2, -z)v_1$$

for $v_i \in \mathbf{V}_{n_i}$. With $v_1 = u_m$ and $v_2 = u_n$, after pairing with u_{m+n}, this gives $A_{mn} = A_{nm}$. Along with the above constraints, this determines all the structure constants A_{mn} except A_{11}, which we normalize to $1/\sqrt{2}$, and A_{13}, which we normalize to 1. Here is a table summarizing the results.

A_{mn}	$n = 0$	$n = 1$	$n = 2$	$n = 3$
$m = 0$	1	1	1	1
$m = 1$	1	$\dfrac{1}{\sqrt{2}}$	$\dfrac{1}{\sqrt{2}}$	1
$m = 2$	1	$\dfrac{1}{\sqrt{2}}$	1	$\dfrac{1}{\sqrt{2}}$
$m = 3$	1	1	$\dfrac{1}{\sqrt{2}}$	$\dfrac{1}{\sqrt{2}}$

DEFINITION. *Define the matrices*

$$B_0 = \begin{bmatrix} 1 & 0 \\ 0 & 1 \end{bmatrix}, \qquad B_\infty = \begin{bmatrix} 0 & 1 \\ 1 & 0 \end{bmatrix}, \qquad B_{\mathbf{i}} = \begin{bmatrix} 1 & \mathbf{i} \\ \mathbf{i} & 1 \end{bmatrix},$$

$$B_{-\mathbf{i}} = \begin{bmatrix} -1 & \mathbf{i} \\ \mathbf{i} & -1 \end{bmatrix}, \qquad B_{-1} = \begin{bmatrix} 1 & -1 \\ 1 & 1 \end{bmatrix}, \qquad B_1 = \begin{bmatrix} 1 & 1 \\ -1 & 1 \end{bmatrix}.$$

THEOREM 15. *With A_{mn} as given in the above table, and with notations as defined above, we have*

$$(1 - x_0^4)^{\frac{1}{4}} z_2^{\frac{1}{8}} \, [G(z_1, z_2)]_0$$

$$\sim (1 - x_\infty^4)^{\frac{1}{4}} z_2^{\frac{1}{8}} \, x_\infty^{-1} B_\infty \, [G(z_1, z_2)]_\infty$$

$$\sim (1 - x_{\mathbf{i}}^4)^{\frac{1}{4}} z_1^{\frac{1}{8}} \, (\mathbf{i}x_{\mathbf{i}} + 1)^{-1} B_{\mathbf{i}} \, [G(z_2, z_1)]_{\mathbf{i}}$$

$$\sim (1 - x_{-\mathbf{i}}^4)^{\frac{1}{4}} z_1^{\frac{1}{8}} \, (\mathbf{i}x_{-\mathbf{i}} - 1)^{-1} B_{-\mathbf{i}} \, [G(z_2, z_1)]_{-\mathbf{i}}$$

$$\sim (1 - x_{-1}^4)^{\frac{1}{4}} z^{\frac{1}{8}} \, (x_{-1} + 1)^{-1} B_{-1} \, [H(z, z_2)]_{-1}$$

$$\sim (1 - x_1^4)^{\frac{1}{4}} z^{\frac{1}{8}} \, (-x_1 + 1)^{-1} B_1 \, [H(z, z_2)]_1$$

where $\sim$ means that these series converge absolutely in their appropriate domains to the same globally defined matrix valued function

$$\begin{bmatrix} x_0 & x_0 & x_0 & x_0 \\ 1 & 1 & 1 & 1 \end{bmatrix}$$

on the four sheeted covering.

Note that if we associate to a matrix $M = \begin{bmatrix} a & b \\ c & d \end{bmatrix}$ a linear fractional transformation $f(x) = \frac{ax+b}{cx+d}$ then we associate to the matrix B_α the linear fractional transformation

$$f_\alpha(x_\alpha) = x_0 = \frac{ax_\alpha + b}{cx_\alpha + d}.$$

6. Inductive formulas

Now let us see how the general two-point functions are determined inductively.

THEOREM 16. *Let $v_i \in \mathbf{V}_{n_i}$, $1 \leq i \leq 4$, be eigenvectors for $L(0)$ with $L(0)v_i = wt(v_i)v_i = |v_i|v_i$. For $i = 1, 2$ let $Y_i = Y(v_i, z_i)$, $\partial_i = \partial/\partial z_i$, and recall the notation*

$$G(v_1, v_2, v_3, v_4; z_1, z_2) = (Y_1 Y_2 v_3, v_4).$$

Then for any $k \in \mathbb{Z}$ we have

$$
\begin{aligned}
G(v_1, v_2, v_3, L(-k)v_4; z_1, z_2) = {} & G(v_1, v_2, L(k)v_3, v_4; z_1, z_2) \\
& + (z_1^{k+1}\partial_1 + z_2^{k+1}\partial_2)G(v_1, v_2, v_3, v_4; z_1, z_2) \\
& + (k+1)(z_1^k|v_1| + z_2^k|v_2|)G(v_1, v_2, v_3, v_4; z_1, z_2) \\
& + \sum_{i \geq 1} \binom{k+1}{i+1} z_1^{k-i} G(L(i)v_1, v_2, v_3, v_4; z_1, z_2) \\
& + \sum_{i \geq 1} \binom{k+1}{i+1} z_2^{k-i} G(v_1, L(i)v_2, v_3, v_4; z_1, z_2).
\end{aligned}
$$

PROOF. We have

$$
\begin{aligned}
G(v_1, v_2, v_3, L(-k)v_4; z_1, z_2) = {} & (L(k)Y_1 Y_2 v_3, v_4) \\
= {} & ([L(k), Y_1]Y_2 v_3, v_4) + (Y_1[L(k), Y_2]v_3, v_4) + (Y_1 Y_2 L(k)v_3, v_4) \\
= {} & \sum_{i \geq 0} \binom{k+1}{i} z_1^{k+1-i}(Y(L(i-1)v_1, z_1)Y_2 v_3, v_4) \\
& + \sum_{i \geq 0} \binom{k+1}{i} z_2^{k+1-i}(Y_1 Y(L(i-1)v_2, z_2)v_3, v_4) \\
& + (Y_1 Y_2 L(k)v_3, v_4)
\end{aligned}
$$

giving the result after separating the $i = 0$ and the $i = 1$ terms and reindexing. ∎

THEOREM 17. *Let $v_i \in \mathbf{V}_{n_i}$, $1 \leq i \leq 4$, be eigenvectors for $L(0)$ with $L(0)v_i = wt(v_i)v_i = |v_i|v_i$. Let $\partial_1 = \partial/\partial z_1$, $\partial = \partial/\partial z$, and recall the notation*

$$H(v_1, v_2, v_3, v_4; z, z_2) = (Y(Y(v_1, z)v_2, z_2)v_3, v_4).$$

Then for any $k \in \mathbb{Z}$ we have

$$
\begin{aligned}
&H(v_1, v_2, v_3, L(-k)v_4; z, z_2) \\
&= H(v_1, v_2, L(k)v_3, v_4; z, z_2) \\
&+ z_2^{k+1} \partial_2 H(v_1, v_2, v_3, v_4; z, z_2) \\
&+ (k+1)(z_2^k |v_2| + (z_2 + z)^k |v_1|) H(v_1, v_2, v_3, v_4; z, z_2) \\
&+ [(z_2 + z)^{k+1} - z_2^{k+1}] \partial H(v_1, v_2, v_3, v_4; z, z_2) \\
&+ \sum_{i \geq 1} \binom{k+1}{i+1} z_2^{k-i} H(v_1, L(i)v_2, v_3, v_4; z, z_2) \\
&+ \sum_{j \geq 1} \binom{k+1}{j+1} (z_2 + z)^{k-j} H(L(j)v_1, v_2, v_3, v_4; z, z_2).
\end{aligned}
$$

PROOF. For brevity, let us write

$$
H = H(v_1, v_2, v_3, v_4; z, z_2), \quad H(L(k)v_3) = H(v_1, v_2, L(k)v_3, v_4; z, z_2),
$$

$$
H(L(j)v_1) = H(L(j)v_1, v_2, v_3, v_4; z, z_2)
$$

and

$$
H(L(i)v_2) = H(v_1, L(i)v_2, v_3, v_4; z, z_2).
$$

We have

$$
\begin{aligned}
&H(v_1, v_2, v_3, L(-k)v_4; z, z_2) \\
&= ([L(k), Y(Y(v_1, z)v_2, z_2)]v_3, v_4) + (Y(Y(v_1, z)v_2, z_2)L(k)v_3, v_4) \\
&= \sum_{i \geq 0} \binom{k+1}{i} z_2^{k+1-i} (Y(L(i-1)Y(v_1, z)v_2, z_2)v_3, v_4) + H(L(k)v_3) \\
&= z_2^{k+1} \partial_2 H + H(L(k)v_3) + \sum_{i \geq 0} \binom{k+1}{i+1} z_2^{k-i} (Y([L(i), Y(v_1, z)]v_2, z_2)v_3, v_4) \\
&\quad + (k+1) z_2^k |v_2| H + \sum_{i \geq 1} \binom{k+1}{i+1} z_2^{k-i} (Y(Y(v_1, z)L(i)v_2, z_2)v_3, v_4) \\
&= z_2^{k+1} \partial_2 H + H(L(k)v_3) \\
&\quad + \sum_{i \geq 0} \binom{k+1}{i+1} z_2^{k-i} \left(z^{i+1} \partial H + (i+1) z^i |v_1| H + \sum_{j \geq 1} \binom{i+1}{j+1} z^{i-j} H(L(j)v_1) \right) \\
&\quad + (k+1) z_2^k |v_2| H + \sum_{i \geq 1} \binom{k+1}{i+1} z_2^{k-i} H(L(i)v_2)
\end{aligned}
$$

$$= z_2^{k+1}\partial_2 H + H(L(k)v_3) + z_2^{k+1}\sum_{i\geq 0}\binom{k+1}{i+1}\left(\frac{z}{z_2}\right)^{i+1}\partial H$$

$$+ |v_1|z_2^k H\sum_{i\geq 0}\binom{k+1}{i+1}(i+1)\left(\frac{z}{z_2}\right)^i$$

$$+ z_2^{k-j}\sum_{j\geq 1}H(L(j)v_1)\sum_{i\geq 0}\binom{k+1}{i+1}\binom{i+1}{j+1}\left(\frac{z}{z_2}\right)^{i-j}$$

$$+ (k+1)z_2^k|v_2|H + \sum_{i\geq 1}\binom{k+1}{i+1}z_2^{k-i}H(L(i)v_2)$$

which gives the result after using

$$\binom{k+1}{i+1}(i+1) = (k+1)\binom{k}{i}$$

and

$$(z_2 + z)^m = z_2^m\left(1+\frac{z}{z_2}\right)^m = z_2^m\sum_{i\geq 0}\binom{m}{i}\left(\frac{z}{z_2}\right)^i.$$

∎

For $1 \leq i \leq 4$ let $v_i \in \mathbf{V}_{n_i}$ with $wt(v_i) = |v_i| = N_i + \Delta_{n_i}$ for $0 \leq N_i \in \mathbb{Z}$. For $n_1, n_2, n_3 \in \{1,3\}$ we have

$$\Delta(n_1, n_2+n_3) = \begin{cases} 0 & \text{if } n_2 + n_3 = 0 \\[2mm] \frac{1}{2} & \text{if } n_2 + n_3 = 2 \end{cases} \quad \text{and} \quad \Delta(n_2, n_3) = \begin{cases} \frac{1}{8} & \text{if } n_2 + n_3 = 0 \\[2mm] -\frac{3}{8} & \text{if } n_2 + n_3 = 2 \end{cases}$$

so that

$$\Delta(n_1, n_2 + n_3) + \Delta(n_2, n_3) = \tfrac{1}{8}$$

for $n_1, n_2, n_3 \in \{1,3\}$. Let $N = N_1 + N_2 + N_3 - N_4$. With $t = z_2/z_1$ and $\Delta = \Delta(n_1, n_2 + n_3)$ we have

$$G(v_1, v_2, v_3, v_4; z_1, z_2) = t^{N_1 - N_4 + \Delta}z_2^{-N - \frac{1}{8}}\Phi(v_1, v_2, v_3, v_4; t)$$

where

$$\Phi(v_1, v_2, v_3, v_4; t) = \sum_{0 \leq k \in \mathbb{Z}}t^k\Phi_k(v_1, v_2, v_3, v_4)$$

is a power series in t. Similarly, with $s = z/z_2 = t^{-1} - 1$ and $\bar{\Delta} = \Delta(n_1 + n_2, n_3)$ we have

$$H(v_1, v_2, v_3, v_4; z, z_2) = s^{N_3 - N_4 + \bar{\Delta}}z^{-N - \frac{1}{8}}\Psi(v_1, v_2, v_3, v_4; s)$$

where

$$\Psi(v_1, v_2, v_3, v_4; s) = \sum_{0 \leq k \in \mathbb{Z}}s^k\Psi_k(v_1, v_2, v_3, v_4)$$

is a power series in s.

We choose the relationship $s^{1/2} = (t^{-1}-1)^{1/2}$ such that we have $(1-x_0^2)/2x_0 = -2x_1/(1 - x_1^2)$. We wish to rewrite the recursions for G and H given in the last two theorems as recursions for the functions

$$\tilde{\Phi}(v_1, v_2, v_3, v_4; x) = (1 - x^4)^{\frac{1}{4}} z_2^{N+\frac{1}{8}} G(v_1, v_2, v_3, v_4; z_1, z_2)$$

$$= (1 - x^4)^{\frac{1}{4}} t^{N_1 - N_4 + \Delta} \Phi(v_1, v_2, v_3, v_4; t)$$

where $x = x_0$ or $x = x_\infty$ and $t^{1/2} = 2x/(1 + x^2)$, and

$$\tilde{\Psi}(v_1, v_2, v_3, v_4; w) = (1 - w^4)^{\frac{1}{4}} z^{N+\frac{1}{8}} H(v_1, v_2, v_3, v_4; z, z_2)$$

$$= (1 - w^4)^{\frac{1}{4}} s^{N_3 - N_4 + \bar{\Delta}} \Psi(v_1, v_2, v_3, v_4; s)$$

where $w = x_1$ or $w = x_{-1}$ and $s^{1/2} = 2w/(1 - w^2)$.

We will later apply the first of these recursions to the functions

$$\tilde{\Phi}(v_2, v_1, v_3, v_4; y) = (1 - y^4)^{\frac{1}{4}} z_1^{N+\frac{1}{8}} G(v_2, v_1, v_3, v_4; z_2, z_1)$$

$$= (1 - y^4)^{\frac{1}{4}} t^{-N_2 + N_4 - \Delta'} \Phi(v_2, v_1, v_3, v_4; t^{-1})$$

where $y = x_{\mathrm{i}}$ or $y = x_{-\mathrm{i}}$ and $\Delta' = \Delta(n_2, n_1 + n_3)$. These will be obtained from the first kind of recursions by switching v_1 with v_2, z_1 with z_2, and n_1 with n_2, and by replacing t by t^{-1}. Define the matrices

$$[\tilde{\Phi}_{top}]_\alpha = (1 - x_\alpha^4)^{\frac{1}{4}} z_2^{\frac{1}{8}} [G(z_1, z_2)]_\alpha$$

for $\alpha = 0$ or $\alpha = \infty$,

$$[\tilde{\Phi}_{top}]_\alpha = (1 - x_\alpha^4)^{\frac{1}{4}} z_1^{\frac{1}{8}} [G(z_2, z_1)]_\alpha$$

for $\alpha = \mathrm{i}$ or $\alpha = -\mathrm{i}$, and

$$[\tilde{\Psi}_{top}]_\alpha = (1 - x_\alpha^4)^{\frac{1}{4}} z^{\frac{1}{8}} [H(z, z_2)]_\alpha$$

for $\alpha = 1$ or $\alpha = -1$. Then our earlier work, which will give the base cases of our inductions, can be written as

$$[\tilde{\Phi}_{top}]_0 \sim x_\infty^{-1} B_\infty \, [\tilde{\Phi}_{top}]_\infty$$

$$\sim (\mathrm{i}x_{\mathrm{i}} + 1)^{-1} B_{\mathrm{i}} \, [\tilde{\Phi}_{top}]_{\mathrm{i}} \sim (\mathrm{i}x_{-\mathrm{i}} - 1)^{-1} B_{-\mathrm{i}} \, [\tilde{\Phi}_{top}]_{-\mathrm{i}}$$

$$\sim (x_{-1} + 1)^{-1} B_{-1} \, [\tilde{\Psi}_{top}]_{-1} \sim (-x_1 + 1)^{-1} B_1 \, [\tilde{\Psi}_{top}]_1$$

The general case requires us to define 2×4 matrices which are analogs of $[\tilde{\Phi}_{top}]_\alpha$ and $[\tilde{\Psi}_{top}]_\alpha$ with the "top" vectors u_{n_i} replaced by general vectors. There is no loss of generality in assuming that these vectors v_i are $L(0)$-eigenvectors with $wt(v_i) = N_i + \Delta_{n_i}$, $1 \le i \le 4$. We must do this in such a way that our induction formulas imply that the above top $\sim$ relations hold for the more general matrices. This means that we need a more general description than the pattern of subscripts defining the matrices $[G(z_1, z_2)]$, $[G(z_2, z_1)]$ and $[H(z, z_2)]$. We will use the isomorphism θ on $\mathbf{CM}(\mathbb{Z})$ defined at the beginning of this paper to give such a description. Assume that $v_1, v_2, v_3 \in \mathbf{V}_1$ and $v_4 \in \mathbf{V}_3$ are $L(0)$-eigenvectors with weights as given above. In the entries of the following matrices we will write the functions $\tilde{\Phi}(v_1, v_2, v_3, v_4; x_\alpha)$ and $\tilde{\Psi}(v_1, v_2, v_3, v_4; x_\alpha)$ more briefly as just $\tilde{\Phi}(v_1, v_2, v_3, v_4)$ and $\tilde{\Psi}(v_1, v_2, v_3, v_4)$. For $\alpha = 0$ or $\alpha = \infty$, define the matrix

$$[\tilde{\Phi}]_\alpha = [\tilde{\Phi}(v_1, v_2, v_3, v_4; x_\alpha)] =$$
$$\begin{bmatrix} \tilde{\Phi}(v_1, v_2, v_3, v_4) & \tilde{\Phi}(\theta v_1, v_2, v_3, \theta v_4) & \tilde{\Phi}(\theta v_1, \theta v_2, \theta v_3, \theta v_4) & \tilde{\Phi}(v_1, \theta v_2, \theta v_3, v_4) \\ \tilde{\Phi}(\theta v_1, \theta v_2, v_3, v_4) & \tilde{\Phi}(v_1, \theta v_2, v_3, \theta v_4) & \tilde{\Phi}(v_1, v_2, \theta v_3, \theta v_4) & \tilde{\Phi}(\theta v_1, v_2, \theta v_3, v_4) \end{bmatrix}.$$

For $\alpha = \mathbf{i}$ or $\alpha = -\mathbf{i}$ define the matrix

$$[\tilde{\Phi}]_\alpha = [\tilde{\Phi}(v_1, v_2, v_3, v_4; x_\alpha)] =$$
$$\begin{bmatrix} \tilde{\Phi}(v_2, v_1, v_3, v_4) & \tilde{\Phi}(\theta v_2, v_1, v_3, \theta v_4) & \tilde{\Phi}(\theta v_2, \theta v_1, \theta v_3, \theta v_4) & \tilde{\Phi}(v_2, \theta v_1, \theta v_3, v_4) \\ \tilde{\Phi}(\theta v_2, \theta v_1, v_3, v_4) & \tilde{\Phi}(v_2, \theta v_1, v_3, \theta v_4) & \tilde{\Phi}(v_2, v_1, \theta v_3, \theta v_4) & \tilde{\Phi}(\theta v_2, v_1, \theta v_3, v_4) \end{bmatrix}.$$

For $\alpha = 1$ or $\alpha = -1$ define the matrix

$$[\tilde{\Psi}]_\alpha = [\tilde{\Psi}(v_1, v_2, v_3, v_4; x_\alpha)] =$$
$$\begin{bmatrix} \tilde{\Psi}(v_1, v_2, v_3, v_4) & \tilde{\Psi}(\theta v_1, \theta v_2, v_3, v_4) & \tilde{\Psi}(v_1, v_2, \theta v_3, \theta v_4) & \tilde{\Psi}(\theta v_1, \theta v_2, \theta v_3, \theta v_4) \\ \tilde{\Psi}(\theta v_1, v_2, v_3, \theta v_4) & \tilde{\Psi}(v_1, \theta v_2, v_3, \theta v_4) & \tilde{\Psi}(\theta v_1, v_2, \theta v_3, v_4) & \tilde{\Psi}(v_1, \theta v_2, \theta v_3, v_4) \end{bmatrix}.$$

In our inductions we will see such matrices with one of the vectors, say v_i, replaced by $L(-k)v_i$. We will denote the above matrices with such a replacement by $[\tilde{\Phi}(L(-k)v_i)]_\alpha$ and $[\tilde{\Psi}(L(-k)v_i)]_\alpha$.

Let $\Phi' = \frac{\partial}{\partial t}\Phi$, $\Psi' = \frac{\partial}{\partial s}\Psi$ and let us use notations such as $\tilde{\Phi}(L(i)v_1) = \tilde{\Phi}(L(i)v_1, v_2, v_3, v_4; x)$ as we did in the proof of Theorem 17. We have

$$\partial_1 G = \frac{N_4 - N_1 - \Delta}{z_1} G - \frac{z_2}{z_1^2} t^{N_1 - N_4 + \Delta} z_2^{-N - \frac{1}{8}} \Phi'$$

$$\partial_2 G = \frac{\Delta - N_2 - N_3 - \frac{1}{8}}{z_2} G + \frac{1}{z_1} t^{N_1 - N_4 + \Delta} z_2^{-N - \frac{1}{8}} \Phi'$$

$$\partial H = \frac{-N_1 - N_2 - \frac{1}{8} + \bar{\Delta}}{z} H + \frac{1}{z_2} s^{N_3 - N_4 + \bar{\Delta}} z^{-N - \frac{1}{8}} \Psi'$$

$$\partial_2 H = \frac{N_4 - N_3 - \bar{\Delta}}{z_2} H - \frac{z}{z_2^2} s^{N_3 - N_4 + \bar{\Delta}} z^{-N - \frac{1}{8}} \Psi'.$$

THEOREM 18. *Define the operator*

$$\mathcal{L}_x = \frac{x(1 + x^2)}{2(1 - x^2)} \frac{d}{dx} + \frac{x^4}{2(1 - x^2)^2} = \frac{x(1 + x^2)}{2(1 - x^2)} \left[\frac{d}{dx} + \frac{x^3}{1 - x^4} \right].$$

Then for any $k \in \mathbb{Z}$ we have

$$\tilde{\Phi}(L(-k)v_4) = \tilde{\Phi}(L(k)v_3) - (N + \tfrac{1}{8})\tilde{\Phi} + (1 - t^{-k})\mathcal{L}_x \tilde{\Phi}$$

$$+ \sum_{i \geq 0} \binom{k+1}{i+1} \left(t^{i-k} \tilde{\Phi}(L(i)v_1) + \tilde{\Phi}(L(i)v_2) \right).$$

PROOF. For any $k \in \mathbb{Z}$ we have

$$
\tilde{\Phi}(L(-k)v_4) = (1 - x^4)^{\frac{1}{4}} z_2^{N-k+\frac{1}{8}} G(v_1, v_2, v_3, L(-k)v_4; z_1, z_2)
$$

$$
= (1 - x^4)^{\frac{1}{4}} z_2^{N-k+\frac{1}{8}} G(v_1, v_2, L(k)v_3, v_4; z_1, z_2)
$$

$$
+ (1 - x^4)^{\frac{1}{4}} z_2^{N-k+\frac{1}{8}} (z_1^{k+1} \partial_1 + z_2^{k+1} \partial_2) G(v_1, v_2, v_3, v_4; z_1, z_2)
$$

$$
+ (1 - x^4)^{\frac{1}{4}} z_2^{N-k+\frac{1}{8}} (k+1)(z_1^k |v_1| + z_2^k |v_2|) G(v_1, v_2, v_3, v_4; z_1, z_2)
$$

$$
+ (1 - x^4)^{\frac{1}{4}} z_2^{N-k+\frac{1}{8}} \sum_{i \geq 1} \binom{k+1}{i+1} z_1^{k-i} G(L(i)v_1, v_2, v_3, v_4; z_1, z_2)
$$

$$
+ (1 - x^4)^{\frac{1}{4}} z_2^{N-k+\frac{1}{8}} \sum_{i \geq 1} \binom{k+1}{i+1} z_2^{k-i} G(v_1, L(i)v_2, v_3, v_4; z_1, z_2)
$$

$$
= \tilde{\Phi}(L(k)v_3) + ((N_4 - N_1 - \Delta)t^{-k} + (\Delta - N_2 - N_3 - \tfrac{1}{8}))\tilde{\Phi}
$$

$$
+ (t - t^{1-k})t^{N_1 - N_4 + \Delta}(1 - x^4)^{\frac{1}{4}} \Phi' + (k+1)(t^{-k}|v_1| + |v_2|)\tilde{\Phi}
$$

$$
+ \sum_{i \geq 1} \binom{k+1}{i+1} t^{i-k} \tilde{\Phi}(L(i)v_1) + \sum_{i \geq 1} \binom{k+1}{i+1} \tilde{\Phi}(L(i)v_2).
$$

Note that

$$
\frac{d\Phi}{dx} = \frac{d\Phi}{dt}\frac{dt}{dx} \qquad \text{and} \qquad \frac{dt}{dx} = \frac{8x(1-x^2)}{(1+x^2)^3} = \frac{8x(1-x^4)}{(1+x^2)^4}
$$

so that

$$
\Phi' = \frac{d\Phi}{dt} = \frac{d\Phi}{dx}\frac{(1+x^2)^4}{8x(1-x^4)}.
$$

From

$$
\frac{d}{dx}\tilde{\Phi}(x) = \frac{d}{dx}\left((1-x^4)^{\frac{1}{4}} t^{N_1 - N_4 + \Delta} \Phi(t)\right)
$$

$$
= \frac{-x^3}{1-x^4}\tilde{\Phi}(x) + \frac{N_1 - N_4 + \Delta}{t}\frac{dt}{dx}\tilde{\Phi}(x) + (1-x^4)^{\frac{1}{4}} t^{N_1 - N_4 + \Delta}\frac{d\Phi}{dx}
$$

we get

$$
(1-x^4)^{\frac{1}{4}} t^{N_1 - N_4 + \Delta}\frac{d\Phi}{dt} = \left(\frac{dt}{dx}\right)^{-1}\frac{d\tilde{\Phi}}{dx} + \frac{x^3}{1-x^4}\left(\frac{dt}{dx}\right)^{-1}\tilde{\Phi} - \frac{N_1 - N_4 + \Delta}{t}\tilde{\Phi}.
$$

This gives us

$$
(t - t^{1-k})t^{N_1 - N_4 + \Delta}(1-x^4)^{\frac{1}{4}} \Phi'
$$

$$
= (1 - t^{-k})\left(\frac{x(1+x^2)}{2(1-x^2)}\frac{d}{dx} + \frac{x^4}{2(1-x^2)^2} + (N_4 - N_1 - \Delta)\right)\tilde{\Phi}
$$

$$
= (1 - t^{-k})(\mathcal{L}_x + (N_4 - N_1 - \Delta))\tilde{\Phi}
$$

which gives the result.

Note that the operator $\mathcal{L}_x$ and multiplication by any integral power of $t^{\frac{1}{2}}$ preserve the ring $\mathcal{R}_x$.

LEMMA 19. *For any $g \in \mathbb{C}(x)$, if $x = u^{-1}$, we have*

$$\mathcal{L}_x(xg(x^{-1})) = u^{-1}\mathcal{L}_u(g(u)).$$

PROOF. We have

$$
\begin{aligned}
\mathcal{L}_x(xg(x^{-1})) &= \frac{x(1+x^2)}{2(1-x^2)}\left(x\frac{dg}{du}\frac{du}{dx} + g(x^{-1}) + \frac{x^4}{1-x^4}g(x^{-1})\right) \\
&= \frac{u^2+1}{2u(u^2-1)}\left(-u\frac{dg}{du} + g(u) + \frac{1}{u^4-1}g(u)\right) \\
&= \frac{1+u^2}{2(1-u^2)}\frac{dg}{du} + \frac{u^3}{2(1-u^2)^2}g(u) \\
&= u^{-1}\mathcal{L}_u(g(u)).
\end{aligned}
$$

∎

LEMMA 20. *For any $g \in \mathbb{C}(x)$, if $x = \frac{\pm y + \mathbf{i}}{\mathbf{i}y \pm 1}$, $t^{1/2} = \frac{2x}{1+x^2}$ and $N \in \mathbb{Z}$, we have*

$$\mathcal{L}_x(t^N(\mathbf{i}y \pm 1)^{-1}g(x)) = \frac{-t^N}{\mathbf{i}y \pm 1}(\mathcal{L}_y - N - \tfrac{1}{8})g\left(\frac{\pm y + \mathbf{i}}{\mathbf{i}y \pm 1}\right).$$

PROOF. We have

$$
\begin{aligned}
&\mathcal{L}_x(t^N(\mathbf{i}y \pm 1)^{-1}g(x)) \\
&= \frac{x(1+x^2)}{2(1-x^2)}\left[\frac{t^N}{\mathbf{i}y \pm 1}\frac{dg}{dy}\frac{dy}{dx} + t^N g(x)\frac{d}{dx}(\mathbf{i}y \pm 1)^{-1}\right. \\
&\qquad\left. + Nt^{N-1}\frac{dt}{dx}(\mathbf{i}y \pm 1)^{-1}g(x) + t^N(\mathbf{i}y \pm 1)^{-1}\frac{x^3}{1-x^4}g(x)\right] \\
&= \frac{x(1+x^2)}{2(1-x^2)}\left[\frac{t^N}{\mathbf{i}y \pm 1}\frac{(\mathbf{i}y \pm 1)^2}{2}\frac{d}{dy} - t^N\frac{\mathbf{i}}{2}\right. \\
&\qquad\left. + Nt^{N-1}2\frac{2x}{1+x^2}\frac{2(1-x^2)}{(1+x^2)^2}(\mathbf{i}y \pm 1)^{-1} + t^N(\mathbf{i}y \pm 1)^{-1}\frac{x^3}{1-x^4}\right]g(x) \\
&= \frac{-t^N}{\mathbf{i}y \pm 1}\frac{y(1+y^2)}{2(1-y^2)}\left[\frac{d}{dy} - \frac{\mathbf{i}}{\mathbf{i}y \pm 1} - \frac{2N(1-y^2)}{y(1+y^2)}\right. \\
&\qquad\left. + \frac{(\pm y + \mathbf{i})^3}{\pm 4y\mathbf{i}(1-y^2)(\mathbf{i}y \pm 1)}\right]g\left(\frac{\pm y + \mathbf{i}}{\mathbf{i}y \pm 1}\right) \\
&= \frac{-t^N}{\mathbf{i}y \pm 1}\frac{y(1+y^2)}{2(1-y^2)}\left[\frac{d}{dy} + \frac{3y^2-1}{4y(1-y^2)} - \frac{2N(1-y^2)}{y(1+y^2)}\right]g\left(\frac{\pm y + \mathbf{i}}{\mathbf{i}y \pm 1}\right)
\end{aligned}
$$

$$= \frac{-t^N}{\mathbf{i}y \pm 1} \frac{y(1+y^2)}{2(1-y^2)} \left[\frac{d}{dy} + \frac{y^3}{1-y^4} - \frac{1-y^2}{4y(1+y^2)} - \frac{2N(1-y^2)}{y(1+y^2)} \right] g\left(\frac{\pm y + \mathbf{i}}{\mathbf{i}y \pm 1} \right)$$

$$= \frac{-t^N}{\mathbf{i}y \pm 1} \frac{y(1+y^2)}{2(1-y^2)} \left[\frac{d}{dy} + \frac{y^3}{1-y^4} - \frac{(1-y^2)(1+8N)}{4y(1+y^2)} \right] g\left(\frac{\pm y + \mathbf{i}}{\mathbf{i}y \pm 1} \right)$$

$$= \frac{-t^N}{\mathbf{i}y \pm 1} \left[\mathcal{L}_y - \frac{y(1+y^2)}{2(1-y^2)} \frac{(1-y^2)(1+8N)}{4y(1+y^2)} \right] g\left(\frac{\pm y + \mathbf{i}}{\mathbf{i}y \pm 1} \right)$$

$$= \frac{-t^N}{\mathbf{i}y \pm 1} (\mathcal{L}_y - N - \tfrac{1}{8}) g\left(\frac{\pm y + \mathbf{i}}{\mathbf{i}y \pm 1} \right)$$

$\blacksquare$

LEMMA 21. *For any* $g \in \mathbb{C}(x)$, *if* $x = \frac{w \pm 1}{\mp w + 1}$, $s^{1/2} = \frac{2w}{1-w^2}$ *and* $N \in \mathbb{Z}$, *we have*

$$\mathcal{L}_x(s^{-N}(\mp w + 1)^{-1} g(x)) = \frac{-s^{-N}}{\mp w + 1}(s^{-1}\mathcal{L}_w - (N + \tfrac{1}{8})(1 + s^{-1})) g\left(\frac{w \pm 1}{\mp w + 1} \right).$$

PROOF. We have

$$\mathcal{L}_x(s^{-N}(\mp w + 1)^{-1} g(x))$$

$$= \frac{x(1+x^2)}{2(1-x^2)} \left[\frac{s^{-N}}{\mp w + 1} \frac{dg}{dw} \frac{dw}{dx} + s^{-N} g(x) \frac{dw}{dx} \frac{d}{dw} \frac{1}{\mp w + 1} \right.$$

$$\left. + \frac{1}{\mp w + 1} g(x) \frac{dw}{dx} \frac{d}{dw} s^{-N} + \frac{s^{-N}}{(\mp w + 1)} \frac{x^3}{(1-x^4)} g(x) \right]$$

$$= \frac{x(1+x^2)}{2(1-x^2)} \left[\frac{s^{-N}}{\mp w + 1} \frac{(\mp w + 1)^2}{2} \frac{d}{dw} \pm \frac{s^{-N}}{2} \right.$$

$$\left. - \frac{1}{\mp w + 1} \frac{(\mp w + 1)^2}{2} N s^{-N} \frac{2(1+w^2)}{w(1-w^2)} + \frac{s^{-N}}{(\mp w + 1)} \frac{x^3}{(1-x^4)} \right] g(x)$$

$$= \frac{x(1+x^2)}{2(1-x^2)} \frac{s^{-N}}{\mp w + 1} \left[\frac{(\mp w + 1)^2}{2} \frac{d}{dw} \pm \frac{\mp w + 1}{2} \right.$$

$$\left. - \frac{N(\mp w + 1)^2(1+w^2)}{w(1-w^2)} + \frac{x^3}{1-x^4} \right] g(x)$$

$$= \frac{x(1+x^2)(\mp w + 1)^2}{4(1-x^2)} \frac{s^{-N}}{\mp w + 1} \left[\frac{d}{dw} \pm \frac{1}{\mp w + 1} \right.$$

$$\left. - \frac{2N(1+w^2)}{w(1-w^2)} + \frac{2x^3}{(1-x^4)(\mp w + 1)^2} \right] g(x)$$

$$= \frac{-s^{-N}}{\mp w + 1} \frac{(1-w^2)(1+w^2)}{8w} \left[\frac{d}{dw} \pm \frac{1}{\mp w + 1} \right.$$

$$\left. - \frac{2N(1+w^2)}{w(1-w^2)} \mp \frac{(w \pm 1)^3}{4(\mp w + 1)w(1+w^2)} \right] g(x)$$

$$= \frac{-s^{-N}}{\mp w + 1} \frac{(1 - w^2)(1 + w^2)}{8w} \left[\frac{d}{dw} + \frac{w^3}{1 - w^4} - \frac{(8N + 1)(1 + w^2)}{4w(1 - w^2)} \right] g(x)$$

$$= \frac{-s^{-N}}{\mp w + 1} \left[\frac{(1 - w^2)(1 + w^2)}{8w} \left(\frac{d}{dw} + \frac{w^3}{1 - w^4} \right) - (N + \tfrac{1}{8}) \frac{(1 + w^2)^2}{4w^2} \right] g(x)$$

$$= \frac{-s^{-N}}{\mp w + 1} \left[s^{-1} \mathcal{L}_w - (N + \tfrac{1}{8})(1 + s^{-1}) \right] g \left(\frac{w \pm 1}{\mp w + 1} \right)$$

$\blacksquare$

THEOREM 22. *For any* $k \in \mathbb{Z}$ *we have*

$$\tilde{\Psi}(L(-k)v_4) =$$
$$\tilde{\Psi}(L(k)v_3) - (N + \tfrac{1}{8})s^{-k-1}[(1 + s)^{k+1} - 1]\tilde{\Psi} + s^{-k-1}[(1 + s)^k - 1]\mathcal{L}_w \tilde{\Psi}$$
$$+ \sum_{i \geq 0} \binom{k + 1}{i + 1} \left[s^{i-k} \tilde{\Psi}(L(i)v_2) + \left(\frac{1 + s}{s} \right)^{k-i} \tilde{\Psi}(L(i)v_1) \right].$$

PROOF. For any $k \in \mathbb{Z}$ we have

$$\tilde{\Psi}(L(-k)v_4)$$
$$= (1 - w^4)^{\frac{1}{4}} z^{N-k+\frac{1}{8}} H(v_1, v_2, v_3, L(-k)v_4; z, z_2)$$
$$= (1 - w^4)^{\frac{1}{4}} z^{N-k+\frac{1}{8}} H(v_1, v_2, L(k)v_3, v_4; z, z_2)$$
$$+ (1 - w^4)^{\frac{1}{4}} z^{N-k+\frac{1}{8}} z_2^{k+1} \partial_2 H(v_1, v_2, v_3, v_4; z, z_2)$$
$$+ (1 - w^4)^{\frac{1}{4}} z^{N-k+\frac{1}{8}} [(z_2 + z)^{k+1} - z_2^{k+1}] \partial H(v_1, v_2, v_3, v_4; z, z_2)$$
$$+ (1 - w^4)^{\frac{1}{4}} z^{N-k+\frac{1}{8}} \sum_{i \geq 0} \binom{k + 1}{i + 1} z_2^{k-i} H(v_1, L(i)v_2, v_3, v_4; z, z_2)$$
$$+ (1 - w^4)^{\frac{1}{4}} z^{N-k+\frac{1}{8}} \sum_{i \geq 0} \binom{k + 1}{i + 1} (z_2 + z)^{k-i} H(L(i)v_1, v_2, v_3, v_4; z, z_2)$$

$$= \tilde{\Psi}(L(k)v_3) + (N_4 - N_3 - \bar{\Delta})s^{-k} \tilde{\Psi} - s^{1-k}(1 - w^4)^{\frac{1}{4}} s^{N_3 - N_4 + \bar{\Delta}} \Psi'$$
$$- (N_1 + N_2 + \tfrac{1}{8} - \bar{\Delta})s^{-k-1}[(1 + s)^{k+1} - 1]\tilde{\Psi}$$
$$+ s^{-k}[(1 + s)^{k+1} - 1]s^{N_3 - N_4 + \bar{\Delta}}(1 - w^4)^{\frac{1}{4}} \Psi'$$
$$+ \sum_{i \geq 0} \binom{k + 1}{i + 1} s^{i-k} \tilde{\Psi}(L(i)v_2) + \sum_{i \geq 0} \binom{k + 1}{i + 1} s^{i-k}(1 + s)^{k-i} \tilde{\Psi}(L(i)v_1)$$

Note that

$$\frac{d\Psi}{dw} = \frac{d\Psi}{ds} \frac{ds}{dw} \qquad \text{and} \qquad \frac{ds}{dw} = \frac{8w(1 + w^2)}{(1 - w^2)^3} = \frac{8w(1 - w^4)}{(1 - w^2)^4}$$

so that

$$\Psi' = \frac{d\Psi}{ds} = \frac{d\Psi}{dw} \frac{(1 - w^2)^4}{8w(1 - w^4)}.$$

From

$$\frac{d}{dw}\tilde{\Psi}(w) = \frac{d}{dw}\left((1 - w^4)^{\frac{1}{4}}s^{N_3 - N_4 + \bar{\Delta}}\Psi(s)\right)$$

$$= \frac{-w^3}{1 - w^4}\tilde{\Psi}(w) + \frac{N_3 - N_4 + \bar{\Delta}}{s}\frac{ds}{dw}\tilde{\Psi}(w) + (1 - w^4)^{\frac{1}{4}}s^{N_3 - N_4 + \bar{\Delta}}\frac{d\Psi}{dw}$$

we get

$$(1 - w^4)^{\frac{1}{4}}s^{N_3 - N_4 + \bar{\Delta}}\frac{d\Psi}{ds} = \left(\frac{ds}{dw}\right)^{-1}\frac{d\tilde{\Psi}}{dw} + \frac{w^3}{1 - w^4}\left(\frac{ds}{dw}\right)^{-1}\tilde{\Psi} - \frac{N_3 - N_4 + \bar{\Delta}}{s}\tilde{\Psi}.$$

The two terms involving Ψ' in the above expression for $\tilde{\Psi}(L(-k)v_4)$ combine to give

$$s^{-k}(1 + s)[(1 + s)^k - 1](1 - w^4)^{\frac{1}{4}}s^{N_3 - N_4 + \bar{\Delta}}\frac{d\Psi}{ds}$$

and since $1 + s = \frac{(1 + w^2)^2}{(1 - w^2)^2}$ this gives us

$$s^{-k}(1 + s)[(1 + s)^k - 1](1 - w^4)^{\frac{1}{4}}s^{N_3 - N_4 + \bar{\Delta}}\Psi'$$

$$= s^{-k}[(1 + s)^k - 1]\left(\frac{1 - w^4}{8w}\frac{d}{dw} + \frac{w^2}{8} - \frac{(N_3 - N_4 + \bar{\Delta})(1 + s)}{s}\right)\tilde{\Psi}$$

$$= s^{-k}[(1 + s)^k - 1]\left(\frac{\mathcal{L}_w}{s} - \frac{(N_3 - N_4 + \bar{\Delta})(1 + s)}{s}\right)\tilde{\Psi}.$$

Substituting this in the earlier expression, after some simplification, we get the result. ∎

Note that the operator $\mathcal{L}_w$ and multiplication by any rational function in $s^{\frac{1}{2}}$ preserve the ring $\mathcal{R}_w$.

Now let us return to the problem of determining the general two-point functions inductively.

THEOREM 23. *For $1 \leq i \leq 4$ let $v_i \in \mathbf{V}_{n_i}$ be eigenvectors for $L(0)$ with $L(0)v_i = wt(v_i)v_i = |v_i|v_i$. For $i = 1, 2$ let $Y_i = Y(v_i, z_i)$, $\partial_i = \partial/\partial z_i$. Then for any $k \in \mathbb{Z}$ we have*

$$G(L(-k)v_1, v_2, v_3, v_4; z_1, z_2)$$

$$= \sum_{i \geq 0}\binom{k + i - 2}{i}z_1^i G(v_1, v_2, v_3, L(k + i)v_4; z_1, z_2)$$

$$+ (-1)^k(z_1 - z_2)^{1-k}\partial_2 G(v_1, v_2, v_3, v_4; z_1, z_2)$$

$$+ (-1)^k \sum_{j \geq 0}(-1)^{j+1}\binom{1 - k}{j + 1}(z_1 - z_2)^{-k-j}G(v_1, L(j)v_2, v_3, v_4; z_1, z_2)$$

$$+ (-1)^k \sum_{i \geq 0}\binom{k + i - 1}{i + 1}z_1^{-k-i}G(v_1, v_2, L(i)v_3, v_4; z_1, z_2)$$

$$+ (-1)^k z_1^{1-k}(G(v_1, v_2, v_3, L(1)v_4; z_1, z_2) - (\partial_1 + \partial_2)G(v_1, v_2, v_3, v_4; z_1, z_2)).$$

PROOF. We have

$$G(L(-k)v_1, v_2, v_3, v_4; z_1, z_2) = (Y(L(-k)v_1, z_1)Y_2v_3, v_4)$$

$$= \sum_{i \geq 0} \binom{k+i-2}{i} z_1^i (L(-k-i)Y_1Y_2v_3, v_4)$$

$$+ (-1)^k \sum_{i \geq 0} \binom{k+i-2}{i} z_1^{-k+1-i} (Y_1L(i-1)Y_2v_3, v_4)$$

$$= \sum_{i \geq 0} \binom{k+i-2}{i} z_1^i G(v_1, v_2, v_3, L(k+i)v_4; z_1, z_2)$$

$$+ (-1)^k \sum_{i \geq 0} \binom{k+i-2}{i} z_1^{-k+1-i} \Big[(Y_1[L(i-1), Y_2]v_3, v_4)$$

$$+ (Y_1Y_2L(i-1)v_3, v_4) \Big]$$

$$= \sum_{i \geq 0} \binom{k+i-2}{i} z_1^i G(v_1, v_2, v_3, L(k+i)v_4; z_1, z_2)$$

$$+ (-1)^k \sum_{i \geq 0} \binom{k+i-2}{i} z_1^{-k+1-i} \sum_{j \geq 0} \binom{i}{j} z_2^{i-j} (Y_1Y(L(j-1)v_2, z_2)v_3, v_4)$$

$$+ (-1)^k \sum_{i \geq 0} \binom{k+i-2}{i} z_1^{-k+1-i} (Y_1Y_2L(i-1)v_3, v_4)$$

$$= \sum_{i \geq 0} \binom{k+i-2}{i} z_1^i G(L(k+i)v_4) + (-1)^k z_1^{1-k} \sum_{i \geq 0} \binom{k+i-2}{i} \left(\frac{z_2}{z_1}\right)^i \partial_2 G$$

$$+ (-1)^k \sum_{j \geq 0} \sum_{i \geq 0} \binom{k+i-2}{i} \binom{i}{j+1} z_1^{1-k} z_2^{-j-1} \left(\frac{z_2}{z_1}\right)^i G(L(j)v_2)$$

$$+ (-1)^k \sum_{i \geq 0} \binom{k+i-1}{i+1} z_1^{-k-i} G(L(i)v_3) + (-1)^k z_1^{1-k} (Y_1Y_2L(-1)v_3, v_4)$$

$$= \sum_{i \geq 0} \binom{k+i-2}{i} z_1^i G(L(k+i)v_4) + (-1)^k z_1^{1-k} \partial_2 G \sum_{i \geq 0} \binom{k+i-2}{i} \left(\frac{z_2}{z_1}\right)^i$$

$$+ (-1)^k \sum_{j \geq 0} z_1^{-j-k} G(L(j)v_2) \sum_{i \geq 0} \binom{k+i-2}{i} \binom{i}{j+1} \left(\frac{z_2}{z_1}\right)^{i-j-1}$$

$$+ (-1)^k \sum_{i \geq 0} \binom{k+i-1}{i+1} z_1^{-k-i} G(L(i)v_3)$$

$$+ (-1)^k z_1^{1-k} (G(L(1)v_4) - ([L(-1), Y_1]Y_2v_3, v_4) - (Y_1[L(-1), Y_2]v_3, v_4)).$$

We may write the last line as

$$(-1)^k z_1^{1-k} (G(L(1)v_4) - (\partial_1 + \partial_2)G).$$

Also, note that

$$\binom{k+i-2}{i} = (-1)^i \binom{1-k}{i}$$

so

$$\sum_{i \geq 0} \binom{k+i-2}{i} \left(\frac{z_2}{z_1}\right)^i = \sum_{i \geq 0} (-1)^i \binom{1-k}{i} \left(\frac{z_2}{z_1}\right)^i = \left(1 - \frac{z_2}{z_1}\right)^{1-k}$$

and therefore,

$$(-1)^k z_1^{1-k} \partial_2 G \sum_{i \geq 0} \binom{k+i-2}{i} \left(\frac{z_2}{z_1}\right)^i = (-1)^k (z_1 - z_2)^{1-k} \partial_2 G.$$

In addition, we have

$$\sum_{i \geq 0} (-1)^i \binom{1-k}{i} \binom{i}{j+1} t^{i-j-1} = \frac{1}{(j+1)!} \partial_t^{j+1} \sum_{i \geq 0} (-1)^i \binom{1-k}{i} t^i$$

$$= \frac{1}{(j+1)!} \partial_t^{j+1} (1-t)^{1-k} = (-1)^{j+1} \binom{1-k}{j+1} (1-t)^{-j-k}$$

so that

$$(-1)^k \sum_{j \geq 0} z_1^{-j-k} G(L(j)v_2) \sum_{i \geq 0} \binom{k+i-2}{i} \binom{i}{j+1} \left(\frac{z_2}{z_1}\right)^{i-j-1}$$

$$= (-1)^k \sum_{j \geq 0} z_1^{-j-k} G(L(j)v_2) \sum_{i \geq 0} (-1)^i \binom{1-k}{i} \binom{i}{j+1} \left(\frac{z_2}{z_1}\right)^{i-j-1}$$

$$= (-1)^k \sum_{j \geq 0} z_1^{-j-k} G(L(j)v_2)(-1)^{j+1} \binom{1-k}{j+1} \left(1 - \frac{z_2}{z_1}\right)^{-j-k}$$

$$= (-1)^k \sum_{j \geq 0} G(L(j)v_2)(-1)^{j+1} \binom{1-k}{j+1} (z_1 - z_2)^{-j-k}$$

which gives the result. $\blacksquare$

THEOREM 24. *For $1 \leq i \leq 4$ let $v_i \in \mathbf{V}_{n_i}$ be eigenvectors for $L(0)$ with $L(0)v_i = wt(v_i)v_i = |v_i|v_i$. For $i = 1, 2$ let $Y_i = Y(v_i, z_i)$, $\partial_i = \partial/\partial z_i$. Then for* *any $k \in \mathbb{Z}$ we have*

$$G(v_1, L(-k)v_2, v_3, v_4; z_1, z_2) = \sum_{i \geq 0} \binom{k+i-2}{i} z_2^i G(v_1, v_2, v_3, L(k+i)v_4; z_1, z_2)$$

$$+ (-1)^k \sum_{i \geq 0} \binom{k+i-1}{i+1} z_2^{-k-i} G(v_1, v_2, L(i)v_3, v_4; z_1, z_2) - (\partial_1 G)(z_1 - z_2)^{1-k}$$

$$+ \sum_{j \geq 0} G(L(j)v_1, v_2, v_3, v_4; z_1, z_2)(-1)^j z_2^{-j-k}.$$

$$\cdot \sum_{i \geq 0} (-1)^i \binom{1-k}{i} \binom{i+j+k-1}{j+1} \left(\frac{z_2}{z_1}\right)^{i+j+k}$$

$$+ (-1)^k z_2^{1-k} (G(v_1, v_2, v_3, L(1)v_4; z_1, z_2) - (\partial_1 + \partial_2)G(v_1, v_2, v_3, v_4; z_1, z_2)).$$

PROOF. We have

$$G(v_1, L(-k)v_2, v_3, v_4; z_1, z_2) = (Y_1 Y(L(-k)v_2, z_2)v_3, v_4)$$

$$= \sum_{i \geq 0} \binom{k+i-2}{i} z_2^i (Y_1 L(-k-i) Y_2 v_3, v_4)$$

$$+ (-1)^k \sum_{i \geq 0} \binom{k+i-2}{i} z_2^{-k+1-i} (Y_1 Y_2 L(i-1) v_3, v_4)$$

$$= \sum_{i \geq 0} \binom{k+i-2}{i} z_2^i (L(-k-i) Y_1 Y_2 v_3, v_4)$$

$$- \sum_{i \geq 0} \binom{k+i-2}{i} z_2^i ([L(-k-i), Y_1] Y_2 v_3, v_4)$$

$$+ (-1)^k \sum_{i \geq 0} \binom{k+i-1}{i+1} z_2^{-k-i} G(L(i)v_3)$$

$$+ (-1)^k z_2^{1-k} (G(L(1)v_4) - (\partial_1 + \partial_2)G)$$

$$= \sum_{i \geq 0} \binom{k+i-2}{i} z_2^i G(L(k+i)v_4)$$

$$- \sum_{i \geq 0} \binom{k+i-2}{i} z_2^i \sum_{j \geq 0} \binom{-k-i+1}{j} z_1^{1-i-j-k} (Y(L(j-1)v_1, z_1) Y_2 v_3, v_4)$$

$$+ (-1)^k \sum_{i \geq 0} \binom{k+i-1}{i+1} z_2^{-k-i} G(L(i)v_3)$$

$$+ (-1)^k z_2^{1-k} (G(L(1)v_4) - (\partial_1 + \partial_2)G).$$

We may rewrite the second line in the last expression as

$$-\sum_{j\geq 0} G(L(j-1)v_1)(-1)^j z_2^{1-j-k}.$$

$$\cdot \sum_{i\geq 0}(-1)^i \binom{1-k}{i}\binom{i+j+k-2}{j}\left(\frac{z_2}{z_1}\right)^{i+j+k-1}$$

$$= -(\partial_1 G)(z_1 - z_2)^{1-k}$$

$$+ \sum_{j\geq 0} G(L(j)v_1)(-1)^j z_2^{-j-k} \sum_{i\geq 0}(-1)^i \binom{1-k}{i}\binom{i+j+k-1}{j+1}\left(\frac{z_2}{z_1}\right)^{i+j+k}$$

giving the result. Note that the $j = 0$ term in the last line simplifies to $(k-1)|v_1|G(z_1 - z_2)^{-k}$. ∎

THEOREM 25. *For any $k \in \mathbb{Z}$ we have*

$$H(L(-k)v_1) =$$

$$\sum_{i\geq 0}(-1)^i \binom{1-k}{i}(z_2 + z)^i H(L(k+i)v_4)$$

$$+ (-1)^k (z_2 + z)^{1-k}(H(L(1)v_4) - \partial_2 H)$$

$$+ (-1)^k \sum_{j\geq 0} H(L(j)v_3) z^{-j-k} \sum_{i\geq 0}\binom{1-k}{i}\binom{i+j+k-1}{j+1}\left(\frac{z}{z_2}\right)^{i+j+k}$$

$$+ (-1)^k \sum_{i\geq 0}\binom{k+i-1}{i+1} z^{-k-i} H(L(i)v_2)$$

$$+ (-1)^k z^{1-k}(\partial_2 H - \partial H)$$

and

$$H(L(-k)v_2) = \sum_{i\geq 0}\binom{k+i-2}{i} z_2^i H(L(k+i)v_4)$$

$$+ (-1)^k z_2^{1-k}(H(L(1)v_4) - \partial_2 H)$$

$$+ (-1)^k \sum_{i\geq 0}\binom{k+i-1}{i+1} z_2^{-k-i} H(L(i)v_3)$$

$$- z^{1-k}\partial H - \sum_{i\geq 0}\binom{1-k}{i+1} z^{-k-i} H(L(i)v_1)$$

THEOREM 26. *For any $k \in \mathbb{Z}$ we have*

$$\tilde{\Phi}(L(-k)v_1) = \sum_{i \geq 0} \binom{k+i-2}{i} t^{-i} \tilde{\Phi}(L(k+i)v_4)$$

$$+ (-1)^k (N + \tfrac{1}{8}) t^{k-1} [1 - (1-t)^{1-k}] \tilde{\Phi}$$

$$+ (-1)^k t^{k-1} (1-t)[(1-t)^{-k} - 1] \mathcal{L}_x \tilde{\Phi}$$

$$+ (-1)^k \sum_{j \geq 0} (-1)^{j+1} \binom{1-k}{j+1} \left(\frac{t}{1-t}\right)^{j+k} \tilde{\Phi}(L(j)v_2)$$

$$+ (-1)^k \sum_{i \geq 0} \binom{k+i-1}{i+1} t^{k+i} \tilde{\Phi}(L(i)v_3)$$

$$+ (-1)^k t^{k-1} \tilde{\Phi}(L(1)v_4)$$

and

$$\tilde{\Phi}(L(-k)v_2) = \sum_{i \geq 0} \binom{k+i-2}{i} \tilde{\Phi}(L(k+i)v_4)$$

$$+ (-1)^k \sum_{i \geq 0} \binom{k+i-1}{i+1} \tilde{\Phi}(L(i)v_3)$$

$$+ [t^k (1-t)^{-k} - (-1)^k](1-t) \mathcal{L}_x \tilde{\Phi}$$

$$+ \sum_{j \geq 0} (-1)^j \tilde{\Phi}(L(j)v_1) \sum_{i \geq 0} (-1)^i \binom{1-k}{i} \binom{i+j+k-1}{j+1} t^{i+j+k}$$

$$+ (-1)^k \tilde{\Phi}(L(1)v_4) + (-1)^k (N + \tfrac{1}{8}) \tilde{\Phi}.$$

THEOREM 27. *For any $k \in \mathbb{Z}$ we have*

$$\tilde{\Psi}(L(-k)v_1) = \sum_{i \geq 0} (-1)^i \binom{1-k}{i} \left(\frac{1+s}{s}\right)^i \tilde{\Psi}(L(k+i)v_4)$$

$$+ (-1)^k \left(\frac{1+s}{s}\right)^{1-k} \tilde{\Psi}(L(1)v_4)$$

$$+ (-1)^k \sum_{j \geq 0} \tilde{\Psi}(L(j)v_3) \sum_{i \geq 0} \binom{1-k}{i} \binom{i+j+k-1}{j+1} s^{i+j+k}$$

$$+ (-1)^k \left[\left(\frac{s}{1+s}\right)^k - 1 \right] \mathcal{L}_w \tilde{\Psi}$$

$$+ (-1)^k \sum_{i \geq 0} \binom{k+i-1}{i+1} \tilde{\Psi}(L(i)v_2) + (-1)^k (N + \tfrac{1}{8}) \tilde{\Psi}$$

and

$$\tilde{\Psi}(L(-k)v_2) = \sum_{i \geq 0} \binom{k+i-2}{i} s^{-i} \tilde{\Psi}(L(k+i)v_4)$$
$$+ (-1)^k s^{k-1} \tilde{\Psi}(L(1)v_4) + (N + \tfrac{1}{8}) \tilde{\Psi}$$
$$+ (-1)^k \sum_{i \geq 0} \binom{k+i-,1}{i+1} s^{k+i} \tilde{\Psi}(L(i)v_3)$$
$$- \sum_{i \geq 0} \binom{1-k}{i+1} \tilde{\Psi}(L(i)v_1) + \frac{1}{1+s}[(-1)^k s^k - 1]\mathcal{L}_w \tilde{\Psi}.$$

THEOREM 28. *(Rationality) For $1 \leq i \leq 4$ let $v_i \in \mathbf{V}_{n_i}$ be eigenvectors for $L(0)$ with $L(0)v_i = wt(v_i)v_i = |v_i|v_i$. Suppose that $n_i \in \{1,3\}$, $n_4 = n_1 + n_2 + n_3$ (mod 4), $wt(v_i) = N_i + \Delta_{n_i}$ and $N = N_1 + N_2 + N_3 - N_4$. Then, after the substitution $(z_2/z_1)^{1/2} = 2x/(1+x^2)$, the series*

$$\tilde{\Phi}(v_1, v_2, v_3, v_4; x) = (1-x^4)^{\frac{1}{4}} z_2^{N + \frac{1}{8}} G(v_1, v_2, v_3, v_4; z_1, z_2)$$

converges absolutely in the domain $|x| < \sqrt{3 - 2\sqrt{2}}$ to a rational function in the ring $\mathcal{R}_x = \mathbb{C}[x, x^{-1}(x^4 - 1)^{-1}]$.

PROOF. We proof the statement by induction on $M = N_1 + N_2 + N_3 + N_4$. The base case, when $N_i = 0$ for $1 \leq i \leq 4$, is given by Theorem 13. The result of Theorem 18 with $k > 0$ inductively reduces the general case to the case where $N_4 = 0$, and with $k < 0$ it reduces further to the case where $N_3 = 0$ also. The two results of Theorem 26 with $k > 0$ further reduce one to the base case. ∎

THEOREM 29. *For any $v_1, v_2, v_3 \in \mathbf{V}_1$, $v_4 \in \mathbf{V}_3$, we have*

$$[\tilde{\Phi}]_0 \sim \frac{1}{x_\infty} B_\infty [\tilde{\Phi}]_\infty.$$

PROOF. It sufficies to prove this for homogeneous vectors v_i with $wt(v_i) = N_i + \Delta_{n_i}$, where $v_i \in \mathbf{V}_{n_i}$. We will prove this using Theorem 18, Theorem 26, and Lemma 19, by induction on $N_1 + N_2 + N_3 + N_4$. The base case, when all $N_i = 0$, has been established already. First note that since $wt(v_i) = wt(\theta(v_i))$, the N in Theorems 18 and 26 is the same for each of the entries in the matrices $[\tilde{\Phi}]_0$ and $[\tilde{\Phi}]_\infty$. Also, $t = 4x_0^2/(1+x_0^2)^2 = 4x_\infty^2/(1+x_\infty^2)^2$, so the t in Theorems 18 and 26 is the same whether x is x_0 or x_∞. First we will do the inductive step which reduces the weight of v_4. Assume the statement is true as stated for all choices of four vectors whose weights add up to be less than or equal to $N_1 + N_2 + N_3 + N_4$. We will show that it is then true with v_4 replaced by

$L(-k)v_4$ for any $0 < k \in \mathbb{Z}$. We have

$$[\tilde{\Phi}(L(-k)v_4)]_0 =$$
$$[\tilde{\Phi}(L(k)v_3)]_0 - (N + \tfrac{1}{8})[\tilde{\Phi}]_0 + (1 - t^{-k})\mathcal{L}_{x_0}[\tilde{\Phi}]_0$$
$$+ \sum_{i \geq 0} \binom{k+1}{i+1} \left(t^{i-k}[\tilde{\Phi}(L(i)v_1)]_0 + [\tilde{\Phi}(L(i)v_2)]_0 \right)$$
$$\sim \frac{1}{x_\infty} B_\infty [\tilde{\Phi}(L(k)v_3)]_\infty - (N + \tfrac{1}{8}) \frac{1}{x_\infty} B_\infty [\tilde{\Phi}]_\infty + (1 - t^{-k}) \mathcal{L}_{x_0} \frac{1}{x_\infty} B_\infty [\tilde{\Phi}]_\infty$$
$$+ \sum_{i \geq 0} \binom{k+1}{i+1} \frac{1}{x_\infty} B_\infty \left(t^{i-k}[\tilde{\Phi}(L(i)v_1)]_\infty + [\tilde{\Phi}(L(i)v_2)]_\infty \right)$$
$$= \frac{1}{x_\infty} B_\infty \left[[\tilde{\Phi}(L(k)v_3)]_\infty - (N + \tfrac{1}{8})[\tilde{\Phi}]_\infty + (1 - t^{-k})\mathcal{L}_{x_\infty}[\tilde{\Phi}]_\infty \right]$$
$$+ \frac{1}{x_\infty} B_\infty \sum_{i \geq 0} \binom{k+1}{i+1} \left(t^{i-k}[\tilde{\Phi}(L(i)v_1)]_\infty + [\tilde{\Phi}(L(i)v_2)]_\infty \right)$$
$$= \frac{1}{x_\infty} B_\infty [\tilde{\Phi}(L(-k)v_4)]_\infty.$$

But the same calculation with $0 > k \in \mathbb{Z}$ shows that the statement is also true with v_3 replaced by $L(k)v_3$.

Applying the two parts of Theorem 26 for $0 < k \in \mathbb{Z}$, one similarly gets that the statement is true with v_i replaced by $L(-k)v_i$, for $i = 1, 2$. Lemma 19 plays a crucial role, giving

$$\mathcal{L}_{x_0} \frac{1}{x_\infty} [\tilde{\Phi}]_\infty = \frac{1}{x_\infty} \mathcal{L}_{x_\infty} [\tilde{\Phi}]_\infty.$$

∎

THEOREM 30. *Let $v_1, v_2, v_3 \in \mathbf{V}_1$ and $v_4 \in \mathbf{V}_3$ with $wt(v_i) = |v_i| = N_i + \Delta_{n_i}$ and $N = N_1 + N_2 + N_3 - N_4$. Then we have*

$$[\tilde{\Phi}]_0 \sim \frac{t^N}{\mathbf{i}x_\mathbf{i} + 1} B_\mathbf{i} [\tilde{\Phi}]_\mathbf{i} \sim \frac{t^N}{\mathbf{i}x_{-\mathbf{i}} - 1} B_{-\mathbf{i}} [\tilde{\Phi}]_{-\mathbf{i}}.$$

PROOF. We will prove this using Theorem 18, Theorem 26, and Lemma 20, by induction on $N_1 + N_2 + N_3 + N_4$. The base case, when all $N_i = 0$, has been established already. Since $wt(v_i) = wt(\theta(v_i))$, the N in Theorems 18 and 26 is the same for each of the entries in the matrices $[\tilde{\Phi}]_0$ and $[\tilde{\Phi}]_{\pm \mathbf{i}}$. First we will do the inductive step which reduces the weight of v_4. Assume the statement is true as stated for all choices of four vectors whose weights add up to be less than or equal to $N_1 + N_2 + N_3 + N_4$. We will show that it is then true with v_4 replaced by $L(-k)v_4$ for any $0 < k \in \mathbb{Z}$. We have

$$
\begin{aligned}
&[\tilde{\Phi}(L(-k)v_4)]_0 \\
&= [\tilde{\Phi}(L(k)v_3)]_0 - (N + \tfrac{1}{8})[\tilde{\Phi}]_0 + (1 - t^{-k})\mathcal{L}_{x_0}[\tilde{\Phi}]_0 \\
&\quad + \sum_{i \geq 0} \binom{k+1}{i+1} \left(t^{i-k}[\tilde{\Phi}(L(i)v_1)]_0 + [\tilde{\Phi}(L(i)v_2)]_0 \right) \\
&\sim \frac{t^{N-k}}{\mathbf{i}x_{\mathbf{i}} + 1} B_{\mathbf{i}}[\tilde{\Phi}(L(k)v_3)]_{\mathbf{i}} - (N + \tfrac{1}{8})\frac{t^N}{\mathbf{i}x_{\mathbf{i}} + 1} B_{\mathbf{i}}[\tilde{\Phi}]_{\mathbf{i}} + (1 - t^{-k})\mathcal{L}_{x_0}\frac{t^N}{\mathbf{i}x_{\mathbf{i}} + 1} B_{\mathbf{i}}[\tilde{\Phi}]_{\mathbf{i}} \\
&\quad + \sum_{i \geq 0} \binom{k+1}{i+1} \frac{t^{N-i}}{\mathbf{i}x_{\mathbf{i}} + 1} B_{\mathbf{i}} \left(t^{i-k}[\tilde{\Phi}(L(i)v_1)]_{\mathbf{i}} + [\tilde{\Phi}(L(i)v_2)]_{\mathbf{i}} \right) \\
&= \frac{t^{N-k}}{\mathbf{i}x_{\mathbf{i}} + 1} B_{\mathbf{i}} \Big[[\tilde{\Phi}(L(k)v_3)]_{\mathbf{i}} - (N + \tfrac{1}{8})t^k[\tilde{\Phi}]_{\mathbf{i}} + (1 - t^{-k}) \left(-t^k(\mathcal{L}_{x_{\mathbf{i}}} - N - \tfrac{1}{8})[\tilde{\Phi}]_{\mathbf{i}} \right) \\
&\quad + \sum_{i \geq 0} \binom{k+1}{i+1} \left([\tilde{\Phi}(L(i)v_1)]_{\mathbf{i}} + t^{k-i}[\tilde{\Phi}(L(i)v_2)]_{\mathbf{i}} \right) \Big] \\
&= \frac{t^{N-k}}{\mathbf{i}x_{\mathbf{i}} + 1} B_{\mathbf{i}} \Big[[\tilde{\Phi}(L(k)v_3)]_{\mathbf{i}} + (1 - t^k)\mathcal{L}_{x_{\mathbf{i}}}[\tilde{\Phi}]_{\mathbf{i}} - (N + \tfrac{1}{8})[\tilde{\Phi}]_{\mathbf{i}} \\
&\quad + \sum_{i \geq 0} \binom{k+1}{i+1} \left([\tilde{\Phi}(L(i)v_1)]_{\mathbf{i}} + t^{k-i}[\tilde{\Phi}(L(i)v_2)]_{\mathbf{i}} \right) \Big] \\
&= \frac{t^{N-k}}{\mathbf{i}x_{\mathbf{i}} + 1} B_{\mathbf{i}}[\tilde{\Phi}(L(-k)v_4)]_{\mathbf{i}}.
\end{aligned}
$$

In the last step we used Theorem 18 with v_1 and v_2 switched, z_1 and z_2 switched, and therefore, t replaced by t^{-1}. The same calculation with $0 > k \in \mathbb{Z}$ shows that the statement is also true with v_3 replaced by $L(k)v_3$.

We now apply the two parts of Theorem 26 for $0 < k \in \mathbb{Z}$ to get that the statement is true with v_i replaced by $L(-k)v_i$, for $i = 1, 2$. Lemma 20 plays a crucial role. We have

$$
\begin{aligned}
&[\tilde{\Phi}(L(-k)v_1)]_0 \\
&= \sum_{i \geq 0} \binom{k+i-2}{i} t^{-i}[\tilde{\Phi}(L(k+i)v_4)]_0 \\
&\quad + (-1)^k(N + \tfrac{1}{8})t^{k-1}[1 - (1-t)^{1-k}][\tilde{\Phi}]_0 \\
&\quad + (-1)^k t^{k-1}(1-t)[(1-t)^{-k} - 1]\mathcal{L}_{x_0}[\tilde{\Phi}]_0 \\
&\quad + (-1)^k \sum_{j \geq 0} (-1)^{j+1} \binom{1-k}{j+1} \left(\frac{t}{1-t} \right)^{j+k} [\tilde{\Phi}(L(j)v_2)]_0 \\
&\quad + (-1)^k \sum_{i \geq 0} \binom{k+i-1}{i+1} t^{k+i}[\tilde{\Phi}(L(i)v_3)]_0 + (-1)^k t^{k-1}[\tilde{\Phi}(L(1)v_4)]_0
\end{aligned}
$$

$$\sim \sum_{i\geq 0} \binom{k+i-2}{i} t^{-i} \frac{t^{N+k+i}}{\mathbf{i}x_{\mathbf{i}}+1} B_{\mathbf{i}}[\tilde{\Phi}(L(k+i)v_4)]_{\mathbf{i}}$$

$$+ (-1)^k (N+\tfrac{1}{8}) t^{k-1}[1-(1-t)^{1-k}] \frac{t^N}{\mathbf{i}x_{\mathbf{i}}+1} B_{\mathbf{i}}[\tilde{\Phi}]_{\mathbf{i}}$$

$$+ (-1)^k t^{k-1}(1-t)[(1-t)^{-k}-1] \mathcal{L}_{x_0} \frac{t^N}{\mathbf{i}x_{\mathbf{i}}+1} B_{\mathbf{i}}[\tilde{\Phi}]_{\mathbf{i}}$$

$$+ (-1)^k \sum_{j\geq 0} (-1)^{j+1} \binom{1-k}{j+1} \left(\frac{t}{1-t}\right)^{j+k} \frac{t^{N-j}}{\mathbf{i}x_{\mathbf{i}}+1} B_{\mathbf{i}}[\tilde{\Phi}(L(j)v_2)]_{\mathbf{i}}$$

$$+ (-1)^k \sum_{i\geq 0} \binom{k+i-1}{i+1} t^{k+i} \frac{t^{N-i}}{\mathbf{i}x_{\mathbf{i}}+1} B_{\mathbf{i}}[\tilde{\Phi}(L(i)v_3)]_{\mathbf{i}}$$

$$+ (-1)^k t^{k-1} \frac{t^{N+1}}{\mathbf{i}x_{\mathbf{i}}+1} B_{\mathbf{i}}[\tilde{\Phi}(L(1)v_4)]_{\mathbf{i}}$$

$$= \sum_{i\geq 0} \binom{k+i-2}{i} \frac{t^{N+k}}{\mathbf{i}x_{\mathbf{i}}+1} B_{\mathbf{i}}[\tilde{\Phi}(L(k+i)v_4)]_{\mathbf{i}}$$

$$+ (-1)^k (N+\tfrac{1}{8}) t^{k-1}[1-(1-t)^{1-k}] \frac{t^N}{\mathbf{i}x_{\mathbf{i}}+1} B_{\mathbf{i}}[\tilde{\Phi}]_{\mathbf{i}}$$

$$+ (-1)^k t^{k-1}(1-t)[(1-t)^{-k}-1] \left[\frac{-t^N}{\mathbf{i}x_{\mathbf{i}}+1} B_{\mathbf{i}}(\mathcal{L}_{x_i}-N-\tfrac{1}{8})\right][\tilde{\Phi}]_{\mathbf{i}}$$

$$+ (-1)^k \sum_{j\geq 0} (-1)^{j+1} \binom{1-k}{j+1} (1-t)^{-j-k} \frac{t^{N+k}}{\mathbf{i}x_{\mathbf{i}}+1} B_{\mathbf{i}}[\tilde{\Phi}(L(j)v_2)]_{\mathbf{i}}$$

$$+ (-1)^k \sum_{i\geq 0} \binom{k+i-1}{i+1} \frac{t^{N+k}}{\mathbf{i}x_{\mathbf{i}}+1} B_{\mathbf{i}}[\tilde{\Phi}(L(i)v_3)]_{\mathbf{i}} + (-1)^k \frac{t^{N+k}}{\mathbf{i}x_{\mathbf{i}}+1} B_{\mathbf{i}}[\tilde{\Phi}(L(1)v_4)]_{\mathbf{i}}$$

$$= \frac{t^{N+k}}{\mathbf{i}x_{\mathbf{i}}+1} B_{\mathbf{i}}\left[\sum_{i\geq 0} \binom{k+i-2}{i} [\tilde{\Phi}(L(k+i)v_4)]_{\mathbf{i}}\right.$$

$$+ (-1)^k (N+\tfrac{1}{8}) t^{-1}[1-(1-t)^{1-k}][\tilde{\Phi}]_{\mathbf{i}}$$

$$- (-1)^k t^{-1}(1-t)[(1-t)^{-k}-1](\mathcal{L}_{x_i}-N-\tfrac{1}{8})[\tilde{\Phi}]_{\mathbf{i}}$$

$$+ (-1)^k \sum_{j\geq 0} (-1)^{j+1} \binom{1-k}{j+1} (1-t)^{-j-k} [\tilde{\Phi}(L(j)v_2)]_{\mathbf{i}}$$

$$\left. + (-1)^k \sum_{i\geq 0} \binom{k+i-1}{i+1} [\tilde{\Phi}(L(i)v_3)]_{\mathbf{i}} + (-1)^k [\tilde{\Phi}(L(1)v_4)]_{\mathbf{i}}\right]$$

$$= \frac{t^{N+k}}{\mathbf{i}x_{\mathbf{i}}+1} B_{\mathbf{i}}\left[\sum_{i\geq 0} \binom{k+i-2}{i} [\tilde{\Phi}(L(k+i)v_4)]_{\mathbf{i}}\right.$$

$$+ (-1)^k (N+\tfrac{1}{8})[\tilde{\Phi}]_{\mathbf{i}} + [t^{-k}(1-t^{-1})^{-k}-(-1)^k](1-t^{-1})\mathcal{L}_{x_i}[\tilde{\Phi}]_{\mathbf{i}}$$

$$+ \sum_{j\geq 0} (-1)^j [\tilde{\Phi}(L(j)v_2)]_{\mathbf{i}} \sum_{i\geq 0} \binom{1-k}{i}\binom{i+j+k-1}{j+1} t^{-i-j-k}$$

$$+ (-1)^k \sum_{i \geq 0} \binom{k+i-1}{i+1} [\tilde{\Phi}(L(i)v_3)]_i + (-1)^k [\tilde{\Phi}(L(1)v_4)]_i \bigg]$$

$$= \frac{t^{N+k}}{ix_i + 1} B_i [\tilde{\Phi}(L(-k)v_1)]_i.$$

In the last step we used the second part of Theorem 26 with v_1 and v_2 switched, z_1 and z_2 switched, and therefore, t replaced by t^{-1}.

The calculation for $[\tilde{\Phi}(L(-k)v_2)]_0$ is similar, completing the inductive proof of the first part of the theorem. The second part, with x_{-i} in place of x_i, B_{-i} in place of B_i, and certain sign changes, is proved in exactly the same way. ∎

THEOREM 31. *Let* $v_1, v_2, v_3 \in \mathbf{V}_1$ *and* $v_4 \in \mathbf{V}_3$ *with* $wt(v_i) = |v_i| = N_i + \Delta_{n_i}$ *and* $N = N_1 + N_2 + N_3 - N_4$. *Then we have*

$$[\tilde{\Phi}]_0 \sim \frac{s^{-N}}{-x_1 + 1} B_1 [\tilde{\Psi}]_1 \sim \frac{s^{-N}}{x_{-1} + 1} B_{-1} [\tilde{\Psi}]_{-1}.$$

PROOF. We will prove this using Theorems 18, 22, 27, and Lemma 21, by induction on $N_1 + N_2 + N_3 + N_4$. The base case, when all $N_i = 0$, has been established already. First note that since $wt(v_i) = wt(\theta(v_i))$, the N in Theorems 18 and 27 is the same for each of the entries in the matrices $[\tilde{\Phi}]_0$ and $[\tilde{\Psi}]_{\pm 1}$. First we will do the inductive step which reduces the weight of v_4. Assume the statement is true as stated for all choices of four vectors whose weights add up to be less than or equal to $N_1 + N_2 + N_3 + N_4$. We will show that it is then true with v_4 replaced by $L(-k)v_4$ for any $0 < k \in \mathbb{Z}$. We have

$$[\tilde{\Phi}(L(-k)v_4)]_0$$
$$= [\tilde{\Phi}(L(k)v_3)]_0 - (N + \tfrac{1}{8})[\tilde{\Phi}]_0 + (1 - t^{-k})\mathcal{L}_{x_0}[\tilde{\Phi}]_0$$
$$+ \sum_{i \geq 0} \binom{k+1}{i+1} \left(t^{i-k}[\tilde{\Phi}(L(i)v_1)]_0 + [\tilde{\Phi}(L(i)v_2)]_0 \right)$$

$$\sim \frac{s^{-N+k}}{-x_1 + 1} B_1 [\tilde{\Psi}(L(k)v_3)]_1 - (N + \tfrac{1}{8}) \frac{s^{-N}}{-x_1 + 1} B_1 [\tilde{\Psi}]_1$$
$$+ (1 - t^{-k})\mathcal{L}_{x_0} \frac{s^{-N}}{-x_1 + 1} B_1 [\tilde{\Psi}]_1$$
$$+ \sum_{i \geq 0} \binom{k+1}{i+1} \frac{s^{-N+i}}{-x_1 + 1} B_1 \left(t^{i-k}[\tilde{\Psi}(L(i)v_1)]_1 + [\tilde{\Psi}(L(i)v_2)]_1 \right)$$

$$= \frac{s^{-N+k}}{-x_1 + 1} B_1 \bigg[[\tilde{\Psi}(L(k)v_3)]_1 - (N + \tfrac{1}{8})s^{-k}[\tilde{\Psi}]_1$$
$$- (1 - t^{-k})s^{-k} \left(s^{-1}\mathcal{L}_{x_1} - (N + \tfrac{1}{8})(1 + s^{-1}) \right) [\tilde{\Psi}]_1$$
$$+ \sum_{i \geq 0} \binom{k+1}{i+1} \left[\left(\frac{s+1}{s} \right)^{k-i} [\tilde{\Psi}(L(i)v_1)]_1 + s^{i-k}[\tilde{\Psi}(L(i)v_2)]_1 \right] \bigg]$$

$$= \frac{s^{-N+k}}{-x_1+1} B_1 \Big[[\tilde{\Psi}(L(k)v_3)]_1 - (N+\tfrac{1}{8})s^{-k}[\tilde{\Psi}]_1 + ((1+s)^k - 1)s^{-k-1}\mathcal{L}_{x_1}[\tilde{\Psi}]_1$$

$$+ (1 - (1+s)^k)s^{-k}(1+s^{-1})(N+\tfrac{1}{8})[\tilde{\Psi}]_1$$

$$+ \sum_{i \geq 0} \binom{k+1}{i+1} \left[\left(\frac{s+1}{s} \right)^{k-i} [\tilde{\Psi}(L(i)v_1)]_1 + s^{i-k}[\tilde{\Psi}(L(i)v_2)]_1 \right] \Big]$$

$$= \frac{s^{-N+k}}{-x_1+1} B_1 \Big[[\tilde{\Psi}(L(k)v_3)]_1 - (N+\tfrac{1}{8})s^{-k-1}((1+s)^{k+1} - 1)[\tilde{\Psi}]_1$$

$$+ s^{-k-1}((1+s)^k - 1)\mathcal{L}_{x_1}[\tilde{\Psi}]_1$$

$$+ \sum_{i \geq 0} \binom{k+1}{i+1} \left[\left(\frac{s+1}{s} \right)^{k-i} [\tilde{\Psi}(L(i)v_1)]_1 + s^{i-k}[\tilde{\Psi}(L(i)v_2)]_1 \right] \Big]$$

$$= \frac{s^{-N+k}}{-x_1+1} B_1 [\tilde{\Psi}(L(-k)v_4)]_1$$

We used Theorem 22 in the last step.

The same calculation with $0 > k \in \mathbb{Z}$ shows that the statement is also true with v_3 replaced by $L(k)v_3$.

Applying the two parts of Theorem 27 for $0 < k \in \mathbb{Z}$, one similarly gets that the statement is true with v_i replaced by $L(-k)v_i$, for $i = 1, 2$. Lemma 21 plays a crucial role. ∎

We are, at last, ready to state the new "matrix" Jacobi-Cauchy Identity which is valid for the $c = \frac{1}{2}$ minimal model. This is the main objective of this paper, but considerable further work remains to be done in order to understand other minimal models and the WZW models. That will be the subject of future investigations.

THEOREM 32. *(Matrix Jacobi-Cauchy Identity) Let* $v_1, v_2, v_3 \in \mathbf{V}_1$ *and* $v_4 \in \mathbf{V}_3$ *with* $wt(v_i) = |v_i| = N_i + \Delta_{n_i}$ *and* $N = N_1 + N_2 + N_3 - N_4$. *Let* $f(x)$ *be any function in* $\mathcal{R}_x$. *Let* C_0 *be a small positively oriented circle with center* $x_0 = 0$ *and for* $\alpha \in \{\infty, 1, -1, \mathbf{i}, -\mathbf{i}\}$ *let* C_α *be the circle obtained from* C_0 *by the appropriate Möbius transformation sending* 0 *to* α. *Then we have*

$$0 = \oint_{C_0} [\Phi]_0 f(x_0) dx_0$$

$$+ \oint_{C_\infty} \frac{-1}{x_\infty^3} B_\infty [\tilde\Phi]_\infty f(1/x_\infty)\ dx_\infty$$

$$+ \oint_{C_i} \frac{2t^N}{(ix_i + 1)^3} B_i [\tilde\Phi]_i f\left(\frac{x_i + i}{ix_i + 1}\right)\ dx_i$$

$$+ \oint_{C_{-i}} \frac{2t^N}{(ix_{-i} - 1)^3} B_{-i} [\tilde\Phi]_{-i} f\left(\frac{-x_{-i} + i}{ix_{-i} - 1}\right)\ dx_{-i}$$

$$+ \oint_{C_1} \frac{2s^{-N}}{(-x_1 + 1)^3} B_1 [\tilde\Psi]_1 f\left(\frac{x_1 + 1}{-x_1 + 1}\right)\ dx_1$$

$$+ \oint_{C_{-1}} \frac{2s^{-N}}{(x_{-1} + 1)^3} B_{-1} [\tilde\Psi]_{-1} f\left(\frac{x_{-1} - 1}{x_{-1} + 1}\right)\ dx_{-1}.$$

PROOF. Apply the Cauchy residue theorem to $f(x)$ times the globally defined matrix valued function $G(x)$ to which each of the expressions in Theorems 29 - 31 converge. It says that the sum of the residues at the six possible poles is zero, that is,

$$0 = \sum_\alpha \oint_{C_\alpha} G(x_0) f(x_0) dx_0.$$

Let

$$x_0 = \mu_\alpha(x_\alpha) = \frac{a_\alpha x_\alpha + b_\alpha}{c_\alpha x_\alpha + d_\alpha}$$

denote the Möbius transformation chosen before Theorem 13 to relate x_0 to x_α. Then the chain rule gives $dx_0 = \frac{a_\alpha d_\alpha - b_\alpha c_\alpha}{(c_\alpha x_\alpha + d_\alpha)^2} dx_\alpha$. For each α we write the corresponding integral with $G(x_0) f(x_0)$ represented by the series in x_α which converges to it in the neighborhood of $x_0 = \alpha$, and we write dx_0 as the appropriate chain rule factor times dx_α. ∎

Just as Corollary 3 was obtained from Theorem 2, we can obtain infinitely many explicit identities for products of components of intertwining operators from Theorem 32 by making explicit choices for the test function $f(x)$. One computes the integral in each term of the Matrix Jacobi-Cauchy Identity (MJCI) by expanding the integrand as a series in the local variable x_α and then finding the residue as the coefficient of x_α^{-1}. The algebra involved is tedious but straightforward, so we will not present it here. The most general function $f(x) \in \mathcal{R}_x$ is a linear combination of functions of the form

$$f(x) = x^{m_0}(1 + ix)^{m_i}(1 - ix)^{m_{-i}}(1 - x)^{m_1}(1 + x)^{m_{-1}}$$

where $m_0, m_i, m_{-i}, m_1, m_{-1} \in \mathbb{Z}$ are five independent parameters. Since the MJCI is linear in $f(x)$, it suffices to compute the identity just for $f(x)$ of that form. In fact, using a few simple combinatorial identities, there is a considerable simplification of the result if we restrict $f(x)$ to be a rational function of the variable t, say $t^r(1-t)^s$. This seems a natural restriction because the correlation functions made from intertwining operators were defined from series in z_1 and z_2 which could be written in terms of t. Only later did we have to express them in terms of the local variables x_α in order to use properties of the hypergeometric functions to relate the functions at the six poles. It makes the local variables seem like a necessary but artificial technical construction, and one might think that the final answer should reflect the fact that there were only three possible poles in the t-plane. This may not be the last word on that subject, but our results seem to show that with that restriction on f there are six terms in the MJCI, but the second row of the matrix yields a trivial identity. We find a slightly simpler result if we restrict the test function to be of the form $f(x) = x^2 t^r(1-t)^s$. We give the results in the next three corollaries.

If we were to write out each entry of each 2×4 matrix in the identity, then the result would not fit across the page, and the pattern would be very similar in each column. So we will just show the first column of the answer in Corollary 33. The other columns are easily obtained from the first one by modifying the pattern of vectors v_1, v_2, v_3, v_4 in the first row and $\theta v_1, \theta v_2, v_3, v_4$ in the second row, as shown in the matrices $[\tilde{\Phi}]_\alpha$ and $[\tilde{\Psi}]_\alpha$ defined before Theorem 18. In Corollaries 34 and 35, since the second row of the matrix yields a trivial identity, we only give the identity coming from the first row.

COROLLARY 33. *Let* $v_1, v_2, v_3 \in \mathbf{V}_1$ *and* $v_4 \in \mathbf{V}_3$ *with* $wt(v_i) = |v_i| = N_i + \Delta_{n_i}$ *and* $N = N_1 + N_2 + N_3 - N_4$. *Let*

$$f(x) = x^{m_0}(1 + \mathbf{i}x)^{m_\mathbf{i}}(1 - \mathbf{i}x)^{m_{-\mathbf{i}}}(1 - x)^{m_1}(1 + x)^{m_{-1}}$$

for $m_0, m_\mathbf{i}, m_{-\mathbf{i}}, m_1, m_{-1} \in \mathbb{Z}$, *and let* $m = m_0 + m_\mathbf{i} + m_{-\mathbf{i}} + m_1 + m_{-1}$. *Define the following constants in terms of those parameters:*

$$A_\infty = (-1)^{m_1 + m_{-i} + 1}(i)^{m_i + m_{-i}},$$

$$A_i = 2^{m_1 + m_i + m_{-i} + 1}(i)^{m_0 + m_i}(1 + i)^{m_{-1} - m_1},$$

$$A_{-i} = 2^{m_{-1} + m_i + m_{-i} + 1}(-1)^{m_0 + m_{-i} + 1}(i)^{m_0 + m_{-i}}(1 + i)^{m_1 - m_{-1}},$$

$$A_1 = 2^{m_1 + m_{-1} + m_{-i} + 1}(-1)^{m_1}(1 + i)^{m_i - m_{-i}},$$

$$A_{-1} = 2^{m_1 + m_{-1} + m_i + 1}(-1)^{m_0}(1 + i)^{m_{-i} - m_i},$$

and for each $k \geq 0$ *define*

$$Q_k = 2(k + N_1 - N_4), \qquad R_k = 2(k - N_1 - N_3), \qquad S_k = 2(k - N_1 - N_2).$$

In the following formula, for each $k \geq 0$, *and for each* $\alpha \in \{0, \infty, \mathbf{i}, -\mathbf{i}, 1, -1\}$, $\sum_\alpha$ *and* $\sum'_\alpha$ *denote summations over all integers* $p_1, p_{-1}, p_\mathbf{i}, p_{-\mathbf{i}} \geq 0$ *such that the sum* $p_1 + p_{-1} + p_\mathbf{i} + p_{-\mathbf{i}}$ *equals a fixed value,* $\hat{p}$, *depending on* k *and some of the given parameters. The fixed values are as follows. For* $\sum_0$, $\hat{p} = -m_0 - 2 - Q_k$, *for* $\sum'_0$, $\hat{p} = -m_0 - 1 - Q_k$, *for* $\sum_\infty$, $\hat{p} = m + 1 - Q_k$, *for* $\sum'_\infty$, $\hat{p} = m + 2 - Q_k$, *for* $\sum_\mathbf{i}$, $\hat{p} = -m_\mathbf{i} - 2 - R_k$, *for* $\sum'_\mathbf{i}$, $\hat{p} = -m_\mathbf{i} - 1 - R_k$, *for* $\sum_{-\mathbf{i}}$, $\hat{p} = -m_{-\mathbf{i}} - 2 - R_k$, *for* $\sum'_{-\mathbf{i}}$, $\hat{p} = -m_{-\mathbf{i}} - 1 - R_k$, *for* $\sum_1$, $\hat{p} = -m_1 - 2 - S_k$, *for* $\sum'_1$, $\hat{p} = -m_1 - 1 - S_k$, for $\sum_{-1}$, $\hat{p} = -m_{-1} - 2 - S_k$, and for $\sum'_{-1}$, $\hat{p} = -m_{-1} - 1 - S_k$. *Within the summations we use the notations*

$$C_1 = \begin{pmatrix} \frac{1}{4} + m_1 \\ p_1 \end{pmatrix}, \ C_{-1} = \begin{pmatrix} \frac{1}{4} + m_{-1} \\ p_{-1} \end{pmatrix}, \ C_i = \begin{pmatrix} \frac{1}{4} + m_i \\ p_i \end{pmatrix}, \ C_{-i} = \begin{pmatrix} \frac{1}{4} + m_{-i} \\ p_{-i} \end{pmatrix},$$

and

$$D = (-1)^{p_1 + p_{-i}} (i)^{p_i + p_{-i}}.$$

To shorten our notation for the correlation functions, we write

$$\Phi_k = \Phi_k(v_1, v_2, v_3, v_4) = \left(Y_{k-N_4+\frac{7}{16}}(v_1) Y_{-k+N_3-\frac{7}{16}}(v_2) v_3, v_4 \right),$$

$$\Phi_k^\theta = \Phi_k(\theta v_1, \theta v_2, v_3, v_4) = \left(Y_{k-N_4-\frac{1}{16}}(\theta v_1) Y_{-k+N_3+\frac{1}{16}}(\theta v_2) v_3, v_4 \right),$$

$$\Phi_k^* = \Phi_k(v_2, v_1, v_3, v_4) = \left(Y_{k-N_4+\frac{7}{16}}(v_2) Y_{-k+N_3-\frac{7}{16}}(v_1) v_3, v_4 \right),$$

$$\Phi_k^{*\theta} = \Phi_k(\theta v_2, \theta v_1, v_3, v_4) = \left(Y_{k-N_4-\frac{1}{16}}(\theta v_2) Y_{-k+N_3+\frac{1}{16}}(\theta v_1) v_3, v_4 \right),$$

$$\Psi_k = \Psi_k(v_1, v_2, v_3, v_4) = \left(Y_{N_3-N_4}(Y_{-k+N_2-\frac{7}{16}}(v_1) v_2) v_3, v_4 \right),$$

$$\Psi_k^\theta = \Psi_k(\theta v_1, v_2, v_3, \theta v_4) = \left(Y_{N_3-N_4}(Y_{-k+N_2+\frac{1}{16}}(\theta v_1) v_2) v_3, \theta v_4 \right).$$

Then we have the identity

$$0 = \sum_{0 \leq k \in \mathbb{Z}} \left(2^{Q_k} \begin{bmatrix} \sum_0 C_1 C_{-1} \left(\begin{smallmatrix} \frac{1}{4}+m_i-Q_k-1 \\ p_i \end{smallmatrix} \right) \left(\begin{smallmatrix} \frac{1}{4}+m_{-i}-Q_k-1 \\ p_{-i} \end{smallmatrix} \right) D2\Phi_k \\ \sum'_0 C_1 C_{-1} \left(\begin{smallmatrix} \frac{1}{4}+m_i-Q_k \\ p_i \end{smallmatrix} \right) \left(\begin{smallmatrix} \frac{1}{4}+m_{-i}-Q_k \\ p_{-i} \end{smallmatrix} \right) D\Phi_k^\theta \end{bmatrix} \right.$$

$$+ 2^{Q_k} A_\infty B_\infty \begin{bmatrix} \sum_\infty C_1 C_{-1} \left(\begin{smallmatrix} \frac{1}{4}+m_{-i}-Q_k-1 \\ p_i \end{smallmatrix} \right) \left(\begin{smallmatrix} \frac{1}{4}+m_i-Q_k-1 \\ p_{-i} \end{smallmatrix} \right) D2\Phi_k \\ \sum'_\infty C_1 C_{-1} \left(\begin{smallmatrix} \frac{1}{4}+m_{-i}-Q_k \\ p_i \end{smallmatrix} \right) \left(\begin{smallmatrix} \frac{1}{4}+m_i-Q_k \\ p_{-i} \end{smallmatrix} \right) D\Phi_k^\theta \end{bmatrix}$$

$$+ 2^{R_k} A_i B_i \begin{bmatrix} \sum_i C_1 C_{-1} \left(\begin{smallmatrix} \frac{1}{4}-m-R_k-4 \\ p_i \end{smallmatrix} \right) \left(\begin{smallmatrix} \frac{1}{4}+m_0-R_k-1 \\ p_{-i} \end{smallmatrix} \right) D2\Phi_k^* \\ \sum'_i C_1 C_{-1} \left(\begin{smallmatrix} \frac{1}{4}-m-R_k-3 \\ p_i \end{smallmatrix} \right) \left(\begin{smallmatrix} \frac{1}{4}+m_0-R_k \\ p_{-i} \end{smallmatrix} \right) D\Phi_k^{*\theta} \end{bmatrix}$$

$$+ 2^{R_k} A_{-i} B_{-i} \begin{bmatrix} \sum_{-i} C_1 C_{-1} \left(\begin{smallmatrix} \frac{1}{4}+m_0-R_k-1 \\ p_i \end{smallmatrix} \right) \left(\begin{smallmatrix} \frac{1}{4}-m-R_k-4 \\ p_{-i} \end{smallmatrix} \right) D2\Phi_k^* \\ \sum'_{-i} C_1 C_{-1} \left(\begin{smallmatrix} \frac{1}{4}+m_0-R_k \\ p_i \end{smallmatrix} \right) \left(\begin{smallmatrix} \frac{1}{4}-m-R_k-3 \\ p_{-i} \end{smallmatrix} \right) D\Phi_k^{*\theta} \end{bmatrix}$$

$$+ 2^{S_k} A_1 B_1 \begin{bmatrix} \sum_1 C_i C_{-i} \left(\begin{smallmatrix} \frac{1}{4}-m-S_k-4 \\ p_1 \end{smallmatrix} \right) \left(\begin{smallmatrix} \frac{1}{4}+m_0-S_k-1 \\ p_{-1} \end{smallmatrix} \right) D\Psi_k \\ \sum'_1 C_i C_{-i} \left(\begin{smallmatrix} \frac{1}{4}-m-S_k-3 \\ p_1 \end{smallmatrix} \right) \left(\begin{smallmatrix} \frac{1}{4}+m_0-S_k \\ p_{-1} \end{smallmatrix} \right) D\Psi_k^\theta \end{bmatrix}$$

$$\left. + 2^{S_k} A_{-1} B_{-1} \begin{bmatrix} \sum_{-1} C_i C_{-i} \left(\begin{smallmatrix} \frac{1}{4}+m_0-S_k-1 \\ p_1 \end{smallmatrix} \right) \left(\begin{smallmatrix} \frac{1}{4}-m-S_k-4 \\ p_{-1} \end{smallmatrix} \right) D\Psi_k \\ \sum'_{-1} C_i C_{-i} \left(\begin{smallmatrix} \frac{1}{4}+m_0-S_k \\ p_1 \end{smallmatrix} \right) \left(\begin{smallmatrix} \frac{1}{4}-m-S_k-3 \\ p_{-1} \end{smallmatrix} \right) D\Psi_k^\theta \end{bmatrix} \right).$$

COROLLARY 34. *Let* $v_1, v_2, v_3 \in \mathbf{V}_1$ *and* $v_4 \in \mathbf{V}_3$ *with* $wt(v_i) = |v_i| = N_i + \Delta_{n_i}$ *and* $N = N_1 + N_2 + N_3 - N_4$. *For any* $r, s \in \mathbb{Z}$, *define*

$$a = \tfrac{1}{4} + 2r - 4, \qquad b = r + s - 1, \qquad c = -s - 1.$$

With notation as in Corollary 33, for any $r, s \in \mathbb{Z}$, *we have*

$$0 = 2^{2r+2(N_1-N_4)} \sum_{0 \leq k \in \mathbb{Z}} 2^{-2k} \Phi_k \sum_{0 \leq q \in \mathbb{Z}} \binom{\tfrac{1}{4} + 2s}{q} \binom{\tfrac{1}{4} - 2(r+s) - Q_k - 1}{-Q_k/2 - r - 1 - q} (-1)^q$$

$$- 2^{2r+2(N_1-N_4)} \sum_{0 \leq k \in \mathbb{Z}} 2^{-2k} \Phi_k^\theta \sum_{0 \leq q \in \mathbb{Z}} \binom{\tfrac{1}{4} + 2s}{q} \binom{\tfrac{1}{4} - 2(r+s) - Q_k}{-Q_k/2 - r + 1 - q} (-1)^q$$

$$+ (-1)^s 2^{-2b-2(N_1+N_3)} \sum_{0 \leq k \in \mathbb{Z}} 2^{-2k} \Phi_k^* \cdot$$

$$\sum_{0 \leq q \in \mathbb{Z}} \binom{\tfrac{1}{4} + 2s}{q} \left[\binom{a - R_k}{b - R_k/2 - q} - 3 \binom{a - R_k}{b - R_k/2 - 1 - q} \right] (-1)^q$$

$$- (-1)^s 2^{-2b-2(N_1+N_3)} \sum_{0 \leq k \in \mathbb{Z}} 2^{-2k} \Phi_k^{*\theta} \cdot$$

$$\sum_{0 \leq q \in \mathbb{Z}} \binom{\tfrac{1}{4} + 2s}{q} \left[\binom{a - R_k + 1}{b - R_k/2 - 1 - q} - 3 \binom{a - R_k + 1}{b - R_k/2 - q} \right] (-1)^q$$

$$+ 2^{-2c-2(N_1+N_2)} \sum_{0 \leq k \in \mathbb{Z}} 2^{-2k} \Psi_k \cdot$$

$$\sum_{0 \leq q \in \mathbb{Z}} \binom{\tfrac{1}{4} - 2(r+s)}{c - S_k/2 - q} \left[\binom{a - S_k}{q} - 3 \binom{a - S_k}{q - 1} \right] (-1)^q$$

$$- 2^{-2c-2(N_1+N_2)} \sum_{0 \leq k \in \mathbb{Z}} 2^{-2k} \Psi_k^\theta \cdot$$

$$\sum_{0 \leq q \in \mathbb{Z}} \binom{\tfrac{1}{4} - 2(r+s)}{c - S_k/2 - q} \left[\binom{a - S_k + 1}{q - 1} - 3 \binom{a - S_k + 1}{q} \right] (-1)^q.$$

COROLLARY 35. *Let* $v_1, v_2, v_3 \in \mathbf{V}_1$ *and* $v_4 \in \mathbf{V}_3$ *with* $wt(v_i) = |v_i| = N_i + \Delta_{n_i}$, $N = N_1 + N_2 + N_3 - N_4$ *and* $\Gamma = N_1 + N_2 + N_3 + N_4$. *With notation as in Corollary 33, for any* $r, s \in \mathbb{Z}$, *we have*

$$\sum_{0 \leq k \in \mathbb{Z}} 2^{-2k} \sum_{0 \leq q \leq k} \binom{\tfrac{1}{4} + 2r + 2(N_1 + N_2)}{q} (-1)^q.$$

$$\left[\binom{\tfrac{1}{4} - 2r + S_k - 1}{k - q} [\Phi_{r+s+\Gamma-k} + (-1)^{r+N_1+N_2+1} \Phi_{-s-1-k}^*] \right.$$

$$\left. - \binom{\tfrac{1}{4} - 2r + S_k}{k - q} [\Phi_{r+s+\Gamma-k}^\theta + (-1)^{r+N_1+N_2+1} \Phi_{-s-1-k}^{*\theta}] \right]$$

$$= \sum_{0 \le k \in \mathbb{Z}} 2^{-2k} \sum_{0 \le q \le k} \binom{\frac{1}{4} + 2s + 2(N_1 + N_3)}{k - q} (-1)^q \cdot$$

$$\left[\binom{\frac{1}{4} - 2s + R_k}{q} \Psi^\theta_{-r-1-k} - \binom{\frac{1}{4} - 2s + R_k - 1}{q} \Psi_{-r-1-k} \right].$$

Tables of Constants from Section 4

$n_1 = 1 \quad n_2 = 1$

n_3	0	1	2	3
a	0	$\frac{5}{4}$	0	$\frac{1}{4}$
b	1	$\frac{3}{4}$	$\frac{2}{3}$	$\frac{3}{4}$
c	0	$\frac{3}{2}$	$\frac{1}{3}$	$\frac{1}{2}$
a'	0	$\frac{5}{4}$	0	$\frac{1}{4}$
b'	1	$\frac{3}{4}$	$\frac{2}{3}$	$\frac{3}{4}$
c'	2	$\frac{3}{2}$	$\frac{4}{3}$	$\frac{3}{2}$
A	$-\frac{3}{8}$	$\frac{1}{2}$	$\frac{1}{8}$	0
B	0	$-\frac{3}{8}$	$\frac{1}{2}$	$\frac{1}{8}$
C	$-\frac{3}{8}$	$-\frac{3}{8}$	$-\frac{3}{8}$	$-\frac{3}{8}$
A'	0	$\frac{1}{2}$	1	$\frac{1}{2}$
B'	$-\frac{3}{8}$	$-\frac{3}{8}$	$-\frac{3}{8}$	$-\frac{3}{8}$
C'	0	$-\frac{3}{8}$	$\frac{1}{2}$	$\frac{1}{8}$

$n_1 = 3 \quad n_2 = 3$

n_3	0	1	2	3
a	0	$\frac{1}{4}$	0	$\frac{5}{4}$
b	1	$\frac{3}{4}$	$\frac{2}{3}$	$\frac{3}{4}$
c	0	$\frac{1}{2}$	$\frac{1}{3}$	$\frac{3}{2}$
a'	0	$\frac{1}{4}$	0	$\frac{5}{4}$
b'	1	$\frac{3}{4}$	$\frac{2}{3}$	$\frac{3}{4}$
c'	2	$\frac{3}{2}$	$\frac{4}{3}$	$\frac{3}{2}$
A	$-\frac{3}{8}$	0	$\frac{1}{8}$	$\frac{1}{2}$
B	0	$\frac{1}{8}$	$\frac{1}{2}$	$-\frac{3}{8}$
C	$-\frac{3}{8}$	$-\frac{3}{8}$	$-\frac{3}{8}$	$-\frac{3}{8}$
A'	0	$\frac{1}{2}$	1	$\frac{1}{2}$
B'	$-\frac{3}{8}$	$-\frac{3}{8}$	$-\frac{3}{8}$	$-\frac{3}{8}$
C'	0	$\frac{1}{8}$	$\frac{1}{2}$	$-\frac{3}{8}$

$n_1 = 1 \quad n_2 = 3$

n_3	0	1	2	3
a	0	$\frac{1}{4}$	-1	$\frac{1}{4}$
b	1	$-\frac{1}{4}$	$-\frac{1}{3}$	$\frac{3}{4}$
c	0	$\frac{1}{2}$	$\frac{1}{3}$	$\frac{3}{2}$
a'	0	$\frac{1}{4}$	-1	$\frac{1}{4}$
b'	1	$\frac{3}{4}$	$-\frac{1}{3}$	$-\frac{1}{4}$
c'	2	$\frac{1}{2}$	$-\frac{2}{3}$	$\frac{1}{2}$
A	$\frac{1}{8}$	0	$-\frac{3}{8}$	$\frac{1}{2}$
B	0	$\frac{1}{8}$	$\frac{1}{2}$	$-\frac{3}{8}$
C	$\frac{1}{8}$	$\frac{1}{8}$	$\frac{1}{8}$	$\frac{1}{8}$
A'	0	0	0	0
B'	$\frac{1}{8}$	$\frac{1}{8}$	$\frac{1}{8}$	$\frac{1}{8}$
C'	0	$-\frac{3}{8}$	$\frac{1}{2}$	$\frac{1}{8}$

$n_1 = 3 \quad n_2 = 1$

n_3	0	1	2	3
a	0	$\frac{1}{4}$	-1	$\frac{1}{4}$
b	1	$\frac{3}{4}$	$-\frac{1}{3}$	$-\frac{1}{4}$
c	0	$\frac{3}{2}$	$\frac{1}{3}$	$\frac{1}{2}$
a'	0	$\frac{1}{4}$	-1	$\frac{1}{4}$
b'	1	$-\frac{1}{4}$	$-\frac{1}{3}$	$\frac{3}{4}$
c'	2	$\frac{1}{2}$	$-\frac{2}{3}$	$\frac{1}{2}$
A	$\frac{1}{8}$	$\frac{1}{2}$	$-\frac{3}{8}$	0
B	0	$-\frac{3}{8}$	$\frac{1}{2}$	$\frac{1}{8}$
C	$\frac{1}{8}$	$\frac{1}{8}$	$\frac{1}{8}$	$\frac{1}{8}$
A'	0	0	0	0
B'	$\frac{1}{8}$	$\frac{1}{8}$	$\frac{1}{8}$	$\frac{1}{8}$
C'	0	$\frac{1}{8}$	$\frac{1}{2}$	$-\frac{3}{8}$

REFERENCES

[BPZ]   A. A. Belavin, A. M. Polyakov, A. B. Zamolodchikov, *Infinite conformal symmetry in two-dimensional quantum field theory*, Nucl. Phys. **B241** (1984), 333–380.

[BMP]   P. Bouwknegt, J. McCarthy, K. Pilch, *Free field approach to 2-dimensional conformal field theories*, Progr. of Theor. Phys., Suppl. No. **102** (1990), 67–135.

[DL]   C. Dong, J. Lepowsky, *Abelian intertwining algebras - a generalization of vertex operator algebras,*, Proc. Symp. Pure Math, 56, Part 2, Algebraic Groups and Their Generalizations (William J. Haboush and Brian J. Parshall, ed.), Amer. Math. Soc., Providence, RI, 1994, pp. 261–293.

[DMZ]   C. Dong, G. Mason, Y. Zhu, *Discrete series of the Virasoro algebra and the moonshine module*, Proc. of Symp. Pure Math., 56, Part 2, Algebraic Groups and Their Generalizations (William J. Haboush and Brian J. Parshall, ed.), Amer. Math. Soc., Providence, RI, 1994, pp. 295–316.

[F]   A. J. Feingold, *Constructions of vertex operator algebras*, Proc. Symp. Pure Math., 56, Part 2, Algebraic Groups and Their Generalizations (William J. Haboush and Brian J. Parshall, ed.), Amer. Math. Soc., Providence, RI, 1994, pp. 317–336.

[FF]   A. J. Feingold, I. B. Frenkel, *Classical affine algebras*, Adv. in Math. **56** (1985), 117–172.

[FFR]   A. J. Feingold, I. B. Frenkel, J. F. X. Ries, *Spinor Construction of Vertex Operator Algebras, Triality and $E_8^{(1)}$*, Contemp. Math., 121, Amer. Math. Soc., Providence, RI, 1991.

[Fr1]   I. B. Frenkel, *Spinor representations of affine Lie algebras*, Proc. Natl. Acad. Sci. USA **77** (1980), 6303–6306.

[Fr2]   _____, *Two constructions of affine Lie algebra representations and boson-fermion correspondence in quantum field theory*, J. Funct. Anal. **44** (1981), 259–327.

[FHL]   I. B. Frenkel, Yi-Zhi Huang, J. Lepowsky, *On axiomatic approaches to vertex operator algebras and modules*, Memoirs Amer. Math. Soc., 104, No. 594, Amer. Math. Soc., Providence, RI, 1993.

[FLM]   I. B. Frenkel, J. Lepowsky, A. Meurman, *Vertex Operator Algebras and the Monster*, Pure and Applied Math., 134, Academic Press, Boston, 1988.

[FZ]   I. B. Frenkel, Y. Zhu, *Vertex operator algebras associated to representations of affine and Virasoro algebras*, Duke Math. J. **66** (1992), 123–168.

[GO]   P. Goddard, D. I. Olive, *Kac-Moody and Virasoro algebras in relation to quantum physics*, Internat. J. Mod. Phys. A **1** (1986), 303–414.

[L]   Y. L. Luke, *The Special Functions and Their Approximations, Vol. 1*, Academic Press, New York, 1969.

[MaS]   G. Mack, V. Schomerus, *Conformal field algebras with quantum symmetry from the theory of superselection sectors*, Commun. Math. Phys. **134** (1990), 139–196.

[MS]   G. Moore and N. Seiberg, *Classical and quantum conformal field theory*, Commun. Math. Phys. **123** (1989), 177–254.

[S]   L. J. Slater, *Generalized Hypergeometric Functions*, Cambridge University Press, Cambridge, Great Britain, 1966.

[TK]   A. Tsuchiya, Y. Kanie, *Vertex operators in conformal field theory on $\mathbb{P}^1$ and monodromy representations of braid group*, Conformal Field Theory and Solvable Lattice Models, Adv. Studies in Pure Math., vol. 16, Academic Press, New York, 1988, pp. 297–372.

[W]   M. D. Weiner, *Bosonic construction of vertex operator para-algebras from symplectic affine Kac-Moody algebras*, Memoirs Amer. Math. Soc. (submitted).

DEPT. OF MATH. SCI., THE STATE UNIVERSITY OF NEW YORK, BINGHAMTON, NEW YORK 13902-6000

MATH. DEPT., EAST CAROLINA UNIVERSITY, GREENVILLE, NORTH CAROLINA 27858-4353

MATH. DEPT., BROOME COMMUNITY COLLEGE, BINGHAMTON, NEW YORK 13902-1017

Contemporary Mathematics
Volume **193**, 1996

THE MCKAY-THOMPSON SERIES ASSOCIATED WITH THE IRREDUCIBLE CHARACTERS OF THE MONSTER

KOICHIRO HARADA AND MONG LUNG LANG

ABSTRACT. Let $\mathbb{V} = \coprod_{h=0}^{\infty} \mathbb{V}_h$ be the graded monster module of the monster simple group $\mathbb{M}$ and let χ_k be an irreducible representation of $\mathbb{M}$. The generating function of c_{hk} (the multiplicity of χ_k in $\mathbb{V}_h$) is determined. Furthermore, the invariance group of the modular function associated with the generating function is also determined in this paper.

1. INTRODUCTION

Let $\mathbb{M}$ be the monster simple group and $\mathbb{V}$ be the monster module of Frenkel-Lepowsky-Meurman [4]. $\mathbb{V}$ is a graded $\mathbb{M}$ module

$$\mathbb{V} = \coprod_{h=0}^{\infty} \mathbb{V}_h$$

such that

$$j(q) - 744 = \sum_{h=0}^{\infty} \dim \mathbb{V}_h q^{h-1}.$$

In particular, $\dim \mathbb{V}_0 = 1$, $\dim \mathbb{V}_1 = 0$, $\dim \mathbb{V}_2 = 196884$, $\dim \mathbb{V}_3 = 21493760$, $\cdots$. Let χ_k, $1 \le k \le 194$, be the irreducible characters of $\mathbb{M}$ (the χ_k's are ordered according to their degrees.) which will often be used to denote the irreducible representations also. For the first few $\mathbb{V}_i$'s, we have the decompositions :

$$\mathbb{V}_0 = \chi_1,$$

$$\mathbb{V}_2 = \chi_1 + \chi_2,$$

$$\mathbb{V}_3 = \chi_1 + \chi_2 + \chi_3.$$

In general, write

$$\mathbb{V}_h = \sum_{k=1}^{194} c_{hk} \chi_k$$

where c_{hk} is the multiplicity of χ_k in $\mathbb{V}_h$. The table of c_{hk} for $0 \le h \le 51$, $1 \le k \le 194$ can be found in McKay-Strauss [7].

We also list some of the multiplicities c_{h1} of the trivial character χ_1 in V_h.

h	0	2	3	4	5	6	7	8	9	10	11 $\cdots$	20 $\cdots$	30 $\cdots$	40 $\cdots$	50
c_{h1}	1	1	1	2	2	4	4	7	8	12	14 $\cdots$	167 $\cdots$	1762 $\cdots$	15913 $\cdots$	129734

1991 *Mathematics Subject Classification.* 20D08; Secondary 11F03.

Key words and phrases. monster, monster module, modular functions, invariance groups.

Let us consider, for each irreducible character χ_k, the generating function :

$$t_{\chi_k}(x) = x^{-1}\sum_{h=0}^{\infty} c_{hk}x^h.$$

The multiplicity c_{hk} can be computed as follows :

$$c_{hk} = \frac{1}{|\mathbb{M}|}\sum_{g\in\mathbb{M}} Tr(g|\mathbb{V}_h)\chi_k(g).$$

Therefore the generating function of c_{hk} is

$$t_{\chi_k}(x) = x^{-1}\sum_{h=0}^{\infty}\sum_{g\in\mathbb{M}}\frac{1}{|\mathbb{M}|}Tr(g|\mathbb{V}_h)\chi_k(g)x^h.$$

If we replace the indeterminate x by $q = e^{2\pi iz}$, $z\in\mathbb{H} = \{z\in\mathbb{C}|\mathrm{Im}(z) > 0\}$ then

$$t_{\chi_k}(q) = \frac{1}{|\mathbb{M}|}\sum_{g\in\mathbb{M}}\chi_k(g)t_g(q)$$

where

$$t_g(q) = q^{-1}\sum_{h=0}^{\infty}Tr(g|\mathbb{V}_h)q^h$$

is the McKay-Thompson series for the element g in $\mathbb{M}$. Thus $t_\chi(q)$ for the irreducible character χ is the weighted sum of the McKay-Thompson series for the elements g of $\mathbb{M}$. Not all $t_\chi(q)$'s are distinct and in fact there are exactly 172 distinct $t_\chi(q)$'s, since

$$t_\chi(q) = t_{\bar\chi}(q)$$

where $\bar\chi$ is the complex conjugate of χ and there are no other equalities among $t_\chi(q)$'s. One of the obvious questions one will raise here will be :

Problem. Determine the invariance group Γ_χ of $t_\chi(q)$.

Here Γ_χ is defined to be :

Definition. $\Gamma_\chi = \{A \in SL_2(\mathbb{R})|t_\chi(Az) = t_\chi(z)\}$.

Let us here review some of the properties of the invariance group Γ_g of the McKay-Thompson series $t_g(z)$ for the element $g \in \mathbb{M}$. (See [3] for notation.)

(0). For $G \subset GL_2^+(\mathbb{R})$, $\bar G$ is the image of G in $PGL_2^+(\mathbb{R})$.

(1). $\Gamma_0(N) = \{A = \begin{pmatrix} a & b \\ c & d \end{pmatrix} \in SL_2(\mathbb{Z})|c \equiv 0 \ (\mathrm{mod}\ N) \}$.

(2). For an exact divisor $e||N$ (i.e. $e|N$ and $\gcd(e, \frac{N}{e}) = 1$) let

$$W_e = \begin{pmatrix} ae & b \\ cN & de \end{pmatrix}, \quad a, b, c, d \in \mathbb{Z}, \quad ade^2 - bcN = e.$$

Then $\bar W_e$ normalizes $\bar\Gamma_0(N)$ and $\bar W_e^2 \in \bar\Gamma_0(N)$.

(3). Let h be a divisor of n. Then $n|h + e, f, \cdots$ is defined to be

$$\begin{pmatrix} \frac{1}{h} & 0 \\ 0 & 1 \end{pmatrix}\langle\Gamma_0(\frac{n}{h}), W_e, W_f, \cdots\rangle\begin{pmatrix} h & 0 \\ 0 & 1 \end{pmatrix}.$$

The statement (4) below was a conjecture in [3] and was proved by Borcherds [1]. Statement (5) follows from (4) by a case by case argument.

(4). For each g in $\mathbb{M}$, Γ_g, the invariance group of $t_g(z)$, is a normal subgroup of index h_g in $n_g|h_g + e_g, f_g, \cdots$, the eigen group of g [3]. Note that for each A in $n_g|h_g + e_g, f_g, \cdots$, $(t_g|A)(z) = \sigma t_g(z)$ where σ is an h_g-th root of unity. We will often use n, h, e, f, etc. instead of n_g, h_g e_g, f_g, etc. for simplicity.

(5). Γ_g contains $\Gamma_0(n_g h_g)$.

For each irreducible character of $\mathbb{M}$, we now define :

Definition. $N_\chi = \mathrm{lcm}\{n_g h_g | g \in \mathbb{M}, \chi(g) \neq 0\}$.

It is obvious that $t_\chi(z)$ is invariant under $\Gamma_0(N_\chi)$. Note that N_χ can be quite large ($N_{\chi_1} = 2^6 3^3 5^2 7 \cdot 11 \cdot 13 \cdot 17 \cdot 19 \cdot 23 \cdot 29 \cdot 31 \cdot 41 \cdot 47 \cdot 59 \cdot 71$) or relatively small ($N_{\chi_{166}} = 2^6 3^2 7 = 4032$).

The purpose of this paper is to show

Theorem. $\Gamma_\chi = \Gamma_0(N_\chi)$.

2. Poles of $t_\chi(z)$

For each cusp c in $\mathbb{Q} \cup \{\infty\}$, we define Φ_c to be the set $\Phi_c = \{g \in \mathbb{M} | c$ is equivalent to ∞ in $\Gamma_g\}$ and decompose $t_\chi(z)$ into :

$$\frac{1}{|\mathbb{M}|} \sum_{g \in \Phi_c} \chi(g) t_g(z) + \frac{1}{|\mathbb{M}|} \sum_{g \notin \Phi_c} \chi(g) t_g(z).$$

The McKay-Thompson series $t_g(z)$ is a generator of the function field of the compact Riemann surface $\Gamma_g \backslash \mathbb{H}^*$ ($\mathbb{H}^* = \mathbb{H} \cup \{\infty\}$) of genus 0 and has a unique pole at ∞ (and at all cusps $c \in \mathbb{Q}$ equivalent to ∞ in Γ_g). Obviously, $\frac{1}{|\mathbb{M}|} \sum_{g \notin \Phi_c} \chi(g) t_g(z)$ is holomorphic at c. Hence, whether c is a pole of $t_\chi(z)$ or not is determined by the singular part of

$$\frac{1}{|\mathbb{M}|} \sum_{g \in \Phi_c} \chi(g) t_g(z)$$

at c. For example, if $c = \infty$, then ∞ is a pole of $t_\chi(z)$ if and only if χ is the trivial character since the singular part of $t_\chi(z)$ at ∞ is given by $\frac{1}{|\mathbb{M}|} \sum_{g \in \mathbb{M}} \chi(g) \frac{1}{q}$ and

$$\sum_{g \in \mathbb{M}} \chi_i(g) = \begin{cases} |\mathbb{M}| & \text{if } i = 1 \\ 0 & \text{if } i \neq 1 \end{cases}$$

Suppose $c \in \mathbb{Q}$. For each g in Φ_c, let $A \in SL_2(\mathbb{Z})$ be chosen so that $A\infty = c$. Then $t_g(Az) = (t_g|A)(z)$ has an expansion in $q = e^{2\pi i z}$ of the form

$$a q^{-\frac{1}{\mu}} + \cdots$$

where $\mu = [\langle \begin{pmatrix} 1 & 1 \\ 0 & 1 \end{pmatrix} \rangle : A^{-1} \Gamma_0(n_g h_g)_c A]$, where the subscript c denotes the stabilizer. We contend that the contribution of the $t_g(z)$ to the singular part of $t_\chi(z)$ is

$$a q^{-\frac{1}{\mu}}.$$

Indeed, by our assumption the cusp c is equivalent to ∞ in Γ_g and so there is $B \in \Gamma_g$ such that $B\infty = c$ and

$$(t_g|B)(q) = t_g(q) = q^{-1} + \sum_{i \geq 0} a_i q^i.$$

The only difference between $(t_g|A)(z)$ and $(t_g|B)(z)$ lies in the power of q and a, hence the contention.

In order to determine whether c is pole of $t_\chi(z)$ or not, one has to :

(1). Determine whether c is equivalent to ∞ in Γ_g or not.

(2). Determine the singular part of $\frac{1}{|M|} \sum_{g \in \Phi_c} \chi(g) t_g(z)$ at c.

We will investigate those questions in the next section.

3. EQUIVALENCE OF CUSPS

In this section, we will study the equivalence of cusps in Γ_g, $g \in \mathbb{M}$.

Lemma 1. *For each exact divisor e of N and for each c such that $\gcd(c,e) = 1$, $\Gamma_0(N)$ admits an Atkin-Lehner involution of the form $W_e = \begin{pmatrix} ae & b \\ cN & de \end{pmatrix}$. Moreover, one can choose either $a = 1$ or $d = 1$ if desired.*

Proof. For each c such that c and e are relatively prime, we have $\gcd(\frac{cN}{e}, e) = 1$. Hence, there exists b and y such that $ye - \frac{bcN}{e} = 1$, or $ye^2 - bcN = e$. The lemma follows by writing y into ad for suitable a and d. $\square$

Lemma 2. *Let $\gcd(a,b) = 1$ and M be nonzero integers. Then there exists a pair of integers x and y satisfying $\gcd(xM, y) = 1$ and $ax + by = 1$.*

Proof. This is a well known fact of the elementary number theory. Let x' and y' be a pair of integral solutions of the equation $ax + by = 1$ and let $M = M_a M_{y'} M'$ be the decomposition of M into a product of coprime factors such that M_a and a, $M_{y'}$ and y', have the same prime factors. It is clear that $y = y' + aM'$ and $x = x' - bM'$ is also a pair of solutions to the equation. Note that $\gcd(x,y) = 1$ since it is a solution of $ax + by = 1$. Furthermore, one has $\gcd(y, M) = \gcd(y' + aM', M_a M_{y'} M')$ $= \gcd(y' + aM', M') = 1$. Therefore x and y is pair of integral solutions of the equation such that $\gcd(xM, y) = 1$. $\square$

Lemma 3. *Let $\frac{x}{y}$, $\gcd(x,y) = 1$, be a rational number. Then $\frac{x}{y}$ is equivalent to some $\frac{x'}{y'}$, $\gcd(x',y') = 1$ in $\Gamma_0(N)$, where $y' = \gcd(N,y)$. Furthermore, if $\frac{x}{y}$ is equivalent to $\frac{x''}{y''}$ with $y''|N$, $y'' > 0$, and $\gcd(x'',y'') = 1$, then $y'' = y'$.*

Proof. Consider the equality

$$\begin{pmatrix} a & b \\ cN & d \end{pmatrix} \frac{x}{y} = \frac{ax + by}{cNx + dy} = \frac{ax + dy}{y'(cx\frac{N}{y'} + d\frac{y}{y'})}$$

Note that $\gcd(x\frac{N}{y'}, \frac{y}{y'}) = 1$, hence the equation

$$c\frac{xN}{y'} + d\frac{y}{y'} = 1$$

is solvable for c, d in $\mathbb{Z}$. Applying Lemma 2, we may assume that c and d are integral solutions of the above equation such that $\gcd(cN, d) = 1$. Let a and b be chosen such that $ad - cbN = 1$. Summarizing, we now conclude that $\frac{x}{y}$ is equivalent to $\frac{ax+by}{y'}$ by $\begin{pmatrix} a & b \\ cN & d \end{pmatrix} \in \Gamma_0(N)$. Since $\gcd(ax+by, y') = \gcd(ax, y') = \gcd(a, y') = 1$, first part of the lemma is proved. As for the second part, suppose

$$\begin{pmatrix} a & b \\ cN & d \end{pmatrix} \frac{x}{y} = \frac{ax+by}{cNx+dy} = \frac{ax+dy}{y'(cx\frac{N}{y'}+d\frac{y}{y'})} = \frac{x''}{y''}.$$

We first note $y'|y''$, since $\gcd(ax+by, y') = 1$. To show $y''|y'$, suppose that y'' possesses a prime power q^t such that y' is not a multiple of q^t, then $q^t|y'(cx\frac{N}{y'}+d\frac{y}{y'})$ implies $q|(cx\frac{N}{y'}+d\frac{y}{y'})$. Since $y''|N$, q is a divisor of $\frac{N}{y'}$, hence $q|d$. This implies that $\gcd(cN, d) \neq 1$, against our choice of c, d. Thus, $y''|y'$ and the second part of the lemma is proved. $\square$

In Lemma 4 and Lemma 5, $G = n|h+e, f, \cdots$ is the eigen group of the invariance group Γ_g.

Lemma 4. *Let $g \in \mathbb{M}$ and let $\Gamma_g \leq G = n|h + e, f, \cdots$ be the invariance group of $t_g(z)$. Then $G\infty = \Gamma_g\infty$.*

Proof. Since $q = \exp(2\pi i z)$ is a local parameter of $t_g(z)$, the stabilizer $(\Gamma_g)_\infty$ of ∞ is generated by $\begin{pmatrix} 1 & 1 \\ 0 & 1 \end{pmatrix}$. As for G, G_∞ is generated by $\begin{pmatrix} 1 & \frac{1}{h} \\ 0 & 1 \end{pmatrix}$. Hence

$$[G : \Gamma_g] = h = [G_\infty : (\Gamma_g)_\infty]$$

Consequently, $G\infty = \Gamma_g\infty$. $\square$

Lemma 5. *Let $g \in \mathbb{M}$ and let $\Gamma_g \leq n|h + e, f, \cdots$ be the invariance group of $t_g(z)$. Then $\frac{x}{y}$, $\gcd(x,y) = 1$, is equivalent to ∞ in Γ_g if and only if*

$$\gcd(\frac{n}{h}, \frac{y}{\gcd(y,h)}) \in \left\{ \frac{n}{h}, \frac{n}{he}, \frac{n}{hf}, \cdots \right\}$$

Proof. To simplify our notation, let $N = \frac{n}{h}$. Choose the Atkin-Lehner involution W_e as described in Lemma 1 with $\gcd(c, N) = 1$. One has $W_e\infty = \frac{a}{c\frac{N}{e}}$, and

$$\gcd(a, c\frac{N}{e}) = 1.$$

By Lemma 3, there exists $\gamma_e \in \Gamma_0(N)$ such that $\gamma_e W_e\infty = \frac{e'}{\frac{N}{e}}$ where $\gcd(e', \frac{N}{e}) = 1$, since $\gcd(c\frac{N}{e}, N) = \frac{N}{e}$. Note that e is an exact divisor of N, hence among the representitives of inequivalent cusps of $\Gamma_0(N)$, there is exactly one and only one cusp

z with denominator $\frac{N}{e}$ (see Harada [5]). Without loss of generality, we may assume that $z = \frac{1}{\frac{N}{e}}$. Therefore, we may assume that γ_e is chosen so that $\gamma_e W_e \infty = \frac{1}{\frac{N}{e}}$. Hence the G-orbit of ∞ can be decomposed into,

$$\begin{pmatrix} \frac{1}{h} & 0 \\ 0 & 1 \end{pmatrix} \Gamma_0(N) \frac{1}{N} \cup \begin{pmatrix} \frac{1}{h} & 0 \\ 0 & 1 \end{pmatrix} \Gamma_0(N) \frac{1}{\frac{N}{e}} \cup \begin{pmatrix} \frac{1}{h} & 0 \\ 0 & 1 \end{pmatrix} \Gamma_0(N) \frac{1}{\frac{N}{f}} \cup \cdots .$$

Hence $\frac{x}{y}$ is equivalent to ∞ in G if and only if

$$\begin{pmatrix} h & 0 \\ 0 & 1 \end{pmatrix} \frac{x}{y} = \frac{\frac{hx}{\gcd(y,h)}}{\frac{y}{\gcd(y,h)}} \in \Gamma_0(N) \frac{1}{N} \cup \Gamma_0(N) \frac{1}{\frac{N}{e}} \cup \Gamma_0(N) \frac{1}{\frac{N}{f}} \cup \cdots ,$$

which is equivalent to, by Lemma 3,

$$\gcd(\frac{y}{\gcd(y,h)}, N) \in \left\{ N, \frac{N}{e}, \frac{N}{f}, \cdots \right\} .$$

$G\infty = \Gamma_g \infty$ as shown in Lemma 4 and so $\frac{x}{y}$ is equivalent to ∞ in Γ_g if and only if

$$\gcd(\frac{y}{\gcd(y,h)}, \frac{n}{h}) \in \left\{ \frac{n}{h}, \frac{n}{he}, \frac{n}{hf}, \cdots \right\} .$$

$\square$

Corollary 6. $0 = \frac{0}{1}$ *is equivalent to* ∞ *in* Γ_g *if and only if* $n = h$ *or* $G = n|h + e, f, \cdots$ *admits the Atkin-Lehner involution* $W_{\frac{n}{h}}$.

Proof. Since $\gcd(1, \frac{n}{h}) = 1$, G must admits an Atkin-Lehner involution W_e such that $\frac{n}{he} = 1$, hence $e = \frac{n}{h}$. $\square$

Let χ be an irreducible character of the monster $\mathbb{M}$. In order to determine the singular part of

$$\frac{1}{|\mathbb{M}|} \sum_{g \in \Phi_c} \chi(g) t_g(z)$$

at the cusp $c = \frac{x}{y}$, where $\gcd(x, y) = 1$ and $y | N_\chi$, it is necessary to find a matrix P_c in $SL_2(\mathbb{R})$ such that $P_c \infty = c$ and determine the q-expansion of

$$\frac{1}{|\mathbb{M}|} \sum_{g \in \Phi_c} \chi(g) (t_g | P_c)(z),$$

which will be called the q-expansion of $t_\chi(z)$ at c. Such a matrix P_c is easy to find and choice is not unique. To ease the computation of the q-expansion of

$$\frac{1}{|\mathbb{M}|} \sum_{g \in \Phi_c} \chi(g) (t_g | P_c)(z),$$

it is necessary to find a good P_c so that the transformation formula of $t_g | P_c$ can be obtained for every g in Φ_c simultaneously. What we shall do is as follows. Namely, we will find a matrix P_c so that one can associate with P_c an upper triangular matrix $U_{c,g}$ such that

$$P_c U_{c,g}^{-1} \in n_g | h_g + e_g, f_g, \cdots .$$

Since $n_g|h_g + e_g, f_g, \cdots$ is the eigen group of Γ_g, elements in $n_g|h_g + e_g, f_g, \cdots$ map t_g to $\sigma_g t_g$ where σ_g is an h_g-th root of unity which depends on g and on some other quantities. Therefore

$$t_g|P_c = \sigma_g t_g|U_{c,g}.$$

A good transformation formula for $t_g|P_c$ is obtained since $U_{c,g}$ is upper trianglar.

Let y_0 be the exact divisor of N_χ such that y and y_0 share the same prime divisors. Then $\gcd(y, x\frac{N_\chi}{y_0}) = 1$ and so there is a matrix $P_c \in SL_2(\mathbb{Z})$ of the form

$$P_c = \begin{pmatrix} x & w \\ y & \frac{zN_\chi}{y_0} \end{pmatrix}.$$

Lemma 4 implies that $\frac{x}{y}$ is equivalent to ∞ in Γ_g if and only if $\frac{x}{y}$ is equivalent to ∞ in the eigen group $n_g|h_g + e_g, f_g, \cdots$ of g and so by Lemma 5, $\frac{x}{y}$ is equivalent to ∞ in Γ_g if and only if $\gcd(\frac{n_g}{h_g}, \frac{y}{\gcd(y,h_g)}) \in \left\{ \frac{n}{h}, \frac{n}{he}, \frac{n}{hf}, \cdots \right\}$. More precisely, $\frac{x}{y}$ is equivalent to ∞ by an element in $n_g|h_g$ if

$$\gcd(\frac{n_g}{h_g}, \frac{y}{\gcd(y,h_g)}) = \frac{n_g}{h_g}$$

and is equivalent to ∞ by an Atkin-Lehner involution W_{e_g} of $n_g|h_g + e_g, f_g, \cdots$ if

$$\frac{n_g}{h_g e_g} = \gcd(\frac{n_g}{h_g}, \frac{y}{\gcd(y,h_g)}) \in \left\{ \frac{n}{he}, \frac{n}{hf}, \cdots \right\}.$$

Lemma 7. *Suppose that $\gcd(\frac{n_g}{h_g}, \frac{y}{\gcd(y,h_g)}) = \frac{n_g}{h_g}$. Let u_g be chosen so that $\frac{yu_g}{h_g} + \frac{zN_\chi \gcd(h_g,y)}{y_0 h_g}$ is an integer. Then*

$$P_c U_{c,g}^{-1} = P_c \begin{pmatrix} \frac{h_g}{\gcd(h_g,y)} & \frac{u_g}{h_g} \\ 0 & \frac{\gcd(h_g,y)}{h_g} \end{pmatrix} \in n_g|h_g$$

where

$$U_{c,g} = \begin{pmatrix} \frac{\gcd(h_g,y)}{h_g} & -\frac{u_g}{h_g} \\ 0 & \frac{h_g}{\gcd(h_g,y)} \end{pmatrix}.$$

Proof. To show the existence of u_g, simply solve the equation

$$\frac{y}{\gcd(h_g,y)} u_g + \frac{N_\chi}{y_0} z \equiv 0 \pmod{\frac{h_g}{\gcd(h_g,y)}}.$$

Then $yu_g + z\frac{N_\chi}{y_0}\gcd(h_g,y) \equiv 0 \pmod{h_g}$ and hence $\frac{yu_g}{h_g} + \frac{zN_\chi \gcd(h_g,y)}{y_0 h_g}$ is an integer.
The matrix

$$P_c \begin{pmatrix} \frac{h_g}{\gcd(h_g,y)} & \frac{u_g}{h_g} \\ 0 & \frac{\gcd(h_g,y)}{h_g} \end{pmatrix} = \begin{pmatrix} \frac{xh_g}{\gcd(h_g,y)} & \frac{xu_g+w\gcd(h_g,y)}{h_g} \\ \frac{yh_g}{\gcd(h_g,y)} & \frac{yu_g}{h_g} + \frac{zN_\chi\gcd(h_g,y)}{y_0 h_g} \end{pmatrix}$$

has the property

(1). $\frac{yu_g}{h_g} + \frac{zN_\chi\gcd(h_g,y)}{y_0 h_g}$ is an integer by our choice of u_g, and,

(2). $\frac{yh_g}{\gcd(h_g,y)}$ is a multiple of n_g since $\gcd(\frac{n_g}{h_g}, \frac{y}{\gcd(y,h_g)}) = \frac{n_g}{h_g}$.
Therefore

$$P_c U_{c,g}^{-1} = P_c \begin{pmatrix} \frac{h_g}{\gcd(h_g,y)} & \frac{u_g}{h_g} \\ 0 & \frac{\gcd(h_g,y)}{h_g} \end{pmatrix} \in n_g | h_g. \quad \square$$

Corollary 8. *Suppose that* $\gcd(\frac{n_g}{h_g}, \frac{y}{\gcd(y,h_g)}) = \frac{n_g}{h_g}$. *Then*

$$t_g | P_c = \sigma_g t_g | U_{c,g} = \sigma_g t_g(U_{c,g}z) = \sigma_g t_g(\frac{\gcd(h_g,y)^2}{h_g{}^2}z - \frac{u_g}{h_g^2}\gcd(h_g,y))$$

where σ_g *is an* h_g-*th root of unity.*

Proof. Since $P_c \begin{pmatrix} \frac{h_g}{\gcd(h_g,y)} & \frac{u_g}{h_g} \\ 0 & \frac{\gcd(h_g,y)}{h_g} \end{pmatrix} \in n_g | h_g$ and $n_g | h_g$ is the eigen group
of $t_g(z)$

$$t_g | P_c \begin{pmatrix} \frac{h_g}{\gcd(h_g,y)} & \frac{u_g}{h_g} \\ 0 & \frac{\gcd(h_g,y)}{h_g} \end{pmatrix} = \sigma_g t_g(z).$$

This completes the proof of the corollary. $\quad \square$

We now consider the case that c is equivalent to ∞ in the eigen group of Γ_g by an Atkin-Lehner involution W_{e_g}.

Lemma 9. *Suppose that* $c = \frac{x}{y}$ *is equivalent to* ∞ *in the eigen group of* Γ_g *by an Atkin-Lehner involution* W_{e_g}. *Let an integer* u_g *be chosen such that* e_g *is a divisor of an integer* $\frac{u_g y}{h_g} + \frac{z N_\chi \gcd(h_g,y)}{h_g y_0}$ *where*

$$e_g = \frac{\frac{n_g}{h_g}}{\gcd(\frac{n_g}{h_g}, \frac{y}{\gcd(y,h_g)})}.$$

Then

$$P_c U_{c,g}^{-1} = P_c \begin{pmatrix} \frac{e_g h_g}{\gcd(h_g,y)} & \frac{u_g}{h_g} \\ 0 & \frac{\gcd(h_g,y)}{h_g} \end{pmatrix} = W_{e_g} \in n_g | h_g + e_g, f_g, \cdots.$$

Furthermore,

$$t_g | P_c = \sigma_g t_g(U_{c,g}z) = \sigma_g t_g(\frac{\gcd(h_g,y)^2}{e_g h_g{}^2}z - \frac{u_g}{e_g h_g^2}\gcd(h_g,y))$$

where σ_g *is an* h_g-*th root of unity.*

Proof. Let us first show that such an u_g exists. We will need u_g such that

$$yu_g + z\frac{N_\chi}{y_0}\gcd(h_g,y) \equiv 0 \pmod{e_g h_g}.$$

This follows from

$$\frac{y}{\gcd(h_g,y)}u_g + z\frac{N_\chi}{y_0} \equiv 0 \pmod{\frac{h_g}{\gcd(h_g,y)}e_g}.$$

Since $\gcd(\frac{n_g}{h_g}, \frac{y}{\gcd(h_g,y)}) = \frac{n_g}{h_g e_g}$ and e_g is an exact divisor of $\frac{n_g}{h_g}$, we see that

$$\gcd(\frac{y}{\gcd(h_g,y)}, e_g) = 1.$$

Therefore $\frac{y}{\gcd(h_g,y)}$ is invertible modulo $\frac{h_g}{\gcd(h_g,y)}e_g$, hence u_g exists as required.
The matrix

$$P_c\begin{pmatrix} \frac{e_g h_g}{\gcd(h_g,y)} & \frac{u_g}{h_g} \\ 0 & \frac{\gcd(h_g,y)}{h_g} \end{pmatrix} = \begin{pmatrix} \frac{x e_g h_g}{\gcd(h_g,y)} & \frac{x u_g + w\gcd(h_g,y)}{h_g} \\ \frac{y e_g h_g}{\gcd(h_g,y)} & \frac{y u_g}{h_g} + \frac{z N_\chi \gcd(h_g,y)}{y_0 h_g} \end{pmatrix}$$

has the property

(1). $\frac{u_g y}{h_g} + \frac{z N_\chi \gcd(h_g,y)}{h_g y_0}$ is a multiple of e_g by choice of u_g, and,

(2). $\frac{y e_g h_g}{\gcd(h_g,y)}$ is a multiple of n_g, since $\gcd(\frac{n_g}{h_g}, \frac{y}{\gcd(y,h_g)}) = \frac{n_g}{h_g e_g}$.

Therefore

$$P_c\begin{pmatrix} \frac{e_g h_g}{\gcd(h_g,y)} & \frac{u_g}{h_g} \\ 0 & \frac{\gcd(h_g,y)}{h_g} \end{pmatrix} = W_{e_g} \in n_g|h_g + e_g, f_g, \cdots.$$

Since c is equivalent to ∞ in the eigen group of Γ_g by W_{e_g}, the transformation
formula follows easily. $\square$

Remark. It is easy to see that Lemma 7 and Corollary 8 are included in
Lemma 9 if $e_g = 1$, in which case every element of $n_g|h_g$ is called an Atkin-Lehner
involution for $e_g = 1$. This abuse of words will be used occasionally for the balance
of the paper.

The singular part of $\frac{1}{|\mathbb{M}|}\sum_{g\in\Phi_c}\chi(g)(t_g|P_c)(z)$ at $z = \infty i$ is now determined by

$$\text{sing}_{P_c}t_\chi = \frac{1}{|\mathbb{M}|}\sum_{g\in\Phi_c}\chi(g)\frac{\sigma_g}{e^{2\pi i U_{c,g}z}}.$$

We give a few examples in the calculation of $\text{sing}_{P_c}t_\chi$'s. Note that the first
example will be used later in the determination of the invariance groups.

Example 1. Suppose $c = \frac{0}{1}$. Then $\text{sing}_{P_0}t_\chi = \frac{1}{|\mathbb{M}|}\sum_{g\in\Phi_c}\chi(g)\frac{\sigma_g}{q^{\frac{1}{n_g h_g}}}$.

Proof. In this case $y = 1$. We may choose $P_0 = \begin{pmatrix} x & w \\ y & \frac{x N_\chi}{y_0} \end{pmatrix} = \begin{pmatrix} 0 & -1 \\ 1 & 0 \end{pmatrix}$
and $u_g = 0$. If the condition of Corollary 8 holds, then $n_g = h_g$ and

$$t_g|P_0 = \sigma_g t_g(\frac{z}{h_g^2}) = \sigma_g t_g(\frac{z}{n_g h_g}).$$

The only $g \in \Phi_0 \subseteq \mathbb{M}$ satisfying $n_g = h_g$ are $1A$ and $3C$. We have

$$t_{1A}|P_0 = t_{1A} \quad \text{and} \quad t_{3C}|P_0 = \sigma_{3C}t_{3C}(\frac{1}{9}z).$$

On the other hand, if the condition of Lemma 9 holds, then $n_g = e_g h_g$ and

$$t_g|P_0 = \sigma_g t_g(\frac{z}{e_g h_g^2}) = \sigma_g t_g(\frac{z}{n_g h_g}).$$

Note that for all the remaining $g \in \Phi_0 \backslash \{1A, 3C\} \subseteq \mathbb{M}$, 0 is equivalent to ∞ in Γ_g by the Atkin-Lehner involution $W_{\frac{n_g}{h_g}}$ and hence the condition of Lemma 9 holds. $\square$

Remark. Example 1 shows that 0 is a pole of the McKay-Thompson series $t_\chi(z)$ for every χ, since $n_g h_g \neq 1$ for $g \neq 1$ and so the coefficient of q^{-1} is nonzero.

Example 2. Let χ be the trivial character and let $c = \frac{1}{3}$. Then $P_{\frac{1}{3}} = \begin{pmatrix} 1 & [\frac{N_0}{81}] \\ 3 & \frac{N_0}{27} \end{pmatrix}$ ($[x]$ is the integral part of x and $\frac{N_0}{27} \equiv 1 \pmod 3$) and $t_{84A}|P_{\frac{1}{3}} = \sigma_{84A} t_{84A}(\frac{1}{56}z + \frac{1}{2})$.

Proof. We know $\Gamma_{84A} < 84|2+$. Since $\gcd(\frac{84}{2}, \frac{3}{\gcd(3,2)}) = 3 = \frac{84}{2e}$, we see that e is 14, and can choose $u_g = -28$.

$$P_{\frac{1}{3}} \begin{pmatrix} 28 & -\frac{28}{2} \\ 0 & \frac{1}{2} \end{pmatrix} = W_{14} \in 84|2+.$$

The rest follows easily. $\square$

Remark. (1). The invariance group Γ_g of the harmonics $n|h + e, f, \cdots$ are not fully determined. (For each g, one can write down a set of generators of the invariance group Γ_g easily. But determining whether or not a given element in $SL_2(\mathbb{R})$ is a word of those generators is nontrivial.) Hence we have to settle for σ_g being an h_g-th root of unity.

(2). $\sigma_g = 1$ if $h_g = 1$.

(3). Let $p \in \{11, 17, 19, 23, 29, 31, 41, 47, 59, 71\}$. Applying Lemma 5, one can prove $\Phi_0 = \Phi_{\frac{x}{p}}$ and $\Phi_{\frac{1}{32}} = \Phi_{\frac{1}{64}}$.

4. INVARIANCE GROUP

In this section, we will determine the invariance group of the McKay-Thompson series t_χ for each irreducible character χ. We know that the modular function $f = t_\chi$ is invariant by $K_f = \Gamma_0(N_\chi)$. We want to know the full invariance group Γ_χ of f in $SL(2, \mathbb{R})$. Since K_f has only finitely many equivalence classes of cusps, the index $[\Gamma_f : K_f]$ must be finite. We will determine Γ_f by considering all possible images of the cusp 0 of K_f under the action of Γ_f. We will argue in a slightly more general setting. Define

$C_f = $ the set of all cusps of K_f, and,
$C_0 = \{c \in C_f | c$ is a pole of $f\}$.

Lemma 10. *We have $\Gamma_f C_f \subseteq C_f$ and $\Gamma_f C_0 \subseteq C_0$.*

Proof. Since $[\Gamma_f : K_f] < \infty$, C_f is also the set of all cusps of Γ_f. The second statement is obvious. $\square$

Lemma 11. *Let f be a modular function and let Γ_f be its invariance group. Suppose that $K_f \le \Gamma_f$. Let $\alpha = \frac{a_1}{c_1}$ ($a_1 \ne 0$) and $\beta = \frac{a_2}{c_2}$ be two inequivalent cusps of K_f. Let*

$$M_1 = \begin{pmatrix} a_1 & b_1 \\ c_1 & d_1 \end{pmatrix} \in SL(2,\mathbb{Z}) \quad and \quad M_2 = \begin{pmatrix} a_2 & b_2 \\ c_2 & d_2 \end{pmatrix} \in SL(2,\mathbb{Z}).$$

Then $\frac{a_1}{c_1}$ and $\frac{a_2}{c_2}$ are equivalent with respect to Γ_f if and only if the q-expansion of $f|_{M_1}$ is derived from that of $f|_{M_2}$ under the substitution $z \to az + b$ for some numbers a and b (if $c_i = 0$, then $a_i = 1$ and $\frac{a_i}{c_i} = \infty$).

Proof. Let $A \in \Gamma_f$ be such that $A\alpha = \beta$. Define the matrix B such that

$$A = M_2 \begin{pmatrix} 1 & 0 \\ -\frac{c_1}{a_1} & 1 \end{pmatrix} B.$$

Since $A\alpha = \beta$, it follows that $B\alpha = \alpha$. Hence

$$B = \begin{pmatrix} 1 & 0 \\ \frac{c_1}{a_1} & 1 \end{pmatrix} \begin{pmatrix} m_{11} & m_{12} \\ 0 & m_{22} \end{pmatrix} \begin{pmatrix} 1 & 0 \\ -\frac{c_1}{a_1} & 1 \end{pmatrix}$$

for some m_{11}, m_{12} and m_{22}. In particular,

$$A = M_2 \begin{pmatrix} m_{11} & m_{12} \\ 0 & m_{22} \end{pmatrix} \begin{pmatrix} 1 & 0 \\ -\frac{c_1}{a_1} & 1 \end{pmatrix}$$

and

$$A\alpha = M_2 \begin{pmatrix} m_{11} & m_{12} \\ 0 & m_{22} \end{pmatrix} \infty.$$

It follows that $f|_{M_1} = f|_{AM_1} =$

$$f|_{M_2} \begin{pmatrix} 1 & 0 \\ -\frac{c_1}{a_1} & 1 \end{pmatrix} BM_1 = f|_{M_2} \begin{pmatrix} m_{11} & m_{12} \\ 0 & m_{22} \end{pmatrix} \begin{pmatrix} a_1 & b_1 \\ 0 & a' \end{pmatrix} = f|_{M_2} \begin{pmatrix} m'_{11} & m'_{12} \\ 0 & m'_{22} \end{pmatrix},$$

for some b_1, a', m'_{11}, m'_{12} and m'_{22}. Consequently, $f|_{M_1}$ is derived from that of $f|_{M_2}$ under the substitution $z \to az + b$ where $a = \frac{m'_{11}}{m'_{22}}$ and $b = \frac{m'_{12}}{m'_{22}}$. Conversely, one sees easily that α and β are equivalent to each other by

$$M_1 \begin{pmatrix} \frac{1}{a} & -\frac{b}{a} \\ 0 & 1 \end{pmatrix} M_2^{-1} \in \Gamma_f.$$

$\square$

The invariance group Γ_f can now be determined as follows :

(1). Determine C_f, the set of all cusps of K_f.
(2). Determine the subset C_0.
(3). Determine the q-expansion of f at all c_i in C_0 by suitable matrices M_i such that $M_i\infty = c_i$.
(4). Apply Lemma 11 and determine the set $E_0 = \{c \in C_0|$ the q-expansion of f at c is derived from the q-expansion of f at 0 under the substitution $z \to az + b\}$

and the set $A_0 = \{A_c \in \Gamma_f, A_c 0 = c\}$. Note that it is sufficient to determine at most one matrix A_c for each representative of inequivalent cusps.

(5). Determine $(\Gamma_f)_0 = \langle B | B = \begin{pmatrix} 1 & 0 \\ m & 1 \end{pmatrix}, m \in r\mathbb{Z}$, for some (fixed) $r \in \mathbb{Q}\rangle$. Note that this can be achieved by investigating the q-expansion of f at 0. Note also that B is of the form given since Γ_f is discrete.

Remark. (1). The McKay-Thompsom series $t_\chi(z)$ has a pole at 0 for every χ as stated in the remark right after Example 1.

(2). Since Lemma 11 applies only when one of the cusp is nonzero, one can not take c to be 0 in (4) above.

(3). One can replace 0 by any cusp and apply our procedure to find the invariance groups.

Lemma 12. $\Gamma_f = \langle K_f, B, A_c, c \in E_0 \rangle$.

Proof. For any $\sigma \in \Gamma_f \backslash \langle K_f, A_c, c \in E_0 \rangle$. Applying Lemma 10, $\sigma 0$ is again a cusp. Hence $\sigma 0$ must be $\langle K_f, A_c, c \in E_0 \rangle$-equivalent to 0. Choose $\delta \in \langle K_f, A_c, c \in E_0 \rangle$ such that $\delta\sigma 0 = 0$. Then $\delta\sigma \in (\Gamma_f)_0$. Hence $\Gamma_f = \langle K_f, B, A_c, c \in E_0 \rangle$ holds. $\square$

Theorem 13. (Helling's Theorem [6]) *The maximal discrete groups of $PSL(2,\mathbb{C})$ commensurable with the modular group $SL(2,\mathbb{Z})$ are just the images of the conjugates of $\Gamma_0(N)+$ for square free N.*

Corollary 14. *For each irreducible character χ of $\mathbb{M}$, the set of prime divisors of the index $[\Gamma_\chi : \Gamma_0(N_\chi)]$ is a subset of $\{2,3,5,7\}$.*

Proof. By Helling's Theorem, any maximal subgroup that contains Γ_χ is a conjugate of some $\Gamma_0(n)+$. Conway has shown in [2] that n must be a divisor of N_χ. Now compare the volumes of the fundamental domains of $SL_2(\mathbb{Z})$, $\Gamma_0(n)$, and $\Gamma_0(N_\chi)$. Noting that the conjugation does not change the volume and that the normalizer of $\Gamma_0(n)$ changes the volume of the fundamental domain by a factor involving only primes 2 and 3, we obtain our lemma since the index

$$[PSL_2(\mathbb{Z}) : \bar\Gamma_0(N)] = \begin{cases} \frac{N^3}{2} \prod_{p|n}(1 - \frac{1}{p^2}) & \text{if } N > 2, \\ 6 & \text{if } N = 2. \end{cases}$$

involves only primes 2, 3, 5, and 7 if $N = N_\chi$ (N_χ can be found in Table 1.) $\square$

Let χ be an irreducible character of $\mathbb{M}$ and Γ_χ be the invariance group of $t_\chi(z)$. We are now ready to prove :

(1). $A_0 = \emptyset$, and,
(2). $(\Gamma_\chi)_0 = (\Gamma_0(N_\chi))_0$.

Lemma 15. *Let χ be an irreducible character of $\mathbb{M}$ and let c be a cusp of $\Gamma_0(N_\chi)$, not equivalent to 0. Then $A_0 = \{A_c | A_c 0 = c\} = \emptyset$.*

Proof. We first recall that the singular part of $\frac{1}{|\mathbb{M}|} \sum_{g \in \Phi_c} \chi(g)(t_g|P_c)(z)$ at $z = \infty i$ is given by

$$\mathrm{sing}_{P_c} t_\chi = \frac{1}{|\mathbb{M}|} \sum_{g \in \Phi_c} \chi(g) \frac{\sigma_g}{e^{2\pi i U_{c,g} z}}.$$

Applying Lemmas 7, 9 and 10 and Corollary 8, we see that it suffices to show that $\mathrm{sing}_{P_0} t_\chi$ can not be derived from $\mathrm{sing}_{P_c} t_\chi$ under the substitution $z \to az + b$ if $c \neq 0$. This is achieved by a *case-by-case* study. We give an example to indicate how the lemma is proved.

Example 3. Let χ be the trivial character of $\mathbb{M}$. Then $A_0 = \emptyset$.

Proof. We first note that for any irreducible character χ of $\mathbb{M}$ and $c \in \mathbb{Q} \cup \{\infty\}$, we have :

(1). The lowest terms in $\mathrm{sing}_{P_c} t_\chi$ and $\mathrm{sing}_{P_0} t_\chi$ are of the form $\frac{r}{q}$ for some number $r \in \mathbb{Q}$, and

(2). Terms in $\mathrm{sing}_{P_c} t_\chi$ and $\mathrm{sing}_{P_0} t_\chi$ are all of the form $\frac{r}{q^{\frac{1}{t}}}$ (Corollary 8, Lemma 9), for some $r \in \mathbb{Q}$, and $t \in \mathbb{N}$.

Since the lowest term in $\mathrm{sing}_{P_0} t_\chi$ is $\frac{r}{q}$, $r \neq 0$, the transformation that sends $\mathrm{sing}_{P_c} t_\chi$ to $\mathrm{sing}_{P_0} t_\chi$ is of the form $z \to az + b$ where a is some positive integer.

Let χ is the trivial character, then by Corollary 8 and Lemma 9, one has

$$\mathrm{sing}_{P_0} t_\chi = \frac{1}{|\mathbb{M}|} \sum_{g \in \Phi_0} \frac{\sigma_g}{q^{\frac{1}{n_g h_g}}} = \frac{1}{|\mathbb{M}|} \left(\frac{1}{q} + \frac{2|\mathbb{M}|}{|C_{\mathbb{M}}(71A)|} \frac{\sigma_{71A}}{q^{\frac{1}{71}}} + \cdots \right),$$

and for any cusp $c = \frac{x}{y}$, $\gcd(x,y) = 1$,

$$\mathrm{sing}_{P_c} t_\chi = \frac{1}{|\mathbb{M}|} \sum_{g \in \Phi_c} \frac{\sigma_g}{e^{2\pi i U_{c,g} z}}.$$

Suppose $\mathrm{sing}_{P_0} t_\chi$ can be derived from $\mathrm{sing}_{P_c} t_\chi$ under $z \to az + b$.

(3). We will show $\gcd(y, 71) = 1$. Suppose false. Then $71|y$ and $71A, 71B \in \Phi_c$. Since $\frac{1}{q^{\frac{1}{71}}}$ appears in $\mathrm{sing}_{P_0} t_\chi$, there exists, by Lemma 11, some g in Φ_c such that

$$\frac{\sigma_g}{e^{2\pi i U_{c,g} z}} | (z \to az + b) = \frac{r}{q^{\frac{1}{71}}} \qquad (*)$$

where r is some constant. Hence

$$\frac{\gcd(h_g, y)^2 a}{e_g h_g^2} = \frac{1}{71}$$

where

$$e_g = \frac{\frac{n_g}{h_g}}{\gcd\left(\frac{n_g}{h_g}, \frac{y}{\gcd(y, h_g)}\right)}.$$

Since a is an integer, this implies that $g = 71A$ or $71B$. Since $\Gamma_{71A} = \Gamma_{71B} = 71+$, we have $n_{71A} = n_{71B} = 71$, $h_{71A} = h_{71B} = 1$, $e_{71A} = e_{71B} = 71$, and $a = 1$. On the other hand, Corollary 8 implies $t_g|P_c = \sigma_g t_g(z - u_g)$ for $g = 71A$ or $71B$. Hence the transformation $(*)$ can not be done. This forces

$$\gcd(y, 71) = 1.$$

(4). Since $\frac{1}{q^{\frac{1}{t}}}$, $t \in \{29, 41, 59, 92, 93, 94, 95, 104, 110, 119\}$ all appear with nonzero coefficients in $\mathrm{sing}_{P_0} t_\chi$, we may similarly conclude that $\gcd(y, p) = 1$ for the other prime divisors of N_0.

Hence $\gcd(y, N_0) = 1$ and c is equivalent to 0. Consequently, $A_0 = \emptyset$.

Lemma 16. $(\Gamma_\chi)_0 = (\Gamma_0(N_\chi)_0$.

Proof. Suppose not. Applying Corollary 14, we see that $(\Gamma_\chi)_0$ contains $B_r = \begin{pmatrix} 1 & 0 \\ \frac{N_\chi}{r} & 1 \end{pmatrix}$ for $r = 2, 3, 5$ or 7. This implies that the cusp ∞ is equivalent to $\frac{r}{N_\chi}$ in Γ_χ. Therefore $\mathrm{sing}_\infty t_\chi$ must be derived from $\mathrm{sing}_{P_{\frac{r}{N_\chi}}} t_\chi$ under the substitution $z \to az + b$. We can now apply an analogous procedure (using $y = \frac{r}{N_\chi}$) as in Example 3 to get a contradiction. □

Remark. One can also prove Lemma 16 by claiming that B_r does not leave $t_\chi(z)$ invariant. Note that it is easy to show the claim since B_r leaves most of the $t_g(z)$ invariant except for those g's such that $n_g h_g$ is not a divisor of $\frac{N_\chi}{r}$.

Combining Lemma 15 and 16, we have :

Theorem 17. *Let χ be an irreducible character of* $\mathbb{M}$. *Then* $\Gamma_\chi = \Gamma_0(N_\chi)$.

N_χ can be found in *Table* 1.

Remark. (1). In Lemma 15 and 16, 0 is a better choice than the other cusps (∞, for example) since among all the $\mathrm{sing}_{P_c} t_\chi$'s, $\mathrm{sing}_{P_0} t_\chi$ is the one that involves most nonzero terms.

(2). N_χ and its prime decomposition is calculated by a software called GAP.

Table   1

χ_i	N_{χ_i}	N_{χ_i} (prime decomposition)
1	2331309585756753201600	$2^6 3^3 5^2 7.11.13.17.19.23.29.31.41.47.59.71$
2	11841091337275200	$2^6 3^3 5^2 7.11.13.17.19.23.29.31.41$
3	437868837806400	$2^6 3^3 5^2 7.11.13.17.19.23.29.47$
4	467584848090400	$2^5 3^3 5^2 7.11.13.17.19.23.41.71$
5	38732026132800	$2^6 3^3 5^2 7.11.13.17.19.47.59$
6	20350725595200	$2^6 3^3 5^2 7.11.13.17.19.31.47$
7	87358471200	$2^5 3^2 5^2 7.11.13.17.23.31$
8	7820482269600	$2^5 3^2 5^2 7.11.13.17.29.31.71$
9	18526958049600	$2^6 3^2 5^2 7.11.13.23.29.41.47$
10	222987885120	$2^6 3^3 5.7.11.13.19.23.59$
11	8490081600	$2^6 3^2 5^2 7.11.13.19.31$
12	19445025600	$2^6 3^2 5^2 7.11.13.19.71$
13	9958865716800	$2^6 3^3 5^2 7.11.13.17.19.23.31$
14	73513400	$2^5 3^3 5.7.11.13.17$
15	2244077793757800	$2^3 3^3 5^2 7.11.13.17.19.23.29.41.47$
16	3749442460305984	$2^6 3^3 13.23.29.31.41.47.59.71$
17	3749442460305984	$2^6 3^3 13.23.29.31.41.47.59.71$
18	726818400	$2^5 3^3 5^2 7.11.19.23$
19	9182927033280	$2^6 3^3 5.7.11.13.19.29.41.47$
20	35703027360	$2^5 3^2 5.7.11.13.17.31.47$
21	98066928960	$2^6 3^3 5.7.11.13.17.23.29$
22	22789166400	$2^6 3^3 5^2 7.11.13.17.31$
23	451392480	$2^5 3^3 5.7.11.23.59$
24	295495200	$2^5 3^2 5^2 7.11.13.41$
25	253955520	$2^6 3^3 5.7.13.17.19$
26	479256378753600	$2^6 3^2 5^2 7.19.31.41.47.59.71$
27	479256378753600	$2^6 3^2 5^2 7.19.31.41.47.59.71$
28	27003936960	$2^6 3^2 5.7.11.13.17.19.29$
29	81995760	$2^4 3^3 5.7.11.17.29$
30	69618669120	$2^6 3^2 5.7.11.13.19.31.41$
31	21416915520	$2^6 3^2 5.7.11.13.17.19.23$
32	214885440	$2^6 3^3 5.7.11.17.19$
33	2882880	$2^6 3^2 5.7.11.13$
34	332640	$2^5 3^3 5.7.11$
35	786240	$2^6 3^3 5.7.13$
36	11147099040	$2^5 3^3 5.7.11.23.31.47$
37	331962190560	$2^5 3^2 5.7.11.13.17.19.23.31$
38	333637920	$2^5 3^3 5.7.11.17.59$
39	845013600	$2^5 3^3 5^2 19.29.71$
40	845013600	$2^5 3^3 5^2 19.29.71$
41	16676856385200	$2^4 3^3 5^2 11.23.31.47.59.71$
42	16676856385200	$2^4 3^3 5^2 11.23.31.47.59.71$
43	186902100	$2^2 3^3 5^2 7.11.29.31$
44	46955594400	$2^5 3.5^2 11.13.41.47.71$

45	46955594400	$2^5 3.5^2 11.13.41.47.71$
46	54880846020	$2^2 3^2 5.7.11.13.17.19.23.41$
47	105386400	$2^5 3^3 5^2 7.17.41$
48	105386400	$2^5 3^3 5^2 7.17.41$
49	49584815280	$2^4 3^3 5.7.11.13.17.19.71$
50	6404580	$2^2 3^2 5.7.13.17.23$
51	12916800	$2^6 3^3 5^2 13.23$
52	646027200	$2^6 3^2 5^2 7.13.17.29$
53	228731328	$2^6 3^2 7.17.47.71$
54	228731328	$2^6 3^2 7.17.47.71$
55	19044013248	$2^6 3^3 13.23.29.31.41$
56	10944013248	$2^6 3^3 13.23.29.31.41$
57	25077360	$2^4 3.5.7.11.23.59$
58	198918720	$2^6 3^3 5.7.11.13.23$
59	19433872080	$2^4 3^3 5.7.23.29.41.47$
60	19433872080	$2^4 3^3 5.7.23.29.41.47$
61	2784600	$2^3 3^2 5^2 7.13.17$
62	245044800	$2^6 3^2 5^2 7.11.13.17$
63	57266969760	$2^5 3^3 5.7.11.13.17.19.41$
64	157477320	$2^3 3^2 5.7.11.13.19.23$
65	818809200	$2^4 3^2 5^2 11.23.29.31$
66	263877213600	$2^5 3^2 5^2 7.11.13.19.41.47$
67	1588466880	$2^6 3^2 5.7.11.13.19.29$
68	33005280	$2^5 3.5.7.11.19.47$
69	937440	$2^5 3^3 5.7.31$
70	32864832	$2^6 3^3 7.11.13.19$
71	182584514112	$2^6 3.7.11.13.17.29.41.47$
72	182584514112	$2^6 3.7.11.13.17.29.41.47$
73	982080	$2^6 3^2 5.11.31$
74	33542208	$2^6 3^3 7.47.59$
75	33542208	$2^6 3^3 7.47.59$
76	7650720	$2^5 3^3 5.7.11.23$
77	931170240	$2^6 3^2 5.7.11.13.17.19$
78	33754921200	$2^4 3^2 5^2 7.11.13.17.19.29$
79	42325920	$2^5 3^2 5.7.13.17.19$
80	4969440	$2^5 3^2 5.7.17.29$
81	63126554400	$2^5 3^2 5^2 7.13.23.59.71$
82	63126554400	$2^5 3^2 5^2 7.13.23.59.71$
83	208304928	$2^5 3^2 13.23.41.59$
84	208304928	$2^5 3^2 13.23.41.59$
85	704223936	$2^6 3^3 13.23.29.47$
86	704223936	$2^6 3^3 13.23.29.47$
87	1235520	$2^6 3^3 5.11.13$
88	3967200	$2^5 3^2 5^2 19.29$
89	11737440	$2^5 3^3 5.11.13.19$
90	11737440	$2^5 3^3 5.11.13.19$
91	2542811040	$2^5 3^2 5.7.11.17.19.71$
92	22102080	$2^6 3.5.7.11.13.23$
93	1441440	$2^5 3^2 5.7.11.13$
94	879840	$2^5 3^2 5.13.47$

95	16576560	$2^4 3^2 5.7.11.13.23$
96	21677040	$2^4 3^2 5.7.11.17.23$
97	5267201940	$2^2 3^2 5.7.11.13.23.31.41$
98	7900200	$2^3 3^3 5^2 7.11.19$
99	1660401600	$2^6 3.5^2 11.13.41.59$
100	1660401600	$2^6 3.5^2 11.13.41.59$
101	932769600	$2^6 3.5^2 7.17.23.71$
102	7601451872175	$3^3 5^2 7.17.19.29.41.59.71$
103	7601451872175	$3^3 5^2 7.17.19.29.41.59.71$
104	6511680	$2^6 3^2 5.7.17.19$
105	5844589984800	$2^5 3^2 5^2 7.19.31.47.59.71$
106	5844589984800	$2^5 3^2 5^2 7.19.31.47.59.71$
107	2434219200	$2^6 3^2 5^2 7.19.31.41$
108	2434219200	$2^6 3^2 5^2 7.19.31.41$
109	280800	$2^5 3^3 5^2 13$
110	947520	$2^6 3^2 5.7.47$
111	1016747424	$2^5 3^3 7.11.17.29.31$
112	386100	$2^2 3^3 5^2 11.13$
113	6568800	$2^5 3.5^2 7.17.23$
114	1374912	$2^6 3^2 7.11.31$
115	151200	$2^5 3^3 5^2 7$
116	92400	$2^4 3.5^2 7.11$
117	411840	$2^6 3^2 5.11.13$
118	19562400	$2^5 3^2 5^2 11.13.19$
119	12524852340	$2^2 3^3 5.7.11.13.17.29.47$
120	41801760	$2^5 3^2 5.7.11.13.29$
121	75698280	$2^3 3^3 5.7.17.19.31$
122	8148853440	$2^6 3^2 5.7.13.17.31.59$
123	864175548600	$2^3 3^3 5^2 7.13.17.31.47.71$
124	3456702194400	$2^5 3^3 5^2 7.13.17.31.47.71$
125	3456702194400	$2^5 3^3 5^2 7.13.17.31.47.71$
126	119700	$2^2 3^2 5^2 7.19$
127	13695552	$2^6 3^2 13.31.59$
128	752016096	$2^5 3^3 13.23.41.71$
129	752016096	$2^5 3^3 13.23.41.71$
130	19320840	$2^3 3^2 5.7.11.17.41$
131	1004683680	$2^5 3^2 5.7.11.13.17.41$
132	164160	$2^6 3^3 5.19$
133	14379596431200	$2^5 3^3 5^2 7.11.13.17.19.29.71$
134	14208480	$2^5 3^3 5.11.13.23$
135	497653200	$2^4 3^3 5^2 11.59.71$
136	497653200	$2^4 3^3 5^2 11.59.71$
137	995276700	$2^2 3^3 5^2 11.23.31.47$
138	5109369408	$2^6 3^3 11.13.23.29.31$
139	514080	$2^5 3^3 5.7.17$
140	59017104226080	$2^5 3^3 5.7.11.13.17.19.29.31.47$
141	1151710560	$2^5 3^2 5.7.11.13.17.47$
142	3767400	$2^3 3^2 5^2 7.13.23$
143	7600320	$2^6 3^2 5.7.13.29$
144	11823840	$2^5 3^3 5.7.17.23$

145	8558550	$2.3^2 5^2 7.11.13.19$
146	664020	$2^2 3^2 5.7.17.31$
147	4320	$2^5 3^3 5$
148	655188534	$2.3^3 7.11.13.17.23.31$
149	102240	$2^5 3^2 5.71$
150	157248	$2^6 3^3 7.13$
151	26489342880	$2^5 3^2 5.7.11.13.17.23.47$
152	93276960	$2^5 3.5.7.17.23.71$
153	13137600	$2^6 3.5^2 7.17.23$
154	428400	$2^4 3^2 5^2 7.17$
155	18221280	$2^5 3.5.7.11.17.29$
156	190072512	$2^6 3^2 7.17.47.59$
157	21176100	$2^2 3^3 5^2 11.23.31$
158	37130940	$2^2 3^3 5.7.11.19.47$
159	852390	$2.3^3 5.7.11.41$
160	184363200	$2^6 3^2 5^2 7.31.59$
161	108803771818560	$2^6 3^2 5.7.13.17.19.23.29.41.47$
162	1657656	$2^3 3^2 7.11.13.23$
163	345290400	$2^5 3^2 5^2 7.13.17.31$
164	90014400	$2^6 3^2 5^2 7.19.47$
165	30240	$2^5 3^3 5.7$
166	4032	$2^6 3^2 7$
167	4062240	$2^5 3^2 5.7.13.31$
168	3204801600	$2^6 3.5^2 7.11.13.23.29$
169	24196995900	$2^2 3^2 5^2 7.11.17.19.23.47$
170	668304	$2^4 3^3 7.13.17$
171	73920	$2^6 3.5.7.11$
172	6983776800	$2^5 3^3 5^2 7.11.13.17.19$
173	32959080	$2^3 3^2 5.7.11.29.41$
174	3115200	$2^6 3.5^2 11.59$
175	48163383908640	$2^5 3^2 5.7.11.13.17.19.31.47.71$
176	427211200	$2^6 5^2 13.19.23.47$
177	858816	$2^6 3^3 7.71$
178	21416915520	$2^6 3^2 5.7.11.13.17.19.23$
179	14400	$2^6 3^2 5^2$
180	14400	$2^6 3^2 5^2$
181	154881891350	$2.3^3 5^2 11.13.19.29.31.47$
182	2009280	$2^6 3.5.7.13.23$
183	6339168	$2^5 3^3 11.23.29$
184	26429760	$2^6 3^3 5.7.19.23$
185	32730048	$2^6 3^3 13.31.47$
186	7425600	$2^6 3.5^2 7.13.17$
187	8237275200	$2^6 3^2 5^2 7.11.17.19.23$
188	15120	$2^4 3^3 5.7$
189	54774720	$2^6 3^2 5.7.11.13.19$
190	27989280	$2^5 3^3 5.11.19.31$
191	34272	$2^5 3^2 7.17$
192	3500640	$2^5 3^2 5.11.13.17$
193	1049200425	$3^3 5^2 7.13.19.29.31$
194	1404480	$2^6 3.5.7.11.19$

References

1. R. E. Borcherds, *Monstrous Moonshine and Monstrous Lie Superalgebras,* Invent. Math. **109** (1992), 405-444.
2. J. H. Conway, *Understanding Groups like* $\Gamma_0(N)$, (preprint).
3. J. H. Conway and S. P. Norton, *Monstrous Moonshine,* Bull. London Math. Soc. **11** (1979), 308-338.
4. I. Frenkel, J. Lepowsky and A. Meurman, *Vertex Operator Algebras and the Monster*, Academic Press, Inc. (1988).
5. K. Harada, *Modular Functions, Modular Forms and Finite Group*, Lecture Notes at The Ohio State University, (1987).
6. H. Helling, *On the Commensurablility Classes of Rational Modular Group*, J. London Math. Soc. **2** (1970), 67-72.
7. J. McKay and H. Strauss, *The q-series of Monstrous Moonshine and the Decomposition of the Head Characters*, Comm. Alg. **18(1)** (1990), 253-278.

DEPARTMENT OF MATHEMATICS, THE OHIO STATE UNIVERSITY, COLUMBUS, OHIO 43210, USA
E-mail address : haradako@math.ohio-state.edu

DEPARTMENT OF MATHEMATICS, NITIONAL UNIVERSITY OF SINGAPORE, SINGA-PORE 0511, REPUBLIC OF SINGAPORE
E-mail address : matlml@leonis.nus.sg

Contemporary Mathematics
Volume **193**, 1996

Some quilts for the Mathieu groups

TIM HSU

May 27, 1995

ABSTRACT. Quilts and T-systems were introduced in Conway and Hsu [**2**], and extensively developed, along with the related concept of monodromy systems, in Hsu [**3**]. T-systems are orbits of a certain action of $\mathbf{B}_3$ (the 3-string braid group) on pairs of elements of a group; monodromy systems are orbits of a closely related action of $\mathbf{B}_3$ on triples of involutions of a group; and quilts are diagrams which classify transitive permutation representations, and therefore orbit structures, of $\mathbf{B}_3$. Examples of quilts of T-systems/monodromy systems for the Mathieu simple groups are presented, and some associated presentations are examined.

1. Braids and monodromy

The 3-string braid group $\mathbf{B}_3$ is given by either of the following presentations:

$$(1) \qquad \langle L,\, R \,|\, LR^{-1}L = R^{-1}LR^{-1} \rangle$$

$$(2) \qquad \langle V,\, E \,|\, V^3 = E^2 \rangle$$

where the braids L and R are shown in Figure 1, $V = R^{-1}L$, and $E = LR^{-1}L$.

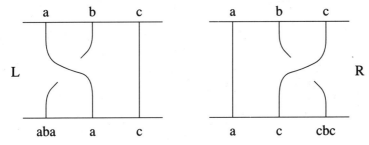

FIGURE 1. The braids L and R

Furthermore, if

$$(3) \qquad Z = E^2 = (LR^{-1}L)^2 = V^3 = (R^{-1}L)^3,$$

then $\mathbf{B}_3/\langle Z \rangle \cong \mathbf{PSL}_2(\mathbf{Z})$, with L, resp. R, being sent to $\begin{pmatrix} 1 & 1 \\ 0 & 1 \end{pmatrix}$, resp. $\begin{pmatrix} 1 & 0 \\ 1 & 1 \end{pmatrix}$.

1991 *Mathematics Subject Classification*. Primary 20D08, 20D60; Secondary 20F05, 20F36, 20H05.

The author was supported in part by an NSF graduate fellowship and DOE GAANN grant #P200A10022.A03.

Let τ be the set of triples of involutions in a group G. The rules

$$
\begin{aligned}
(a,b,c)L &\mapsto (aba,a,c) \quad \text{and} \\
(a,b,c)R &\mapsto (a,c,cbc)
\end{aligned}
$$

(4)

define an action of $\mathbf{B}_3$ on τ, as can be verified using presentation (1). We call the action defined by (4) the *monodromy action on triples*. Figure 1 serves as a visual mnemonic for the monodromy action, in that the involution on the undercrossing string is conjugated by the involution on the overcrossing string. As we shall see in Section 3, if these involutions are 6-transpositions, this action will often have a certain "genus 0" property, and so Norton [6] proposed examining this action as a possible explanation for Moonshine. (See also Mason [4].) We call an orbit of the monodromy action a *monodromy system*.

We can reduce the monodromy action to pairs of elements of G in the following way. For a triple (a,b,c), let $\alpha = ab$ and $\beta = bc$. As the reader may verify, the rules in (4) induce

$$
\begin{aligned}
(\alpha,\beta)\,L^i &\mapsto (\alpha,\alpha^i\beta) \quad \text{and} \\
(\alpha,\beta)\,R^i &\mapsto (\alpha\beta^i,\beta).
\end{aligned}
$$

(5)

These new rules define an action of $\mathbf{B}_3$ on pairs of elements of G. We call this action the *reduced monodromy action*. We call an orbit of the reduced action a *T-system*, and we note that any monodromy system has an associated T-system. As it turns out, since $\mathrm{stab}(a,b,c) = \mathrm{stab}(ab,bc)$ (Theorem 8.2.5 of Hsu [3]), we lose nothing by considering only the reduced action.

Note that if we fix a T-system T, it follows from (5) that for any pairs (α,β) and (γ,δ) in T, $\langle \alpha,\beta \rangle = \langle \gamma,\delta \rangle$. Therefore, when studying a single T-system T, we may as well assume that for any pair $(\alpha,\beta) \in T$, $G = \langle \alpha,\beta \rangle$. When this assumption holds for T, we say that T is a T-system *for* G (as opposed to a T-system with elements in G.)

REMARK 1.1. Note that up to conjugacy, the monodromy action just permutes a, b, and c. In particular, if the involutions in a monodromy system are all in a given conjugacy class C, then for any pair (α,β) in the associated T-system, α and β are both the product of two elements in C.

REMARK 1.2. As the reader may verify, $(a,b,c)Z = (a^{cba}, b^{cba}, c^{cba})$. In other words, the action of Z conjugates a triple as a whole, and so up to conjugacy, considering either the monodromy or reduced monodromy action, Z acts trivially.

2. Quilts

We define some terms arising from permutation representations of $\mathbf{B}_3$. Let Ω be a finite set on which $\mathbf{B}_3$ acts transitively. A Z-orbit of Ω is an orbit of the subgroup $\langle Z \rangle$ of $\mathbf{B}_3$. It follows from transitivity and the centrality of Z in $\mathbf{B}_3$ that each Z-orbit of Ω has the same size, say M. We call M the *modulus* of Ω. Note that there is a natural action of $\mathbf{PSL}_2(\mathbf{Z})$ on the Z-orbits of Ω. The structure of this action as a permutation representation of $\mathbf{PSL}_2(\mathbf{Z})$ is called the *modular structure* of Ω.

A *modular quilt* is a diagram, drawn on a connected oriented surface, which represents a transitive permutation representation of $\mathbf{PSL}_2(\mathbf{Z})$. A modular quilt is made up of triangular "puzzle pieces" called *seams* (Figure 2). The vertical line at the top of the seam in Figure 2 is called a *hash mark*, and the object at the bottom of the seam is called a *dot*.

FIGURE 2. A seam

Seams are assembled in a modular quilt by the following rules. At each hash mark of Q, either 2 or 1 seams meet in what is known as an *edge* (Figure 3(a)) or a *collapsed edge* (Figure 3(b)), respectively; and at each dot of Q, either 3 or 1 seams meet in a *vertex* (Figure 3(c)) or a *collapsed vertex* (Figure 3(d)), respectively.

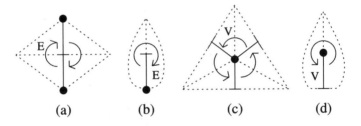

(a)   (b)   (c)   (d)

FIGURE 3. Edges and vertices

Each modular quilt Q then determines a transitive permutation representation of $\mathbf{PSL}_2(\mathbf{Z}) \cong \langle E, V \mid 1 = E^2 = V^3 \rangle$ on the seams of Q, since the rules indicated by the arrows in Figure 3 assign a permutation of order dividing 2 to E and a permutation of order dividing 3 to V. It is not hard to see that any transitive permutation representation of $\mathbf{PSL}_2(\mathbf{Z})$ has an associated modular quilt, and it follows that modular quilts are equivalent to transitive permutation representations of $\mathbf{PSL}_2(\mathbf{Z})$.

Note that the elements $L = EV^{-1}$ and $R = EV^{-2}$ act as shown in Figure 4. Now, if we draw a modular quilt with the dotted lines omitted, the remaining parts of the diagram divide the surface up into open areas called *patches*. From Figure 4, we see that patches correspond 1-to-1 with $\langle L \rangle$-orbits, and that the size of an $\langle L \rangle$-orbit is the number of sides (the *size*) of its associated patch. (See Figures 5–11 for examples of what patches look like. The number on each patch is its size.)

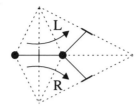

FIGURE 4. Action of L and R

REMARK 2.1. If Q is a finite modular quilt, and $\Gamma \subseteq \mathbf{PSL}_2(\mathbf{Z})$ is a point stabilizer of the transitive permutation representation induced by Q, then Q is a picture of the quotient of the complex upper half-plane by Γ, with patches corresponding to cusps, and collapsed edges and vertices corresponding to elliptic fixed points. See Conway and Hsu [2] or section 2.6 of Hsu [3] for details.

Having defined modular quilts, we can now define *quilts* to be certain suitably annotated modular quilts. Let Ω be a finite set on which $\mathbf{B}_3$ acts transitively, and let M be the modulus of Ω. From the above discussion, we know that the modular structure of Ω is represented by some modular quilt Q, with the seams of Q corresponding to the Z-orbits of Ω. It turns out that if Q is suitably decorated with integers mod M, then the full structure of Ω can be recovered from this decorated version of Q. In analogy with our discussion of modular quilts, it can then be shown that quilts are equivalent to transitive permutation representations of $\mathbf{B}_3$ (Theorem 2.3.13 of Hsu [3]).

In particular, quilts can be used to classify T-systems, since structurally, T-systems are just particular examples of transitive permutation representations of $\mathbf{B}_3$. Moreover, quilts are designed so that many of their geometric features correspond to interesting features of T-systems. For example, let Q be the quilt of a T-system T. Up to conjugacy (see Remark 1.2), the seams of Q correspond to pairs in T, and the patches of Q correspond to elements of G which are part of a pair in T. To elaborate on the latter relation, suppose a patch P corresponds to an element α, again up to conjugacy. It turns out that the size of P always divides the order of α, and is often equal to the order of α. For later use, we say that P is a *collapsed patch* if the size of P is less than the order of α; otherwise, we say that P is an *uncollapsed patch*.

REMARK 2.2. Our repeated caveat of "up to conjugacy" is used to avoid discussing some tricky issues in the rigorous definition of quilts. For versions of the above material without this caveat, see Conway and Hsu [2] or chapter 3 of Hsu [3].

Since the structure of T-systems is completely captured by quilts, much information about T-systems can be derived from the geometry of quilts. For instance, the following results (Theorems 5.1.4 and 5.1.6 of Hsu [3]) are especially relevant to this paper.

THEOREM 2.3. *Let Q be a quilt with s seams and modulus M, and suppose Q*

has no collapsed patches. Then M divides s. Furthermore, if Q has no collapsed vertices, M divides $s/2$, and if Q has no collapsed edges, M divides $s/3$.

THEOREM 2.4. *A quilt of genus 0 which has no collapsed patches is determined by its modular structure and its modulus.*

CONVENTIONS. Since all of the quilts in this paper happen to have genus 0 and no collapsed patches, each of these quilts is determined by its modular structure and its modulus, and so this is the only information we describe. We draw a quilt of genus 0 by omitting a "point at infinity" and treating the diagram as if it were planar. We also omit drawing dotted lines. The symbol

$$(6) \qquad\qquad M \cdot S(a^i, b^j, c^k, \dots)$$

denotes a quilt of modulus M, S seams, genus 0, and i patches of size a, j patches of size b, k patches of size $c \dots$. We sometimes abbreviate (6) to $M \cdot S$. Note that the permutation representation of $\mathbf{B}_3$ associated with $M \cdot S$ has order MS, and the induced modular representation has order S.

3. 6-transposition T-systems

Let G be a group which has a 6-transposition class C (e.g., ATLAS [1] class $2A$ in the Monster). If a, b, and c are in C, then the quilt of the T-system T obtained from (a, b, c) is likely to have genus 0, as we see from the following argument of Norton [6]. First of all, since the corresponding elements ab, bc, ac, etc., have order ≤ 6 (Remark 1.1), the quilt of T has all patch sizes ≤ 6. A trivalent polyhedron whose faces have at most 6 sides has genus either 0 or 1, with the latter occurring only when all of the faces have 6 sides. It is therefore reasonable to expect the quilt of a 6-transposition T-system to have genus 0, and in fact, all of the ones in this paper do. (There are, however, several known examples of 6-transposition T-systems with quilts of genus 1. For instance, Norton [5] has discovered a series of three such T-systems coming from class $2A$ in the Monster, with triples of involutions generating $U_3(5)$, the Held group, and the Monster itself.)

For the Mathieu simple groups M_{11}, M_{12}, $M_{21} \cong L_3(4)$, M_{22}, M_{23}, and M_{24}, ATLAS class $2A$ is a class of 6-transpositions, and class $2B$ of M_{12} is as well. All of the 6-transposition T-systems for the Mathieu groups have been found; Table 1 gives the number of such systems, up to isomorphism, arising from each 6-transposition class in a Mathieu group. (Class $2A$ is assumed, unless otherwise indicated.) For brevity, we show the quilts of only two T-systems coming from each 6-transposition class of a Mathieu group (Figures 5–11). Many other examples of quilts of 6-transposition T-systems for finite simple groups can be found in Chapter 9 of Hsu [3].

REMARK 3.1. We note that the $11 \cdot 22$ quilt for M_{12} is a very special case here, as it turns out to be the quilt of a T-system which can arise from either class $2A$ or class $2B$. That is, even though $2A$ and $2B$ are not equivalent by

118   TIM HSU

Class	M_{11}	M_{12} (2A)	M_{12} (2B)	M_{21}	M_{22}	M_{23}	M_{24}
T-systems	2	4	3	3	9	6	3

TABLE 1. Number of 6-transposition T-systems for the Mathieu groups

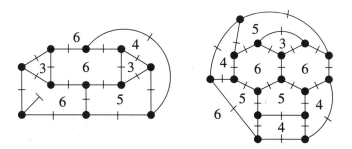

FIGURE 5. M_{11}: $11 \cdot 33(3^2, 4, 5, 6^3)$ and $8 \cdot 48(3, 4^3, 5^3, 6^3)$

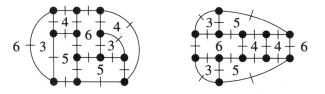

FIGURE 6. M_{12} (2A): $6 \cdot 36(3^2, 4^2, 5^2, 6^2)A$ and $6 \cdot 36(3^2, 4^2, 5^2, 6^2)B$

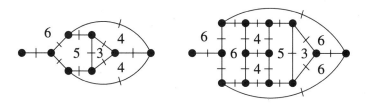

FIGURE 7. M_{12} (2B): $11 \cdot 22(3, 4^2, 5, 6)$ and $10 \cdot 40(3, 4^2, 5, 6^4)$

FIGURE 8. M_{21}: $7 \cdot 21(3, 4^2, 5^2)$ and $5 \cdot 30(3, 4^3, 5^3)$

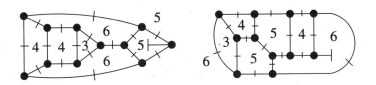

FIGURE 9. M_{22}: $11 \cdot 33(3, 4^2, 5^2, 6^2)A$ and $11 \cdot 33(3, 4^2, 5^2, 6^2)B$

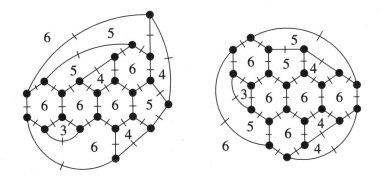

FIGURE 10. M_{23}: $11 \cdot 66(3, 4^3, 5^3, 6^6)A$ and $11 \cdot 66(3, 4^3, 5^3, 6^6)B$

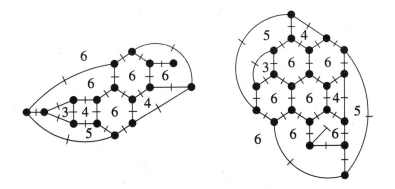

FIGURE 11. M_{24}: $23 \cdot 46(3, 4^2, 5, 6^5)$ and $21 \cdot 63(3, 4^2, 5^2, 6^7)$

any automorphism of M_{12}, they both share one particular manner of generating M_{12}. See Section 4 for one result of this coincidence.

4. Presentations from quilts

When there is a T-system T for a group G, such that T has quilt Q, by abuse of terminology, we say that Q is a quilt for G. As it turns out, for Q to be a quilt for G, it is necessary for G to satisfy the relations of a certain presentation arising naturally from Q. The group $G(Q)$ given by this presentation is called the *group of* Q. We can think of $G(Q)$ as the largest group for which Q is a quilt. Table 2 lists the groups of the quilts in Figures 5–11. In each case, $G(Q)$ is generated by α and β, with defining relations as shown. (The symbol $\rightleftharpoons$ means "commutes with.")

Now, let G be a group generated by α and β, and let Q be the quilt of the T-system of (α, β). Since Q is a quilt for G, G is a homomorphic image of $G(Q)$, and so one might hope that for interesting G, such as simple G or perfect G, $G(Q)$ is (almost) equal to G. Specifically, we can ask the following about our examples from the Mathieu groups:

QUESTION 4.1. *In each of the rows of Table 2, how close is $G(Q_i)$ to G?*

In general, $G(Q)$ will sometimes be close to G. For instance, the following has been computed with GAP [7]:

$$(7) \qquad\qquad G(Q_1) \cong 3^6 : M_{11}$$
$$(8) \qquad\qquad G(Q_3) \cong G(Q_4) \cong M_{12}$$
$$(9) \qquad\qquad G(Q_6) \cong 2.M_{12}$$
$$(10) \qquad\qquad G(Q_8) \cong (2^2 \times 3).M_{21}$$

$G(Q_1)$ is the split extension of the ternary Golay 12-code by M_{11} (acting transitively on 12 letters), and $G(Q_6)$ and $G(Q_8)$ are the indicated Schur central extensions.

Unfortunately, it appears that in most cases, even if G is simple, $G(Q)$ is much bigger than G. For instance, it is easy to show that $M_{12} \times M_{12}$ is a homomorphic image of $G(Q_5)$ (Conway and Hsu [2]), with projection on one factor arising from a class $2A$ monodromy system, and projection on the other arising from a class $2B$ monodromy system. In fact, it seems probable that $G(Q_5)$ is actually infinite. See Chapter 10 of Hsu [3] for other results in the same vein.

Q_i	Name	G	Defining relations for $G(Q_i)$
Q_1	$11 \cdot 33$	M_{11}	$1 = \alpha^6 = \beta^5 = (\alpha\beta)^6 = (\alpha^2\beta)^3 = (\alpha^{-1}\beta)^6 = (\alpha\beta^2)^3$ $= (\alpha\beta^3)^4$
Q_2	$8 \cdot 48$	M_{11}	$1 = \alpha^6 = \beta^6 = (\alpha\beta)^5 = (\alpha^2\beta)^3 = (\alpha^3\beta)^6 = (\alpha^4\beta)^5$ $= (\alpha^{-1}\beta)^4 = (\alpha\beta^2)^4 = (\alpha\beta^3)^5 = (\alpha\beta^4)^4$
Q_3	$6 \cdot 36A$	M_{12}	$1 = \alpha^5 = \beta^5 = (\alpha\beta)^6 = (\alpha^2\beta)^4 = (\alpha^3\beta)^3 = (\alpha^{-1}\beta)^6$ $= (\alpha\beta^2)^3 = (\alpha\beta^3)^4$
Q_4	$6 \cdot 36B$	M_{12}	$1 = \alpha^6 = \beta^4 = (\alpha\beta)^5 = (\alpha^2\beta)^3 = (\alpha^3\beta)^6 = (\alpha^4\beta)^3$ $= (\alpha^{-1}\beta)^5 = (\alpha\beta^2)^4$
Q_5	$11 \cdot 22$	M_{12}	$1 = \alpha^4 = \beta^4 = (\alpha\beta)^3 = (\alpha^2\beta)^5 = (\alpha^{-1}\beta)^6$
Q_6	$10 \cdot 40$	M_{12}	$1 = \alpha^6 = \beta^6 = (\alpha\beta)^3 = (\alpha^2\beta)^5 = (\alpha^3\beta)^4 = (\alpha^4\beta)^6$ $= (\alpha^{-1}\beta)^6 = (\alpha\beta^3)^4; \qquad \alpha, \beta \rightleftharpoons (\beta^{-1}\alpha\beta\alpha^{-1})^5$
Q_7	$7 \cdot 21$	M_{21}	$1 = \alpha^4 = \beta^4 = (\alpha\beta)^3 = (\alpha^2\beta)^5 = (\alpha^{-1}\beta)^5$
Q_8	$5 \cdot 30$	M_{21}	$1 = \alpha^5 = \beta^5 = (\alpha\beta)^3 = (\alpha^2\beta)^5 = (\alpha^3\beta)^4 = (\alpha^{-1}\beta)^4$ $= (\alpha\beta^3)^4$
Q_9	$11 \cdot 33A$	M_{22}	$1 = \alpha^6 = \beta^6 = (\alpha\beta)^5 = (\alpha^2\beta)^5 = (\alpha^3\beta)^4 = (\alpha^4\beta)^4$ $= (\alpha^{-1}\beta)^3$
Q_{10}	$11 \cdot 33B$	M_{22}	$1 = \alpha^5 = \beta^5 = (\alpha\beta)^4 = (\alpha^2\beta)^3 = (\alpha^{-1}\beta)^6 = (\alpha\beta^2)^6$ $= (\alpha\beta^3)^4$
Q_{11}	$11 \cdot 66A$	M_{23}	$1 = \alpha^6 = \beta^6 = (\alpha\beta)^4 = (\alpha^2\beta)^5 = (\alpha^3\beta)^6 = (\alpha^4\beta)^3$ $= (\alpha^{-1}\beta)^6 = (\alpha\beta^2)^6 = (\alpha\beta^3)^5 = (\alpha\beta^4)^4$ $= (\alpha^3\beta\alpha^2\beta)^5 = (\alpha\beta^2\alpha\beta^3)^4 = (\alpha\beta^3\alpha\beta^4)^6$
Q_{12}	$11 \cdot 66B$	M_{23}	$1 = \alpha^6 = \beta^6 = (\alpha\beta)^5 = (\alpha^2\beta)^6 = (\alpha^3\beta)^3 = (\alpha^4\beta)^5$ $= (\alpha^{-1}\beta)^6 = (\alpha\beta^2)^4 = (\alpha\beta^3)^6 = (\alpha\beta^4)^4$ $= (\alpha^2\beta\alpha\beta)^5 = (\alpha^{-1}\beta\alpha^4\beta)^6 = (\alpha\beta^3\alpha\beta^4)^4$
Q_{13}	$23 \cdot 46$	M_{24}	$1 = \alpha^5 = \beta^6 = (\alpha\beta)^6 = (\alpha^2\beta)^4 = (\alpha^3\beta)^3 = (\alpha^{-1}\beta)^6$ $= (\alpha\beta^2)^4 = (\alpha\beta^3)^6 = (\alpha\beta^4)^6$
Q_{14}	$21 \cdot 63$	M_{24}	$1 = \alpha^6 = \beta^6 = (\alpha\beta)^4 = (\alpha^2\beta)^5 = (\alpha^3\beta)^3 = (\alpha^4\beta)^6$ $= (\alpha^{-1}\beta)^6 = (\alpha\beta^2)^6 = (\alpha\beta^3)^5 = (\alpha\beta^4)^4$ $= (\alpha^{-1}\beta\alpha^4\beta)^6 = (\alpha\beta^3\alpha\beta^4)^6$

TABLE 2. Quilt groups for Mathieu examples

References

1. J. H. Conway, R. T. Curtis, S. P. Norton, R. A. Parker, and R. A. Wilson, *ATLAS of finite groups*, Oxford Univ. Press, 1985.
2. J. H. Conway and T. M. Hsu, *Quilts and T-systems*, to appear in J. Alg., 1995.
3. T. M. Hsu, *Quilts, T-systems, and the combinatorics of Fuchsian groups*, Ph.D. thesis, Princeton Univ., 1994.

4. G. Mason, *Remarks on moonshine and orbifolds*, Groups, Combinatorics, and Geometry (M. W. Liebeck and J. Saxl, eds.), Cambridge Univ. Press, 1992, pp. 108–120.
5. S. P. Norton, personal communication.
6. _____, *Generalized moonshine*, Proc. Sympos. Pure Math., vol. 47, AMS, 1987, pp. 208–209.
7. M. Schönert et al., *GAP: Groups, algorithms and programming*, Lehrstuhl D für Mathematik, RWTH Aachen, April 1992, Version 3.1.

DEPARTMENT OF MATHEMATICS, PRINCETON UNIVERSITY, PRINCETON, NJ 08544
E-mail address: timhsu@math.princeton.edu

Contemporary Mathematics
Volume **193**, 1996

A nonmeromorphic extension of the moonshine module vertex operator algebra

YI-ZHI HUANG

ABSTRACT. We describe a natural structure of an abelian intertwining algebra (in the sense of Dong and Lepowsky) on the direct sum of the untwisted vertex operator algebra constructed from the Leech lattice and its (unique) irreducible twisted module. When restricting ourselves to the moonshine module, we obtain a new and conceptual proof that the moonshine module has a natural structure of a vertex operator algebra. This abelian intertwining algebra also contains an irreducible twisted module for the moonshine module with respect to the obvious involution. In addition, it contains a vertex operator superalgebra and a twisted module for this vertex operator superalgebra with respect to the involution which is the identity on the even subspace and is -1 on the odd subspace. It also gives the superconformal structures observed by Dixon, Ginsparg and Harvey.

The relation between the modular function $J(q) = j(q) - 744$ and dimensions of certain representations of the Monster was first noticed by McKay and Thompson (see [**Th**]). Based on these observations, McKay and Thompson conjectured the existence of a natural ($\mathbb{Z}$-graded) infinite-dimensional representation of the Monster group such that its graded dimension as a $\mathbb{Z}$-graded vector space is equal to $J(q)$. In the famous paper [**CN**] by Conway and Norton, remarkable numerology between McKay-Thompson series (graded traces of elements of the Monster acting on the conjectured infinite-dimensional representation of the Monster) and modular functions was collected, and surprising conjectures about those modular functions occured were presented. The Monster was constructed by Griess [**G**] later but the mysterious connection between the Monster and modular functions was still not expained. In [**FLM1**], Frenkel, Lepowsky and Meurman constructed a natural infinite-dimensional representation of the Monster (called the moonshine module and denoted $V^\natural$) using the method of vertex

1991 *Mathematics Subject Classification.* Primary 17B69; Secondary 17B68, 81T40.

This research is supported in part by NSF grant DMS-9301020 and by DIMACS, an NSF Science and Technology Center funded under contract STC-88-09648.

operators. The moonshine module constructed by them provided a remarkable conceptual framework towards the understanding of monstrous moonshine. In particular, some of the numerology and conjectures in [**CN**] were expained and proved in [**FLM1**]. Motivated partly by [**FLM1**], Borcherds [**B1**] developed a general theory of vertex operators based on an even lattice. From this general theory, he axiomized the notion of "vertex algebra" and using the results announced in [**FLM1**], he stated that the moonshine module $V^\natural$ has a structure of such an algebra. In [**FLM2**], Frenkel, Lepowsky and Meurman proved that the moonshine module $V^\natural$ has a natural structure of vertex operator algebra and the Monster is the automorphism group of this vertex operator algebra. Their proof is very involved and uses triality and some results in group theory. On the other hand, $V^\natural$ can be viewed as a substructure of a $\mathbb{Z}_2$-orbifold conformal field theory. Using techniques developed in string theory, Dolan, Goddard and Montague [**DGM1**] gave another proof that $V^\natural$ has a natural structure of vertex operator algebra. Their proof works for a class of $\mathbb{Z}_2$-orbifold theories and thus allows them to give a further interpretation of Frenkel-Lepowsky-Meurman's triality [**DGM2**] [**DGM3**] (see also [**L**]). But their proof is still very technical and complicated.

Using the moonshine module $V^\natural$ constructed by Frenkel, Lepowsky and Meurman, the no-ghost theorem in string theory and the theory of generalized Kac-Moody algebras (or Borcherds algebras), Borcherds [**B2**] completed the proof of the monstrous moonshine conjecture in [**CN**]. Part of Borcherds' proof has been simplified recently by Jurisich [**J**] and by Jurisich, Lepowsky and Wilson [**JLW**]. But the last step of Borcherds' proof uses some case by case identification which is conceptually unsatisfactory. Also it seems that the methods used in the proof cannot be used to prove the generalized moonshine conjecture [**N**]. In [**Tu1**], [**Tu2**] and [**Tu3**], Tuite showed that the monstrous moonshine conjecture, and the generalized moonshine conjecture in some special cases, can be understood by using Frenkel-Lepowsky-Meurman's uniqueness conjecture on $V^\natural$ and some conjectures in the mathematically yet-to-be-established orbifold conformal field theory. In particular, he pointed out that nonmeromorphic operator algebras for orbifold theories of the Leech lattice theory and the moonshine module, which are the foundation of his idea, are still to be constructed, even in the original simplest $\mathbb{Z}_2$-orbifold case. The importance of these nonmeromorphic operator algebras is that they give the whole genus-zero chiral parts of the orbifold conformal theories. Also the automorphism groups of these nonmeromorphic operator algebras might be of interest.

In this paper, we describe a construction of the nonmeromorphic extension of $V^\natural$ in the original simplest $\mathbb{Z}_2$-orbifold case. The main tools which we use are the decompositions of the untwisted vertex operator algebra associated to the Leech lattice and its irreducible twisted module into direct sums of irreducible modules for a tensor product of the Virasoro vertex operator algebra with central charge $\frac{1}{2}$ obtained by Dong, Mason and Zhu ([**DMZ**], [**D3**]), and the theory of tensor products of modules for a vertex operator algebra developed by Lepowsky and the author ([**HL1**]–[**HL5**], [**H2**]). Precisely speaking, we describe a natural structure of an abelian intertwining algebra (in the sense of Dong and Lepowsky [**DL**]) on the direct sum of the untwisted vertex operator algebra constructed from the Leech lattice and its (unique) irreducible twisted module with respect to an involution induced from a reflection of the Leech lattice (see Theorem 3.8). Our construction and proof are conceptual, that is, every step is natural in the theory of vertex operator algebras. In particular, when restricting ourselves to the moonshine module, we obtain a new and conceptual proof that the moonshine module has a natural structure of vertex operator algebra. This abelian intertwining algebra also contains a (unique) irreducible twisted module (which has also been obtained by Dong and Mason using a different method) for the moonshine module with respect to the obvious involution, a vertex operator superalgebra and a twisted module for this vertex operator superalgebra with respect to the involution which is the identity on the even subspace and is -1 on the odd subspace. This abelian intertwining algebra also gives the superconformal structures observed by Dixon, Ginsparg and Harvey [**DGH**]. We define a superconformal vertex operator algebra to be an abelian intertwining algebra of a certain type equipped with an element which together with the Virasoro element generates a super-Virasoro algebra. Then the abelian intertwining algebra structure above together with any one of the superconformal structures of Dixon, Ginsparg and Harvey is a superconformal vertex operator algebra.

Note that the tensor product theory can be applied to modules for any rational vertex operator algebra satisfying certain conditions described in [**H2**] and [**HL4**] (see also Subsection 2.2 below). In the present paper, the results in [**DMZ**] and [**D3**] are used to show that the tensor product theory can be used and to calculate the fusion rules. Thus for other orbifold theories and conformal field theories, if the conditions to use the tensor product theory are satisfied and the fusion rules can be calculated, we can also construct the nonmeromorphic operator algebra in the same way. In particular, it might be possible to prove the result of [**DM**] using the tensor product theory.

Since the present paper uses almost all basic concepts and results in the algebraic theory of vertex operator algebras, it is impossible to give a complete exposition to those basic materials which we need. Therefore we assume that the reader is familiar with the basic notions and results in the theory of vertex operator algebras. For details, see [**FLM1**] and [**FHL**]. For more material on the axiomatic aspects of the theory of vertex operator algebras, we shall refer the reader to the appropriate references. We shall, however, review briefly the results on the moonshine module and other related results which we need. We shall also give a brief account of the part which we need of the tensor product theory of modules for a vertex operator algebra.

This paper is organized as follows: Section 1 is a review of the constructions of and results on the moonshine module and related structures. Section 2 is a review of a certain part of the tensor product theory for modules for a vertex operator algebra. The construction of the nonmeromorphic extension is sketched in Section 3. Some consequences of this nonmeromorphic extension, including the vertex operator algebra structure on the moonshine module, the twisted module for the moonshine module, a vertex operator superalgebra and its twisted module in this extension, and the superconformal structures of Dixon, Ginsparg and Harvey, are described in Section 4.

The details of the proofs of the results described in this paper will be published elsewhere.

Acknowledgement. I would like to thank Chongying Dong for discussions and Jim Lepowsky for discussions, comments and especially explanations of results in [**FLM1**] and [**FLM2**]. I would also like to thank Wanglai Li for pointing out a mistake in the description of the nonmeromorphic extension in an earlier version of this paper.

1. Leech lattice theory and the moonshine module

In this section we review briefly the constructions of the untwisted vertex operator algebra, associated to the Leech lattice, its irreducible twisted module and the moonshine module $V^\natural$. We also review some results on these algebras and modules and related structures. For more details, the reader is referred to [**FLM2**], [**DMZ**], [**D1**], [**D2**], [**D3**].

1.1. The Golay code $\mathcal{C}$. Let $\Omega = \{S_1, \ldots, S_n\}$ be a finite set. A *(binary linear) code* on Ω is a subspace of the vector space $\mathcal{P}(\Omega) = \coprod_{i=1}^n \mathbb{Z}_2 S_i$ over $\mathbb{Z}_2$ spanned by elements of Ω. Any element S of $\mathcal{P}(\Omega)$ is a linear combination of $S_1, \ldots, S_n$. The number of nonzero coefficients is called the *weight* of S and is denoted as $|S|$. A code $\mathcal{S}$ is said to be of *type* II if $n \in 4\mathbb{Z}$, $|S| \in 4\mathbb{Z}$ for all $S \in \mathcal{S}$ and $S_1 + \cdots + S_n \in \mathcal{S}$. The usual dot product for a vector space with a basis gives a natural nonsingular symmetric bilinear form on $\mathcal{P}(\Omega)$. The annihilator of a code $\mathcal{S}$ in $\mathcal{P}(\Omega)$ with respect to this bilinear form is again a code. It is called the *dual code* of $\mathcal{S}$ and is denoted $\mathcal{S}^\circ$. A code is called *self-dual* if it is equal to its dual code.

THEOREM 1.1. *There is a self-dual code of type II on a 24-element set such that it has no elements of weight 4. It is unique up to isomorphism.*

The code in this theorem is called the *Golay code* and is denoted $\mathcal{C}$.

1.2. The Leech lattice Λ. A *(rational) lattice of rank* $n \in \mathbb{N}$ is a rank n free abelian group L equipped with a rational-valued symmetric $\mathbb{Z}$-bilinear form $\langle \cdot, \cdot \rangle$. A lattice is *nondegenerate* if its form is nondegenerate.

Let L be a lattice. For $m \in \mathbb{Q}$, we set $L_m = \{\alpha \in L \mid \langle \alpha, \alpha \rangle = m\}$. The lattice L is said to be *even* if $L_m = 0$ for any $m \in \mathbb{Q}$ which is not an even integer. The lattice L is said to be *integral* if the form is integral-valued and to be *positive definite* if the form is positive definite. Even lattices are integral. Let $L_\mathbb{Q} = L \otimes_\mathbb{Z} \mathbb{Q}$. Then $L_\mathbb{Q}$ is an n-dimensional vector space over $\mathbb{Q}$ in which L is embedded and the form on L is extended to a symmetric $\mathbb{Q}$-bilinear form on $L_\mathbb{Q}$. The lattice is nondegenerate if and only if this form on $L_\mathbb{Q}$ is nondegenerate. A lattice may be equivalently defined as the $\mathbb{Z}$-span of a basis of a finite-dimensional rational vector space equipped with a symmetric bilinear form. The dual of L is the set $L^\circ = \{\alpha \in L_\mathbb{Q} \mid \langle \alpha, L \rangle \subset \mathbb{Z}\}$. This set is a lattice if and if L is nondegenerate, and in this case, L° has as a base the dual base of a given base. The lattice L is said to be *self-dual* if $L = L^\circ$. This is equivalent to L being integral and *unimodular*, which means that $|\det(\langle \alpha_i, \alpha_j \rangle)| = 1$.

Recall that the Golay code $\mathcal{C}$ is defined on a 24-element set Ω. Let $\mathfrak{h} = \coprod_{k \in \Omega} \mathbb{C}\alpha_k$ be a vector space with basis $\{\alpha_k \mid k \in \Omega\}$ and provided $\mathfrak{h}$ with the symmetric bilinear form $\langle \cdot, \cdot \rangle$ such that $\langle \alpha_k, \alpha_l \rangle = 2\delta_{kl}$ for $k, l \in \Omega$. For $S \subset \Omega$, set $\alpha_S = \sum_{k \in S} \alpha_k$. For any fixed element k_0 of Ω, the subset

$$
\begin{aligned}
\Lambda &= \sum_{C \in \mathcal{C}} \mathbb{Z}\frac{1}{2}\alpha_C + \sum_{k \in \Omega} \mathbb{Z}(\frac{1}{4}\alpha_\Omega - \alpha_k) \\
&= \sum_{C \in \mathcal{C}} \mathbb{Z}\frac{1}{2}\alpha_C + \sum_{k,l} \mathbb{Z}(\alpha_k + \alpha_l) + \mathbb{Z}(\frac{1}{4}\alpha_\Omega - \alpha_{k_0}),
\end{aligned}
$$

of $\mathfrak{h}$, equipped with the restriction to Λ of the form on $\mathfrak{h}$, is a lattice. This lattice is the *Leech lattice*.

THEOREM 1.2. *The Leech lattice Λ is an even unimodular lattice such that $\Lambda_2 = \emptyset$. It is unique up to isometry.*

The Leech lattice is generated by Λ_4. It is easy to see that the elements $\pm\alpha_k \pm \alpha_l$, $k, l \in \Omega$, $k \neq l$, are in Λ_4. Obviously, $\theta : \Lambda \to \Lambda, \alpha \mapsto -\alpha$ is an isometry of the Leech lattice such that $\theta^2 = 1$.

1.3. The untwisted vertex operator algebra V_Λ. Let $\mathfrak{h}$ be the same vector space as in Subsection 1.2. We view $\mathfrak{h}$ as an abelian Lie algebra and consider the $\mathbb{Z}$-graded untwisted affine Lie algebra $\hat{\mathfrak{h}} = \coprod_{n \in \mathbb{Z}} \mathfrak{h} \otimes t^n \oplus \mathbb{C}c \oplus \mathbb{C}d$, its Heisenberg subalgebra $\hat{\mathfrak{h}}_\mathbb{Z} = \coprod_{n \in \mathbb{Z}, n \neq 0} \mathfrak{h} \otimes t^n \oplus \mathbb{C}c$ and the subalgebra $\hat{\mathfrak{h}}_\mathbb{Z}^- = \coprod_{n < 0} \mathfrak{h} \otimes t^n$. The symmetric algebra $S(\hat{\mathfrak{h}}_\mathbb{Z}^-)$ over $\hat{\mathfrak{h}}_\mathbb{Z}^-$ is a $\mathbb{Z}$-graded $\hat{\mathfrak{h}}_\mathbb{Z}$-irreducible $\mathfrak{h}$-module. Let $\hat{\Lambda}$ be a central extension of Λ by a cyclic group $\langle \kappa \rangle$ of order 2. We denote the projection from $\hat{\Lambda}$ to Λ by $^-$. Define the faithful character $\chi : \langle \kappa \rangle \to \mathbb{C}^\times$ by $\chi(\kappa) = -1$. Denote by $\mathbb{C}_\chi$ the one-dimensional space $\mathbb{C}$ viewed as a $\langle \kappa \rangle$-module on which $\langle \kappa \rangle$ acts according to χ and denote by $\mathbb{C}\{\Lambda\}$ the induced $\hat{\Lambda}$-module $\mathbb{C}[\hat{\Lambda}] \otimes_{\mathbb{C}[\langle \kappa \rangle]} \mathbb{C}_\chi$ (where $\mathbb{C}[\hat{\Lambda}]$ and $\mathbb{C}[\langle \kappa \rangle]$ are the group algebras of $\hat{\Lambda}$ and $\langle \kappa \rangle$, respectively). Set $V_\Lambda = S(\hat{\mathfrak{h}}_\mathbb{Z}^-) \otimes \mathbb{C}\{\Lambda\}$. We regard $S(\hat{\mathfrak{h}}_\mathbb{Z}^-)$ as the trivial $\hat{\Lambda}$-module and V_Λ as the corresponding tensor product $\hat{\Lambda}$-module. View $\mathbb{C}\{\Lambda\}$ as a trivial $\hat{\mathfrak{h}}_\mathbb{Z}$-module and for $\alpha \in \mathfrak{h}$, define $\alpha(0) : \mathbb{C}\{\Lambda\} \to \mathbb{C}\{\Lambda\}$ by $\alpha(0)(a \otimes 1) = \langle \alpha, \bar{a} \rangle(a \otimes 1)$ for any $a \in \hat{\Lambda}$. Also define $x^\alpha \in (\text{End } \mathbb{C}\{\Lambda\})[x, x^{-1}]$ for $\alpha \in \Lambda$ by $x^\alpha(a \otimes 1) = x^{\langle \alpha, \bar{a} \rangle}(a \otimes 1)$ for any $a \in \hat{\Lambda}$. Give $\mathbb{C}\{\Lambda\}$ a $\mathbb{Z}$-gradation (*weight*) by wt $a \otimes 1 = \frac{1}{2}\langle \bar{a}, \bar{a} \rangle$ for all $a \in \hat{\Lambda}$. Then V_Λ has a $\mathbb{Z}$-gradation (*weight*) obtained from the tensor product gradation. Let $d \in \tilde{\mathfrak{h}}$ act as the weight operators on $S(\hat{\mathfrak{h}}_\mathbb{Z}^-)$ and on $\mathbb{C}\{\Lambda\}$ and give V_Λ the tensor product $\tilde{\mathfrak{h}}$-module structure. We denote the action of $\alpha \otimes t^n$ by $\alpha(n)$ for $\alpha \in \mathfrak{h}$ and denote the element $1 \otimes (a \otimes 1) \in V_\Lambda$ by $\iota(a)$ for $a \in \hat{\Lambda}$.

For $\alpha \in \mathfrak{h}$, let $\alpha(x) = \sum_{n \in \mathbb{Z}} \alpha(n) x^{-n-1}$. For any $a \in \hat{\Lambda}$, we define

$$
\begin{aligned}
Y_{V_\Lambda}(\iota(a), x) &= {}^\circ_\circ e^{\int \bar{a}(x)} {}^\circ_\circ \\
&= {}^\circ_\circ \exp\left(-\sum_{n<0} \frac{\bar{a}(n)}{n} x^{-n} \right) \exp\left(-\sum_{n>0} \frac{\bar{a}(n)}{n} x^{-n} \right) a x^{\bar{a}} {}^\circ_\circ
\end{aligned}
$$

(the *(untwisted) vertex operator associated to* $\iota(a)$) of $(\text{End } V_\Lambda)[[x, x^{-1}]]$, where the normal ordering is defined by

$$
\begin{aligned}
{}^\circ_\circ \alpha_1(m)\alpha_2(n) {}^\circ_\circ &= {}^\circ_\circ \alpha_2(n)\alpha_1(m) {}^\circ_\circ = \begin{cases} \alpha_1(m)\alpha_2(n) & m \le n, \\ \alpha_2(n)\alpha_1(m) & m \ge n, \end{cases} \\
{}^\circ_\circ \alpha(m) a {}^\circ_\circ &= {}^\circ_\circ a\alpha(m) {}^\circ_\circ = a\alpha(m), \\
{}^\circ_\circ x^\alpha a {}^\circ_\circ &= {}^\circ_\circ a x^\alpha {}^\circ_\circ = a x^\alpha
\end{aligned}
$$

for $\alpha_1, \alpha_2, \alpha \in \mathfrak{h}$, $m, n \in \mathbb{Z}$, $\alpha \in \Lambda$ and $a \in \hat{\Lambda}$. For $v = \alpha_1(-n_1) \cdots \alpha_k(n_k) \cdot \iota(a)$, we define the *(untwisted) vertex operator* associated to v to be

$$
\begin{aligned}
Y_{V_\Lambda}(v, x) = {}^\circ_\circ &\left(\frac{1}{(n_1 - 1)!} \frac{d^{n_1 - 1}\alpha_1(x)}{dx^{n_1 - 1}} \right) \cdots \\
&\cdot \left(\frac{1}{(n_k - 1)!} \frac{d^{n_k - 1}\alpha_k(x)}{dx^{n_k - 1}} \right) Y_{V_\Lambda}(\iota(a), x) {}^\circ_\circ.
\end{aligned}
$$

Extending Y_{V_Λ} by linearity, we obtain a linear map $V_\Lambda \to (\text{End } V_\Lambda)[[x, x^{-1}]]$, $v \to Y_{V_\Lambda}(v, x)$. Let $\{h_1, \ldots, h_{24}\}$ be an orthonormal basis of $\mathfrak{h}$. Consider $\omega = \frac{1}{2} \sum_{i=1}^{24} h_i(-1)^2$. The following theorem is a special case of a theorem due to Borcherds [**B1**]; see [**FLM2**]:

THEOREM 1.3. *The quadruple* $(V_\Lambda, Y_{V_\Lambda}, \iota(1), \omega)$ *is a vertex operator algebra with central charge (or rank) equal to 24.*

When there is no confusion, we shall use Y to denote the vertex operator map Y_{V_Λ}.

The following theorem is a special case of a theorem due to Dong [**D1**]:

THEOREM 1.4. *Any irreducible* V_Λ*-module is isomorphic to* V_Λ *as a* V_Λ*-module and any* V_Λ*-module is a finite sum of copies of* V_Λ *as a* V_Λ*-module.*

On $\mathbb{C}\{\Lambda\}$ there is a unique positive definite hermitian form $(\cdot, \cdot)_{\mathbb{C}\{\Lambda\}}$ (see [**FLM2**]) such that

$$
(a \otimes 1, b \otimes 1)_{\mathbb{C}\{\Lambda\}} = \begin{cases} 0 & \bar{a} \ne \bar{b}, \\ 1 & a = b. \end{cases}
$$

It is easy to see that on V_Λ there is a unique bilinear form $\langle \cdot, \cdot \rangle_{V_\Lambda}$ such that

$$
\begin{aligned}
\langle \iota(a), \iota(b) \rangle_{V_\Lambda} &= (a \otimes 1, b \otimes 1)_{\mathbb{C}\{\Lambda\}}, \\
\langle d \cdot u, v \rangle_{V_\Lambda} &= \langle u, d \cdot v \rangle_{V_\Lambda}, \\
\langle \alpha(n) \cdot u, v \rangle_{V_\Lambda} &= \langle u, \alpha(n) \cdot v \rangle_{V_\Lambda}.
\end{aligned}
$$

Recall the isometry θ of the Leech lattice. For $v = \alpha_1(-n_1) \cdots \alpha_k(-n_k) \cdot \iota(a) \in V_\Lambda$, we define $\theta(v) = (-1)^k \alpha_1(-n_1) \cdots \alpha_k(-n_k) \cdot \iota(a^{-1})$. Using linearity, we obtain a linear map $\theta : V_\Lambda \to V_\Lambda$. It is clear that $\theta^2 = 1$ and θ is an automorphism of the vertex operator algebra V_Λ. Thus $V_\Lambda = V_\Lambda^+ + V_\Lambda^-$ where $V_\Lambda^\pm$ are the eigenspace of θ with eigenvalue ± 1. The subspace V_Λ^+ is a vertex operator algebra and both $V_\Lambda^\pm$ are irreducible V_Λ^+-modules.

1.4. The twisted module V_Λ^T for V_Λ. Consider the $\mathbb{Z} + \frac{1}{2}$-graded twisted affine Lie algebra $\tilde{\mathfrak{h}}[-1] = \coprod_{n \in \mathbb{Z} + \frac{1}{2}} \mathfrak{h} \otimes t^n \oplus \mathbb{C}c \oplus \mathbb{C}d$, its Heisenberg subalgebra $\hat{\mathfrak{h}}_{\mathbb{Z} + \frac{1}{2}} = \coprod_{n \in \mathbb{Z} + \frac{1}{2}} \mathfrak{h} \otimes t^n \oplus \mathbb{C}c$ and the subalgebra $\hat{\mathfrak{h}}_{\mathbb{Z} + \frac{1}{2}}^- = \coprod_{n < 0} \mathfrak{h} \otimes t^n$. The symmetric algebra $S(\hat{\mathfrak{h}}_{\mathbb{Z} + \frac{1}{2}}^-)$ over $\hat{\mathfrak{h}}_{\mathbb{Z} + \frac{1}{2}}^-$ is a $\mathbb{Z} + \frac{1}{2}$-graded $\hat{\mathfrak{h}}_{\mathbb{Z} + \frac{1}{2}}$-irreducible $\tilde{\mathfrak{h}}[-1]$-module.

Let $K = \{a^2 \kappa^{\langle \bar{a}, \bar{a} \rangle / 2} \mid a \in \hat{\Lambda}\}$. Then K is a central subgroup of $\hat{\Lambda}$. The following result is proved in [**FLM2**]:

THEOREM 1.5. *The quotient group $\hat{\Lambda}/K$ has a unique (up to equivalence) irreducible module T such that $\kappa K \to -1$ on T. Moreover, the corresponding representation of $\hat{\Lambda}/K$ is the unique faithful irreducible representation and $\dim T = 2^{12}$. To construct T, let Φ be any subgroup of Λ such that $2\Lambda \subset \Phi \subset \Lambda$, $|\Phi/2\Lambda| = 2^{12}$ and $\frac{1}{2}\langle \alpha, \alpha \rangle \in 2\mathbb{Z}$ for any $\alpha \in \Phi$. Then the preimage $\hat{\Phi}$ of Φ under the homomorphism $^- : \hat{\Lambda} \to \Lambda$ is a maximal subgroup of $\hat{\Lambda}$ and $\hat{\Phi}/K$ is an elementary abelian 2-group. Let $\Psi : \hat{\Phi}/K \to \mathbb{C}^\times$ be any homomorphism such that $\Psi(\kappa K) = -1$ and denote by $\mathbb{C}_\Psi$ the one-dimensional $\hat{\Phi}$-module with the corresponding character. Then viewed as a $\hat{\Lambda}$-module*

$$
\begin{aligned}
T &= \mathbb{C}[\hat{\Lambda}] \otimes_{\mathbb{C}[\hat{\Phi}]} \mathbb{C}_\Psi \\
&\simeq \mathbb{C}[\Lambda/\Phi] \quad \text{(linearly)}.
\end{aligned}
$$

For any $a \in \hat{\Lambda}$, the element $a \otimes 1 \in T$ is denoted by $t(a)$.

Set $V_\Lambda^T = S(\hat{\mathfrak{h}}_{\mathbb{Z} + \frac{1}{2}}^-) \otimes T$. We view T as a $\hat{\Lambda}$-module and regard $S(\hat{\mathfrak{h}}_{\mathbb{Z} + \frac{1}{2}}^-)$ as the trivial $\hat{\Lambda}$-module and V_Λ as the corresponding tensor product $\hat{\Lambda}$-module. Give T a $\mathbb{Z} + \frac{1}{2}$-gradation (*weight*) by wt $t = \frac{24}{16} = \frac{3}{2}$ for all $t \in T$. Then V_Λ^T has a $\mathbb{Z} + \frac{1}{2}$-gradation (*weight*) obtained from the tensor product gradation. Let $d \in \tilde{\mathfrak{h}}$ act as the weight operators on $S(\hat{\mathfrak{h}}_{\mathbb{Z} + \frac{1}{2}}^-)$ and on T. View T as a trivial $\hat{\mathfrak{h}}$-module and give V_Λ the tensor product $\tilde{\mathfrak{h}}[-1]$-module structure. We denote the action of $\alpha \otimes t^n$ for $\alpha \in \mathfrak{h}$ and $n \in \mathbb{Z} + \frac{1}{2}$ by $\alpha(n)$.

For any $\alpha \in \mathfrak{h}$, let $\alpha(x) = \sum_{n \in \mathbb{Z} + \frac{1}{2}} \alpha(n) x^{-n-1}$. (Note that though we use the same notation as in Subsection 1.3, $\alpha(x)$ in Subsection 1.3 acts on a different space.) For any $a \in \hat{\Lambda}$, we define the *twisted vertex operator*

$$
\begin{aligned}
Y_{V_\Lambda^T}(\iota(a), x) &= 2^{-\langle \bar{a}, \bar{a} \rangle} \, {}^\circ_\circ e^{\int \bar{a}(x)} \, {}^\circ_\circ a x^{-\langle \bar{a}, \bar{a} \rangle / 2} \\
&= 2^{-\langle \bar{a}, \bar{a} \rangle} \exp\left(\sum_{n<0} \frac{\bar{a}(n+\frac{1}{2})}{n+\frac{1}{2}} x^{-(n+\frac{1}{2})} \right) \cdot \\
&\qquad \cdot \exp\left(\sum_{n \geq 0} \frac{\bar{a}(n+\frac{1}{2})}{n+\frac{1}{2}} x^{-(n+\frac{1}{2})} \right) a x^{-\langle \bar{a}, \bar{a} \rangle / 2}
\end{aligned}
$$

of $(\text{End } V_\Lambda^T)[[x^{\frac{1}{2}}, x^{-\frac{1}{2}}]]$, where the normal ordering is defined by

$$
{}^\circ_\circ \alpha_1(m) \alpha_2(n) {}^\circ_\circ = {}^\circ_\circ \alpha_2(n) \alpha_1(m) {}^\circ_\circ = \begin{cases} \alpha_1(m)\alpha_2(n) & m \leq n, \\ \alpha_2(n)\alpha_1(m) & m \geq n, \end{cases}
$$

for $\alpha_1, \alpha_2 \in \mathfrak{h}$, $m, n \in \mathbb{Z}$. For $v = \alpha_1(-n_1) \cdots \alpha_k(-n_k) \cdot \iota(a) \in V_\Lambda$, we define

$$
\begin{aligned}
Y_0(v, x) &= {}^\circ_\circ \left(\frac{1}{(n_1 - 1)!} \frac{d^{n_1 - 1} \alpha_1(x)}{dx^{n_1 - 1}} \right) \cdots \\
&\qquad \cdot \left(\frac{1}{(n_k - 1)!} \frac{d^{n_k - 1} \alpha_k(x)}{dx^{n_k - 1}} \right) Y_{V_\Lambda^T}(\iota(a), x) {}^\circ_\circ.
\end{aligned}
$$

Let c_{mn} be the complex numbers determined by the formula

$$
\sum_{n,m \geq 0} c_{mn} x^m y^n = -\log\left(\frac{(1+x)^{1/2} + (1+y)^{1/2}}{2} \right).
$$

We define the *twisted vertex operator associated to v* to be

$$
Y_{V_\Lambda^T}(v, x) = Y_0\left(\exp\left(\sum_{m,n \geq 0} \sum_{i=1}^{24} c_{mn} h_i(m) h_i(n) x^{-m-n} \right) v, x \right)
$$

where as in Subsection 1.3 $\{h_1, \ldots, h_{24}\}$ is an orthogonal basis for $\mathfrak{h}$. The following theorem is a special case of a theorem due Frenkel, Lepowsky and Meurman [**FLM2**]:

THEOREM 1.6. *The pair $(V_\Lambda^T, Y_{V_\Lambda^T})$ is a θ-twisted V_Λ-module.*

When there is no confusion, we shall use Y to denote the vertex operator map $Y_{V_\Lambda^T}$.

The following theorem is a special case of a theorem due to Dong [**D2**]:

THEOREM 1.7. *Any irreducible θ-twisted V_Λ-module is isomorphic to V_Λ^T and any θ-twisted V_Λ-module is a finite sum of copies of V_Λ^T.*

On T there is a unique positive definite hermitian form $(\cdot, \cdot)_T$ (see [**FLM2**]) such that

$$
(t(a), t(b))_T = \begin{cases} 0 & a\hat{\Phi} \neq b\hat{\Phi}, \\ 1 & a = b. \end{cases}
$$

Thus on V_Λ there is a unique bilinear form $\langle \cdot, \cdot \rangle_{V_\Lambda^T}$ such that

$$
\begin{aligned}
\langle 1 \otimes t(a), 1 \otimes t(b) \rangle_{V_\Lambda^T} &= (t(a), t(b))_T, \\
\langle d \cdot u, v \rangle_{V_\Lambda^T} &= \langle u, d \cdot v \rangle_{V_\Lambda^T}, \\
\langle \alpha(n) \cdot u, v \rangle_{V_\Lambda^T} &= \langle u, \alpha(-n) \cdot v \rangle_{V_\Lambda^T}.
\end{aligned}
$$

For $w = \alpha_1(-n_1) \cdots \alpha_k(-n_k) \otimes t \in V_\Lambda^T$, we define

$$
\theta(w) = (-1)^{k+1} \alpha_1(-n_1) \cdots \alpha_k(-n_k) \otimes t.
$$

Using linearity, we obtain a linear map $\theta : V_\Lambda^T \to V_\Lambda^T$. It is clear that $\theta^2 = 1$ and θ is an automorphism of the twisted V_Λ-module V_Λ^T. Thus $V_\Lambda^T = (V_\Lambda^T)^+ + (V_\Lambda^T)^-$ where $(V_\Lambda^T)^\pm$ are the eigenspace of θ with eigenvalue ± 1. Both $(V_\Lambda^T)^+$ and $(V_\Lambda^T)^-$ are irreducible V_Λ^+-modules.

1.5. The moonshine module $V^\natural$. The *moonshine module* is defined to be $V^\natural = V_\Lambda^+ \oplus (V_\Lambda^T)^+$. It is not difficult to show that the generating function of the dimensions of the homogeneous subspaces of $V^\natural$ is equal to $qJ(q) = q(j(q) - 744)$. The following result is established in [**FLM2**]:

THEOREM 1.8. *The $\mathbb{Z}$-graded space $V^\natural$ has a natural vertex operator algebra structure and its automorphism group is the Monster.*

The proof of Theorem 1.8 in [**FLM2**] uses triality and some results in group theory. Though the proof of the theorem above is involved, there is a direct and natural way to define the vertex operator map $Y_{V^\natural}$ as was carried out in [**FHL**] for the sum of an arbitrary vertex operator algebra and an arbitrary $\mathbb{Z}$-graded module for the vertex operator algebra. We recall this construction in the case of the moonshine module here: For $u, v \in V_\Lambda^+$, $Y_{V^\natural}(u, x)v = Y_{V_\Lambda}(u, x)v$; for $u \in V_\Lambda^+$, $v \in (V_\Lambda^T)^+$, $Y_{V^\natural}(u, x)v = Y_{V_\Lambda^T}(u, x)v$; for $u \in (V_\Lambda^T)^+$, $v \in V_\Lambda^+$, $Y_{V^\natural}(u, x)v = e^{xL(-1)} Y_{V_\Lambda^T}(v, -x)u$; for $u, v \in (V_\Lambda^T)^+$, $Y_{V^\natural}(u, x)v$ is defined by

$$
\langle w, Y_{V^\natural}(u, x)v \rangle_{V_\Lambda} = \langle Y_{V_\Lambda^T}(w, -x^{-1}) e^{xL(1)} (-x^2)^{-L(0)} u, e^{x^{-1}L(1)} v \rangle_{V_\Lambda^T}
$$

for all $w \in V_\Lambda$. The vacuum of $V^\natural$ is $\iota(1)$ and the Virasoro element is ω.

We now discuss the decompositions of V_Λ^+ and its modules into direct sums of modules for a tensor product of the Virasoro vertex operator algebra of central charge $\frac{1}{2}$. Let $L(\frac{1}{2}, 0)$ be the rational Virasoro vertex operator algebra of central charge $\frac{1}{2}$ and $L(\frac{1}{2}, i)$, $i = 0, \frac{1}{2}, \frac{1}{16}$, the irreducible $L(\frac{1}{2}, 0)$-modules (see [**FZ**] and [**DMZ**]). The following result is proved by Dong, Mason and Zhu [**DMZ**]:

THEOREM 1.9. *There exist $\omega_i \in V_\Lambda^+$, $i = 1, \ldots, 48$, such that $\omega = \omega_1 + \cdots + \omega_{48}$ and the vertex operator subalgebra L of V_Λ^+ generated by ω_i, $i = 1, \ldots, 48$, is isomorphic to $L(\frac{1}{2}, 0)^{\otimes 48}$. In particular, for $W = V_\Lambda^\pm, (V_\Lambda^T)^\pm$, $(W, Y_W|_{L \otimes W})$ are $L(\frac{1}{2}, 0)^{\otimes 48}$-modules.*

A vector in an $L(\frac{1}{2}, 0)^{48}$-module W is a *lowest weight vector* if it is a lowest weight vector when W is regarded as an $L(\frac{1}{2}, 0)$-module for any one of the 48 vertex operator subalgebras $L(\frac{1}{2}, 0)$. The *lowest weight* of a lowest vector v is defined in the obvious way. It is an array of 48 complex numbers. For any homogeneous subspace $W_{(n)}$ of an $L(\frac{1}{2}, 0)^{48}$-module W, we denote the subspace spanned by all lowest weight vectors in $W_{(n)}$ by $W_{(n)}^l$. Let a, b, c be nonnegative integers such that $a + b + c = 48$. A lowest weight vector $v \in W$ with lowest weight $(h_1, \ldots, h_{48})$ is called a vector of type (a, b, c) if $\#\{h_i \mid h_i = 0\} = a$, $\#\{h_i \mid h_i = \frac{1}{2}\} = b$ and $\#\{h_i \mid h_i = \frac{1}{16}\} = c$. We write $W_{(n)}^l = \coprod_{a,b,c} m_{a,b,c}(a, b, c)$ which means that there are $m_{a,b,c}$ linearly independent vectors of type (a, b, c) in $W_{(n)}$. In [**DMZ**], the following information on the lowest weight vectors in V_Λ and in V_Λ^T is obtained:

THEOREM 1.10. *We have the following decompositions:*

$$
\begin{aligned}
(V_\Lambda)_{(1)}^l &= 24(46, 2, 0), \\
(V_\Lambda^T)_{(3/2)}^l &= 2^{12}(24, 0, 24), \\
(V_\Lambda^T)_{(2)}^l &= 24 \cdot 2^{12}(23, 1, 24).
\end{aligned}
$$

The following information on the decompositions of $V_\Lambda^\pm, (V_\Lambda^T)^\pm$ into direct sums of irreducible $L(\frac{1}{2}, 0)^{\otimes 48}$-modules can be found in [**D3**]:

THEOREM 1.11. *Let $W = V_\Lambda^\pm, (V_\Lambda^T)^\pm$. As an $L(\frac{1}{2}, 0)^{\otimes 48}$-module,*

$$
W = \coprod_{h_i = 0, \frac{1}{2}, \frac{1}{16}} c_{h_1 \cdots h_{48}} L(\frac{1}{2}, h_1) \otimes \cdots \otimes L(\frac{1}{2}, h_{48}).
$$

If $W = V_\Lambda^\pm$ and $c_{h_1 \cdots h_{48}} \neq 0$ for $h_i \in \{0, \frac{1}{2}, \frac{1}{16}\}$, $i = 1, \ldots, 48$, then

$$
(h_{2j-1}, h_{2j}) \in \{(0, 0), (0, \frac{1}{2}), (\frac{1}{2}, 0), (\frac{1}{2}, \frac{1}{2}), (\frac{1}{16}, \frac{1}{16})\},
$$

$1 \leq j \leq 24$. If $W = (V_\Lambda^T)^\pm$ and $c_{h_1 \cdots h_{48}} \neq 0$ for $h_i \in \{0, \frac{1}{2}, \frac{1}{16}\}$, $i = 1, \ldots, 48$, then

$$
(h_{2j-1}, h_{2j}) \in \{(0, \frac{1}{16}), (\frac{1}{16}, 0), (\frac{1}{2}, \frac{1}{16}), (\frac{1}{16}, \frac{1}{2})\},
$$

$1 \leq j \leq 24$. Let $x_1^i, \ldots, x_{48}^i$, $i = 1, \ldots, 24$, be given by $(x_{2j-1}^i, x_{2j}^i) = (0, 0)$, $j = 1, \ldots, 24$, $j \neq i$, and $(x_{2j-1}^j, x_{2j}^j) = (\frac{1}{2}, \frac{1}{2})$, $j = 1, \ldots, 24$. When $W = V_\Lambda^-$, the multiplicities $c_{x_1^i \cdots x_{48}^i} = 1$, $i = 1, \ldots, 24$.

The following results are due to Dong [**D3**]:

THEOREM 1.12. *The vertex operator algebra V_Λ^+ has only the four irreducible modules $V_\Lambda^\pm, (V_\Lambda^T)^\pm$ (up to isomorphism) and any V_Λ^+-module is a finite sum of irreducible modules.*

THEOREM 1.13. *The moonshine module vertex operator algebra $V^\natural$ has only one irreducible module, $V^\natural$ itself, (up to isomorphisms) and any $V^\natural$-module is a finite sum of irreducible modules.*

2. Tensor products of modules for a vertex operator algebra

In this section we summarize the basic concepts and constructions in the theory of tensor products of modules for a vertex operator algebra and those results (mainly the associativity) which we need in this paper. Details can be found in [**HL2**]–[**HL3**], [**HL5**], [**H2**]. This theory is initiated in [**HL1**]. For the complete picture of the tensor product theory, see [**HL4**].

2.1. The definition, some properties and two constructions of $P(z)$-tensor products. Let $(V, Y, \mathbf{1}, \omega)$ be a vertex operator algebra and (W, Y) a V-module. For any $v \in V$ and $n \in \mathbb{Z}$, there is a well-defined natural action of v_n on $\overline{W}$. Moreover, for fixed $v \in V$, any infinite linear combination of the v_n of the form $\sum_{n<N} a_n v_n$ $(a_n \in \mathbb{C})$ acts on $\overline{W}$ in a well-defined way.

Fix $z \in \mathbb{C}^\times$ and let (W_1, Y_1), (W_2, Y_2) and (W_3, Y_3) be V-modules. A $P(z)$-*intertwining map of type* $\binom{W_3}{W_1 W_2}$ is a linear map $F : W_1 \otimes W_2 \to \overline{W}_3$ satisfying the condition

$$x_0^{-1}\delta\left(\frac{x_1 - z}{x_0}\right) Y_3(v, x_1)F(w_{(1)} \otimes w_{(2)}) =$$

$$= z^{-1}\delta\left(\frac{x_1 - x_0}{z}\right) F(Y_1(v, x_0)w_{(1)} \otimes w_{(2)})$$

$$+ x_0^{-1}\delta\left(\frac{z - x_1}{-x_0}\right) F(w_{(1)} \otimes Y_2(v, x_1)w_{(2)})$$

for $v \in V$, $w_{(1)} \in W_1$, $w_{(2)} \in W_2$. The vector space of $P(z)$-intertwining maps of type $\binom{W_3}{W_1 W_2}$ is denoted by $\mathcal{M}[P(z)]_{W_1 W_2}^{W_3}$. A $P(z)$-*product of W_1 and W_2 is* a V-module (W_3, Y_3) equipped with a $P(z)$-intertwining map F of type $\binom{W_3}{W_1 W_2}$ and is denoted by $(W_3, Y_3; F)$ (or simply by (W_3, F)). Let $(W_4, Y_4; G)$ be another $P(z)$-product of W_1 and W_2. A *morphism* from $(W_3, Y_3; F)$ to $(W_4, Y_4; G)$ is a module map η from W_3 to W_4 such that $G = \overline{\eta} \circ F$, where $\overline{\eta}$ is the map from $\overline{W}_3$ to $\overline{W}_4$ uniquely extending η.

A $P(z)$-*tensor product of W_1 and W_2* is a $P(z)$-product

$$(W_1 \boxtimes_{P(z)} W_2, Y_{P(z)}; \boxtimes_{P(z)})$$

such that for any $P(z)$-product $(W_3, Y_3; F)$, there is a unique morphism from

$$(W_1 \boxtimes_{P(z)} W_2, Y_{P(z)}; \boxtimes_{P(z)})$$

to $(W_3, Y_3; F)$. The V-module $(W_1 \boxtimes_{P(z)} W_2, Y_{P(z)})$ is called a $P(z)$-*tensor product module of W_1 and W_2. A $P(z)$-tensor product is unique up to isomorphism.*

We have the following properties:

PROPOSITION 2.1. *Let* $\log z = \log|z| + i \arg z$ *such that* $0 \leq \arg z < 2\pi$ *and* $l_p(z) = \log z + 2\pi p i$, $p \in \mathbb{Z}$. *For any value* $p \in \mathbb{Z}$, *we have an isomorphism from the vector space* $\mathcal{V}^{W_3}_{W_1 W_2}$ *of intertwining operators of type* $\binom{W_3}{W_1 W_2}$ *to the vector space* $\mathcal{M}[P(z)]^{W_3}_{W_1 W_2}$. *This isomorphism takes an intertwining operator of the type* $\binom{W_3}{W_1 W_2}$ *to the* $P(z)$-*intertwining map of the same type obtained from the intertwining operator by substituting the complex powers of* $e^{l_p(z)}$ *for the complex powers of the formal variable.*

PROPOSITION 2.2. *Suppose that* $W_1 \boxtimes_{P(z)} W_2$ *exists. We have a natural isomorphism*

$$\operatorname{Hom}_V(W_1 \boxtimes_{P(z)} W_2, W_3) \quad \overset{\sim}{\to} \quad \mathcal{M}[P(z)]^{W_3}_{W_1 W_2}$$
$$\eta \quad \mapsto \quad \overline{\eta} \circ \boxtimes_{P(z)}. \tag{2.1}$$

PROPOSITION 2.3. *Let* $U_1, \ldots, U_k$, $W_1, \ldots, W_l$ *be* V-*modules and suppose that each* $U_i \boxtimes_{P(z)} W_j$ *exists. Then* $(\coprod_i U_i) \boxtimes_{P(z)} (\coprod_j W_j)$ *exists and there is a natural isomorphism*

$$\left(\coprod_i U_i \right) \boxtimes_{P(z)} \left(\coprod_j W_j \right) \overset{\sim}{\to} \coprod_{i,j} U_i \boxtimes_{P(z)} W_j. \tag{2.2}$$

We consider the following special but important class of vertex operator algebras: A vertex operator algebra V is *rational* if it satisfies the following conditions:

(i) There are only finitely many irreducible V-modules (up to equivalence).

(ii) Every V-module is completely reducible (and is in particular a *finite* direct sum of irreducible modules).

(iii) All the fusion rules (the dimensions of spaces of intertwining operators) for V are finite (for triples of irreducible modules and hence arbitrary modules).

PROPOSITION 2.4. *Let* V *be rational and let* W_1, W_2 *be* V-*modules. Then*

$$(W_1 \boxtimes_{P(z)} W_2, Y_{P(z)}; \boxtimes_{P(z)})$$

exists and the $P(z)$-*tensor product module* $W_1 \boxtimes_{P(z)} W_2$ *of* W_1 *and* W_2 *is isomorphic to the* V-*module* $\coprod_{i=1}^k (\mathcal{V}^{M_i}_{W_1 W_2})^* \otimes M_i$ *where* $\{M_1, \ldots, M_k\}$ *is a set of representatives of the equivalence classes of irreducible* V-*modules.*

We now describe the constructions of a $P(z)$-tensor product of two modules. For two V-modules (W_1, Y_1) and (W_2, Y_2), we define an action of

$$V \otimes \iota_+ \mathbb{C}[t, t^{-1}, (z^{-1} - t)^{-1}]$$

on $(W_1 \otimes W_2)^*$ (where as in [**FLM2**] and [**HL2**], ι_+ denotes the operation of expansion of a rational function of t in the direction of positive powers of t), that is, a linear map

$$\tau_{P(z)} : V \otimes \iota_+ \mathbb{C}[t, t^{-1}, (z^{-1} - t)^{-1}] \to \text{End}\, (W_1 \otimes W_2)^*, \qquad (2.3)$$

by

$$\left(\tau_{P(z)} \left(x_0^{-1} \delta \left(\frac{x_1^{-1} - z}{x_0} \right) Y_t(v, x_1) \right) \lambda \right) (w_{(1)} \otimes w_{(2)})$$

$$= z^{-1} \delta \left(\frac{x_1^{-1} - x_0}{z} \right) \lambda(Y_1(e^{x_1 L(1)} (-x_1^{-2})^{L(0)} v, x_0) w_{(1)} \otimes w_{(2)})$$

$$+ x_0^{-1} \delta \left(\frac{z - x_1^{-1}}{-x_0} \right) \lambda(w_{(1)} \otimes Y_2^*(v, x_1) w_{(2)}) \qquad (2.4)$$

for $v \in V$, $\lambda \in (W_1 \otimes W_2)^*$, $w_{(1)} \in W_1$, $w_{(2)} \in W_2$, where

$$Y_t(v, x) = v \otimes x^{-1} \delta \left(\frac{t}{x} \right). \qquad (2.5)$$

There is an obvious action of

$$V \otimes \iota_+ \mathbb{C}[t, t^{-1}, (z^{-1} - t)^{-1}]$$

on any V-module. We have:

PROPOSITION 2.5. *Under the natural isomorphism*

$$\text{Hom}(W_3', (W_1 \otimes W_2)^*) \xrightarrow{\sim} \text{Hom}(W_1 \otimes W_2, \overline{W}_3), \qquad (2.6)$$

the maps in $\text{Hom}(W_3', (W_1 \otimes W_2)^*)$ *intertwining the two actions of*

$$V \otimes \iota_+ \mathbb{C}[t, t^{-1}, (z^{-1} - t)^{-1}]$$

on W_3' *and* $(W_1 \otimes W_2)^*$ *correspond exactly to the* $P(z)$-*intertwining maps of type* $\binom{W_3}{W_1 W_2}$.

Write

$$Y_{P(z)}'(v, x) = \tau_{P(z)}(Y_t(v, x)) \qquad (2.7)$$

and

$$Y_{P(z)}'(\omega, x) = \sum_{n \in \mathbb{Z}} L_{P(z)}'(n) x^{-n-2}. \qquad (2.8)$$

We call the eigenspaces of the operator $L_{P(z)}'(0)$ the *weight subspaces* or *homogeneous subspaces* of $(W_1 \otimes W_2)^*$, and we have the corresponding notions of *weight vector* (or *homogeneous vector*) and *weight*.

Let W be a subspace of $(W_1 \otimes W_2)^*$. We say that W is *compatible for* $\tau_{P(z)}$ if every element of W satisfies the following nontrivial and subtle condition (called the *compatibility condition*) on $\lambda \in (W_1 \otimes W_2)^*$: The formal Laurent series $Y'_{P(z)}(v, x_0)\lambda$ involves only finitely many negative powers of x_0 and

$$\tau_{P(z)} \left(x_0^{-1}\delta \left(\frac{x_1^{-1} - z}{x_0} \right) Y_t(v, x_1) \right) \lambda =$$

$$= x_0^{-1}\delta \left(\frac{x_1^{-1} - z}{x_0} \right) Y'_{P(z)}(v, x_1)\lambda \quad \text{for all } v \in V. \tag{2.9}$$

Also, we say that W is ($\mathbb{C}$-)*graded* if it is $\mathbb{C}$-graded by its weight subspaces, and that W is a V-*module* (respectively, *generalized module*) if W is graded and is a module (respectively, generalized module, see [**HL1**] and [**HL2**]) when equipped with this grading and with the action of $Y'_{P(z)}(\cdot, x)$. The weight subspace of a subspace W with weight $n \in \mathbb{C}$ will be denoted $W_{(n)}$.

Define

$$W_1 \boxdot_{P(z)} W_2 = \sum_{W \in \mathcal{W}_{P(z)}} W = \bigcup_{W \in \mathcal{W}_{P(z)}} W \subset (W_1 \otimes W_2)^*, \tag{2.10}$$

where $\mathcal{W}_{P(z)}$ is the set of all compatible modules for $\tau_{P(z)}$ in $(W_1 \otimes W_2)^*$. We have:

PROPOSITION 2.6. *Let V be a rational vertex operator algebra and W_1, W_2 V-modules. Then $(W_1 \boxdot_{P(z)} W_2, Y'_{P(z)}|_{V \otimes W_1 \boxdot_{P(z)} W_2})$ is a module.*

Now we assume that V is rational. In this case, we define a V-module $W_1 \boxtimes_{P(z)} W_2$ by

$$W_1 \boxtimes_{P(z)} W_2 = (W_1 \boxdot_{P(z)} W_2)' \tag{2.11}$$

and we write the corresponding action as $Y_{P(z)}$. Applying Proposition 2.5 to the special module $W_3 = W_1 \boxtimes_{P(z)} W_2$ and the identity map $W_3' \to W_1 \boxdot_{P(z)} W_2$, we obtain a canonical $P(z)$-intertwining map of type $\binom{W_1 \boxtimes_{P(z)} W_2}{W_1 W_2}$, which we denote

$$\boxtimes_{P(z)} : W_1 \otimes W_2 \quad \to \quad \overline{W_1 \boxtimes_{P(z)} W_2}$$

$$w_{(1)} \otimes w_{(2)} \quad \mapsto \quad w_{(1)} \boxtimes_{P(z)} w_{(2)}. \tag{2.12}$$

We have:

PROPOSITION 2.7. *The $P(z)$-product $(W_1 \boxtimes_{P(z)} W_2, Y_{P(z)}; \boxtimes_{P(z)})$ is a $P(z)$-tensor product of W_1 and W_2.*

Observe that any element of $W_1 \boxdot_{P(z)} W_2$ is an element λ of $(W_1 \otimes W_2)^*$ satisfying:

The compatibility condition:
   (a) The *lower truncation condition*: For all $v \in V$, the formal Laurent series $Y'_{P(z)}(v, x)\lambda$ involves only finitely many negative powers of x.
   (b) The formula (2.9) holds.

The local grading-restriction condition:

(a) The *grading condition*: λ is a (finite) sum of weight vectors of $(W_1 \otimes W_2)^*$.

(b) Let W_λ be the smallest subspace of $(W_1 \otimes W_2)^*$ containing λ and stable under the component operators $\tau_{P(z)}(v \otimes t^n)$ of the operators $Y'_{P(z)}(v, x)$ for $v \in V$, $n \in \mathbb{Z}$. Then the weight spaces $(W_\lambda)_{(n)}$, $n \in \mathbb{C}$, of the (graded) space W_λ have the properties

$$\dim\,(W_\lambda)_{(n)} < \infty \quad \text{for } n \in \mathbb{C}, \tag{2.13}$$

$$(W_\lambda)_{(n)} = 0 \quad \text{for } n \text{ whose real part is sufficiently small.} \tag{2.14}$$

We have another construction of $W_1 \boxtimes_{P(z)} W_2$ using these conditions:

THEOREM 2.8. *The subspace of $(W_1 \otimes W_2)^*$ consisting of the elements satisfying the compatibility condition and the local grading-restriction condition, equipped with $Y'_{P(z)}$, is a generalized module and is equal to $W_1 \boxtimes_{P(z)} W_2$.*

The following result follows immediately from Proposition 2.6, the theorem above and the definition of $W_1 \boxtimes_{P(z)} W_2$:

COROLLARY 2.9. *Let V be a rational vertex operator algebra and W_1, W_2 two V-modules. Then the contragredient module of the module $W_1 \boxtimes_{P(z)} W_2$, equipped with the $P(z)$-intertwining map $\boxtimes_{P(z)}$, is a $P(z)$-tensor product of W_1 and W_2 equal to the structure $(W_1 \boxtimes_{P(z)} W_2, Y_{P(z)}; \boxtimes_{P(z)})$ constructed above.*

2.2. The associativity. Given any V-modules W_1, W_2, W_3, W_4 and W_5, let $\mathcal{Y}_1$, $\mathcal{Y}_2$, $\mathcal{Y}_3$ and $\mathcal{Y}_4$ be intertwining operators of type $\binom{W_4}{W_1 W_5}$, $\binom{W_5}{W_2 W_3}$, $\binom{W_5}{W_1 W_2}$ and $\binom{W_4}{W_5 W_3}$, respectively. Consider the following conditions for the product of $\mathcal{Y}_1$ and $\mathcal{Y}_2$ and for the iterate of $\mathcal{Y}_3$ and $\mathcal{Y}_4$, respectively:

Convergence and extension property for products: There exists an integer N (depending only on $\mathcal{Y}_1$ and $\mathcal{Y}_2$), and for any $w_{(1)} \in W_1$, $w_{(2)} \in W_2$, $w_{(3)} \in W_3$, $w'_{(4)} \in W'_4$, there exist $j \in \mathbb{N}$, $r_i, s_i \in \mathbb{R}$, $i = 1, \ldots, j$, and analytic functions $f_i(z)$ on $|z| < 1$, $i = 1, \ldots, j$, satisfying

$$\Re(\text{wt } w_{(1)} + \text{wt } w_{(2)} + s_i) > N, \quad i = 1, \ldots, j, \tag{2.15}$$

such that

$$\langle w'_{(4)}, \mathcal{Y}_1(w_{(1)}, x_2)\mathcal{Y}_2(w_{(2)}, x_2)w_{(3)}\rangle_{W_4}\Big|_{x_1^n = e^{n \log z_1},\, x_2^n = e^{n \log z_2},\, n \in \mathbb{C}}$$

$$\tag{2.16}$$

is convergent when $|z_1| > |z_2| > 0$ and can be analytically extended to the multi-valued analytic function

$$\sum_{i=1}^{j} z_2^{r_i}(z_1 - z_2)^{s_i} f_i\left(\frac{z_1 - z_2}{z_2}\right) \tag{2.17}$$

when $|z_2| > |z_1 - z_2| > 0$.

Convergence and extension property for iterates: There exists an integer $\tilde{N}$ (depending only on $\mathcal{Y}_3$ and $\mathcal{Y}_4$), and for any $w_{(1)} \in W_1$, $w_{(2)} \in W_2$, $w_{(3)} \in W_3$, $w'_{(4)} \in W'_4$, there exist $k \in \mathbb{N}$, $\tilde{r}_i, \tilde{s}_i \in \mathbb{R}$, $i = 1, \ldots, k$, and analytic functions $\tilde{f}_i(z)$ on $|z| < 1$, $i = 1, \ldots, k$, satisfying

$$\Re(\text{wt } w_{(2)} + \text{wt } w_{(3)} + s_i) > \tilde{N}, \quad i = 1, \ldots, k, \qquad (2.18)$$

such that

$$\langle w'_{(4)}, \mathcal{Y}_4(\mathcal{Y}_3(w_{(1)}, x_0)w_{(2)}, x_2)w_{(3)}\rangle_{W_4}\bigg|_{x_0^n = e^{n \log(z_1 - z_2)},\, x_2^n = e^{n \log z_2},\, n \in \mathbb{C}}$$

$$(2.19)$$

is convergent when $|z_2| > |z_1 - z_2| > 0$ and can be analytically extended to the multi-valued analytic function

$$\sum_{i=1}^{k} z_1^{\tilde{r}_i} z_2^{\tilde{s}_i} \tilde{f}_i\left(\frac{z_2}{z_1}\right) \qquad (2.20)$$

when $|z_1| > |z_2| > 0$.

If for any V-modules W_1, W_2, W_3, W_4 and W_5 and any intertwining operators $\mathcal{Y}_1$ and $\mathcal{Y}_2$ of the types as above, the convergence and extension property for products holds, we say that *the products of the intertwining operators for V have the convergence and extension property*. Similarly we can define what *the iterates of the intertwining operators for V have the convergence and extension property* means.

If a generalized V-module $W = \coprod_{n \in \mathbb{C}} W_{(n)}$ satisfying the condition that $W_{(n)} = 0$ for n whose real part is sufficiently small, we say that W is *lower-truncated*.

Assume that the products and the iterates of the intertwining operators for V are convergent. Let W_1, W_2 and W_3 be three V-modules, $w_{(1)} \in W_1$, $w_{(2)} \in W_2$ and $w_{(3)} \in W_3$ and $z_1, z_2 \in \mathbb{C}$ satisfying $|z_1| > |z_2| > |z_1 - z_2| > 0$. By Proposition 2.1, any $P(z)$-intertwining maps (for $z = z_1, z_2, z_1 - z_2$) can be obtained from certain intertwining operators by substituting complex powers of $e^{\log z}$ for the complex powers of the formal variable x. Thus $w_{(1)} \boxtimes_{P(z_1)} (w_{(2)} \boxtimes_{P(z_2)} w_{(3)})$ (or $(w_{(1)} \boxtimes_{P(z_1 - z_2)} w_{(2)}) \boxtimes_{P(z_2)} w_{(3)}$) is a product (or an iterate) of two intertwining operators evaluated at $w_{(1)} \otimes w_{(2)} \otimes w_{(3)}$ and with the complex powers of the formal variables replaced by the complex powers of $e^{\log z_1}$ and of $e^{\log z_2}$ (or by the complex powers of $e^{\log(z_1 - z_2)}$ and of $e^{\log z_2}$). By assumption, $w_{(1)} \boxtimes_{P(z_1)} (w_{(2)} \boxtimes_{P(z_2)} w_{(3)})$ (or $(w_{(1)} \boxtimes_{P(z_1 - z_2)} w_{(2)}) \boxtimes_{P(z_2)} w_{(3)}$) is a well-defined element of $\overline{W_1 \boxtimes_{P(z_1)} (W_2 \boxtimes_{P(z_2)} W_3)}$ (or of $\overline{(W_1 \boxtimes_{P(z_1 - z_2)} W_2) \boxtimes_{P(z_2)} W_3}$). The following result is proved in [**H2**]:

THEOREM 2.10. *Assume that V is a rational vertex operator algebra and all irreducible V-modules are $\mathbb{R}$-graded. Also assume that V satisfies the following conditions:*

(i) *Every finitely-generated lower-truncated generalized V-module is a V-module.*

(ii) *The products or the iterates of the intertwining operators for V have the convergence and extension property.*

Then for any V-modules W_1, W_2 and W_3 and any complex numbers z_1 and z_2 satisfying $|z_1| > |z_2| > |z_1 - z_2| > 0$; there exists a unique isomorphism $\mathcal{A}_{P(z_1),P(z_2)}^{P(z_1-z_2),P(z_2)}$ from $W_1 \boxtimes_{P(z_1)} (W_2 \boxtimes_{P(z_2)} W_3)$ to $(W_1 \boxtimes_{P(z_1-z_2)} W_2) \boxtimes_{P(z_2)} W_3$ such that for any $w_{(1)} \in W_1$, $w_{(2)} \in W_2$ and $w_{(3)} \in W_3$,

$$\overline{\mathcal{A}}_{P(z_1),P(z_2)}^{P(z_1-z_2),P(z_2)}(w_{(1)} \boxtimes_{P(z_1)} (w_{(2)} \boxtimes_{P(z_2)} w_{(3)}))$$
$$= (w_{(1)} \boxtimes_{P(z_1-z_2)} w_{(2)}) \boxtimes_{P(z_2)} w_{(3)},$$

where

$$\overline{\mathcal{A}}_{P(z_1),P(z_2)}^{P(z_1-z_2),P(z_2)} : \overline{W_1 \boxtimes_{P(z_1)} (W_2 \boxtimes_{P(z_2)} W_3)} \to \overline{(W_1 \boxtimes_{P(z_1-z_2)} W_2) \boxtimes_{P(z_2)} W_3}$$

is the unique extension of $\mathcal{A}_{P(z_1),P(z_2)}^{P(z_1-z_2),P(z_2)}$.

3. The nonmeromorphic extension of $V^\natural$

We sketch the construction the nonmeromorphic extension of $V^\natural$ in this section. This nonmeromorphic extension is in fact the algebra of all intertwining operators for the vertex operator algebra V_Λ^+. We first verify that the conditions for the tensor product theory reviewed in Section 2 are satisfied by V_Λ^+. Then we calculate the fusion rules for V_Λ^+. The nonmeromorphic extension is obtained using the fusion rules and the tensor product theory. The details of the proofs of the results stated in Subsections 3.1 and 3.2 are given in [**H3**].

3.1. Modules for the Virasoro vertex operator algebras and their tensor products.

To prove that V_Λ^+ satisfies the conditions in Theorem 2.10, we shall use the results in [**DMZ**] and [**D3**]. So we first have to prove that the tensor product theory can be applied to the vertex operator algebra $L(\frac{1}{2}, 0)$. The rationality of $L(\frac{1}{2}, 0)$ is proved in [**DMZ**]. In general, the rationality of the Virasoro vertex operator algebra $L(c_{p,q}, 0)$ of central charge $c_{p,q} = 1 - 6\frac{(p-q)^2}{pq}$ is proved in [**W**] for an arbitrary pair p, q of relatively prime positive integers larger than 1. We have the following result for $L(c_{p,q}, 0)$:

PROPOSITION 3.1. *Let p, q be a pair of relatively prime positive integers larger than 1. Then we have:*

(i) *Every finitely-generated lower-truncated generalized $L(c_{p,q}, 0)$-module is a module.*

(ii) *The products of the intertwining operators for $L(c_{p,q}, 0)$ have the convergence and extension property.*

Sketch of the proof   The first conclusion is an easy exercise on the representations of the Virasoro algebra. The second conclusion is proved using the representation theory of the Virasoro algebra and the differential equations of Belavin, Polyakov and Zamolodchikov (BPZ equations) for correlation functions in the minimal models [**BPZ**]. We show that

$$\langle w'_{(4)}, \mathcal{Y}_1(w_{(1)}, x_1)\mathcal{Y}_2(w_{(2)}, x_2)w_{(3)}\rangle\Big|_{x_1^n = e^{n\log z_1}, x_2^n = e^{n\log z_2}, n\in\mathbb{C}} \tag{3.1}$$

satisfies a system of BPZ equations when $w_{(1)}, w_{(2)}, w_{(3)}$ and $w'_{(4)}$ are all lowest weight vectors. The BPZ equation has only regular singular points. Thus using the theory of equations of regular singular points (see, for example, Appendix B of [**K**]) and the definition of intertwining operator, we can show that (3.1) in this case is convergent when $|z_1| > |z_2| > 0$ and can be analytically extended to a function of the form (2.17) when $|z_2| > |z_1 - z_2| > 0$. This together with the brackets of $L(n)$, $n \in \mathbb{C}$, with the intertwining operators and the $L(-1)$-derivative property for the intertwining operators shows that the products of the intertwining operators for $L(c_{p,q}, 0)$ have the convergence and extension property. $\square$

Next we discuss tensor products of $L(c_{p,q}, 0)$.

LEMMA 3.2. *Let n be a positive integer, (p_i, q_i), $i = 1, \ldots, n$, n pairs of relatively prime positive integers larger than 1, $V = L(c_{p_1,q_1}, 0) \otimes \cdots \otimes L(c_{p_n,q_n}, 0)$, $W_i = L(c_{p_1,q_1}, h_1^{(i)}) \otimes \cdots \otimes L(c_{p_n,q_n}, h_n^{(i)})$, $i = 1, 2, 3$, irreducible V-modules and $\mathcal{Y}$ an intertwining operator of type $\binom{W_3}{W_1 W_2}$. Then there exist intertwining operators $\mathcal{Y}_i$ of type $\binom{L(c_{p_i,q_i}, h_i^{(3)})}{L(c_{p_i,q_i}, h_i^{(1)})L(c_{p_i,q_i}, h_i^{(2)})}$, $i = 1, \ldots, n$, such that*

$$\mathcal{Y} = \mathcal{Y}_1 \otimes \cdots \otimes \mathcal{Y}_n.$$

This lemma is an easy consequence of the result in [**DMZ**] expressing the fusion rules of V in terms of the fusion rules of $L(c_{p_i,q_i}, 0)$, $i = 1, \ldots, n$. It can also be proved directly using the special properties of the Virasoro vertex operator algebras. Using Proposition 3.1, Lemma 3.2 and the methods used to prove results on modules for tensor products of a vertex operator algebra in [**FHL**], we obtain:

PROPOSITION 3.3. *For any positive integer n and any n pairs (p_i, q_i) of relatively prime positive integers larger than 1, $i = 1, \ldots, n$, we have:*

(i) *Every finitely-generated lower-truncated generalized $L(c_{p_1,q_1}, 0) \otimes \cdots \otimes L(c_{p_n,q_n}, 0)$-module is a module.*

(ii) *The products of the intertwining operators for*

$$L(c_{p_1,q_1}, 0) \otimes \cdots \otimes L(c_{p_n,q_n}, 0)$$

have the convergence and extension property.

3.2. A class of vertex operator algebras and the associativity of tensor products. Let n be a positive integer, (p_i, q_i), $i = 1, \ldots, n$, n pairs of relatively prime positive integers larger than 1. A vertex operator algebra V is said to be in the class $\mathcal{C}_{p_1, q_1; \ldots; p_n, q_n}$ if V has a vertex operator subalgebra isomorphic to $L(c_{p_1, q_1}, 0) \otimes \cdots \otimes L(c_{p_n, q_n}, 0)$. The work of Dong, Mason and Zhu [**DMZ**] shows that V_Λ^+ is in the class $\mathcal{C}_{3,4; \ldots; 3,4}$ with $n = 48$.

Using Proposition 3.3, we can prove:

PROPOSITION 3.4. *Let n be a positive integer, (p_i, q_i), $i = 1, \ldots, n$, n pairs of relatively prime positive integers larger than 1, and let V be a vertex operator algebra in the class $\mathcal{C}_{p_1, q_1; \ldots; p_n, q_n}$. Then we have:*

(i) *Every finitely-generated lower-truncated generalized V-module is a module.*

(ii) *The products of the intertwining operators for V have the convergence and extension property.*

The proof of the second conclusion is easy. The proof of the first conclusion is more subtle than what it seems to be since a finitely-generated generalized V-module is not obviously finitely-generated as a generalized $L(c_{p_1, q_1}, 0) \otimes \cdots \otimes L(c_{p_n, q_n}, 0)$-module.

3.3. The fusion rules for V_Λ^+. We first quote a result proved in [**FHL**] and [**HL3**]:

PROPOSITION 3.5. *Let V be a vertex operator algebra and W_1, W_2, W_3 three V-modules. Then for any permutation σ of three letters, $\mathcal{N}_{W_{\sigma(1)} W_{\sigma(2)}}^{W'_{\sigma(3)}} = \mathcal{N}_{W_1 W_2}^{W'_3}$. In particular, if W_1, W_2 and W_3 are all self-dual, that is, are isomorphic to their contragredient modules W'_1, W'_2 and W'_3, respectively, then $\mathcal{N}_{W_{\sigma(1)} W_{\sigma(2)}}^{W_{\sigma(3)}} = \mathcal{N}_{W_1 W_2}^{W_3}$.*

In the special case that one of the three modules is the adjoint module and the other two are irreducible, it is easy to prove:

PROPOSITION 3.6. *Let V be a vertex operator algebra and W_1, W_2 irreducible V-modules contragredient to themselves. Then the fusion rules $\mathcal{N}_{W_1 W_2}^V$, $\mathcal{N}_{V W_1}^{W_2}$ and $\mathcal{N}_{W_1 V}^{W_2}$ are equal to 0 if W_1 and W_2 are not isomorphic and are equal to 1 if they are isomorphic.*

For $i, j \in \mathbb{Z}_2$, let

$$M^{i,j} = \begin{cases} V_\Lambda^+, & i = j = 0, \\ V_\Lambda^-, & i = 1, j = 0, \\ (V_\Lambda^T)^+, & i = 0, j = 1, \\ (V_\Lambda^T)^-, & i = j = 1. \end{cases}$$

THEOREM 3.7. *The fusion rules for V_Λ^+ is*

$$\mathcal{N}_{M^{i_1,j_1}M^{i_2,j_2}}^{M^{i_3,j_3}} = \begin{cases} 1, & i_3 = i_1 + i_2, j_3 = j_1 + j_2, \\ 0, & otherwise. \end{cases}$$

Sketch of the proof The proof uses Proposition 3.5, Proposition 3.6, Theorem 1.9, Theorem 1.10, Theorem 1.11 and the tensor product theory described in Section 2, especially the associativity for $P(\cdot)$-tensor products. Using the first half of Theorem 1.11, we can prove $\mathcal{N}_{M^{i_1,1}M^{i_2,1}}^{M^{i_3,1}} = 0$ and $\mathcal{N}_{M^{i_1,0}M^{i_2,0}}^{M^{i_3,1}} = 0$. These fusion rules together with Proposition 3.5 and Proposition 3.6 reduces the problem to the calculations of $\mathcal{N}_{V_\Lambda^-(V_\Lambda^T)^+}^{(V_\Lambda^T)^-}$, $\mathcal{N}_{V_\Lambda^-(V_\Lambda^T)^-}^{(V_\Lambda^T)^-}$, $\mathcal{N}_{V_\Lambda^-(V_\Lambda^T)^+}^{(V_\Lambda^T)^+}$ and $\mathcal{N}_{V_\Lambda^- V_\Lambda^-}^{V_\Lambda^-}$. The calculations of these fusion rules use more details on the lowest weight vectors with respect to $L(\frac{1}{2},0)^{\otimes 48}$ in $V_\Lambda^\pm$ and $(V_\Lambda^T)^\pm$ given in Theorem 1.10 and the associativity of the intertwining operators which is in fact equivalent to the associativity of the $P(\cdot)$-tensor products. By Theorem 1.12, we know that V_Λ^+ is rational and the four irreducible V_Λ^+-modules are all $\mathbb{R}$-graded. The rationality of V_Λ^+, the fact that all V_Λ^+-modules are $\mathbb{R}$-graded and Proposition 3.4 guarantee that we can use the tensor product theory, especially the associativity. □

3.4. The nonmeromorphic extension. Let

$$W^\natural = V_\Lambda \oplus V_\Lambda^+ = \coprod_{(i,j)\in\mathbb{Z}_2\oplus\mathbb{Z}_2} M^{i,j}.$$

We define a vertex operator map $Y_{W^\natural}: W^\natural \otimes W^\natural \to W^\natural[[x^{\frac{1}{2}}, x^{-\frac{1}{2}}]]$ as follows: For $u,v \in V_\Lambda$, $Y_{W^\natural}(u,x)v = Y_{V_\Lambda}(u,v)$; for $u \in V_\Lambda$, $v \in V_\Lambda^T$, $Y_{W^\natural}(u,x)v = Y_{V_\Lambda^T}(u,x)v$; for $u \in V_\Lambda^T$, $v \in V_\Lambda$, $Y_{W^\natural}(u,x)v = e^{xL(-1)}Y_{V_\Lambda^T}(v,e^{\pi i}x)u$; for $u,v \in V_\Lambda^T$, $Y_{W^\natural}(u,x)v$ is defined by

$$\langle w, Y_{W^\natural}(u,x)v\rangle_{V_\Lambda} = \langle Y_{V_\Lambda^T}(w, e^{\pi i}x^{-1})e^{xL(1)}(e^{\pi i}x^2)^{-L(0)}u, e^{x^{-1}L(1)}v\rangle_{V_\Lambda^T}$$

for all $w \in V_\Lambda$. We have a (nonsymmetric) nondegenerate $\{1,-1\}$-valued $\mathbb{Z}$-bilinear form Ω_{SU} (the subscript SU means super, see Subsection 4.2) on the finite abelian group $\mathbb{Z}_2 \oplus \mathbb{Z}_2$ determined uniquely by $\Omega_{SU}((1,0),(1,0)) = 1$, $\Omega_{SU}((1,0),(0,1)) = -1$, $\Omega_{SU}((0,1),(1,0)) = 1$, $\Omega_{SU}((0,1),(0,1)) = 1$, and the bilinearity. To formulate the main result of this paper, we need the notions of abelian intertwining algebra, whose definition, examples and axiomatic properties can be found in [**DL**]. In the definition of abelian intertwining algebra, part of the data is an abelian group G and a normalized abelian 3-cocycle (F,Ω) for the abelian group G with values in $\mathbb{C}^\times$, where F is a normalized 3-cocycle for G as a group. In this paper, G is $\mathbb{Z}_2 \oplus \mathbb{Z}_2$, F is trivial (denoted by 1) and Ω is equal to Ω_{SU} defined above.

THEOREM 3.8. *The structure* $(W^\natural, Y_{W^\natural}, \mathbf{1}, \omega, 2, \mathbb{Z}_2 \oplus \mathbb{Z}_2, 1, \Omega_{SU})$ *is an abelian intertwining algebra of central charge* 24.

Sketch of the proof The proof uses the fusion rules in Theorem 3.7 and the tensor product theory described in Section 2, especially the associativity. We already know that the tensor product theory can be applied to V_Λ^+. From [**H2**], we know that the associativity of $P(\cdot)$-tensor products is equivalent to the associativity of the intertwining operators. So in this case, we have the associativity for intertwining operators. By the fusion rules, for every ordered triple of irreducible V_Λ^+-modules, any two intertwining operators of the type specified by this triple are linearly dependent. On the other hand, $Y_{W^\natural}$ restricted to any ordered triple of irreducible V_Λ^+-modules is an intertwining operator and is nonzero if the fusion rule is nonzero. Thus the associativity for intertwining operators gives the associativity for $Y_{W^\natural}$. It can be verified directly that $Y_{W^\natural}$ satisfies a version of the skew symmetry for vertex operators. Combining the associativity and the skew symmetry of $Y_{W^\natural}$, we obtain the commutativity of $Y_{W^\natural}$. It can be shown that if the fusion algebra for a rational vertex operator algebra satisfying the conditions in Theorem 2.10 is the group algebra of an abelian group, the products and the iterates of intertwining operators among irreducible modules must be appropriate expansions of generalized rational functions (see [**DL**] for the meaning of generalized rational functions). In our case, the fusion algebra is the group algebra of the abelian group $\mathbb{Z}_2 \oplus \mathbb{Z}_2$. Thus we have the generalized rationalities of both products and iterates for $Y_{W^\natural}$. $\square$

4. Applications

We give applications of the results obtained in the preceding section. A special case of Theorem 3.8 gives a new and conceptual proof that $V^\natural$ has a natural vertex operator algebra structure. Other special cases give an irreducible twisted module for the moonshine module with respect to the obvious involution, a vertex operator superalgebra and a twisted module for this vertex operator superalgebra with respect to the involution which is the identity on the even subspace and is -1 on the odd subspace. We also use Theorem 3.8 to construct the superconformal structures of Dixon, Ginsparg and Harvey rigorously.

4.1. The moonshine module and its twisted module. When we restrict ourselves to the moonshine module $V^\natural = V_\Lambda^+ \oplus (V_\Lambda^T)^+ \subset W^\natural$, we immediately obtain the following:

THEOREM 4.1. *The quadruple $(V^\natural, Y_{V^\natural}, \mathbf{1}, \omega)$ is a vertex operator algebra.*

Remark 4.2. Unlike the proof of this theorem given in [**FLM2**], our proof does not use triality or any result in group theory. It can be proved without using triality or group theory that any automorphism of the Griess algebra can be extended to an automorphism of the vertex operator algebra $V^\natural$ (this was also observed by Dong). Thus our proof (or any proof without using triality or group theory, for example, the one given in [**DGM1**]) makes the fact that the automorphism group of the Griess algebra and the automorphism group of the vertex operator algebra $V^\natural$ are isomorphic, to be logically independent of triality and group theory. This independence allows us to obtain another proof of the theorem saying that the Monster is the (full) automorphism group of the vertex operator algebra $V^\natural$ based on the the theorem saying that the Monster is the (full) automorphism group of the Griess algebra proved by Griess [**G**] and Tits [**Ti1**] [**Ti2**], simplified by Conway [**C**] and Tits [**Ti2**] and understood conceptually by Frenkel, Lepowsky and Meurman using the theory of vertex operators and triality [**FLM1**] [**FLM2**].

Another immediate consequence is on the irreducible twisted module for $V^\natural$:

THEOREM 4.3. *Let $\tau : V^\natural \to V^\natural$ be an involution defined by $\tau(v) = v$ if $v \in V_\Lambda^+$ and $\tau(v) = -v$ if $v \in (V_\Lambda^T)^+$. Then the pair*

$$(V_\Lambda^- \oplus (V_\Lambda^T)^-, Y_{W^\natural}|_{V^\natural \otimes (V_\Lambda^- \oplus (V_\Lambda^T)^-)})$$

is an irreducible τ-twisted module for $V^\natural$. Any irreducible τ-twisted module is isomorphic to this one.

Theorem 4.3 is also proved by Dong and Mason using a different method.

Remark 4.4. Note that if we are only interested in Theorem 4.1 or Theorem 4.3, we can prove them using only parts of the fusion rules which we calculated for V_Λ^+.

4.2. The superconformal structures of Dixon, Ginsparg and Harvey. The following result observed first by Dixon, Ginsparg and Harvey is proved using Theorem 3.8 and some concrete calculations of twisted vertex operators:

THEOREM 4.5. *For any $t \in T$ satisfying $\langle 1 \otimes t, 1 \otimes t \rangle_{V_\Lambda^T} = 1$ (for example, $t(a)$ for any $a \in \hat{\Lambda}$), let $Y_{W^\natural}(2(1 \otimes t), x) = \sum_{n \in \frac{1}{2}\mathbb{Z}} G(n) x^{-n - \frac{3}{2}}$. Then the operators $L(n)$, $n \in \mathbb{Z}$ and $G(n)$, $n \in \frac{1}{2}\mathbb{Z}$, satisfies the super-Virasoro relations:*

$$[L(m), L(n)] = (m - n)L(m + n) + \frac{\hat{c}_{W^\natural}}{8}(m^3 - m)\delta_{m+n,0}, \qquad (4.1)$$

$$[L(m), G(n)] = \left(\frac{m}{2} - n\right) G(m + n), \qquad (4.2)$$

$$\{G(m), G(n)\} = 2L(m + n) + \frac{\hat{c}_{W^\natural}}{2}\left(m^2 - \frac{1}{4}\right)\delta_{m+n,0} \qquad (4.3)$$

where the super-central charge $\hat{c}_{W^\natural} = 16$ and $\{\cdot, \cdot\}$ denotes the anti-bracket.

To summarize the superconformal structures on $W^\natural$, we need the following notions:

DEFINITION 4.6. An *(N=1) Neveu-Schwarz type superconformal vertex operator algebra of super-central charge (or super-rank) $\hat{c}$ is a vertex operator superalgebra $(V, Y, \mathbf{1}, \omega)$* (see, for example, [**FFR**] or [**DL**]), equipped with an element $\xi \in V$ such that the components of $Y(x^{L(0)}\omega, x)$ and $Y(x^{L(0)}\xi, x)$ satisfies the super-Virasoro relations (4.1)–(4.3) with $\hat{c}_{W^\natural}$ replaced by $\hat{c}$.

In [**KW**], (N=1) Neveu-Schwarz type superconformal vertex operator algebras are studied and are called "N=1 (NS-type) vertex operator superalgebras." The (N=1) Neveu-Schwarz type superconformal vertex operator algebra just defined is denoted by $(V, Y, \mathbf{1}, \omega, \xi)$ or simply by V. *Homomorphisms, isomorphisms and automorphisms* of (N=1) Neveu-Schwarz type superconformal vertex operator algebras are defined in the obvious way.

Let $(V, Y, \mathbf{1}, \omega, \xi)$ be a Neveu-Schwarz type superconformal vertex operator algebra of super-central charge $\hat{c}$. A module for V is defined to be a module for V as a vertex operator superalgebra. We define a linear isomorphism σ of V by linearity and

$$\sigma(v) = \begin{cases} v, & v \in V^0, \\ -v, & v \in V^1, \end{cases} \tag{4.4}$$

where V^0 and V^1 are even and odd subspaces of V, respectively. It is clear that σ is an involution and an automorphism of the Neveu-Schwarz type superconformal vertex operator algebra V. Therefore it is natural to consider σ-twisted V-modules. See [**FFR**] for a definition of σ-twisted V-modules. Following physicists' terminology, we also call an (untwisted) module for V a *Neveu-Schwarz sector for V* and a σ-twisted module for V a *Ramond sector for V*.

DEFINITION 4.7. An *(N=1) superconformal vertex operator algebra of super-central charge* (or *super-rank*) $\hat{c}$ is a abelian intertwining algebra

$$(W, Y_W, \mathbf{1}, \omega, 2, \mathbb{Z}_2 \oplus \mathbb{Z}_2, 1, \Omega_{SU})$$

(where Ω_{SU} is the $\mathbb{Z}$-bilinear form on $\mathbb{Z}_2 \oplus \mathbb{Z}_2$ defined in Subsection 3.4), equipped with an element $\xi \in W$ such that the components of $Y(\omega, x)$ and $Y(\xi, x)$ satisfies the super-Virasoro relations (4.1)–(4.3) with $\hat{c}_{W^\natural}$ replaced by $\hat{c}$. The element ξ is called the *Neveu-Schwarz-Ramond element*. Let $W = \coprod_{(i,j) \in \mathbb{Z}_2 \oplus \mathbb{Z}_2} W^{i,j}$. Then $W_{NS} = W^{0,0} \oplus W^{1,1}$ is called the *Neveu-Schwarz sector* and $W_R = W^{0,1} \oplus W^{1,0}$ is called the *Ramond sector*.

The abelian intertwining algebra underlying a superconformal vertex operator algebra can be described using its substructures as follows:

PROPOSITION 4.8. *Let $(W, Y_W, \mathbf{1}, \omega, 2, \mathbb{Z}_2 \oplus \mathbb{Z}_2, 1, \Omega_{SU})$ be an abelian intertwining algebra of central charge $\frac{3}{2}\hat{c}$. Then we have:*

(i) *The $\mathbb{Z}$-graded vector spaces $W^{0,0}$, $W^{0,0} \oplus W^{0,1}$, $W^{0,0} \oplus W^{1,0}$ with the restrictions of Y_W as the vertex operator maps, the vacuum $\mathbf{1}$ and the Virasoro element ω are vertex operator algebras of central charge $\frac{3}{2}\hat{c}$ and $W_{NS} = W^{0,0} \oplus W^{1,1}$ is a Neveu-Schwarz type superconformal vertex operator algebra of super-central charge $\hat{c}$.*

(ii) *The $\frac{1}{2}\mathbb{Z}$-graded vector spaces $W^{i,j}$, $i, j \in \mathbb{Z}_2$, are modules for $W^{0,0}$.*

(iii) *The $\frac{1}{2}\mathbb{Z}$-graded vector spaces $W^{1,0} \oplus W^{1,1}$ and $W^{0,1} \oplus W^{1,1}$ with the restrictions of Y_W as the vertex operator maps are twisted modules for*

$W^{0,0} \oplus W^{0,1}$ and $W^{0,0} \oplus W^{1,0}$, respectively, with respect to the obvious involutions.

(iv) The $\frac{1}{2}\mathbb{Z}$-graded vector spaces $W_R = W^{0,1} \oplus W^{1,0}$ with with the restrictions of Y_W as the vertex operator maps is a σ-twisted module for W_{NS}.

Conversely, let $W = \coprod_{i,j \in \mathbb{Z}_2} W^{i,j}$ where $W^{i,j}$, $i,j \in \mathbb{Z}_2$, are four $\frac{1}{2}\mathbb{Z}$-graded vector spaces be equipped with a vertex operator map $Y_W : W \otimes W \to W\{x\}$ and two distinguished elements $\mathbf{1}$ and ω such that (i)–(iv) above hold. Then

$$(W, Y_W, \mathbf{1}, \omega, 2, \mathbb{Z}_2 \oplus \mathbb{Z}_2, 1, \Omega_{SU})$$

is an abelian intertwining algebra.

We can summarize the main result and the main applications of the present paper as follows:

THEOREM 4.9. For any $t \in T$ satisfying $\langle 1 \otimes t, 1 \otimes t \rangle_{V_\Lambda^T} = 1$ (for example, $t(a)$ for any $a \in \hat{\Lambda}$), $(W^\natural, Y_{W^\natural}, \mathbf{1}, \omega, 2, \mathbb{Z}_2 \oplus \mathbb{Z}_2, \Omega_{SU}, 2(1 \otimes t))$ is a superconformal vertex operator algebra of super-central charge $\hat{c}_{W^\natural} = 16$. In particular, we have:

(i) The moonshine module $V^\natural = V_\Lambda^+ \oplus (V_\Lambda^T)^+$ with the restriction of $Y_{W^\natural}$ as the vertex operator map, the vacuum $\mathbf{1}$ and the Virasoro element ω is a vertex operator algebras of central charge $\frac{3}{2}\hat{c}_{W^\natural} = 24$ and $W_{NS}^\natural = V_\Lambda^+ \oplus (V_\Lambda^T)^-$ with the restriction of $Y_{W^\natural}$ as the vertex operator map is a Neveu-Schwarz type superconformal vertex operator algebra of super-central charge $\hat{c}_{W^\natural} = 16$.

(ii) The $\frac{1}{2}\mathbb{Z}$-graded vector spaces $V_\Lambda^- \oplus (V_\Lambda^T)^-$ with the restriction of $Y_{W^\natural}$ as the vertex operator map is a τ-twisted modules for $V^\natural$.

(iii) The $\frac{1}{2}\mathbb{Z}$-graded vector spaces $W_R^\natural = V_\Lambda^- \oplus (V_\Lambda^T)^+$ with the restriction of $Y_{W^\natural}$ as the vertex operator map is a σ-twisted module for $W_{NS}^\natural$.

REFERENCES

[BPZ] A. A. Belavin, A. M. Polyakov and A. B. Zamolodchikov, Infinite conformal symmetries in two-dimensional quantum field theory, *Nucl. Phys.* **B241** (1984), 333–380.

[B1]   R. E. Borcherds, Vertex algebras, Kac-Moody algebras, and the Monster, *Proc. Natl. Acad. Sci. USA* **83** (1986), 3068–3071.

[B2]   R. E. Borcherds, Monstrous moonshine and monstrous Lie superalgebras, *Invent. Math.* **109** (1992), 405–444.

[C]   J. H. Conway, A simple construction for the Fischer-Griess monster group, *Invent. Math.* **79** (1985), 513–540.

[CN]   J. H. Conway and S. P. Norton, Monstrous moonshine, *Bull. London Math. Soc.* **11** (1979), 308–339.

[DGH] L. Dixon, P. Ginsparg and J. A. Harvey, Beauty and the beast: superconformal symmetry in a Monster module, *Comm. Math. Phys.* **119** (1988), 221–241.

[DGM1] L. Dolan, P. Goddard and P. Montague, Conformal field theory of twisted vertex operators, *Nucl. Phys.* **B338** (1990), 529–601.

[DGM2] L. Dolan, P. Goddard and P. Montague, Conformal field theory, triality and the Monster group, *Phys. Lett.* **B236**(1990), 165–172.

[DGM3] L. Dolan, P. Goddard and P. Montague, Conformal field theories, representations and lattice constructions, to appear.

[D1]   C. Dong, Vertex algebras associated with even lattices, *J. Algebra* **161** (1993), 245–265.

[D2]   C. Dong, Twisted modules for vertex algebras associated with even lattices, *J. Algebra* **165** (1994), 91–112.

[D3]   C. Dong, Representations of the moonshine module vertex operator algebra, in: *Mathematical Aspects of Conformal and Topological Field Theories and Quantum Groups, Proc. Joint Summer Research Conference, Mount Holyoke, 1992*, ed. P. Sally, M. Flato, J. Lepowsky, N. Reshetikhin and G. Zuckerman, Contemporary Math., Vol. 175, Amer. Math. Soc., Providence, 1994, 27–36.

[DL]   C. Dong and J. Lepowsky, *Generalized Vertex Algebras and Relative Vertex Operators*, Progress in Math., Vol. 112, Birkhäuser, Boston, 1993.

[DM]   C. Dong and G. Mason, On the construction of the moonshine module as a $\mathbb{Z}_p$-orbifold, in: *Mathematical Aspects of Conformal and Topological Field Theories and Quantum Groups, Proc. Joint Summer Research Conference, Mount Holyoke, 1992*, ed. P. Sally, M. Flato, J. Lepowsky, N. Reshetikhin and G. Zuckerman, Contemporary Math., Vol. 175, Amer. Math. Soc., Providence, 1994.

[DMZ]   C. Dong, G. Mason and Y. Zhu, Discrete series of the Virasoro algebra and the moonshine module, in: *Algebraic Groups and Their Generalizations: Quantum and Infinite-Dimensional Methods*, ed. William J. Haboush and Brian J. Parshall, Proc. Symp. Pure. Math., American Math. Soc., Providence, 1994, Vol. 56, Part 2, 295–316.

[FFR]   A. J. Feingold, I. B. Frenkel, J. F. X. Ries, *Spinor construction of vertex operator algebras, triality and $E_8^{(1)}$*, Contemperary Math., Vol. 121, Amer. Math. Soc., Providence, 1991.

[FHL]   I. B. Frenkel, Y.-Z. Huang and J. Lepowsky, On axiomatic approaches to vertex operator algebras and modules, preprint, 1989; *Memoirs Amer. Math. Soc.* **104**, 1993.

[FLM1] I. B. Frenkel, J. Lepowsky and A. Meurman, A natural representation of the Fischer-Griess Monster with the modular function J as character, *Proc. Natl. Acad. Sci. USA* **81** (1984), 3256–3260.

[FLM2] I. B. Frenkel, J. Lepowsky and A. Meurman, *Vertex Operator Algebras and the Monster*, Pure and Appl. Math., Vol. 134, Academic Press, Boston, 1988.

[FZ]   I. B. Frenkel and Y. Zhu, Vertex operator algebras associated to representations of affine and Virasoro algebras, *Duke Math. J.* **66** (1992), 123–168.

[G]   R. L. Griess, Jr., The Friendly Giant, *Invent. Math.* **69** (1982), 1–102.

[H1]   Y.-Z. Huang, Operadic formulation of topological vertex algebras and Gerstenhaber or Batalin-Vilkovisky algebras, *Comm. Math. Phys.*, **164** (1994), 105–144.

[H2]   Y.-Z. Huang, A theory of tensor product for module categories for a vertex operator algebra, IV, *J. Pure Appl. Alg.*, to appear.

[H3]   Y.-Z. Huang, Virasoro vertex operator algebras, the (nonmeromorphic) operator product expansion and the tensor product theory, to appear.

[HL1]   Y.-Z. Huang and J. Lepowsky, Toward a theory of tensor products for representations of a vertex operator algebra, in: *Proc. 20th International Conference on Differential Geometric Methods in Theoretical Physics, New York, 1991*, ed. S. Catto and A. Rocha, World Scientific, Singapore, 1992, Vol. 1, 344–354.

[HL2]   Y.-Z. Huang and J. Lepowsky, A theory of tensor product for module categories for a vertex operator algebra, I, *Selecta Mathematica*, to appear.

[HL3]   Y.-Z. Huang and J. Lepowsky, A theory of tensor product for module categories for a vertex operator algebra, II, *Selecta Mathematica*, to appear.

[HL4]   Y.-Z. Huang and J. Lepowsky, Tensor products of modules for a vertex operator algebras and vertex tensor categories, in: *Lie Theory and Geometry, in honor of Bertram Kostant*, ed. J.-L. Brylinski, R. Brylinski, V. Guillemin, V. Kac, Birkhäuser, Boston, 1994, 349–383.

[HL5]   Y.-Z. Huang and J. Lepowsky, A theory of tensor product for module categories for a vertex operator algebra, III, *J. Pure Appl. Alg.*, to appear.

[J]   E. Jurisich, Generalized Kac-Moody algebras and their relations to free Lie algebras, Ph.D. thesis, Rutgers University, 1994.

[JLW]   E. Jurisich, J. Lepowsky and R. L. Wilson, Realization of the Monster Lie algebra,

Selecta Mathematica, to appear.

[L]   J. Lepowsky, Remarks on vertex operator algebras and moonshine, in: *Proc. 20th International Conference on Differential Geometric Methods in Theoretical Physics, New York, 1991*, ed. S. Catto and A. Rocha, World Scientific, Singapore, 1992, Vol. 1, 362–370.

[K]   A. W. Knapp, *Representation theory of semisimple groups, an overview based on examples*, Princeton Mathematical Series, Vol. 36, Princeton University Press, Princeton, 1986.

[KW]   V. Kac and W. Wang, Vertex operator superalgebras and their representations, to appear.

[LZ]   B. Lian and G. Zuckerman, New perspectives on the BRST-algebraic structure of string theory, *Comm. Math. Phys.* **154** (1993), 613–646.

[N]   S. P. Norton, Generalized moonshine, *Proc. Symp. Pure Math.* **47** (1987), 209–210.

[Th]   J. G. Thompson, Some numerology between the Fischer-Griess Monster and elliptic modular functions, *Bull. London Math. Soc.* **11** (1979), 352–353.

[Ti1]   J. Tits, Résumé de cours, Annuaire du Collège de France, 1982–1983, 89–102.

[Ti2]   J. Tits, On R. Griess' "Friendly Giant," *invent. Math.* **78** (1984), 491–499.

[Tu1]   M. P. Tuite, Monstrous moonshine from orbifolds, *Comm. Math. Phys.* **146** (1992), 277–309

[Tu2]   M. P. Tuite, On the relationship between monstrous moonshine and the uniqueness of the moonshine module, *Comm. Math. Phys.*, to appear.

[Tu3]   M. P. Tuite, Generalised moonshine and abelian orbifold constructions, in: *Moonshine, the Monster and Related Topics, Proc. Joint Summer Research Conference, Mount Holyoke, 1994*, ed. C. Dong and G. Mason, Contemporary Math., Amer. Math. Soc., Providence, to appear.

[W]   W. Wang, Rationality of Virasoro vertex operator algebras, *International Mathematics Research Notices* (in *Duke Math. J.*) **7** (1993), 197–211.

DEPARTMENT OF MATHEMATICS, UNIVERSITY OF PENNSYLVANIA, PHILADELPHIA, PA 19104 AND DEPARTMENT OF MATHEMATICS, RUTGERS UNIVERSITY, NEW BRUNSWICK, NJ 08903 (CURRENT ADDRESS)

E-mail address: yzhuang@math.rutgers.edu

Contemporary Mathematics
Volume **193**, 1996

ON THE BUEKENHOUT – FISCHER
GEOMETRY OF THE MONSTER

A.A. IVANOV

ABSTRACT. The 3-local geometry $\mathcal{G}_3(M)$ of the Monster group M was introduced in [3] as a locally dual polar space of the group $O_8^-(3)$. Independently a rank 3 truncation of $\mathcal{G}_3(M)$ was described in [23] as a minimal 3-local parabolic geometry of M. More recently, $\mathcal{G}_3(M)$ appeared in the construction of the Moonshine module $V^\natural$ as a Z_3-orbifold in [6]. We discuss the relationship between $\mathcal{G}_3(M)$ and a more familiar 2-local parabolic geometry $\mathcal{G}_2(M)$ of the Monster. The geometry $\mathcal{G}_2(M)$ is quite explicitly involved in the construction of $V^\natural$ as a Z_2-orbifold in [7].

1. THE GEOMETRY $\mathcal{G}_2(M)$ AND $V^\natural$ AS A Z_2-ORBIFOLD.

Let $\mathcal{G}_2(M)$ be the 2-local minimal parabolic geometry of the Monster group M which corresponds to the following "tilde" diagram:

$$\mathcal{G}_2(M) : \quad \underset{2}{\circ}\!\!-\!\!-\!\!-\!\!\underset{2}{\circ}\!\!-\!\!-\!\!-\!\!\underset{2}{\circ}\!\!-\!\!-\!\!-\!\!\underset{2}{\circ}\!\!\overset{\sim}{=\!=\!=}\!\!\underset{2}{\circ}.$$

We always assume that in diagrams the types increase from left to right starting with 1. The elements of types 1, 2 and 3 will be called points, lines and planes, respectively. The points in $\mathcal{G}_2(M)$ are the central involutions in M. The centralizer $C(z)$ of such an involution z has the structure $2^{1+24}.Co_1$, that is, $C(z)$ is an extension of an extraspecial group Q_z of order 2^{25} by the Conway group Co_1. As a $GF(2)$-module for Co_1 the quotient $Q_z/\langle z \rangle$ is isomorphic to $\bar{\Lambda} = \Lambda/2\Lambda$ where Λ is the Leech lattice. It is known [8] that $\bar{C} = C(z)/\langle z \rangle$ is a non-split extension of $\bar{\Lambda}$ by Co_1. Two points z and u are collinear in $\mathcal{G}_2(M)$ (are incident to a common line) if and only if $u \in Q_z$ (equivalently if $z \in Q_u$). The line l containing z and u contains their product and can be identified with the triple $\{z, u, zu\}$. In general for $1 \leq i \leq 5$ the elements of type i in $\mathcal{G}_2(M)$ are subgroups of order 2^i in M whose nontrivial elements are pairwise collinear central involutions. The incidence is by inclusion. Notice that only a proper subclass of such subgroups of order 2^5 correspond to elements of type 5.

The geometry $\mathcal{G}_2(M)$ was first constructed in [23] from the maximal 2-local parabolic geometry of the Monster described in [22]. These two geometries share the sets of points, lines and planes. In what follows we will deal with these types only so it does not matter which of the two geometries we have started with.

1991 *Mathematics Subject Classification.* Primary: 20B25, 20D05, 20D08.

Key words and phrases. Monster, diagram geometry.

The research was carried out while the author was visiting the University of Cambridge under a SERC grant

Consider in $\mathcal{G}_2(M)$ a point, a line and a plane which are pairwise incident. Let C, N and L, respectively be the stabilizers of these elements in M. We already know that $C \cong 2^{1+24}.Co_1$, furthermore, $N \cong 2^{2+11+22}.(M_{24} \times S_3)$ and $L \cong 2^{3+6+12+18}.(3 \cdot S_6 \times L_3(2))$. The following result was proved in [11] (see also [10]) as a contribution to a joint project with S.V. Shpectorov on classification of flag-transitive tilde geometries [16].

Theorem 1. *Let $\mathcal{A}$ be the amalgam consisting of the subgroups C, N and L. Then M is the unique group which contains $\mathcal{A}$ and is generated by the elements of the amalgam.* □

Let Γ be the collinearity graph of $\mathcal{G}_2(M)$ that is a graph on the set of points where two points are adjacent if they are collinear. Theorem 1 is equivalent to the claim that the fundamental group of Γ is generated by the geometrical triangles (all whose vertices are incident to a common plane). The latter claim is a reformulation of the simple connectedness of $\mathcal{G}_2(M)$ (as well as of the maximal parabolic Ronan – Smith geometry of the Monster).

Theorem 1 has also played an important role in the proof of the famous conjecture of J.H. Conway on the Y-presentation of the Monster (cf. [4], [19], [14]).

In [7] the Moonshine module $V^\natural$ was constructed as a Z_2-orbifold obtained from the torus R^{24}/Λ where Λ is the Leech lattice. It follows more or less directly from the construction that $V^\natural$ admits an action of the group C. Then it was shown that C is a proper subgroup of Aut $V^\natural$ by constructing an automorphism σ (called triality) which together with a subgroup of C generates the group N. Finally, in order to verify a certain property (the Jacobi identity) of the algebra structure defined on $V^\natural$, it was shown that a subgroups from C and a subgroup from N generate in Aut $V^\natural$ the group L. The identification in [7] of the subgroup in Aut $V^\natural$ generated by C and N appeals to [9].

Let B be a 196 883-dimensional vector space over complex numbers and let $G = GL(B)$. It was shown in [25] that there exists at most one isomorphism ϕ into G of the amalgam $\mathcal{A}_0$ consisting of the subgroups C and N. In the original construction of the Monster in [9] such an isomorphism ϕ was shown to exist. When proving that $\phi(\mathcal{A}_0)$ generates the Monster group inside G, it was crucial to show that $\phi(\mathcal{A}_0)$ preserves on B a structure which is now called the Griess algebra.

Motivated by an intention to develop the theory of the Monster starting with its Y-presentation, the following result was proved in [13].

Theorem 2. *Let E be the element of $\mathcal{G}_2(M)$ stabilized by L (that is a subgroup of order 2^3 in the Monster). Let ϕ be the unique isomorphism of the amalgam $\mathcal{A}_0$ (consisting of C and N) into $G = GL(B)$. Then the subgroup in G generated by $\phi(N_C(E))$ and $\phi(N_N(E))$ is isomorphic to L.* □

The proof of Theorem 2 relies only on analysis of $\mathcal{A}_0$ and ϕ, so it is independent on the existence of the Monster.

Theorem 2 implies that as soon as an isomorphism into G of the rank 2 amalgam $\mathcal{A}_0$ is constructed, we obtain an isomorphism of the rank 3 amalgam $\mathcal{A}$. By Theorem 1 the amalgam $\mathcal{A}$ can only generate the Monster. So it looks like we obtain an independent existence proof of the Monster. But this is not quite true since Theorem 1 was proved via splitting into geometrical triangles the cycles in the collinearity graph Γ of $\mathcal{G}_2(M)$. So the proof relies on existence of the Monster.

In order to obtain an independent existence proof for the Monster along these lines one should produce an independent proof of Theorem 1. A similar strategy was realized in [15] for the fourth sporadic simple group of Janko J_4. For J_4 no structure similar to the Griess algebra is known although it is a flag-transitive automorphism group of a Petersen type geometry whose simple connectedness was proved in [12].

2. The geometry $\mathcal{G}_3(M)$ and $V^\natural$ as a Z_3-orbifold.

Let $\mathcal{G}_3(M)$ be the 3-local geometry of the Monster group introduced in [3] and belonging to the following diagram.

$$\mathcal{G}_3(M): \quad \underset{3}{\circ}\!\!-\!\!-\!\!-\!\!\underset{3}{\circ}\!\!=\!\!=\!\!\underset{9}{\circ}\overset{c^*}{-\!\!-\!\!-}\underset{1}{\circ}.$$

The points of $\mathcal{G}_3(M)$ are subgroups of order 3 in M generated by central 3-elements. The normalizer N_u of such a subgroup u has the structure $3^{1+12}.2 \cdot Suz.2$, that is, N_u is an extension of an extraspecial group Q_u of order 3^{13} by a 2-fold central cover of the Suzuki sporadic simple group with an outer automorphism of order 2 adjoint. The group $2 \cdot 3 \cdot Suz.2$ is the automorphism group of the complex Leech lattice and $Q_u/\langle u \rangle$ is canonically isomorphic to this lattice reduced modulo 3. Two points u and t are collinear if $t \in Q_u$ (equivalently if $u \in Q_t$). For $i = 2$ and 3 the elements of type i in $\mathcal{G}_3(M)$ are elementary abelian subgroups of order 3^i in M whose nontrivial cyclic subgroups are pairwise collinear points (incident to this element). The elements of type 4 in $\mathcal{G}_3(M)$ are called quadrics. Let D be a quadric. Then D is a self-centralized elementary abelian subgroup of order 3^8 in M. The normalizer of D in M induces on it the natural action of $O_8^-(3).2$ (an extension of the simple orthogonal group by an automorphism of order 2). In particular there is a nonsingular quadratic form of minus type on D preserved by the normalizer. The elements of $\mathcal{G}_3(M)$ incident to D are the totally isotropic subspaces with respect to the invariant quadratic form.

The residue in $\mathcal{G}_3(M)$ of a quadric is a classical polar space of the group $O_8^-(3)$. The point graph of this polar space is strongly regular with parameters $v = 1\,066$, $k = 336$, $\lambda = 92$ and $\mu = 112$. The multiplicities of the nontrivial eigenvalues are 246 and 819. By the general theory these numbers are degrees of non-principal irreducible constituents of the rank 3 action of $O_8^-(3)$ on the graph.

The residue in $\mathcal{G}_3(M)$ of an element of type 1 is the geometry $\mathcal{G}(Suz)$ which belongs to the diagram

$$\mathcal{G}(Suz): \quad \underset{1}{\circ}\overset{c}{-\!\!-\!\!-}\underset{9}{\circ}\!\!=\!\!=\!\!\underset{3}{\circ}$$

and can be described as follows. The group $S \cong Suz.2$ contains a conjugacy class Θ of order 3 subgroups such that for $x \in \Theta$ we have $S(x) \equiv N_S(x) \cong 3 \cdot U_4(3).2^2$. The subgroup $S(x)$ acting on Θ has 5 orbits with lengths 1, 280, 486, 8\,505 and 13\,608. Let Θ denote also the graph of valency 280 invariant under the action of S on Θ. Then two subgroups are adjacent in Θ if and only if they commute. The graph is known as the Patterson graph (cf. [2], Section 13.7). The Patterson graph is distance-transitive with the intersection matrix

$$
\begin{pmatrix}
0 & 1 & 0 & 0 & 0 \\
280 & 36 & 8 & 0 & 0 \\
0 & 243 & 128 & 90 & 0 \\
0 & 0 & 144 & 180 & 280 \\
0 & 0 & 0 & 10 & 0
\end{pmatrix}
$$

and locally it is the point graph Σ of the generalized quadrangle of order $(9,3)$, naturally associated with the group $U_4(3)$. Recall that Σ has 280 vertices and 112 cliques of size 10 (corresponding to lines of the generalized quadrangle). Two distinct cliques intersect in at most one vertex and every vertex is in exactly 4 cliques. This means that a maximal set of pairwise commuting subgroups from Θ (a maximal clique) has size 11. Such a clique generates an elementary subgroup of order 3^5 containing no subgroups from Θ outside the clique and with the normalizer in S isomorphic to $3^5 : (M_{11} \times 2)$. Now points, lines and planes in $\mathcal{G}(Suz)$ are vertices, edges and maximal cliques of Θ, respectively. It was shown in [27] that $N_u/\langle u \rangle$ supports the universal group-addmissible embedding of $\mathcal{G}(Suz)$.

Let us turn back to the 3-local geometry $\mathcal{G}_3(M)$ of the Monster group M. Let $\{u_1, u_2, u_3, u_4\}$ be maximal set of pairwise incident elements (a maximal flag) in $\mathcal{G}_3(M)$, where u_i is of type i. Then u_i is a subgroup of order 3^m in the Monster, where $m = 1, 2, 3$ and 8 for $i = 1, 2, 3$ and 4, respectively. Let P_i be the stabilizer of u_i in M (that is $P_i = N_M(u_i)$). Then

$$P_1 \cong 3^{1+12}.2 \cdot Suz.2; \qquad P_2 \cong 3^{2+5+10}.(GL_2(3) \times M_{11});$$

$$P_3 \cong 3^{3+8+6}.2^3.(L_3(3) \times 2); \qquad P_4 \cong 3^8.O_8^-(3).2.$$

Let $\mathcal{B}$ be the amalgam consisting of the Monster subgroups P_1, P_2, P_3 and P_4. We conjecture that $\mathcal{G}_3(M)$ is simply connected or, equivalently, that M is the only group which contains $\mathcal{B}$ and is generated by its elements. Since the polar space of $O_8^-(3)$ is simply connected, the group P_4 is the only one which contains the amalgam of the subgroups $P_4 \cap P_i$ for $1 \le i \le 3$ and is generated by the elements of the amalgam. This implies that P_4 can be dropped from the amalgam $\mathcal{B}$ and reconstructed back from its intersections with the P_i for $1 \le i \le 3$.

In [6] the Moonshine module was constructed as a Z_3-orbifold $V_3^\natural$ also starting with the torus R^{24}/Λ. From the construction the invariance of $V_3^\natural$ under P_1 is visible. An additional automorphism was constructed which generates P_2 together with an appropriate subgroup of P_1. It was also shown that a subgroup of P_1 and a subgroup of P_2 generate in the automorphism group of $V_3^\natural$ the subgroup P_3. In view of the above remark concerning P_4 this implies that Aut $V_3^\natural$ contains the amalgam $\mathcal{B}$. On the identification step of $V_3^\natural$ with the Moonshine module it was shown, appealing to the classification of finite simple groups, that $\mathcal{B}$ generates M in Aut $V_3^\natural$. If the simple connectedness proof for $\mathcal{G}_3(M)$ would be available the reference to the classification could be skipped.

The simple connectedness question for $\mathcal{G}_3(M)$ is also crucial for completing the classification of flag-transitive extended dual polar spaces [26].

It worth mentioning (cf. [3] and [23]) that $\mathcal{G}_3(M)$ contains as a subgeometry another extended dual polar space, a geometry $\mathcal{G}_3(Fi_{24})$ which admits the Fischer group Fi_{24} as a flag-transitive automorphism group and belongs to the following diagram:

$$\mathcal{G}_3(Fi_{24}): \quad \underset{3}{\circ}\!-\!-\!-\!-\!\underset{3}{\circ}\!=\!=\!\underset{3}{\circ}\overset{c^*}{-\!-\!-\!}\underset{1}{\circ}.$$

The subgeometry can be described as follows. Let v be an order 3 subgroup from $O_3(P_1) \cong 3^{1+12}$ which is not a point of $\mathcal{G}_3(M)$ (that is v and u_1 are not conjugated in M). Then $N_M(v) \cong 3 \cdot F_{24}$ and the subgeometry is formed by the elements of $\mathcal{G}_3(M)$ commuting with v. It was shown recently by G. Stroth and the author that $\mathcal{G}_3(Fi_{24})$ is simply connected [17].

3. A 3-LOCAL GEOMETRY $\mathcal{G}(Co_1)$.

A 3-local geometry $\mathcal{G}(Co_1)$ was mentioned in [3] more or less independently of the discussion of $\mathcal{G}_3(M)$. As we will see in the next section, $\mathcal{G}(Co_1)$ is in fact closely related to $\mathcal{G}_3(M)$. Namely, a 2^{24}-fold cover of $\mathcal{G}(Co_1)$ is a kind of subgeometry in $\mathcal{G}_3(M)$.

Let G be the first Conway group Co_1 and Δ be the conjugacy class of order 3 subgroups in G such that for $w \in \Delta$ we have $G(w) \equiv N_G(w) \cong 3 \cdot Suz.2$ (a maximal subgroup of G). The action of G on Δ is described by the following lemma (cf. [20] and Lemma 49.8 in [1]).

Lemma 3. *The group* $G \cong Co_1$ *induces on* Δ *a rank 5 action of degree* $1\,545\,600$. *For* $w \in \Delta$ *the stabilizer* $G(w)$ *acting on* Δ *has five orbits* $\Delta_i(w)$, $0 \leq i \leq 4$ *of lengths* 1, 22 880, 405 405, 1 111 968 *and* 5 346 *with stabilizers* $3 \cdot Suz.2$, $3^2.U_4(3).2^2$, $2^{1+6}_- U_4(2).2$, $J_2.2$ *and* $G_4(2).2$, *respectively. For* $w_i \in \Delta_i$ *the subgroup in* G *generated by* w *and* w_i *is isomorphic to* Z_3, $Z_3 \times Z_3$, $SL_2(3)$, A_5 *and* A_4 *for* $i = 0, 1, 2, 3$ *and* 4, *respectively.* $\square$

In what follows let Δ denote also the graph on Δ in which two vertices w and v are adjacent if and only if $v \in \Delta_1(w)$ (equivalently if $w \in \Delta_1(v)$). Notice that the edges of Δ are all pairs of commuting subgroups from Δ. The intersection matrix of Δ is given below (cf. [20]). The (j,i)-th entry in this matrix shows the number of vertices in $\Delta_{j-1}(w)$ adjacent in Δ to a fixed vertex from $\Delta_{i-1}(w)$ for $1 \leq i,j \leq 5$.

$$\begin{pmatrix} 0 & 1 & 0 & 0 & 0 \\ 22880 & 280 & 480 & 280 & 2080 \\ 0 & 8505 & 5120 & 6300 & 0 \\ 0 & 13608 & 17280 & 16200 & 20800 \\ 0 & 486 & 0 & 100 & 0 \end{pmatrix}.$$

Let us consider the action of $G(w)$ on $\Delta_1(w)$. Clearly w is in the kernel and the action is isomorphic to that of $Suz.2$ by conjugation on a class of order 3 subgroups with normalizer isomorphic to $3 \cdot U_4(3).2^2$. From the intersection matrix of Δ we see that the subgraph of Δ induced by $\Delta_1(w)$ has valency 280 and by the previous section it must be isomorphic to the Patterson graph. This implies that the maximal cliques in Δ have size 12 and two of them have at most 3 common vertices. Define $\mathcal{G}(Co_1)$ to be a rank 4 geometry whose elements of type 1, 2, 3 and

4 are the intersections of size 1, 2, 3 and 12 of (non-necessarily distinct) maximal cliques in Δ. Notices that the elements of type 4 are cliques themselves and for $1 \leq i \leq 3$ every subset of i pairwise adjacent vertices of Δ is an element of type i. It follows directly from the above discussion that $\mathcal{G}(Co_1)$ belongs to the following diagram:

$$\mathcal{G}(Co_1): \quad \underset{1}{\circ}\!\!-\!\!-\!\!-\!\!\underset{1}{\circ}\overset{c}{-\!\!-\!\!-}\underset{9}{\circ}\!\!-\!\!-\!\!-\underset{3}{\circ}.$$

Let $\{w_1, w_2, w_3, w_4\}$ be a maximal flag in $\mathcal{G}(Co_1)$ where w_i is of type i. The vertices of Δ (considered as order 3 subgroups of G) contained in the clique w_i, generate in G an elementary abelian subgroup of order 3^n where $n =$ 1, 2, 3 and 6 for $i =$ 1, 2, 3 and 4, respectively. This subgroup with be identified with w_i. Let G_i be the stabilizer of w_i in G (that is $G_i = N_G(w_i)$). Then

$$G_1 \cong 3 \cdot Suz.2; \qquad G_2 \cong 3^2.U_4(3).2^2;$$

$$G_3 \cong 3^{3+4}.[2^3].S_4.S_3; \quad G_4 \cong 3^6.2 \cdot M_{12}.$$

For $1 \leq i, j \leq 4$ the subgroup w_j is contained in G_i and $G_i \cap G_j = N_{G_i}(w_j)$. Moreover the residue of w_i in $\mathcal{G}(Co_1)$ consists of all G_i-conjugates of w_j for $j \neq i$ with the incidence relation given by inclusion. Finally notice that there is a Sylow 3-subgroup of G which contains w_i for $1 \leq i \leq 4$.

The most important question about a geometry is the one about its universal covering. The geometry $\mathcal{G}(Co_1)$ is the only one in a large family of sporadic geometries for which the question was left unanswered in [21]. For the best of our knowledge it is still open. The construction given below shows that $\mathcal{G}(Co_1)$ possesses a 2^{24}-fold covering.

Let Λ be the Leech lattice. Then $\bar{\Lambda} = \Lambda/2\Lambda$ is an irreducible $GF(2)$-module for Co_1. Let $\tilde{G}$ be a non-split extension of $\bar{\Lambda}$ by Co_1. At least one isomorphism type of such extensions exists. Namely, we can take $\tilde{G}$ to be the quotient $\tilde{C} = C(z)/\langle z \rangle$ where $C(z)$ is the centralizer in the Monster of a central involution z. We are going to define a geometry $\mathcal{G}(\tilde{G})$ acted on flag-transitively by $\tilde{G}$ and possessing a covering onto $\mathcal{G}(Co_1)$. Let S be a Sylow 3-subgroup of G which contains w_i for $1 \leq i \leq 4$. Let $\tilde{S}$ be a Sylow 3-subgroup of $\phi^{-1}(S)$ where ϕ is the natural homomorphism of $\tilde{G}$ onto $G \cong Co_1$. Then ϕ induces an isomorphism ψ of $\tilde{S}$ onto S. Put $\tilde{w}_i = \psi^{-1}(w_i)$ and define the elements of type i in $\mathcal{G}(\tilde{G})$ to be all $\tilde{G}$-conjugates of $\tilde{w}_i$, $1 \leq i \leq 4$. Let the incidence relation be given by inclusion. Then it is easy to see that $\tilde{G}$ acts flag-transitively on $\mathcal{G}(\tilde{G})$ and ϕ induces a morphism of $\mathcal{G}(\tilde{G})$ onto $\mathcal{G}(Co_1)$ which we denote by the same letter ϕ.

Lemma 4. $\phi : \mathcal{G}(\tilde{G}) \to \mathcal{G}(Co_1)$ is a covering of geometries.

Proof. We have to show that (a) $\mathcal{G}(\tilde{G})$ is connected and (b) the restriction of ϕ to the residue of any element $\tilde{w}$ in $\mathcal{G}(\tilde{G})$ is an isomorphism onto the residue of $\phi(\tilde{w})$ in $\mathcal{G}(Co_1)$.

(a) Let H be the subgroup in $\tilde{G}$ generated by all the elements from the connected component of $\mathcal{G}(\tilde{G})$ containing $\tilde{w}_i$ for $1 \leq i \leq 4$. Then the restriction of ϕ to H is surjective. Since $\tilde{G}$ is a non-split extension and $\bar{\Lambda}$ is irreducible under the action of Co_1 this implies that H coincides with the whole $\tilde{G}$.

(b) Since $\tilde{G}$ acts flag-transitively on $\mathcal{G}(\tilde{G})$ we can assume that $\tilde{w} = \tilde{w}_i$ for $i = 1$, 2, 3 or 4. It is well-known [5] that w_1 acts fixed-points freely on $\bar{\Lambda}$. Since w_1 is contained in w_i for $2 \leq i \leq 4$ we conclude that $C_{\bar{\Lambda}}(\tilde{w}_i) = 1$. This and Frattini argument imply that ϕ restricted to $N_{\tilde{G}}(\tilde{w}_i)$ is an isomorphism onto $N_G(w_i)$. Let $\tilde{t}$ be an element incident to $\tilde{w}_i$ in $\mathcal{G}(\tilde{G})$. The definition of the incidence relation implies that either $\tilde{t} \leq \tilde{w}_i$ or $\tilde{w}_i \leq \tilde{t}$ and in any case $\tilde{t}$ is contained in $N_{\tilde{G}}(\tilde{w}_i)$. Clearly $t = \phi(\tilde{t})$ is incident to w_i in $\mathcal{G}(Co_1)$ and since $N_{\tilde{G}}(\tilde{w}_i)$ maps isomorphically onto $N_G(w_i)$, we conclude that $\tilde{t}$ is the only preimage of t in $N_{\tilde{G}}(\tilde{w}_i)$. Now it is clear that two distinct elements incident to $\tilde{w}_i$ are incident if and only if their images under ϕ do. $\square$

As above let $\tilde{C} \cong C(z)/\langle z \rangle$ where $C(z)$ is the centralizer in the Monster of a central involution z. We propose the following.

Conjecture 5. *The covering $\mathcal{G}(\tilde{C}) \to \mathcal{G}(Co_1)$ is universal.*

Notice that a proof of the conjecture would imply that $\tilde{C}$ is the only non-split extension of $\bar{\Lambda}$ by Co_1. As far as I know the second cohomology group of $\bar{\Lambda}$ has not been computed.

To prove the conjecture it might be useful to work with a subgeometry in $\mathcal{G}(\tilde{C})$, stabilized by $2^2 \cdot U_6(2).S_3$ and belonging to the following diagram:

$$\underset{1}{\circ}\!\!-\!\!-\!\!-\!\!\underset{1}{\circ}\!\!-\!\!\overset{c}{-}\!\!-\!\!\underset{3}{\circ}\!\!=\!\!\!=\!\!\underset{3}{\circ}.$$

This subgeometry is simply connected as proved in [18].

The fact that w_1 acts fixed-points freely on $\bar{\Lambda}$ and Lemma 3 enable us to make the following observation.

Lemma 6. *Two points in $\mathcal{G}(\tilde{C})$ are collinear if and only if they commute and the collinearity graph of $\mathcal{G}(\tilde{C})$ is locally Patterson.* $\square$

4. FROM $\mathcal{G}_3(M)$ TO $\mathcal{G}_2(M)$.

In this section we discuss a possible approach to the simple connectedness proof for $\mathcal{G}_3(M)$. The geometry $\mathcal{G}_3(M)$ is simply connected if and only if the Monster M is the unique group which contains the amalgam $\mathcal{B}$ of maximal parabolic subgroups corresponding to the action of M on $\mathcal{G}_3(M)$ (in other words that M is the unique completion of $\mathcal{B}$). One could try to show that the universal completion of $\mathcal{B}$ must contain and be generated by the amalgam $\mathcal{A}$ corresponding to the action of M on the 2-local geometry $\mathcal{G}_2(M)$. By Theorem 1 this would immediately imply that the universal completion is isomorphic to M. In the remainder of the section we reduce the reconstruction of the subgroup C in the universal completion of $\mathcal{B}$ to Conjecture 5, that is to the simple connectedness question for $\mathcal{G}(\tilde{C})$.

As above let $\mathcal{B} = \{P_1, P_2, P_3, P_4\}$ be the amalgam of maximal parabolic subgroups corresponding to the action of M on $\mathcal{G}_3(M)$. Let H be either the Monster group or the universal completion of $\mathcal{B}$ and $\mathcal{H}$ be the geometry locally isomorphic to $\mathcal{G}_3(M)$ on which H induces a flag-transitive action. This means that the elements of type i in $\mathcal{H}$ are the (right) cosets of P_i in H; two elements are incident if the corresponding cosets intersect nontrivially.

Let Σ be the collinearity graph of $\mathcal{H}$. For a vertex $x \in \Sigma$ its stabilizer in H is an H-conjugate of $P_1 \cong 3^{1+12}.2 \cdot Suz.2$.

Let $x \in \Sigma$ and let $H(x)$ be the stabilizer of x in H. Let $\Sigma_1(x)$ be the set of points collinear to x and let $\Sigma_2(x)$ be the set of points not collinear to x but incident with x to a common quadric (an element of type 4). Let Θ be the Patterson graph naturally associated with the residue of x in $\mathcal{H}$. Then $H(x)$ acts on Θ with $O_{3,2}(H(x))$ being in the kernel. Moreover, the lines incident to x are the maximal cliques of Θ while the quadrics incident to x are the vertices of Θ. Clearly for a point $y \in \Sigma_1(x)$ there is a unique line incident both to x and y. By Proposition 3.4 in [3] in the case $H = M$ for $z \in \Sigma_2(x)$ there is a unique quadric incident to both z and x. Clearly the same property must hold in the universal cover of $\mathcal{G}_3(M)$ as well. So the size of $\Sigma_1(x)$ is 3 times the number of lines passing through x (that is $3 \times 232\,960$) and the size of $\Sigma_2(x)$ is 729 times the number of quadrics incident to x (that is $729 \times 22\,880$). A line l (incident to x) is incident to 11 quadrics (also incident to x) and the common stabilizer in H of x and l induces on these 11 quadrics the natural action of M_{11}. These quadrics correspond to the vertices of Θ contained in the clique which corresponds to l.

Let τ be an involution from $O_{3,2}(H(x))$. Then τ maps onto the unique non-trivial central element of the factor $H(x)/O_3(H(x)) \cong 2 \cdot Suz.2$. This factor acts irreducibly on the module $O_3(H(x))/Z(O_3(H(x))) \cong 3^{12}$. Hence τ inverts every element of this module. This and Frattini argument imply that $C_{H(x)}(\tau)/\langle\tau\rangle \cong 3 \cdot Suz.2$. We are going to study the fixed points of τ on Σ. Notice that τ centralizes $Z(O_3(H(x)))$. Let us determine the fixed points of τ on $\Sigma_1(x)$ and $\Sigma_2(x)$. Let l be a line incident to x and let $\{x, y_1, y_2, y_3\}$ be the points incident to l. The subgroups $Z(O_3(H(t)))$ for the points t incident to l, generate an elementary subgroup $Z(l)$ of order 9 and $Z(l)/Z(O_3(H(x)))$ is a cyclic subgroup in the module $O_3(H(x))/Z(O_3(H(x)))$ which must be inverted by τ. Without loss of generality we conclude that τ permutes y_2 and y_3 and stabilizes y_1, inverting $Z(O_3(H(y_1)))$. Thus on every line incident to x, τ stabilizes a single point distinct from x. Clearly $C_{H(x)}(\tau)$ induces on all these points the action of $Suz.2$ as on the cliques of Θ (on the cosets of $3^5 : (M_{11} \times 2)$).

Let D be a quadric incident to x. The stabilizer $H(D)$ of D in H induces on the set of $1\,066$ points incident to D a rank 3 action of $O_8^-(3).2$. Since τ acts trivially on the residue of x, $\tau \in H(D)$. The action of $H(x)$ on the quadrics incident to x shows that the centralizer of τ in $H(D) \cap H(x)$ is isomorphic to $2.3^2 \cdot U_4(3).2^2$. Using the character table of $O_8^-(3)$ (available for instance in [24]) and the degrees of irreducible constituents of its action on the points incident to D (cf. Section 2), we find out that the centralizer of τ in $H(D)$ is twice larger than in $H(D) \cap H(x)$. Hence τ stabilizes exactly one point, say z, incident to D and not collinear to x (that is $z \in \Sigma_2(x)$) and there is an element in $H(D)$ which permutes x and z.

Let Φ_1 be the orbit of $C_H(\tau)$ on points fixed by τ which contains x (and z) and let Φ_2 be the analogous orbit containing the points collinear to x.

Lemma 7. *Let $y \in \Phi_2$. Then the stabilizer of y in $C_H(\tau)$ is isomorphic to $2 \times 3^6 : 2 \cdot M_{12}$. There are 12 points in Φ_1 collinear to y and the above stabilizer induces on these points the natural action of M_{12}.*

Proof. The stabilizer we are after is $C_{H(y)}(\tau)$. We can assume that y is collinear to x. We have observed that τ inverts $Z(O_3(H(y)))$ and hence τ is contained in $H(y) - O^2(H(y))$. We know that the centralizer of τ in $H(x) \cap H(y)$ is isomorphic to $3.2.3^5 : (M_{11} \times 2)$. From [5] we see that the centralizer in $H(y)/O_{3,2}(H(y)) \cong Suz.2$ of the image $\bar{\tau}$ of τ is $2 \times Aut(M_{12})$. Moreover, both preimages τ_1 and τ_2 of $\bar{\tau}$ in $H(y)/O_3(H(y)) \cong 2 \cdot Suz.2$ are conjugated in the latter group and hence their centralizers there are isomorphic to $2 \times 2 \cdot M_{12}$. Finally, since the product $\tau_1 \times \tau_2$ acts fixed-points freely on the module $O_3(H(y))/Z(O_3(H(y)))$, we conclude that the centralizer of τ_i (and also of τ) in this module is 6-dimensional. So the result follows. □

Let Ξ be the graph on Φ_1 invariant under the action of $C_H(\tau)$ in which x is adjacent to the vertices from $\Phi_1 \cap \Sigma_2(x)$. Recall that there is a natural correspondence between the set $\Xi(x)$ of vertices adjacent to x in Ξ and the vertex set of the Patterson graph Θ.

Let $y \in \Phi_2$ and suppose that x and y are incident to a common line l. Let $\Xi(y)$ denote the set of points from Φ_1 collinear to y. By Lemma 7 we know that $\Xi(y)$ contains 12 points. Let D be a quadric incident to l and let z be the point from $\Sigma_2(x)$ incident to D and fixed by τ. Clearly z is collinear to y. Since there are 11 choices for such D and z, the conclusion is that the points from $\Xi(y)$ are pairwise adjacent in Ξ. On the other hand the points from $\Xi(y) - \{x\}$ correspond to a clique of the Patterson graph Θ. Hence adjacent vertices in Θ correspond to adjacent vertices in $\Xi(x)$. It is quite straightforward to show that Ξ is actually locally Patterson graph.

Let H be the Monster group. Then $C_H(\tau) \cong 2^{1+24}.Co_1$ and Ξ is the collinearity graph $\tilde{\Delta}$ of the geometry $\mathcal{G}(\tilde{C})$. If H is the universal completion of $\mathcal{B}$ then each connected component Ξ_c of Ξ is the collinearity graph of a covering of $\mathcal{G}(\tilde{C})$ (compare Lemma 6). If Conjecture 5 would be true then Ξ_c would be isomorphic to $\tilde{\Delta}$ and we could reconstruct C in H as the stabilizer of Ξ_c in $C_H(\tau)$.

REFERENCES

1. M. Aschbacher, *Sporadic Groups*, Cambridge Univ. Press, Cambridge, 1994.
2. A. Brouwer, A. Cohen and A. Neumaier, *Distance Regular Graphs*, Springer, Berlin, 1989.
3. F. Buekenhout and B. Fischer, *A locally dual polar space for the Monster*, unpublished manuscript dated around 1983.
4. J.H. Conway, *Y_{555} and all that*, Groups, Combinatorics and Geometry, London Math. Soc. Lect. Notes vol. 165 (M. Liebeck and J. Saxl, eds.), Cambridge Univ. Press, Cambridge, 1992, pp. 22–23.
5. J.H. Conway, R.T. Curtis, S.P. Norton, R.A. Parker and R.A. Wilson, *Atlas of Finite Groups*, Clarendon Press, Oxford, 1985.
6. C. Dong and G. Mason, *The construction of the Moonshine module as a Z_p-orbifold*, Preprint 1993.
7. I. Frenkel, J. Lepowsky and P. Meurman, *Vertex Operator Algebras and the Monster*, Academic Press, Boston, 1988.
8. R.L. Griess, *Automorphisms of extra special groups and nonvanishing degree 2 cohomology*, Pacific J. Math **28** (1973), 355–421.
9. R.L. Griess, *The Friendly Giant*, Invent. Math. **69** (1982), 1–102.
10. A.A. Ivanov, *Geometric presentations of groups with an application to the Monster*, Proc. ICM-90 Kyoto, Japan August 1990, Springer Verlag, 1991, pp. 385–395.
11. A.A. Ivanov, *A geometric characterization of the Monster*, Groups, Combinatorics and Geometry, London Math. Soc. Lect. Notes vol. 165 (M. Liebeck and J. Saxl, eds.), Cambridge

Univ. Press, Cambridge, 1992, pp. 46–62.

12. A.A. Ivanov, *A presentation for J_4*, Proc. London Math. Soc. **64** (1992), 369–396.

13. A.A. Ivanov, *Constructing the Monster via its Y-presentation*, Combinatorics, Paul Erdös is Eighty, vol. 1, Bolyai Soc. Math. Studies, Budapest, 1993, pp. 253–270.

14. A.A. Ivanov, *Presenting the Baby Monster*, J. Algebra **163** (1994), 88–108.

15. A.A. Ivanov and U. Meierfrankenfeld, *A computer free construction of J_4*, (submitted), Proc. London Math. Soc.

16. A.A. Ivanov and S.V. Shpectorov, *The flag-transitive tilde and Petersen type geometries are all known*, Bull. Amer. Math. Soc. **31** (1994), 173–184.

17. A.A. Ivanov and G. Stroth, *Simple connectedness of the 3-local geometry of F_{24}*, in preparation.

18. T. Meixner, *Some polar towers*, Europ. J. Combin. **12** (1991), 397–416.

19. S.P. Norton, *Constructing the Monster*, Groups, Combinatorics and Geometry, London Math. Soc. Lect. Notes vol. 165 (M. Liebeck and J. Saxl, eds.), Cambridge Univ. Press, Cambridge, 1992, pp. 63–76.

20. C.E. Praeger and L.H. Soicher, *Low rank representations and graphs for sporadic groups*, Preprint 1994.

21. M.A. Ronan, *Coverings of certain finite geometries*, Finite Geometries and Designs, Cambridge Univ. Press, Cambridge, 1981, pp. 316–331.

22. M.A. Ronan and S.D. Smith, *2-local geometries for some sporadic groups*, Proc. Symp. Pure Math. **37** (1980), 283–289.

23. M.A. Ronan and G. Stroth, *Minimal parabolic geometries for the sporadic groups*, Europ. J. Combin. **5** (1984), 59–91.

24. M. Schönert et al., *GAP: Groups, Algorithms and Programming*, Lehrstuhl D für Mathematik, RWTH Aachen, 1993.

25. J.G. Thompson, *Uniqueness of the Fischer–Griess Monster*, Bull. London Math. Soc. **11** (1979), 340–346.

26. S. Yoshiara, *On some extended dual polar spaces,* I, Europ. J. Combin. **15** (1994), 73–86.

27. S. Yoshiara, *Embeddings of flag-transitive classical locally polar geometries of rank 3*, Geom. Dedic. **43** (1992), 121-165.

INSTITUTE FOR SYSTEM ANALYSIS, RUSSIAN ACADEMY OF SCIENCES, 9, PROSPECT 60 LET OKTYABRYA, 117312 MOSCOW, RUSSIA AND DEPARTMENT OF MATHEMATICS, IMPERIAL COLLEGE, 180 QUEEN'S GATE, LONDON SW7 2BZ, ENGLAND, UK

E-mail address: ivanov@cs.vniisi.msk.su

Contemporary Mathematics
Volume **193**, 1996

Operads of Moduli Spaces and Algebraic Structures in Conformal Field Theory

TAKASHI KIMURA

November 30, 1994

ABSTRACT. In the first part of this paper, we construct the cochain complex associated to vertex algebras in terms of operads of moduli spaces. We do so by extending the results of Kimura and Voronov [**23**] where the cohomology of a conformal field theory was defined using ideas from the theory of operads due to Ginzburg and Kapranov [**13**]. In the second part of this paper, we explicate the origins of various algebraic structures that have arisen in topological conformal field theory from geometric structures on the underlying moduli spaces. We begin with the origins of homotopy algebraic structures (homotopy Lie algebras [**21**] and homotopy associative algebras) that are associated to the so-called BRST complex of a topological conformal field theory. Finally, we discuss the Batalin-Vilkovisky algebra structure on BRST cohomology noticed by Lian and Zuckerman [**29**] which arises as representations of homology classes on moduli spaces of Riemann spheres with holomorphically embedded disks, an interpretation due to Getzler [**8**]. We then use results by Harer [**14**], [**15**] on the stability of the homology of the mapping class group to show that there are a finite number of independent operations on the BRST cohomology of a given degree despite the contributions from the higher genus moduli spaces.

1. Introduction

The notion of a conformal field theory which originated in physics has become a part of mathematics in two different ways. The first approach, due to Frenkel, Lepowsky, and Meurman (see [**7**]) is through the notion of a vertex operator algebra which is a rigorization of the notion of a chiral algebra. The moonshine module constructed in [**6**] was shown to have the structure of a vertex operator algebra which has the Monster group as its automorphism group. Another

---

1991 *Mathematics Subject Classification*. Primary 57R19, 53Z05; Secondary 81T40, 55S20, 17B69.

Research was supported in part by an NSF Postdoctoral Research Fellowship

common example of a vertex operator algebra is given by the irreducible highest
weight representation of an affine Lie algebra associated to a simple Lie algebra
which is induced from the trivial representation of this simple Lie algebra. The
latter vertex operator algebra is the one which is relevant to the three manifold
invariants and link invariants in Chern-Simons theory [38] and and study of the
Knizhnik-Zamolodchikov equation. This formulation has the advantage the it is
quite explicitly algebraic and is readily accessible to computation. The second
approach, due to Segal [35], defines a conformal field theory as a morphism from
a category of moduli spaces of Riemann surfaces with holomorphically embedded
disks (extended by a power of the determinant line bundle) to the category of vector spaces. This approach is the rigorization of the so-called operator formalism
from physics. This formulation has the advantage that the role of the geometry
of the moduli spaces is manifest and it has a differential geometric flavor to it.
Recently, Lepowsky and Huang [19] have shown that a vertex operator algebra
gives rise to a (holomorphic) tree level conformal field theory where tree level
means that only the moduli spaces of Riemann spheres are considered. This is a
powerful result as it allows us to use differential geometric tools to analyze vertex
operator algebras. In particular, geometric structures on the moduli spaces of
Riemann spheres give rise to algebraic structures on vertex operator algebras.

Tree level conformal field theories can be most readily described using the
notion of an operad which is an object that parametrizes operations on a vector
space. Operads were originally introduced by May [30] in order to study the
homotopy type of iterated loop spaces but have recently been used to describe
moduli spaces and algebraic structures that have arisen in mathematical physics
such as conformal field theory [21, 8, 18] and topological gravity [27, 22, 9].
The moduli spaces of Riemann spheres with holomorphically embedded disks,
$\mathcal{P}$, form an operad with composition maps induced by sewing and the action of
the permutation groups on the disks. A tree level conformal field theory is a
vector space which is a (projective) representation of the operad $\mathcal{P}$. The result
by Huang and Lepowsky in this context is that a vertex operator algebra is a
(holomorphic) representation (or algebra) over a (partial) operad, $\mathfrak{P}$, containing
$\mathcal{P}$. This reformulation of conformal field theories and vertex operator algebras
in terms of operads is useful as it allows us to apply results from the theory of
operads to conformal field theories and vertex operator algebras.

In the first section of this paper we introduce the notion of an operad, $\mathcal{O}$,
and its representations (or $\mathcal{O}$-algebras) in analogy with a (semi)group and its
representations. We give examples of operads whose representations are associative algebras and commutative, associative algebras and realize them as moduli
spaces of circles and spheres with embedded intervals and disks, respectively.
Next, we introduce a suboperad of $\mathcal{P}$ which is homotopic to it called the framed
little disks operad $\mathcal{F}$ [8]. We then embed $\mathcal{P}$ in the larger partial operad $\mathfrak{P}$.
Similarly, we embed $\mathcal{F}$ into a partial suboperad $\mathfrak{F}$. A vertex algebra is a vertex
operator algebra where some of the axioms involving the Virasoro algebra have

been weakened. It can be regarded as a (holomorphic) $\mathfrak{F}$-algebra [18].

The second section of the paper is devoted to a construction of the cohomology theory associated to $\mathfrak{F}$-algebras extending the construction by Kimura and Voronov [23] where the notion of the cohomology of an algebra over a quadratic operad with values in a module is introduced and applied to obtain the cochain complex associated to a conformal field theory . (Their construction used the notion of Koszul duality from Drinfeld [4], Ginzburg and Kapranov [13] and Kontsevich [26].) We need to generalize their construction because it does not apply to partial operads. We show that $\mathfrak{F}$ is a quadratic partial operad and construct the cochain complex associated to algebras over it. Furthermore, we show that inequivalent first order deformations of $\mathfrak{F}$-algebras which keep the action of the complex affine transformations fixed are parametrized by its second cohomology with values in itself. This cohomology theory is important because it provides a way to characterize invariants associated to $\mathfrak{F}$-algebras. A computation of this cohomology would give information about the moduli space of all $\mathfrak{F}$-algebras. This space is closely related to the space of all vertex operator algebras and conformal field theories which is of great interest in related theories such as mirror symmetry [27].

The third section of the paper is devoted to exploring the geometrical origins of algebraic structures in topological conformal field theory. A (tree level $c = 0$) topological conformal field theory is an algebra over singular chains on the moduli spaces of Riemann spheres with holomorphically embedded disks where the representation space is itself a complex called the BRST complex. Kimura, Stasheff, and Voronov [21] proved that a relative subcomplex of the BRST complex of a topological conformal field theory is a homotopy Lie (or L_∞) algebra. This structure was observed in the physics literature by Zwiebach [41] in the context of closed string field theory. A homotopy Lie algebra is a complex with n-ary operations, one for each n, which induces at the level of cohomology the structure of a Lie algebra. Hinich and Schechtman showed that homotopy Lie algebras are algebras over a certain operad of trees [17]. The geometric origins of the homotopy Lie algebra structure in topological conformal field theory can be found in the canonical stratification of a real blowup, introduced in [21], of a compactification of the moduli space of (stable) Riemann spheres with punctures due to Deligne-Knudsen-Mumford [3, 20, 24]. The latter space is has strata that are naturally indexed by trees, as observed by Beilinson and Ginzburg [2], and these strata can be pulled up to obtain a stratification of the real blowup. This stratification gives rise to a filtration of this space whose associated spectral sequence has a top row that is precisely the homotopy Lie operad of Hinich and Schechtman where the trees are precisely those which index the strata of the moduli space. A result of minimal area metrics by Wolf and Zwiebach [40] is then used to construct a morphism from chains on this moduli space to chains on $\mathcal{P}$ (so-called string vertices) which induces the structure of a homotopy Lie algebra on a subcomplex of the BRST complex. Next, we show that the BRST complex is itself a homotopy associative (or A_∞) algebra with a binary operation which is commutative up to homotopy by realizing the operad responsible for A_∞-algebra as a real version of the Deligne-Knudsen-Mumford compactification of the moduli space of configurations of points on $\mathbb{RP}^1$ and constructing a morphism from this operad to $\mathcal{P}$. This A_∞ structure is expected to be closely related to structures discovered on the BRST complex by Lian and Zuckerman [29] in the context of topological chiral algebras. Furthermore, this structure appears to be a close cousin to the similar structure of a commutative, homotopy associative algebra which appears in topological gravity [22]. In fact, a tree level topological gravity can be understood as an algebra over the operad of homology groups of the Deligne-Knudsen-Mumford compactifications.

In the last section of this paper, we discuss algebraic structures that appear on the BRST cohomology of a topological conformal field theory. Lian and Zuckerman [29] observed using chiral algebra methods that BRST cohomology is endowed with the structure of a Batalin-Vilkovisky algebra which is a commutative, associative algebra with a Poisson bracket and a unary operation which behaves like an odd second order differential operator. Getzler [8] realized that this structure occurs because BRST cohomology is itself an algebra over the homology operad of $\mathcal{P}$ but $\mathcal{P}$ contains the suboperad $\mathcal{F}$, the little disks operad, which is homotopy equivalent to it. Getzler proved that a Batalin-Vilkovisky algebra is nothing more than an algebra over the operad of homology groups of the framed little disks operad. Finally, we discuss topological conformal field theories which need not be tree level and study the operations induced on BRST cohomology from the moduli spaces of higher genus. Unlike the genus zero case, the homology of the higher genus moduli spaces is not generated from compositions of the homology of the moduli space associated to a "pair of pants, cylinders, and caps." Therefore, in principle, there could be an infinite number of independent operations on its BRST cohomology of a given degree. The homology groups of these higher genus moduli spaces are isomorphic to the group homology of the corresponding mapping class group. Harer [14] proved that the group homology of the mapping class group stabilizes for large genus. As a consequence, we prove that there are only a finite number of independent operations of a given degree in a topological conformal field theory and that, for a given degree, the operations arising from a high genus moduli spaces are obtained from those form lower genus moduli spaces composed with the commutative multiplication in the Batalin-Vilkovisky algebra and an invariant bilinear form.

2. A pedestrian guide to operads and moduli spaces

Henceforth, we shall assume for simplicity that all vector spaces are over $\mathbb{C}$. We will see that operads may be regarded as generalizations of (semi)groups (or associative algebras) and their representations parametrize multilinear operations on a vector space.

2.1. Operads and their representations. Let G be a semigroup with unit. In other words, there is a multiplication map $G \times G \rightarrow G$ taking $(g, g') \mapsto g \circ g'$ which is associative and a unit element I in G such that $g \circ I = I \circ g = g$ for all g in G. On the other hand, given a vector space V, the space of linear maps on V, $\mathcal{E}nd\,(V)$, possesses a composition map $\mathcal{E}nd\,(V) \otimes \mathcal{E}nd\,(V) \rightarrow \mathcal{E}nd\,(V)$ taking $f \otimes f' \mapsto f \circ f'$ and a unit element, id_V (the identity map on V). V is a (nontrivial) representation of G on V if there is a morphism $m : G \rightarrow \mathcal{E}nd\,(V)$, i.e. m preserves the unit elements and $m_{g \circ g'} = m_g \circ m_{g'}$. When G is a group, it is natural to replace the endomorphisms of V by automorphisms because of the existence of inverse elements and a morphism of these structures defines the representation of a group.

In the above, we could have replaced the semigroup G by A, an associative algebra with unit, and demanded that the morphism $A \rightarrow \mathcal{E}nd\,(V)$ be linear to obtain the definition of a representation of A subject to the nontriviality condition that the unit element act as the identity on V. For definiteness, assume that we are always dealing with associative algebras with a unit.

If V is a representation of A then V may be regarded as a vector space with a collection of unary operations parametrized by elements in A which satisfy certain compatibility conditions. Suppose, instead, that we are given a vector space V which is endowed with a collection of multilinear operations which satisfy certain relations, e.g. V could be an associative algebra with the bilinear multiplication map. A semigroup will not parametrize such operations on V. An operad $\mathcal{O}$ is a generalization of a semigroup such that if V is a representation of $\mathcal{O}$ then the elements in $\mathcal{O}$ parametrize the multilinear operations on V.

Operads are best understood by looking at an example. The *endomorphism operad* of a vector space V, denoted by $\mathcal{E}nd_V = \{\,\mathcal{E}nd_V\,(n)\,\}_{n \geq 1}$, is a collection of spaces $\mathcal{E}nd_V\,(n) = \mathrm{Hom}(V^{\otimes n}, V)$, each with the action of the permutation group, S_n, which acts by permuting the factors in the tensor product (twisted by the sign representation when V is graded), and with composition maps between them, i.e. the composition, $f \circ_i f'$, of two elements f in $\mathcal{O}(n)$ and f' in $\mathcal{O}(n')$ for $n \geq 1$ and $n' \geq 0$ is given by

$$(f \circ_i f')(v_1 \otimes \cdots \otimes v_{n+n'-1})$$

$$= f(v_1, \otimes \cdots \otimes v_{i-1} \otimes f'(v_i \otimes \cdots v_{i+n'-1}) \otimes \cdots \otimes v_{n+n'-1})$$

for all $i = 1, \ldots, n$ (with an insertion of the usual sign when V is graded), and the permutation groups act equivariantly with respect to the compositions. These composition maps satisfy certain "associativity" conditions. If f, f', and f'' are elements in $\mathcal{O}(n)$, $\mathcal{O}(n')$, and $\mathcal{O}(n'')$ respectively then for any i and j the two ways of obtaining an element in $\mathcal{O}(n + n' + n'' - 2)$ of the form $f(\ldots, f'(\ldots), \ldots, f''(\ldots), \ldots)$, where f' is inserted in the ith slot of f and f'' is inserted in the jth slot of f, via compositions of f, f', and f'', are the same. Similarly, for any i and j, the two ways of obtaining an element in $\mathcal{O}(n + n' + n'' - 2)$ of the form $f(\ldots, f'(\ldots, f''(\ldots), \ldots), \ldots)$ where f' is inserted in the ith slot of f and f'' is inserted in the jth slot of f' via compositions of f, f', and f'' are the same. Finally, there is an element I in $\mathcal{E}nd_V(1)$, the identity map $I : V \to V$, which is a unit with respect to the composition maps. This structure can be formalised in the following way.

An *operad* $\mathcal{O} = \{\mathcal{O}(n)\}_{n \geq 0}$ *with unit* is a collection of objects (topological spaces, vector spaces, complexes, etc. – elements of any symmetric monoidal category) such that each $\mathcal{O}(n)$ has an action of S_n, the permutation group on n elements (S_0 is contains only the identity), and a collection of operations for $n \geq 1$ and $1 \leq i \leq n$, $\mathcal{O}(n) \times \mathcal{O}(n') \to \mathcal{O}(n + n' - 1)$ given by $(f, f') \mapsto f \circ_i f'$ satisfying

   (i) if $f \in \mathcal{O}(n)$, $f' \in \mathcal{O}(n')$, and $f'' \in \mathcal{O}(n'')$ where $1 \leq i < j \leq n$, $n', n'' \geq 0$, and $n \geq 2$ then

$$(2.1) \qquad (f \circ_i f') \circ_{j+n'-1} f'' = (f \circ_j f'') \circ_i f'$$

   where a factor of $(-1)^{|f'||f''|}$ should be inserted on the right hand side if $\mathcal{O}$ is an operad of graded vector spaces,

   (ii) if $f \in \mathcal{O}(n)$, $f' \in \mathcal{O}(n')$, and $f'' \in \mathcal{O}(n'')$ where $n, n' \geq 1$, $n'' \geq 0$, and $i = 1, \ldots, n$ and $j = 1, \ldots, n'$ then

$$(f \circ_i f') \circ_{i+j-1} f'' = f \circ_i (f' \circ_j f''),$$

   (iii) the composition maps are equivariant under the action of the permutation groups,

   (iv) there exists an element I in $\mathcal{O}(1)$ called the *unit* such that for all f in $\mathcal{O}(n)$ and $i = 1, \ldots, n$,

$$I \circ_1 f = f = f \circ_i I$$

In particular, these axioms imply that $\mathcal{O}(1)$ is an associative algebra with unit (in the linear category) so, given an associative algebra with unit A, we can define an operad by setting $\mathcal{O}(1) = A$ and $\mathcal{O}(n)$ to be 0 for all $n \neq 1$. For definiteness, assume that $\mathcal{O}$ is in the linear category.

Let $\mathcal{O}$ be an operad. A vector space V is said to be a *representation of $\mathcal{O}$* or an *$\mathcal{O}$-algebra* if there is a morphism of operads $m : \mathcal{O} \to \mathcal{E}nd_V$, *i.e.* there exists a map $m : \mathcal{O}(n) \to \mathcal{E}nd_V(n)$ for each $n \geq 0$ such that m preserves the unit, is equivariant under the actions of the permutation groups, *i.e.*

$$m_{\sigma\Sigma}(v_1, \ldots, v_n) = m_\Sigma(v_{\sigma^{-1}(1)}, \ldots, v_{\sigma^{-1}(n)})$$

for all $v_1, \ldots, v_n$ in V, σ in S_n, and Σ in $\mathcal{O}(n)$, and satisfies

$$m_{\Sigma \circ_i \Sigma'} = m_\Sigma \circ_i m_{\Sigma'}$$

for all Σ in $\mathcal{O}(n)$, Σ' in $\mathcal{O}(n')$ with $n \geq 1$, and $i = 1, \ldots, n$. In particular, V is a representation of the associative algebra $\mathcal{O}(1)$.

The permutation group in an $\mathcal{O}$-algebra, V, provides information about commutativity of the operations. For example, if Σ is a fixed point in $\mathcal{O}(n)$ under the action of S_n then the n-ary operation, m_Σ, will be totally symmetric.

This definition of an operad may be generalized by relaxing the condition that the unit element exists. Much of what follows can be generalized to this more general setting.

There is a natural notion of a V-module, M, where V is an $\mathcal{O}$-algebra. The composition maps of a V-module M are essentially required to satisfy the axioms of the composition maps in an $\mathcal{O}$-algebra by replacing the V associated to one input and the output by M, and demanding that the axioms of an algebra hold (see [13]) except that one does not allow compositions between elements in M. As usual, the axioms of a V-module M are those which make the algebra V into a module over itself.

2.2. Operads of moduli spaces.

In this paper, we shall be primarily concerned with the case where $\mathcal{O}$ is an operad of manifolds which arise as moduli spaces of some geometric objects or linear operads derived from such operads. The simplest example of such an operad is the *commutative operad (with unit)* $\mathcal{C}om$ which can be regarded as follows. Let S^2 be the oriented two sphere. Let $\mathcal{C}om(n)$ be the moduli space of configurations of $(n + 1)$ distinct, ordered, embedded disks in S^2 where any two such configurations are regarded as equivalent if they are related by an orientation preserving diffeomorphism of S^2. The permutation group acts on $\mathcal{C}om(n)$ by permuting the ordering on the first n maps. The composition maps $\mathcal{C}om(n) \times \mathcal{C}om(n') \rightarrow \mathcal{C}om(n + n' - 1)$ taking $(\Sigma, \Sigma') \mapsto \Sigma \circ_i \Sigma'$ are obtained by cutting out the $(n' + 1)$st disk of Σ' and the ith disk of Σ and then sewing them together for all $i = 1, \ldots, n$. This structure makes $\mathcal{C}om$ into an operad. Because the diffeomorphism group is so large, $\mathcal{C}om(n)$ consists of one point for each n and the permutation group acts trivially. In fact, a $\mathcal{C}om$-algebra, V, is equivalent to saying that V is endowed with the structure of a commutative, associative algebra with unit. V is said to be a *(tree level) two dimensional topological field theory* where tree level means that no higher genus surfaces appear. A similar operad $\mathcal{AS}$, called the *associative operad (with unit)* has an nth component which can be regarded as the moduli space of configurations of $(n + 1)$, distinct, ordered, intervals in the oriented circle but now quotienting by the action of orientation preserving diffeomorphisms of the circle. The composition maps and the action of the permutation groups are defined similarly to the previous case. $\mathcal{AS}(n)$ contains $n!$ elements upon which S_n acts simply transitively. An $\mathcal{AS}$-algebra, V, is nothing more than the structure of an associative algebra with unit on V. V is said to be a *one dimensional topological field theory*.

One of the simplest operads of nontrivial moduli spaces is the *framed little disks operad*, $\mathcal{F}$, which plays an important role in what follows. Let D be the (closed) unit disk about the origin in $\mathbb{C}$. Let $\mathcal{F}(n)$ be the space of configurations of n distinct, ordered embeddings $(f_1, \ldots, f_n)$ of D into D such that for all $i = 1, \ldots, n$, $f_i(z) = a_i z + z_i$ where a_i belongs to $\mathbb{C}^\times$, a nonzero complex number, and z_i belongs to $\mathbb{C}$, *i.e.* each f_i is a composition of translations, rotations, and dilations. Furthermore, the images of D must have mutually disjoint interiors. The collection $\mathcal{F} = \{ \mathcal{F}(n) \}$ is an operad with the permutation group acting by reordering the embedded disks, and the composition maps $\mathcal{F}(n) \times \mathcal{F}(n') \to \mathcal{F}(n + n' - 1)$ taking $(\Sigma, \Sigma') \mapsto \Sigma \circ_i \Sigma'$ are defined by taking the large disk of Σ', shrinking it down to size, rotating appropriately and then inserting it into the ith disk of Σ (see Figure 1). More explicitly, let $(z_1, a_1, \ldots, z_n, a_n)$ and $(z'_1, a'_1, \ldots, z'_{n'}, a'_{n'})$ denote elements in $\mathcal{F}(n)$ and $\mathcal{F}(n')$, respectively, where for $n' \geq 1$, z_i and z'_i belong to $\mathbb{C}$ denoting the center of the ith disks and a_i and a'_i belong to $\mathbb{C}^\times$ denoting the rotation and dilations of the ith disk then

$$(2.2) \quad (z_1, a_1, \ldots, z_n, a_n) \circ_i (z'_1, a'_1, \ldots, z'_{n'}, a'_{n'}) =$$
$$(z_1, a_1, \ldots, z_{i-1}, a_{i-1}, z_i + a_i z'_1, a'_1 a_i, z_i + a_i z'_2, a'_2 a_i, \ldots, z_i + a_i z'_{n'}, a'_{n'} a_i, \ldots, z_n, a_n).$$

If Ω is the sole element in $\mathcal{F}(0)$ then

$$(2.3) \qquad (z_1, a_1, \ldots, z_n, a_n) \circ_i \Omega = (z_1, a_1, \ldots, \hat{z}_i, \hat{a}_i, \ldots, z_n, a_n).$$

$\mathcal{F}$ is an operad of complex manifolds since $\mathcal{F}(n)$ is an $2n$-complex dimensional manifold (with corners) and the composition maps are holomorphic.

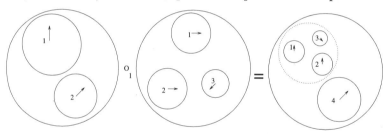

FIGURE 1. The composition map $\circ_1 : \mathcal{F}(2) \times \mathcal{F}(3) \to \mathcal{F}(4)$

2.3. Conformal field theories. Let $\mathbb{CP}^1$ be the Riemann sphere which can be identified with $\mathbb{C} \cup \{\infty\}$ in a standard coordinate z. Let $\mathcal{P}(n)$ be the moduli space of configurations of $(n + 1)$-distinct, ordered biholomorphic embeddings of D into $\mathbb{CP}^1$ whose images have mutually disjoint interiors and where any two such configurations are identified if they are related by a biholomorphic map, *i.e.* a complex projective transformation, where $\mathrm{PSL}(2, \mathbb{C})$ acts upon $\mathbb{CP}^1$ by

$$(2.4) \qquad z \mapsto \frac{az + b}{cz + d}, \ \forall\, a, b, c, d \in \mathbb{C} \text{ where } \det \begin{pmatrix} a & b \\ c & d \end{pmatrix} = 1.$$

$\mathcal{P}$ forms an operad of complex manifolds where the permutation group acts by permuting the first n embeddings, the unit element I in $\mathcal{P}(1)$ being $\mathbb{CP}^1$ with the standard unit disks about 0 and ∞, and the composition maps $\mathcal{P}(n) \times \mathcal{P}(n') \to$

$\mathcal{P}(n + n' - 1)$ taking $(\Sigma, \Sigma') \mapsto \Sigma \circ_i \Sigma'$ are defined by cutting out the $(n+1)$st disk of Σ', the ith disk of Σ and then sewing.

The framed little disks operad $\mathcal{F}$ is a suboperad of $\mathcal{P}$ where the inclusion maps $\mathcal{F}(n) \hookrightarrow \mathcal{P}(n)$ are defined by identifying the big disk of a point in $\mathcal{F}(n)$ with the unit disk D in $\mathbb{C} \subseteq \mathbb{CP}^1$ giving rise to n embedded unit disks in $\mathbb{CP}^1$ and where the $(n+1)$st unit disk is given by the unit disk in the standard chart about ∞ in $\mathbb{CP}^1$. This map is injective and a morphism of operads. In fact, this map is a homotopy equivalence – a fact that will prove to be useful when we discuss topological conformal field theory.

A *(tree level $c = 0$) conformal field theory* , V, is a $\mathcal{P}$-algebra, *i.e.* V is a topological vector space with a smooth morphism of operads $m : \mathcal{P} \to \mathcal{E}nd_V$. The map $m : \mathcal{P}(n) \to \mathcal{E}nd_V(n)$ is called the *n-point correlation function* of V. An important class of conformal field theories are *holomorphic conformal field theories* which are conformal field theories where m is holomorphic. As in the topological case, tree level means that only genus zero Riemann surfaces appear in the definition of $\mathcal{P}$. A general tree level conformal field theory is a collection of smooth maps $m : \mathcal{P} \to \mathcal{E}nd_V$ which are equivariant under the action of the permutation group and such that compositions of elements in $\mathcal{P}$ are mapped into compositions of their images in $\mathcal{E}nd_V$ up to a projective factor. In particular, V is a projective $\mathcal{P}(1)$ module and, as a consequence, is a projective representation of the Virasoro algebra with central charge c. Thus, if m is a morphism of operads then $c = 0$. In general, V can be regarded as an honest algebra over a larger operad involving determinant line bundles which fibers over $\mathcal{P}$ (see [**35**]). For simplicity, we shall restrict to $c = 0$ conformal field theories although much of what follows may be generalized.

2.4. Vertex operator algebras and partial operads.

As shown by Huang and Lepowsky, [**19**], $(c = 0)$ vertex operator algebras may be regarded as holomorphic algebras over a partial operad which contains $\mathcal{P}$ as a suboperad. Vertex operator algebras thus provide an important class of examples of holomorphic conformal field theories.

Henceforth, assume that all coordinate charts on $\mathbb{CP}^1$ which appear in the definition of our moduli spaces have an image containing the origin in $\mathbb{C}$. The preimage of the origin is said to be the *center of this coordinate*.

Let $\mathfrak{P}(n)$ be the moduli space of configurations of $(n+1)$-distinct, ordered holomorphic coordinates (no two of which have coinciding centers) on $\mathbb{CP}^1$ where any two such configurations are identified if they are related by a complex projective transformation. The collection, $\mathfrak{P}$, satisfies all the axioms of an operad except that one cannot always compose elements in $\mathfrak{P}$. Such a collection satisfying the axioms of an operad whenever compositions are defined is called a *partial operad* [**19**]. The partial operad $\mathfrak{P}$ contains $\mathcal{P}$ as a suboperad since holomorphically embedded unit disks may be regarded as holomorphic coordinates. Since both $\mathcal{P}$ and $\mathfrak{P}$ are operads of complex manifolds, the composition maps of

$\mathfrak{P}$ are uniquely determined by analytic continuation of the composition maps of $\mathcal{P}$.

V is said to be $\mathfrak{P}$-algebra if there is a morphism of partial operads $\mathfrak{P} \to \mathcal{E}nd_V$. We shall call holomorphic $\mathfrak{P}$-algebras V, *(c = 0) vertex operator algebras* although as shown by Huang and Lepowsky [19], an actual vertex operator algebra is a bit more subtle. The subtlties are of three types. The first are constraints on the form of the representation of the Virasoro algebra (the $\mathcal{P}(1)$-module structure). The second is about the structure of poles which the correlation functions can have as the centers of charts coincide. The third is that the endomorphism operad is replaced by the space of linear maps on the vertex operator algebra into a somewhat larger space where compositions are defined whenever compositions are defined on $\mathfrak{P}$. We refer the reader to [19] for the details. It is straightforward to generalize our constructions to their situation.

In any case, since $\mathcal{P}$ is a suboperad of $\mathfrak{P}$, vertex operator algebras provide an important class of examples of holomorphic conformal field theories.

2.5. Vertex algebras. Just as the operad $\mathcal{P}$ can be enlarged to the partial operad $\mathfrak{P}$, the framed little disks operad $\mathcal{F}$ can be enlarged to a partial operad $\mathfrak{F}$ [18] which is a partial suboperad of $\mathfrak{P}$.

Let $\mathfrak{F}(n)$ be the space of configurations of n distinct, ordered holomorphically embedded disks, $(f_1, \dots, f_n)$, in $\mathbb{C}$ where the ith map is $f_i(z) = a_i z + z_i$ for z_i in $\mathbb{C}$, a_i in $\mathbb{C}^\times$, and $z_i \neq z_j$ if $i \neq j$. Denote elements in $\mathfrak{F}(n)$ by $(z_1, a_1, \dots, z_n, a_n)$ for all $n \geq 1$ as in $\mathcal{F}(n)$ and denote the sole element in $\mathcal{F}(0)$ by Ω. Now, $\mathfrak{F}(n)$ can be regarded as a $2n$ dimensional complex submanifold of $\mathfrak{P}(n)$ by setting the $(n+1)$st chart to be the standard holomorphic chart about ∞ in $\mathbb{CP}^1$. In fact, $\mathfrak{F}$ is a partial suboperad of $\mathfrak{P}$ with the composition maps given by equation 2.2. $\mathfrak{F}$ is called the *framed little disks partial operad*.

Observe that $\mathfrak{F}$ has a natural partial suboperad $\mathcal{C}$ where $\mathcal{C}(n)$ consists of those points in $\mathfrak{F}(n)$ of the form $(z_1, 1, z_2, 1, \dots, z_n, 1)$. $\mathcal{C}(n)$ is the space of configurations of n distinct, ordered points in $\mathbb{C}$. As complex manifolds, $\mathfrak{F}(n)$ can be written as $\mathcal{C}(n) \times (\mathbb{C}^\times)^n$. In particular, $\mathfrak{F}(0)$ consists of one element denoted by Ω while $\mathfrak{F}(1)$ is the complex Lie group which is the semidirect product of $\mathbb{C}$ with $\mathbb{C}^\times$ which has a Lie algebra which generates (complex) translations L_{-1} and dilations L_0 in the Virasoro algebra. Observe that for all $n \geq 3$, every element in $\mathfrak{F}(n)$ can be decomposed into $(n-1)$ elements of $\mathfrak{F}(2)$.

A *vertex algebra* is an object which satisfies all of the axioms of a vertex operator algebra except that the Virasoro algebra is replaced by the Lie algebra of $\mathfrak{F}(1)$. As in the case of the vertex operator algebra, we shall regard a vertex algebra V as a holomorphic $\mathfrak{F}$-algebra (see [18], [19]).

3. The cohomology of vertex algebras

In this section, we examine first order deformations of an $\mathfrak{F}$-algebra, V, which leave the action of the Lie group $\mathfrak{F}(1)$ fixed. We then generalize the construction of Kimura and Voronov [**23**] where the cochain complex associated to algebras over quadratic operads with values in a module is defined and applied to conformal field theories. The generalization is necessary because $\mathfrak{F}$ is not an operad but merely a partial operad. The second cohomology group of V with values in itself parametrizes inequivalent first order deformations of V. All objects will, unless otherwise stated, be assumed to be smooth (or holomorphic if desired) in what follows.

3.1. The structure of $\mathfrak{F}$-algebras. Henceforth, denote $\mathfrak{F}(0)$ by C, $\mathfrak{F}(1)$ by K, and $\mathfrak{F}(2)$ by E. Let V be an $\mathfrak{F}$-algebra with the following components. Let $e : C \to \mathcal{E}nd_V(0)$ be the image of Ω in V, $\rho : K \to \mathcal{E}nd_V(1)$ be the action of the group K on V, and $m : E \to \mathcal{E}nd_V(2)$ give rise to the set of binary operations. All other operations on the $\mathfrak{F}$-algebra, V, comes from compositions of e, ρ, and m since $\mathfrak{F}$ is generated by C, K, and E. If V is a vertex algebra the vector e can be identified with the *vacuum vector* in V, ρ is given by

$$(3.1) \qquad \rho_{(z,a)}(v) = Y(a^{-L_0}v, z)\, e,$$

and m is given by

$$(3.2)$$

$$m_{(z_1,a_1,z_2,a_2)}(v_1 \otimes v_2) = \begin{cases} Y((a_1)^{-L_0}v_1, z_1)Y((a_2)^{-L_0}v_2, z_2)e, & \text{if } |z_1| > |z_2| \\ Y((a_2)^{-L_0}v_2, z_2)Y((a_1)^{-L_0}v_1, z_1)e, & \text{if } |z_2| > |z_1| \end{cases}$$

and everywhere else m is defined by analytic continuation (see [**18**]).

The relations between (e, ρ, m) arise from the images of the composition maps (see equation 2.2) between C, K, and E. They are given explicitly by the following where (z_1, a_1), (z_1', a_1') belong to K and (z_1, a_1, z_2, a_2), (z_1', a_1', z_2', a_2') belong to $\mathfrak{F}(2)$ keeping in mind that the composition maps are undefined if the resulting expression contains charts with coinciding centers:

$$(3.3) \qquad\qquad\qquad \rho_{(0,1)} \;=\; \mathrm{id}_V,$$

$$(3.4) \qquad\qquad \rho_{(z_1,a_1)} \circ_1 \rho_{(z_2,a_2)} \;=\; \rho_{(z_1+a_1 z_2,\, a_2)},$$

$$(3.5) \qquad\qquad\qquad \rho_{(z_1,a_1)} \circ_1 e \;=\; e,$$

$$(3.6) \qquad\qquad m_{(z_1,a_1,z_2,a_2)} \circ_1 e \;=\; \rho_{(z_2,a_2)},$$

$$(3.7) \qquad\qquad m_{(z_1,a_1,z_2,a_2)} \circ_2 e \;=\; \rho_{(z_1,a_1)},$$

$$(3.8) \qquad m_{(z_1,a_1,z_2,a_2)} \circ_1 \rho_{(z_1',a_1')} \;=\; m_{(z_1,\, a_1 z_1',\, a_1 a_1',\, z_2,\, a_2)},$$

$$(3.9) \qquad m_{(z_1,a_1,z_2,a_2)} \circ_2 \rho_{(z_1',a_1')} \;=\; m_{(z_1,\, a_1,\, z_2+a_2 z_1',\, a_2 a_1')},$$

$$(3.10) \qquad \rho_{(z_1',a_1')} \circ_1 m_{(z_1,a_1,z_2,a_2)} \;=\; m_{(z_1'+a_1' z_1,\, a_1' a_1,\, z_1'+a_1' z_2,\, a_1' a_2)}.$$

There is one additional relation called *the associativity condition* which may be written in the following way. Let Σ belong to $\mathfrak{F}(3)$. Such a surface may be decomposed into two elements (*pairs of pants*) in $\mathfrak{F}(2)$ in three ways. The first is to decompose Σ so that one of the pants contains only the first two punctures. The second is to decompose Σ so that one of the pants contains only the last two punctures. The third is to decompose Σ so that one of the pants contains only the first and third punctures. These three decompositions are parametrized by the three binary 3-trees (see figure 2) by placing pairs of pants at each vertex which sew together in the indicated manner to become Σ. The associativity conditions may be written as

$$(3.11) \qquad (m \circ m)_{(\Sigma, T_i)} = (m \circ m)_{(\Sigma, T_j)}$$

where (Σ, T_i) is any decomposition of Σ as indicated by the tree T_i.

FIGURE 2. The three binary 3-trees T_1, T_2, and T_3

3.2. Deformations of $\mathfrak{F}$-algebras. Consider a first order deformation of an $\mathfrak{F}$-algebra, V, (which fixes the K-module structure) given by $e \mapsto e' = e + t\epsilon + o(t^2)$, and $m \mapsto m' = m + t\alpha + o(t^2)$ where $\epsilon : C \to \mathcal{E}nd_V(0)$ is K-equivariant, and $\alpha : E \to \mathcal{E}nd_V(2)$ is (K, K^2) and S_2 equivariant. Keeping terms of order t, we obtain the following conditions on (ϵ, α) to be a first order deformation of V:

$$(3.12) \qquad \rho_{(z,a)} \circ_1 \epsilon = 0,$$

$$(3.13) \qquad \alpha_{(z_1,a_1,z_2,a_2)} \circ_1 e + m_{(z_1,a_1,z_2,a_2)} \circ_1 \epsilon = 0,$$

$$(3.14) \qquad \alpha_{(z_1,a_1,z_2,a_2)} \circ_2 e + m_{(z_1,a_1,z_2,a_2)} \circ_2 \epsilon = 0,$$

$$(3.15) \qquad \alpha_{(z_1,a_1,z_2,a_2)} \circ_1 \rho_{(z_1',a_1')} = \alpha_{(z_1,a_1 z_1',z_1 a_1',z_2,a_2)},$$

$$(3.16) \qquad \alpha_{(z_1,a_1,z_2,a_2)} \circ_2 \rho_{(z_1',a_1')} = \alpha_{(z_1,a_1,z_2+a_2 z_1',a_2 a_1')},$$

$$(3.17) \qquad \rho_{(z_1',a_1')} \circ_1 \alpha_{(z_1,a_1,z_2,a_2)} = \alpha_{(z_1'+a_1' z_1,a_1' a_1,z_1'+a_1' z_2,a_1' a_2)},$$

and the associativity condition yields

$$(3.18) \qquad (\alpha \circ m + m \circ \alpha)_{(\Sigma,T_i)} = (\alpha \circ m + m \circ \alpha)_{(\Sigma,T_j)}$$

for all $i = 1, 2, 3$.

It may be that, up to first order, such a deformation is an automorphism of V, *i.e.* suppose that there exists an isomorphism $\Phi : V \to V$ such that

$$(3.19) \qquad \Phi \circ_1 m = m' \circ (\Phi \otimes \Phi),$$

$$(3.20) \qquad \Phi \circ_1 e = e'.$$

Setting $\Phi = \mathrm{id}_V + t\beta$ where $\beta : V \to V$ is a linear map, plugging in for (e', ρ', m') and keeping terms up to first order in t, we obtain for all Σ in $\mathfrak{F}(2)$,

$$(3.21) \qquad \alpha_\Sigma = m_\Sigma \circ_2 \beta - \beta \circ_1 m_\Sigma + m_\Sigma \circ_1 \beta,$$

$$(3.22) \qquad \epsilon = \beta \circ_1 e.$$

We have just shown the following.

THEOREM 1. Let V be an $\mathfrak{F}$-algebra. Inequivalent first order deformations of V keeping the K-module structure fixed are parametrized by the space of pairs (ϵ, α) satisfying equations 3.12 to 3.18 quotiented by those which are obtained from β satisfying equations 3.21 and 3.22 where $\epsilon : \mathfrak{F}(0) \to \mathcal{E}nd_V(0)$ is left K-equivariant, $\alpha : \mathfrak{F}(2) \to \mathcal{E}nd_V(2)$ is (K, K^2) and S_2 equivariant, and β belongs to $\mathcal{E}nd_V(1)$.

If V is an $\mathfrak{F}$-algebra then it is certainly an algebra over the partial suboperad, $\widetilde{\mathfrak{F}}$, which is the same as $\mathfrak{F}$ except that $\mathfrak{F}(0)$ is the empty set. It turns out that the space of inequivalent first order deformations of V as an $\mathfrak{F}$-algebra is isomorphic to the space of inequivalent first order deformations of V as an $\widetilde{\mathfrak{F}}$-algebra.

To be more precise, an $\widetilde{\mathfrak{F}}$-algebra, V, is determined by the binary map m and its inequivalent first order deformations are given by those (K, K^2) and S_2-equivariant maps $\alpha : \mathfrak{F}(2) \to \mathcal{E}nd_V(2)$ satisfying equation 3.18 modulo those for which a linear map $\beta : V \to V$ exists satisfying 3.21.

THEOREM 2. Let V be an $\mathfrak{F}$-algebra. The space of inequivalent first order deformations of V which keep the K-module structure fixed is isomorphic to the space of inequivalent first order deformations of V as an algebra over $\widetilde{\mathfrak{F}}$.

PROOF. Let $[\alpha]$ be a nontrivial deformation of V as an $\widetilde{\mathfrak{F}}$-algebra, i.e. $\alpha : \mathfrak{F}(2) \to \mathcal{E}nd_V(2)$ which is (K, K^2) and S_2 equivariant satisfying equations 3.12 to 3.18 where any two such are related if they differ by some β satisfying equation 3.21 and 3.22. Let $\epsilon = -\alpha_\Sigma(e, e)$ for any Σ in $\mathfrak{F}(2)$. A straightforward computation shows that the pair (α, ϵ) defines a nontrivial deformation of V as an $\mathfrak{F}$-algebra. The converse is obvious. $\square$

3.3. $\mathfrak{F}$ as a quadratic partial operad.

We will now show that the partial operad $\mathfrak{F}$ is quadratic in analogy with the notion of a quadratic operad as introduced by Ginzburg and Kapranov [13].

For now, let $\mathfrak{F}$ be the operad which has an empty $\mathfrak{F}(0)$ component. It will prove convenient to let $\mathfrak{F}(n)$ be the vector space generated by the points in the framed little disks partial operad so that K is the linear combination of 2-punctured spheres and E the linear combination of 3-punctured spheres.

Recall that a binary n-tree is a tree with its n incoming edges labeled by the integers $\{1, \ldots, n\}$ such that all its vertices have two inputs and one output. Let $F[E](n)$ be the free operad generated by the S_2 and (K, K^2)-module, E, i.e.

$$F[E](n) = \bigoplus_{\text{binary } n\text{-trees, } T} E[T]$$

where $E[T] = \bigotimes_{v \in T} E[\mathrm{In}v]$ where $\mathrm{In}v$ denotes the set of incoming to a vertex v in the tree T. For example, an element in $E[T]$ where T is a binary 3-tree can be regarded as linear combinations of equivalence classes of triples $[T, \Sigma, \Sigma']$ where Σ and Σ' are pairs of pants which are associated to the top and bottom vertices of T, respectively, with punctures associated to the incoming edges to that vertex (see figure 3). (We have already seen that trees arise naturally in decomposing punctured spheres.) Two triples are said to be equivalent if they are related by the following action of K

$$(T, \Sigma \circ_1 k, \Sigma') \sim (T, \Sigma, k \circ \Sigma')$$

where k belongs to K and Σ, Σ' belong to E. Geometrically, this action of K should be regarded as sliding a two punctured sphere along an internal edge of T from the output of one pair of pants at one bounding vertex to the corresponding input of the other pair of pants.

FIGURE 3. An element in $\mathfrak{F}(3)$, a 3-tree, and a decomposition

$F[E]$ is an operad with composition maps which are given by grafting the root of one tree onto the proper branch of the other.

We shall now define a closely related partial operad $F(E)$. Let $F(E)(n)$ be the subspace of $F[E](n)$ generated by n-trees decorated with pairs of pants at each vertex such that all of these pants can be composed together as indicated by the tree T to obtain an element in $\mathfrak{F}(n)$. Thus, generators of $F(E)(n)$ can be identified with pairs (Σ, T) where Σ is an element in $\mathfrak{F}(n)$ and T is a binary n-tree since there is a unique decomposition of Σ into pairs of pants as indicated by T up to appropriate actions of K (see figure 3). Redefine composition maps $F(E)(n) \otimes F(E)(n') \to F(E)(n + n' - 1)$ taking $(\Sigma, T) \otimes (\Sigma', T') \mapsto (\Sigma, T) \circ_i (\Sigma', T') = (\Sigma \circ_i \Sigma', T \circ_i T')$ where $T \circ_i T'$ is the tree obtained by grafting the root of T' onto the ith input to T. This composition is defined only if $\Sigma \circ_i \Sigma'$ is defined. These composition maps make $F(E)$ into a partial operad.

Let R be the subspace of $F(E)(3)$ consisting of linear combinations of the elements $(\Sigma, T_1) - (\Sigma, T_2)$, $(\Sigma, T_1) - (\Sigma, T_3)$, and $(\Sigma, T_2) - (\Sigma, T_3)$ where Σ belongs to $\mathfrak{F}(n)$ and T_1, T_2, and T_3 are the three binary 3-trees in figure 2. The space R is the *space of relations in $F(E)(3)$* since if (R) is the ideal in $F(E)$ generated by R, then we have the isomorphism of partial operads $\mathfrak{F} = F(E)/(R)$. In analogy to the case of operads, such a partial operad is said to be a *quadratic partial operad*.

3.4. The cochain complex. It will be useful to fix some terminology about trees. Let T be an n-tree. A *terminal edge of* T is an edge which is an input to the tree. Denote the space of all terminal edges of T by $\text{Term}(T)$.

Let $\det(T)$ be the top exterior power of the vector space generated by the internal edges of T, *i.e.* edges which are not inputs to the tree T. Let $\mathcal{U}(n) = \bigoplus_{\text{binary } n\text{-trees } T} E(T) \otimes \det(T)$. A 1-*ternary binary* n-*tree* is a tree with n-inputs which has exact one ternary vertex (a vertex with three inputs), all other vertices being binary vertices. Let T be a 1-ternary tree with ternary vertex u and binary vertices v, then let

(3.23)

$$(R \otimes \det)(T) = (R \otimes \det)(\text{In}u) \otimes \left[\bigotimes_{\text{binary } v \in T} E(\text{In}v) \otimes \det(\text{In}v) \right] \otimes \det(\text{Term}(T))$$

where $(R \otimes \det)(\text{In}u)$ is exactly R except that it is labeled by incoming edges of the vertex u and is twisted by det. Finally, let $\langle R \rangle(n) = \bigcap_{\text{1-ternary } n \text{ trees } T}(R \otimes \det)(T)$, elements of which can be written in terms of linear combinations of binary trees by rewriting the ternary vertex in terms of linear combinations of binary ones.

Let V be an $\mathfrak{F}$-algebra and M a V-module, then the *cochain complex for the cohomology of* V *with values in* M has n-th term given, for all $n \geq 1$, by

(3.24) $C_{\mathfrak{F}}^n(V, M) = \text{Hom}_K(\mathcal{U}(n) \otimes_{S_n, K^n} V^{\otimes n}, M)/(\langle R \rangle(n) \otimes_{K^n} V^{\otimes n})^{\perp}.$

The differential $d : C_{\mathfrak{F}}^n(V, M) \to C_{\mathfrak{F}}^{n+1}(V, M)$ is induced from a map on the numerators of $C_{\mathfrak{F}}^n(V, M)$ which preserves the denominator. Let us denote the cohomology of this complex by $H_{\mathfrak{F}}^{\bullet}(V, M)$.

The differential $d : C^1 = \text{Hom}_K(V, M) \to C^2 = \text{Hom}(E \otimes_{S_2, K} V \otimes V, M)$ of a 1-cochain β is given by

(3.25) $(d\beta)_{\Sigma}(v_1, v_2) = m_{\Sigma}(v_1, \beta(v_2)) - \beta(m_{\Sigma}(v_1, v_2)) + m_{(\sigma\Sigma)}(v_2, \beta(v_1))$

where σ is the transposition in S_2, v_1, v_2 belong to V, Σ belongs to E, m denotes the module map $E \otimes V \otimes M \to M$ or the binary operation in V, $E \otimes V \otimes V \to V$ as is appropriate.

In order to describe the differential $d : C^n \to C^{n+1}$ for $n \geq 2$, we need some additional notions. Let $\text{In}_r T$ denote the set of terminal edges of T which are attached to the root, r, of the tree T. An *extremal vertex of* T is a vertex such that all of its inputs are terminal edges. Let $\text{Ext } T$ denote the set of extremal vertices of T. The differential of an n-cochain α is given by

$$
(d\alpha)(\Sigma, T, e_T; v_1, v_2, \ldots, v_{n+1})
$$
$$
= \sum_{w \in \mathrm{Ext}\, T} \alpha(\Sigma/w, T/w, e_{T/w}; m_{\Sigma^w}(v_i, v_j), v_1, \ldots, \hat{v}_i, \ldots, \hat{v}_j, \ldots, v_{n+1})
$$
$$
+ \sum_{e \in \mathrm{In}_r T} m_{\Sigma^e}(v_i, \alpha(\Sigma/e, T/r, e_{T/r}; v_1, \ldots, \hat{v}_i, \ldots. v_{n+1})),
$$

where (Σ, T) denotes a decomposition of an $(n+1)$ punctured sphere into pairs of pants each of which is located at a vertex of the binary $(n+1)$ tree, T. The first summation runs over extremal vertices w of T with i and j being the labels of the vertices coming into w, Σ^w is the pair of pants at the vertex w, $(\Sigma/w, T/w)$ is what remains of (Σ, T) after removing Σ^w from Σ and T/w is the tree T from which we have cut away the branches above w. The punctures in Σ/w are ordered so that the puncture created by removing Σ^w is the first and then the remaining punctures follow retaining their original ordering in Σ. The order of the inputs to the tree T/w are given similarly. $e_{T/w}$ is just the wedge product of the internal edges of T/w in the same order they were in e_T. The puncture of Σ^w which was originally labeled by i on Σ is relabeled 1, while the other input puncture is relabeled by 2. Similarly, the second summation runs over terminal edges e going into the root of (Σ, T), if any, where i is the label at the input of the edge e, Σ^e is the pair of pants located at the root and $(\Sigma/e, T/r)$ is what remains of Σ after removing Σ^e from (Σ, T). Finally, T/r is the tree with the root and its inputs removed, the punctures of Σ^e retain the same ordering that they had in Σ, the puncture labeled with i on Σ is labeled by a 1 on Σ/e, and $e_{T/r}$ is the product of the corresponding edges of T in the same order.

THEOREM 3. For any $\mathfrak{F}$-algebra, V, and V-module, M, $(C_{\mathfrak{F}}^\bullet(V, M), d)$ is a complex.

PROOF. The proof that $d^2 = 0$ in [23] for the cohomology of an algebra over a quadratic operad with values in a module carries over to $\mathfrak{F}$-algebras and its modules over $\mathfrak{F}$ because in the above, we only decompose the element (Σ, T) in $F(E)(n)$ in the construction of the differential and, furthermore, $F(E)(n)$ is defined so that pairs of pants located at vertices bounding an edge can always be sewn together. $\square$

THEOREM 4. The inequivalent first order deformations of an $\mathfrak{F}$-algebra, V, which leave the K-module structure on V fixed are parametrized by $H_{\mathfrak{F}}^2(V, V)$.

PROOF. The 2-cocycle condition $d\alpha = 0$ is nothing more than equation 3.18 while the fact that $\alpha = d\beta$ for some 1-cochain β is nothing more than equation 3.21. $\square$

When V is an $\mathfrak{F}$-algebra with a trivial K-module structure then V is a commutative associative algebra with unit, that is, a (tree level) two dimensional topological field theory, and the complex $C_{\bar{3}}^{\bullet}$ is precisely the Harrison cochain complex which is the natural complex associated to such algebras.

We expect this complex to be a prototype for the complex associated to $\mathcal{F}$-algebras, the important case for vertex operator algebras. We expect the two complexes to share many common features.

4. Topological conformal field theory

In this section, we will present some results from so-called (tree level $c = 0$) topological conformal field theories and topological operator vertex algebras. A topological conformal field theory is a conformal field theory which has differential form valued correlation functions that are suitably compatible. Integration of these forms over singular chains on our moduli spaces associates an operation to each singular chain. Since the space of singular chains on an operad of topological spaces is itself an operad, a topological conformal field theory is an algebra over the singular chains on $\mathcal{P}$. *Unless otherwise specified, all chains will be over* $\mathbb{R}$.

4.1. Operads of complexes. Let $\mathcal{O}$ be an operad of topological spaces, i.e. $\mathcal{O}(n)$ is a topological space for all n. Let $C_{\bullet}(\mathcal{O}) = \{C_{\bullet}(\mathcal{O}(n))\}$ where $C_{\bullet}(\mathcal{O}(n))$ is the complex of singular chains on $\mathcal{O}(n)$. The composition maps and the action of the permutation groups on $\mathcal{O}$ induce the same on $C_{\bullet}(\mathcal{O})$ making it into an operad of complexes or differential graded vector spaces. That is, for all n, $C_{\bullet}(\mathcal{O}(n))$ is a complex graded by $\mathbb{Z}$ with a differential ∂ : $C_p(\mathcal{O}(n)) \rightarrow C_{p-1}(\mathcal{O}(n))$ where $C_p(\mathcal{O}(n))$ is an S_n-module for all p and n (the module action commutes with the differential), and the composition maps $C_p(\mathcal{O}(n)) \otimes C_{p'}(\mathcal{O}(n')) \rightarrow C_{p+p'}(\mathcal{O}(n+n'-1))$ taking $c \otimes c' \mapsto c \circ_i c'$ are chain maps which satisfy the axioms of an operad.

Let (V, Q) be a complex with differential $Q : V^g \rightarrow V^{g-1}$ for all g in $\mathbb{Z}$. The *endomorphism operad of* (V, Q), $\mathcal{E}nd_V$, is defined as before but with one modification. Let $\mathcal{E}nd_V{}^g(n)$ denote the space of maps in $\mathcal{E}nd_V(n)$ which have degree g.

The complex (V, Q) *is said to be an algebra over* $C_{\bullet}(\mathcal{O})$ if there is a morphism of operads $m : \mathcal{O} \rightarrow \mathcal{E}nd_V$, *i.e.* the collection of chain maps m which preserve the action of the permutation group, compositions, and the grading.

A *(tree level $c = 0$) topological conformal field theory* is a complex (V, Q) with a collection of differential forms $\{\omega_n\}$ such that $\omega_n = \omega_n^0 + \omega_n^1 + \omega_n^2 + \cdots$ satisfying the degree conditions

$$\omega_n^g \in \Omega^g(\mathcal{P}(n)) \otimes \mathcal{E}nd_V{}^{-g}(n),$$

the *descent equations*

$$d\omega_n = Q\omega_n,$$

the *equivariance under composition maps*

$$(\circ_i)^* \omega_{n+n'-1} = \omega_n \circ_i \omega_{n'},$$

for all $i = 1, \ldots, n$ where $(\circ_i)^*$ is the pullback of the composition map $\circ_i :$ $\mathcal{P}(n) \times \mathcal{P}(n') \to \mathcal{P}(n + n' - 1)$, and $\circ_i$ on the right hand side is the composition map in the endomorphism operad, and the *equivariance under the permuation groups*

$$\sigma^* \omega_n = \omega_n \circ \sigma,$$

for all σ in S_n where σ^* is the pullback of the action of S_n on $\mathcal{P}(n)$ and $\circ \sigma$ denotes the action of σ on $V^{\otimes n}$. A *holomorphic topological conformal field theory* is a topological conformal field theory where the differential forms are all holomorphic.

Given a topological conformal field theory (V, Q) and a p-chain c on $\mathcal{P}(n)$, let $\mu_c = \int_c \omega_n$, an element of degree p in $\mathcal{E}nd_V(n)$. This gives rise to a collection of maps $C_\bullet(\mathcal{O}) \to \mathcal{E}nd_V$. The defining conditions of a topological conformal field theory are precisely those which insure that (V, Q) is a $C_\bullet(\mathcal{P})$-algebra.

A topological conformal field theory is, as the name suggests, a conformal field theory since points on $\mathcal{P}(n)$ are zero chains on $\mathcal{P}(n)$ and the axioms of a topological conformal field theory insure that evaluation of these points on $\{\omega_n^0\}$ gives a morphism of operads $m : \mathcal{P} \to \mathcal{E}nd_V$. (We assume that the unit element in $\mathcal{P}$ gets mapped to the identity map in $\mathcal{E}nd_V$.)

If (V, Q) is a topological conformal field theory, Q is called the *BRST operator* and the negative of the grading on V is called the *ghost (or fermion) number* in the physics literature. The complex (V, Q) is called the *BRST complex* and its cohomology $H_\bullet(V)$ is called the *BRST cohomology*.

The BRST complex (V, Q) has an important subcomplex (V_{rel}, Q) called the *semirelative BRST complex* with cohomology $H_\bullet(V_{rel})$, the *semirelative BRST cohomology*. The Lie group $U(1) \subseteq \mathcal{P}(1)$ consists of the points in $\mathcal{P}(1)$ which have the standard chart about ∞ and a chart about 0 which is the standard one rotated by multiplication by an element of $U(1)$. Let $D = \int_{U(1)} \omega_1$ where $U(1)$ is now regarded as a 1-chain in $\mathcal{P}(1)$ then let V_{rel} be the kernel of D. Since D commutes with Q, V_{rel} is a subcomplex of V. An alternate, and perhaps more familar, definition of V_{rel} can be obtained as follows. The morphism $m : \mathcal{P}(1) \to \mathcal{E}nd_V(1)$ induces an action of $\mathfrak{u}(1)$, the Lie algebra of $U(1)$, on V. Let X be a generator of $\mathfrak{u}(1)$ then it acts upon V by a degree 0 map traditionally denoted by L_0^-. Let $b_0^- = \omega_1^1(X)$, a degree 1 endomorphism of V. It is straightforward to show from the definition of a topological conformal field theory that

$$
\begin{aligned}
(b_0^-)^2 &= 0 \\
Q b_0^- + b_0^- Q &= L_0^- \\
Q L_0^- - L_0^- Q &= 0.
\end{aligned}
$$

V_{rel} is then the intersection of the kernels of L_0^- and b_0^-.[1] We will see that this subcomplex will play an important role in what follows.

4.2. Algebraic structures up to homotopy in topological conformal field theory.

In this section, we will show that if (V, Q) is a topological conformal field theory then (V, Q) is a homotopy associative (or A_∞) algebra and a subcomplex (V_{rel}, Q) is a homotopy Lie (or L_∞) algebra. We obtain these structures by analyzing the canonical stratification of the Deligne-Knudsen-Mumford compactification of the moduli space of Riemann spheres with n distinct, ordered punctures and its real analog. Both rely upon the construction of a morphism between those moduli spaces and $\mathcal{P}$ given by the solution to a minimal area metric problem due to Wolf and Zwiebach [40].

Our first result concerns an algebraic structure called a homotopy Lie algebra.

DEFINITION 1. An L_∞ (or homotopy) Lie algebra is a complex V with a differential $Q : V^g \to V^{g-1}$ endowed with, for each $n \geq 2$, a degree $(2n-3)$, graded symmetric n-ary operation, $\underbrace{[\cdot, \dots, \cdot]}_{n}$, satisfying

$$(4.1) \quad Q[v_1, \dots, v_n] + \sum_{i=1}^{n} \epsilon(i)[v_1, \dots, Qv_i, \dots, v_n]$$

$$= \sum_{\substack{k+l=n+1 \\ k,l \geq 2}} \sum_{\substack{\text{unshuffles } \sigma: \\ \{1,2,\dots,n\}=I_1 \cup I_2, \\ I_1=\{i_1,\dots,i_k\}, \; I_2=\{j_1,\dots,j_{l-1}\}}} \epsilon(\sigma)[[v_{i_1}, \dots, v_{i_k}], v_{j_1}, \dots, v_{j_{l-1}}],$$

where $v_1, \dots, v_n$ are elements in V with definite degree, $\epsilon(i) = (-1)^{|v_1|+\dots+|v_{i-1}|}$ is the usual sign obtained by sliding Q by $v_1, \cdots, v_{i-1}$, $\epsilon(\sigma)$ is the sign obtained by sliding the elements v_i by the v_j's in the unshuffle of $v_1, \dots, v_n$. Finally, an unshuffle in S_n is a permutation taking $(1, \dots, n) \mapsto (i_1, \dots, i_p, j_1, \dots, j_{n-p})$ for some p such that the sequences $(i_1, \dots, i_p)$ and $(j_1, \dots, j_{n-p})$ are both increasing.

We have chosen conventions of grading and signs to agree with the physics literature, [41] [39], although an alternate but equivalent choice is given in [28].

When $n = 2$, equation 4.1 says that the binary bracket $[\cdot, \cdot]$ preserves the differential. When $n = 3$, the right hand side is the usual sum in the Jacobi identity while the left hand side is something which is Q-exact when acting upon Q-cocycles involving the homotopy $[\cdot, \cdot, \cdot]$. These two conditions insure that the cohomology of the a homotopy Lie algebra is a Lie algebra with respect to the binary bracket (up to signs due to our choice of conventions). For higher n, equation 4.1 should be regarded as a generalization of the Jacobi identity up to homotopy.

---

[1]Strictly speaking, these two descriptions are equivalent if D has trivial cohomology. This is the case in many examples.

This structure arises in topological conformal field theory from the geometry of certain moduli spaces. Let $\mathcal{M}(n)$ be the moduli space of Riemann spheres with $(n + 1)$ ordered, distinct punctures where any two such configurations are regarded to be equivalent if they are related by the action of $\mathrm{PSL}(2, \mathbb{C})$, i.e. $\mathcal{M}(n) = ((\mathbb{CP}^1)^{n+1} - \Delta)/\mathrm{PSL}(2, \mathbb{C})$ where $\Delta = \{(z_1, \ldots, z_{n+1}) \in (\mathbb{CP}^1)^{n+1} \mid z_i = z_j$ for some $i \neq j\}$, the set of diagonals.

Although the collection of moduli spaces $\mathcal{M} = \{\mathcal{M}(n)\}$ does not form an operad, a natural compactification of $\mathcal{M}(n)$ due to Deligne-Knudsen-Mumford [3, 20, 24], $\overline{\mathcal{M}}(n)$, for all $n \geq 2$, forms an operad of complex manifolds $\overline{\mathcal{M}} = \{\overline{\mathcal{M}}(n)\}_{n \geq 2}$. Here, $\overline{\mathcal{M}}(n)$ is the moduli space of stable genus 0 curves with $(n+1)$ punctures or, equivalently, the moduli space of noded Riemann spheres with $(n+1)$ punctures. Recall that a noded Riemann sphere with $(n+1)$ punctures is a connected complex space where each point has a neighborhood isomorphic to either $\{|z| < \epsilon\}$ or $\{zw = 0 \mid |z| < \epsilon, |w| < \epsilon\}$ (nodes), with $(n + 1)$ punctures none of which is a node, so that each component of the complement of the nodes and punctures is a Riemann sphere with at least three holes. A punctured Riemann sphere acquires a node by undergoing "mitosis," i.e. the sphere forms a long, thin neck away from the punctures which looks locally like $zw = \epsilon$, as ϵ goes to 0 resulting in two punctured spheres attached at a node. Both $\mathcal{M}(n)$ and $\overline{\mathcal{M}}(n)$ are smooth complex algebraic manifolds of complex dimension $n - 2$. The moduli space $\mathcal{M}(n)$ of nonsingular curves is an open submanifold in the projective manifold $\overline{\mathcal{M}}(n)$ and the complement, formed by all the curves with nodes, is a divisor.

S_n acts on $\overline{\mathcal{M}}(n)$ by reordering the first n-punctures and the composition maps $\overline{\mathcal{M}}(n) \times \overline{\mathcal{M}}(n') \to \overline{\mathcal{M}}(n + n' - 1)$ taking $(\Sigma, \Sigma') \mapsto \Sigma \circ_i \Sigma'$ are given by attaching the $(n' + 1)$st puncture of Σ' to the ith puncture of Σ thereby creating a curve with a new double point.

$\overline{\mathcal{M}}(n)$ is stratified by smooth, connected locally closed algebraic subvarieties each of which is indexed by a tree since any stable curve can be obtained by attaching spheres together and the combinatorics of this attaching process, as observed by Beilinson and Ginzburg [2], can be encoded in a tree (see figure 4).

There is another compactification, $\underline{\mathcal{M}}(n)$, of $\mathcal{M}(n)$ for all $n \geq 2$, introduced by Kimura, Stasheff, and Voronov in [21], which is a smooth manifold with corners. It is analogous to a similar real compactification of configuration space considered by Kontsevich [25], Axelrod and Singer [1] and Getzler and Jones [10].

A stable complex curve of genus 0 with $(n+1)$-punctured is said to be *decorated with relative phase parameters at its double points* if the sum of the arguments of germs of holomorphic coordinates on each irreducible component is specified at each double point. This additional data may be regarded as a pair of tangent directions (nonzero covectors up to rescaling by $\mathbb{R}^+$) along the positive real axes, one for each irreducible component on either side of a double point, identifying any pair if they are related by simultaneous rotation by an element of $U(1)$ at each double point. Alternately, $\underline{\mathcal{M}}(n)$ may be regarded as the moduli space of noded Riemann spheres with $(n+1)$ punctures with each node decorated with a phase. Here, we allow mitosis of a Riemann sphere such that the neck looks like $zw = re^{i\theta}$ as r goes to 0 keeping θ fixed, i.e. $\underline{\mathcal{M}}(n)$ is the real blowup of $\overline{\mathcal{M}}(n)$ along the irreducible components of the divisor. $\underline{\mathcal{M}}(n)$ is a real analytic manifold with corners with interior $\mathcal{M}(n)$. The collection $\underline{\mathcal{M}} = \{\,\underline{\mathcal{M}}(n)\,\}_{n\geq 2}$ is not an operad but, the space of chains on $\underline{\mathcal{M}}$, $C_\bullet(\underline{\mathcal{M}})$, do form an operad, after a shift in degree, by using a transfer [21]. We show that (V_{rel}, Q) is an algebra over $C_\bullet(\underline{\mathcal{M}})$. The statification on $\overline{\mathcal{M}}(n)$ is pulled up to one on $\underline{\mathcal{M}}(n)$ via the canonical projection. The spectral sequence associated to the filtration of $\underline{\mathcal{M}}(n)$ arising from this stratification has a top row which is exactly the homotopy Lie operad which was described by Hinich and Schechtman [17] purely in terms of trees. These trees are precisely the ones which index the stratification. By using the result of Wolf and Zwiebach on minimal area metrics [40] to construct a morphism of operads between $C_\bullet(\underline{\mathcal{M}}(n)) \to C_\bullet(\mathcal{P}(n))$, Kimura, Stasheff, and Voronov [21] proved the following.

THEOREM 5. Let (V, Q) be a topological conformal field theory then V_{rel} is a homotopy Lie (or L_∞) algebra. If (V, Q) is a holomorphic topological conformal field theory, then the n-ary brackets vanish if $n \geq 3$.

There is another moduli space which is a real version of $\mathcal{M}(n)$. Consider the real projective space $\mathbb{RP}^1$ which is diffeomorphic to S^1. It can be identified with $\mathbb{R} \cup \{\infty\}$ in a standard chart x. Let $\mathbb{R}^2$ be oriented so that the inclusion map of the real subset $\mathbb{R}^2 \hookrightarrow \mathbb{C}^2$ is orientation preserving. It induces an orientation preserving map $\mathbb{RP}^1 \hookrightarrow \mathbb{CP}^1$ mapping the oriented circle into the Riemann sphere. In analogy with the complex case, let $\mathcal{M}'(n)$ be the moduli space of oriented circles with $(n+1)$ ordered punctures, i.e. $\mathcal{M}'(n) = ((\mathbb{RP}^1)^{n+1} \setminus \Delta)/\operatorname{PSL}(2, \mathbb{R})$ where $\operatorname{PSL}(2, \mathbb{R})$ acts on $\mathbb{RP}^1$ via equation 2.4 with z replaced by the coordinate x and where Δ is the set of diagonals. There is a compactification of $\overline{\mathcal{M}}'(n)$ for $n \geq 2$ which is the real analog of the Deligne-Knudsen-Mumford compactification, the *moduli space of (oriented) circles with nodes and n punctures*, $\overline{\mathcal{M}}'(n)$ where a sphere with node is defined in analogy with the complex case. $\overline{\mathcal{M}}'(n)$ is a smooth manifold with corners with real dimension $n - 2$ and contains $\mathcal{M}(n)$ as an open submanifold. The collection $\overline{\mathcal{M}}' = \{\,\overline{\mathcal{M}}'(n)\,\}_{n\geq 2}$ forms an operad of smooth manifolds with corners where the action of the permutation group and the composition maps are defined in analogy with the complex case. Unlike the complex case, $\overline{\mathcal{M}}'(n)$ has $n!$ components which are permuted into each other under the action of S_n. As in the complex case, this space has a natural stratification with strata indexed by trees.

The operad $\overline{\mathcal{M}}$ and $\overline{\mathcal{M}}'$ are closely related as there is a morphism of operads $\overline{\mathcal{M}}' \hookrightarrow \overline{\mathcal{M}}$ which is induced from the an inclusion map as a real subset $\mathbb{RP}^1 \hookrightarrow \mathbb{CP}^1$.

DEFINITION 2. An A_∞ *(or strongly homotopy associative) algebra*, V, is a complex $Q : V^g \to V^{g-1}$ endowed with a collection of n-ary (linear) operations

$\{ m_n : V^{\otimes n} \to V \}_{n \geq 2}$ with m_n having degree $n - 2$ satisfying

$$(4.2) \quad Q(m_n(v_1, \dots, v_n)) - (-1)^n \sum_{k=1}^{n} (-1)^{\epsilon(k)} m_n(v_1, \dots, Qv_k, \dots, v_n)$$

$$= \sum_{r,s,k} (-1)^{k(s-1)+sn} (m_r \circ_k m_s)(v_1, \dots, v_n)$$

where the summation runs over r, s, k satisfying $r + s = n + 1$, $1 \leq k \leq r$, $2 \leq r < n$ for all $v_1, \dots, v_n$ in V and $n \geq 2$, and $\epsilon(k)$ denotes the sign obtained by sliding Q by $v_1, \dots, v_{k-1}$.

When $n = 2$, equation 4.2 merely says that the binary operation m_2 commutes with the differential. When $n = 3$, the equation says that when acting upon cocycles, the difference between the two ways of associating m_2 is exact. Together, they insure that the homology of V is an associative algebra under the binary operations m_2 but at the level of cochains, m_2 is only associative up to the homotopy given by the ternary operation m_3. Equation 4.2 may be regarded as a generalization of the associativity relation (up to homotopy) on the n-ary operations m_n which are closely related to Massey products in algebraic topology.

Stasheff [36], [37] introduced an operad (although he did not use this language) of cell complexes each component of which is the product of a convex polytope, called an *associahedron*, with the permutation group. He showed that an A_∞-algebra is nothing more than an algebra over the cellular chains on this operad. His operad is isomorphic to $\overline{\mathcal{M}}'$ [10] and its cellular decomposition comes from the canonical stratification of $\overline{\mathcal{M}}'$ which is analogous to the stratification of $\overline{\mathcal{M}}$.

It turns out that there is a family of morphisms $\overline{\mathcal{M}}' \to \mathcal{P}$. Therefore a topological conformal field theory (V, Q) is an algebra over chains on $\overline{\mathcal{M}}'$ and, therefore, an A_∞-algebra. In order to construct this morphism, we need to consider moduli spaces of stable genus 0 curves with $(n+1)$ punctures decorated with additional data at both its punctures and nodes.

Let $\mathcal{N}(n)$ be the moduli space of Riemann spheres with $(n + 1)$ punctures where each puncture is decorated by the argument of a germ of a holomorphic coordinate, *i.e.* there is a tangent direction associated to each puncture. For all $n \geq 2$, $\mathcal{N}(n)$ has a compactification $\underline{\mathcal{N}}(n)$ which fibers over $\underline{\mathcal{M}}(n)$ with fiber $(S^1)^{n+1}$. $\underline{\mathcal{N}}(n)$ is defined as the moduli space of stable n-punctured complex curves of genus 0 decorated with relative phase parameters at each double point and with an argument of a germ of holomorphic coordinate each puncture. The collection $\underline{\mathcal{N}} = \{ \underline{\mathcal{N}}(n) \}_{n \geq 2}$ is an operad of smooth manifolds. Because $\mathbb{RP}^1$ is oriented, every puncture on $\mathbb{RP}^1$ comes with a canonical tangent direction which gets mapped to a tangent direction in $\mathbb{CP}^1$ under the inclusion $\mathbb{RP}^1 \hookrightarrow \mathbb{CP}^1$. This induces a morphism of operads $\overline{\mathcal{M}}' \to \underline{\mathcal{N}}$ induced from the inclusion. Kimura, Stasheff, and Voronov show in [21] that the solution to the minimal area metric problem by Wolf and Zwiebach can be used to construct a family of morphisms $\underline{\mathcal{N}} \to \mathcal{P}$ (called *string vertices*). Zwiebach and Wolf observed that every conformal class of a Riemann sphere with punctures with negative Euler characteristic can be endowed with a unique metric compatible with its complex structure which solves a minimal area problem for metrics on the punctured Riemann sphere subject to the constraint that the length of all homotopically nontrivial closed curves on the punctured sphere is greater than or equal to 2π. This minimal area metric is such that closed homotopically nontrivial geodesics may not intersect. Therefore, a punctured sphere naturally decomposes into flat cylinders foliated by closed geodesics with circumference 2π. The distance between the boundary components of such a cylinder is called the *height* of the cylinder. In particular, any cylinder containing a puncture has infinite height. In the case of a point in $\underline{N}_{n+1}$ with double points, a minimal area metric is assigned to each irreducible component minus its punctures and double points. A cylinder in an irreducible component which does not contain a puncture or a double point is said to be an *internal cylinder*.

We will now construct a family of morphisms of operads $s : \underline{\mathcal{N}}(n) \to \mathcal{P}(n)$ parametrized by a real number $l > \pi$ and a smooth monotonically increasing function $f_l : [2\pi, \infty) \to \mathbb{R}$ such that $f_l(2\pi) = 2\pi$ and $\lim_{x \to +\infty} f_l(x) = 2l$. Given a sphere with nodes, Σ in $\underline{\mathcal{N}}(n)$, each irreducible component may be endowed with its minimal area metric where each component is regarded as a punctured sphere by cutting away its punctures and nodes. Now, replace all internal flat cylinders of every irreducible component with height h greater than 2π with a flat cylinder with height $f_l(h)$. The result is still a minimal area metric. This metric together with the tangent directions at each puncture endows the irreducible component with a holomorphically embedded unit disk about each puncture and double point. Finally, $s(\Sigma)$ in $\mathcal{P}(n)$ is obtained by sewing together each irreducible component on each side of a double point using these charts (see figure 5 from [21]). Therefore, there is a morphism of operads $\overline{\mathcal{M}}' \to \mathcal{P}$. Since $\underline{\mathcal{N}}(2)$ is connected, this implies that the resulting binary operation on any topological conformal field theory is (graded) commutative up to homotopy. This proves the following theorem.

THEOREM 6. Let (V, Q) be a topological conformal field theory, then V is an A_∞ algebra with a binary operation which is (graded) commutative up to homotopy.

We expect this structure to be closely related to similar structures noticed on BRST by Lian and Zuckerman in [29]. Essentially they found a binary operation on the BRST complex which was commutative up to homotopy which gives rise to the commutative multiplication at the level of BRST cohomology.

This A_∞-algebra structure also appears to be closely related to the structure of a so-called commutative homotopy associative algebra that appears in topological gravity [22]. An explicit connection between the two would be enlightening.

4.3. Algebraic structures on BRST cohomology. The operad, $C_\mathcal{P}$, of singular chains on $\mathcal{P}$ induces the structure of an operad on $H_\bullet(\mathcal{P}) = \{ H_\bullet(\mathcal{P}(n)) \}$ where $H_\bullet(\mathcal{P}(n))$ is the homology of $\mathcal{P}(n)$ since all of the relevant structures in $C_\bullet(\mathcal{P})$ commute with the differentials. The homology operad $H_\bullet(\mathcal{P})$ is an operad in the category of graded vector spaces.

Let (V, Q) be a (tree level $c = 0$) topological conformal field theory. Since the morphism $m : C_\bullet(\mathcal{P}) \to \mathcal{E}nd_V$ is a chain map, it induces a morphism of operads $\mu : H_\bullet(\mathcal{P}) \to \mathcal{E}nd_{H_\bullet(V)}$. In other words, the BRST cohomology, $H_\bullet(V)$, of a topological conformal field theory is an $H_\bullet(\mathcal{P})$-algebra. Recall that the morphism of operads $\mathcal{F} \hookrightarrow \mathcal{P}$ is a homotopy equivalence which induces an isomorphism of operads between $H_\bullet(\mathcal{F})$ and $H_\bullet(\mathcal{P})$. Therefore, the BRST cohomology of a topological conformal field theory, $H_\bullet(V)$, is a $H_\bullet(\mathcal{F})$-algebra.

Let W be a graded commutative associative algebra with unit with respect to the multiplication $W \otimes W \to W$ via $(w_1, w_2) \to w_1 w_2$ which has degree 0. Furthermore suppose that W is endowed with a bilinear operation $[\cdot, \cdot] : W \otimes W \to W$ of degree 1 satisfying

$$[w_1, w_2] = -(-1)^{(|w_1|-1)(|w_2|-1)}[w_2, w_1],$$

$$[w_1, [w_2, w_3]] = [[w_1, w_2], w_3] + (-1)^{(|w_1|-1)(|w_2|-1)}[w_2, [w_1, w_3]],$$

and

$$[w_1, w_2 w_3] = [w_1, w_2]w_3 + (-1)^{|w_2|(|w_1|-1)}w_2[w_1, w_3]$$

for all w_1, w_2, w_3 in W. The vector space W endowed with such operations is said to be a *Gerstenhaber algebra with unit*. It is essentially a Poisson (super)algebra with funny signs which arise from the fact that the bracket carries a degree. The space of polyvectors on a smooth manifold under the Schouten bracket forms a natural geometric example of a Gerstenhaber algebra.

Gerstenhaber algebras are associated to certain special topological operads. Let $\mathcal{D} = \{ \mathcal{D}(n) \}$ be the suboperad of $\mathcal{F}$ consisting of those points which are also in the configuration space C_n. This is called the *little disks operad*. F. Cohen [5] proved that the category of algebras over the operad $H_\bullet(\mathcal{D})$ is isomorphic to the category of Gerstenhaber algebras.

A *Batalin-Vilkovisky algebra*, W, is a Gerstenhaber algebra endowed with a degree 1 unary operation $\Delta : W \to W$ such that $\Delta^2 = 0$ and

$$[w_1, w_2] = (-1)^{|w_1|}(\Delta(w_1 w_2) - (\Delta w_1)w_2 - (-1)^{|w_1|}w_1 \Delta w_2)$$

for all w_1 and w_2 in W.

The main result for topological conformal field theory is the following due to Geztler [8].

THEOREM 7. *The category of $H_\bullet(\mathcal{F})$-algebras is isomorphic to the category of Batalin-Vilkovisky algebras.*

COROLLARY 4.1. *If (V, Q) is a topological conformal field theory, then $H_\bullet(V)$ (BRST cohomology) is a Batalin-Vilkovisky algebra.*

It was Lian and Zuckerman [**29**] who first observed the Batalin-Vilkovisky algebra structure on the BRST cohomology of a holomorphic topological conformal field theory, namely, in the context of semi-infinite cohomology of the Virasoro algebra with values in a $c = 26$ vertex operator algebra which is itself a $c = 0$ vertex operator (super)algebra. They also pointed out that their result generalized to so-called topological chiral algebras. Getzler [**8**] was the first to realize the geometric origins of this structure. Huang [**18**] showed that Batalin-Vilkovisky algebras arise in the context of topological vertex algebras, which are (holomorphic) algebras over chains on $\mathfrak{F}$. Kimura, Stasheff, and Voronov [**21**] showed explicitly the form of the operations of a Batalin-Vilkovisky algebra in the context of a nonholomorphic topological conformal field theory.

The multiplication is induced by any point in $\mathcal{F}(2)$ regarded as a 0-cohomology class in $H_0(\mathcal{F}(2)) \cong \mathbb{C}$. The bracket is induced (up to a sign) by the fundamental class of S^1 via $H_1(S^1) \cong H_1(\mathcal{D}(2)) \hookrightarrow H_1(\mathcal{F}(2))$. The image, $[C]$, satisfies

$$[v_1, v_2] = (-1)^{|v_1|} \mu_{[C]}(v_1 \otimes v_2)$$

for all v_1 and v_2 in $H_\bullet(V)$. Finally, the operation Δ is induced from the fundamental class of S^1 in $H_1(S^1) \cong H_1(\mathcal{F}(1)) \cong \mathbb{C}$.

4.4. Higher genus topological conformal field theories. The formalism of operads must be generalized to the formalism of PROPS in order to accomodate conformal field theories beyond tree level, *i.e.* conformal field theories involving higher genus Riemann surfaces. We shall do so by following Segal [**35**], although it can be described using modular operads which were introduced by Getzler and Kapranov [**12**]. Topological conformal field theories beyond tree level can be defined in a similar fashion and this gives rise to a morphism from the homology of the moduli space of Riemann surfaces with holomorphically embedded disks to the BRST cohomology of the topological conformal field theory. In this section, we use the results by Harer [**14**] on the stability of the homology of the mapping class group to prove that the number of independent operations of a given degree on the BRST cohomology of a $c = 0$ topological conformal field theory is finite, something which, unlike the genus zero case, is nontrivial for higher genus theories.

Let $\mathcal{P}_{g,r}$ denote the moduli space of configurations of r distinct holomorphically embedded closed unit disks (with mutually disjoint interiors) in a compact, connected, genus g Riemann surface. Let $\mathcal{P}_g(m, n)$ be the same as $\mathcal{P}_{g,m+n}$ as a space but where m of the disks (called outgoing disks) are labeled by $1, \dots, m$ and the remaining n (called incoming disks) are labeled by $1, \dots, n$. $\mathcal{P}_g(m, n)$ has the natural action of $S_m \times S_n$ and has composition maps $\mathcal{P}_g(m, n) \times \mathcal{P}_{g'}(m', n') \to \mathcal{P}_{g+g'}(m + m' - 1, n + n' - 1)$ denoted by $(\Sigma, \Sigma') \mapsto \Sigma \circ_{(i,j)} \Sigma'$ where the latter is obtained by sewing the jth output of Σ' with the ith input of Σ. There is also another type of sewing operation $\mathcal{P}_g(m, n) \to \mathcal{P}_{g+1}(m - 1, n - 1)$ taking $\Sigma \mapsto \tau_{(i,j)}\Sigma$ which is Σ with outgoing disk j sewn with the incoming disk i.

(Strictly speaking, the domain of $\tau_{(i,j)}$ are those Riemann surfaces with disks whose boundaries do not overlap.) The operad $\mathcal{P}$ is really just the restriction of the above to $\mathcal{P}(n) = \mathcal{P}_0(1, n)$ with composition maps $\Sigma \circ_i \Sigma' = \Sigma \circ_{(i, n'+1)} \Sigma'$ where Σ' belongs to $\mathcal{P}(n')$. $\mathcal{P} = \{\, \mathcal{P}_g(m, n)\, \}$ contains distinguished elements I, $\hat{I}$, and $\hat{I}^{-1}$ in $\mathcal{P}_0(1, 1)$, $\mathcal{P}_0(0, 2)$, and $\mathcal{P}_0(2, 0)$ respectively which all correspond to the same point in $\mathcal{P}_{0,2}$, namely, the sphere with standard charts about 0 and ∞.

The space of endomorphisms associated to a vector space V endowed with a (graded) symmetric nondegenerate bilinear form $\langle \cdot, \cdot \rangle$, $\mathcal{E}nd_V(m, n)$, is a subspace of $\mathrm{Hom}(V^{\otimes n}, V^{\otimes m})$ with the actions of the permutation group and composition maps defined similarly. This subspace consists of those maps f such that $\tau_{(i,j)} f$ exists where $\tau_{(i,j)} f$ corresponds to taking the trace in the ith output and jth input of f. A $c = 0$ conformal field theory, V, is a (smooth) morphism (of PROPs) from $\mathcal{P} \to \mathcal{E}nd_V$ taking $\hat{I}$ to the bilinear form on V and $\hat{I}^{-1}$ to its inverse. A $c = 0$ topological conformal field theory, (V, Q), can be defined in terms of differential forms as before but we shall regard it here as a morphism from the space of singular chains on $\mathcal{P}$, $C_\bullet(\mathcal{P}) \to \mathcal{E}nd_V$ which induces a morphism from the homology groups $H_\bullet(\mathcal{P}) \to \mathcal{E}nd_{H_\bullet(V)}$. The bilinear form induces the same on the BRST cohomology, $H_\bullet(V)$.

Given a $c = 0$ topological conformal field theory, what can one say about the algebraic structure on the BRST cohomology $H_\bullet(V)$ beyond the fact that it is a Batalin-Vilkovisky algebra? Getzler in [8] observed that the bilinear form on $H_\bullet(V)$ is invariant with respect to the associative multiplication (making $H_\bullet(V)$ into a Frobenius algebra) and the unary operation Δ respects the bilinear form. This information can be described in terms of cyclic operads without going beyond tree level [11], however. To obtain information about operations from higher genus moduli spaces, we need to know more about the homology of $\mathcal{P}$.

The moduli space $\mathcal{P}_{g,r}$ is homotopy equivalent to $\mathcal{N}_{g,r}$, the moduli space of connected, compact genus g Riemann surfaces with r, distinct, punctures decorated with a tangent direction at each puncture in analogy with the definition of $\mathcal{N}(r)$ (in fact, $\mathcal{N}_{0,r+1} = \mathcal{N}(r)$). The projection map $\mathcal{P}_{g,r} \to \mathcal{N}_{g,r}$ is given by replacing every embedded disk of a point in $\mathcal{P}_{g,r}$ by its center and a tangent direction pointing along its positive real axis. This map is a homotopy equivalence so that the homology of $\mathcal{P}_{g,r}$ and of $\mathcal{N}_{g,r}$ are isomorphic. It is known [16] that when $2g + r > 2$, i.e. the corresponding surface has negative Euler characteristic, $\mathcal{N}_{g,r}$ deformation retracts onto a cell complex of dimension $4g - 5$, if $r = 0$, $4g + 2r - 4$, if $g > 0$ and $r > 0$, and $2r - 3$, if $g = 0$. In particular, this insures that the operations induced from homology classes of $\mathcal{P}_{g,r}$ have, at most, the degrees equal to the dimension of these cell complexes.

Unfortunately, not much is known about the homology of $\mathcal{N}_{g,r}$ although it has been computed in special cases (see, for example, [15]). However, its homology is essentially the group homology of the associated mapping class group. Recall that if $F_{g,r}$ is a compact, orientable surface of genus g with r boundary components, the *mapping class group of* $F_{g,r}$, $\Gamma_{g,r}$, is the group of isotopy classes of orientation preserving diffeomorphisms on $F_{g,r}$ which restrict to the identity on the boundary. (Alternately, $\Gamma_{g,r}$ may be regarded as isotopy classes of orientation preserving diffeomorphisms on an orientable, genus g surface with r punctures and each puncture decorated with a tangent direction which fixes the decorations.) The moduli space $\mathcal{N}_{g,r}$ is the quotient of the corresponding Teichmüller space by $\Gamma_{g,r}$ which acts properly discontinuously (and freely when $r > 0$). Since Teichmüller space is homeomorphic to $\mathbb{R}^{6g-6+3r}$, we obtain isomorphisms

$$H_k(\mathcal{N}_{g,r}, \mathbb{Z}) \simeq H_k(\Gamma_{g,r}, \mathbb{Z})$$

when $r > 0$, and

$$H_k(\mathcal{N}_{g,0}, \mathbb{Q}) \simeq H_k(\Gamma_{g,0}, \mathbb{Q}).$$

Since $H_\bullet(\Gamma_{g,r})$, $H_\bullet(\mathcal{N}_{g,r})$, and $H_\bullet(\mathcal{P}_{g,r})$ are isomorphic (at least rationally) and the latter have natural composition maps between them (we can break up the embedded disks into an incoming and outgoing set) induced from sewing together Riemann surfaces, it is natural to expect that the group homology itself possess natural composition maps. This turns out to be the case. Consider sewing together the surface $F_{g,r}$ with $F_{g',r'}$ along chosen boundary components to obtain the surface $F_{g+g',r+r'-2}$. This operation induces maps between the mapping class groups which induce composition maps on their homology groups $H_\bullet(\Gamma_{g,r}) \otimes H_\bullet(\Gamma_{g',r'}) \rightarrow H_\bullet(\Gamma_{g+g',r+r'-2})$. Similarly, by sewing together two boundary components on a given surface $F_{g,r}$ to obtain the surface $F_{g+1,r-2}$, one obtains a map on homology $H_\bullet(\Gamma_{g,r}) \rightarrow H_\bullet(\Gamma_{g+1,r-2})$. By keeping track of incoming and outgoing boundary components, these composition maps between the group homology groups $H_\bullet(\Gamma)$ are, in rational cohomology, isomorphic to the composition maps between the homology groups, $H_\bullet(\mathcal{P})$. Studying operations on the BRST cohomology of a topological conformal field theory is thus rationally equivalent to studying algebras over the homology of the mapping class group.

Although the mapping class group is not an arithmetic group, it does possess some of the properties of arithmetic groups. Perhaps the most striking property of such groups is that their group homologies stabilize, *e.g.* the groups $H_k(\mathrm{GL}_n(\mathbb{Z}); \mathbb{Q})$ are isomorphic for large enough values of n. Harer showed [**14**] that the homology of the mapping class group stabilizes for large enough values of the genus g.

To be more precise, let $\Phi : F_{g,r} \rightarrow F_{g,r+1}$ for $r \geq 1$ be the map corresponding to sewing one leg of a pair of pants onto a boundary component of $F_{g,r}$. Let $\Psi : F_{g,r} \rightarrow F_{g+1,r-1}$ for $r \geq 2$ be the map correponding sewing two legs of a pair of pants onto $F_{g,r}$. Finally, let $\eta : F_{g,r} \rightarrow F_{g+1,r-2}$ for $r \geq 2$ be the map sewing two boundary components of $F_{g,r}$ together. All these maps induce the obvious homomorphisms on the mapping class group.

THEOREM 8. The induced map

(4.3) $$\Phi_* : H_k(\Gamma_{g,r}, \mathbb{Z}) \to H_k(\Gamma_{g,r+1}, \mathbb{Z})$$

is an isomorphism when $k > 1$, $g \geq 3k - 2$, and $r \geq 1$ or when $k = 1$, $g \geq 2$, and $r \geq 1$. The induced map

(4.4) $$\Psi_* : H_k(\Gamma_{g,r}, \mathbb{Z}) \to H_k(\Gamma_{g+1,r-1}, \mathbb{Z})$$

is an isomorphism when $k > 1$, $g \geq 3k - 1$, and $r \geq 2$ or when $k = 1$, $g \geq 3$, or $r \geq 2$. Finally, the induced map

(4.5) $$\eta_* : H_k(\Gamma_{g,r}, \mathbb{Z}) \to H_k(\Gamma_{g+1,r-2}, \mathbb{Z})$$

is an isomorphism when $g \geq 3k$ and $r \geq 2$. When $\mathbb{Z}$ is replaced by $\mathbb{Q}$, the η_* extends to an isomorphism $g \geq 3k$ and $r \geq 2$.

By combining these maps, Harer obtained the following.

COROLLARY 4.2. $H_k(\Gamma_{g,r}, \mathbb{Q})$ is independent of g and r when $g \geq 3k+1$ and $H_k(\Gamma_{g,r}, \mathbb{Z})$ is independent of g and r when $g \geq 3k$.

Returning to topological conformal field theory, observe that any operation on $H_\bullet(V)$ with m inputs and n outputs can be converted to an operation with $(m + n - 1)$ inputs and 1 output (an $m + n - 1$-ary operation) provided that $m + n \geq 1$ by using the bilinear form on $H_\bullet(V)$. Although there are à priori an infinite number of independent operations of a given degree on BRST cohomology because of the operations that arise from higher genera moduli spaces, Harer's stability theorem implies that for a given k, when the genus is large enough, the kth homology group of $\mathcal{P}$ stabilizes. Furthermore, the maps Φ_* and Ψ_* are induced by sewing in pairs of pants, a 0 cohomology class of $\mathcal{P}_{0,3}$, onto the appropriate Riemann surfaces. This corresponds to composing the operation associated to a k homology class on the moduli space of curves with a large genus with the commutative binary product in the proper slots. This commutative binary product is the dot product from the Batalin-Vilkovisky algebra which is induced from the class in $H_0(\mathcal{P}(1,2))$ corresponding to a point. The other map η_* corresponds to contracting two of the inputs of an operation corresponding to a k homology class on the moduli space with a large genus with the bilinear form, which is induced from a class in $H_0(\mathcal{P}_2)$ in the proper slots. This proves the following.

THEOREM 9. Let (V, Q) be a $c = 0$ topological conformal field theory. The BRST cohomology $H_\bullet(V)$ is endowed with a finite number of independent operations with degree k for each k.

It is known that the stable rational cohomology groups contain a polynomial algebra with generators κ_i in H^{2i} for $i \geq 1$ [31, 32] which is conjectured to be the entire stable homology of $\mathcal{N}_{g,r}$ [33]. It would be interesting to understand the algebraic structures on BRST cohomology arising from these classes.

Acknowledgments. We would like to thank José Figueroa-O'Farrill, Yi-Zhi Huang, and Jim Stasheff for useful conversations and to Alexander Voronov for catching some typos in the final draft. Special thanks to John Harer for explaining his work to us.

REFERENCES

1. Axelrod, S., Singer, I.M.: *Chern-Simons perturbation theory II.* J. Diff. Geom. **39**, 173–213 (1994), `hep-th/9304087`
2. Beilinson, A., Ginzburg, V.: *Infinitesimal structure of moduli spaces of G-bundles.* Internat. Math. Research Notices **4**, 63–74 (1992)
3. Deligne, P.: *Resumé des premièrs exposés de A. Grothendieck. In :* SGA 7. Lecture Notes in Math., vol. 288, pp. 1–24. Berlin, Heidelberg, New York: Springer-Verlag 1972
4. Drinfel'd, V. G.: Letter to V. Schechtman, September 18, 1988.
5. Cohen, F.R.: *Artin's braid groups, classical homotopy theory and sundry other curiosities.* Contemp. Math. **78**, 167–206 (1988)
6. Frenkel I. B., Lepowsky J. and Meurman, A., *A natural representation of the Fischer-Griess monster with the modular function J as character,* Proc. Natl. Acad. Sci. USA **81** (1984), 3256-3260.
7. Frenkel I. B., Lepowsky J. and Meurman, A., *Vertex operator algebras and the Monster,* Pure and Appl. Math., Vol. 134, Academic Press, Boston, 1988.
8. Getzler, E.: *Batalin-Vilkovisky algebras and two-dimensional topological field theories.* Commun. Math. Phys. **159**, 265-285 (1994), `hep-th/9212043`
9. Getzler, E.: *Two-dimensional topological gravity and equivariant cohomology.* Preprint. Department of Mathematics, MIT 1993, hep-th/9305013
10. Getzler, E., Jones, J.D.S.: *Operads, homotopy algebra and iterated integrals for double loop spaces.* Preprint. Department of Mathematics, MIT March 1994, `hep-th/9403055`
11. Getzler, E., Kapranov, M. M.: *Cyclic operads and cyclic homology.* Preprint. MSRI 1993
12. Getzler, E., Kapranov, M. M.: *Modular operads,* August 1994, `dg-ga/9408003`
13. Ginzburg, V., Kapranov, M. M.: *Koszul duality for operads.* Preprint. Northwestern University 1993
14. Harer, J.: *Stability of the homology of the mapping class groups of orientable surfaces.* Annals of Math., **121** (1985) 215-249.
15. Harer, J.: *The third homology group of the moduli space of curves.* Duke Math. Journal, **63** (1991) 25-55.
16. Harer, J.: *The virtual cohomological dimension of the mapping class group of an orientable surface,* Invent. Math. **84** (1986) 157-176.
17. Hinich, V., Schechtman, V.: *Homotopy Lie algebras.* Preprint. SUNY, Stony Brook 1992
18. Huang, Y.-Z.: *Operadic formulation of topological vertex algebras and Gerstenhaber or Batalin-Vilkovisky algebras.* Commun. Math. Phys. **164**, 105–144 (1994), `hep-th/9306021`
19. Huang, Y.-Z., Lepowsky, J.: *Vertex operator algebras and operads,* The Gelfand mathematical seminars, 1990–1992, Birkhäuser, Boston, 1993, pp. 145–161, `hep-th/9301009`.
20. Keel, S.: *Intersection theory of moduli space of stable n-pointed curves of genus zero.* Trans. Amer. Math. Soc. **330**, 545–574 (1992)
21. Kimura, T., Stasheff, J., and Voronov, A. A. *On operad structures of moduli spaces and string theory,* Preprint 936, RIMS, Kyoto University, Kyoto, Japan, July 1993, `hep-th/9307114`. Commun. Math. Phys. (to appear)
22. Kimura, T., Stasheff, J., and Voronov, A. A. *Homology of moduli spaces of curves and commutative homotopy algebras,* Preprint, University of Pennsylvania and University of North Carolina. February 1995.
23. Kimura, T., Voronov, A. A.: *The cohomology of algebras over moduli spaces,* Proceedings of the 1994 Texel Conference on Moduli spaces of curves, edited by R. Dijkgraaf, G. van der Geer, and C. Faber (submitted to), `hep-th/9410108`.
24. Knudsen, F.F.: *The projectivity of moduli spaces of stable curves, II: the stacks $M_{g,n}$.* Math. Scand. **52**, 161–199 (1983)

25. Kontsevich, M.: *Feynman diagrams and low-dimensional topology*. Preprint. Max-Planck-Insitut für Mathematik, Bonn 1993

26. M. Kontsevich, *Formal (non)-commutative symplectic geometry*, The Gelfand mathematical seminars, 1990-1992 (L. Corwin, I. Gelfand, and J. Lepowsky, eds.), Birkhauser, 1993, pp. 173–187.

27. M. Kontsevich, Yu. I. Manin, *Gromov-Witten classes, quantum cohomology, and enumerative geometry*, Preprint MPI/93, Max-Planck-Institut für Mathematik, Bonn, February 1994, `hep-th/9402147`.

28. Lada, T., Stasheff, J.D.: *Introduction to sh Lie algebras for physicists*. Preprint UNC-MATH-92/2. University of North Carolina, Chapel Hill September 1992, `hep-th/9209099`

29. Lian, B.H., Zuckerman, G. J.: *New perspectives on the BRST-algebraic structure of string theory*. Commun. Math. Phys. **154**, 613–646 (1993), `hep-th/9211072`

30. May, J.P.: *The geometry of iterated loop spaces*. Lecture Notes in Math., vol. 271. Berlin, Heidelberg, New York: Springer-Verlag 1972

31. Miller, E.: *The homology of the mapping class group*, J. Diff. Geom. **24** (1986), 1-14.

32. Morita, S. *Characteristic classes of surface bundles I*, preprint, 1986.

33. Mumford, D.: *Towards an enumerative geometry on the moduli space of curves* in *Arithmetic and Geometry, Vol. II*, Birkhauser, Boston, 1983, 271-328.

34. Penkava, M., Schwarz, A.: *On some algebraic structures arising in string theory*. Preprint UCD-92-03. University of California, Davis 1993, `hep-th/9212072`

35. Segal, G.: *The definition of conformal field theory*. Preprint. Oxford

36. Stasheff, J.D.: *On the homotopy associativity of H-spaces,I*. Trans. Amer. Math. Soc. **108**, 275–292 (1963)

37. Stasheff, J.D.: *On the homotopy associativity of H-spaces, II*. Trans. Amer. Math. Soc. **108**, 293–312 (1963)

38. Witten, E.: *Quantum field theory and the Jones polynomial*. Commun. Math. Phys. **121** (1989) 351-399.

39. Witten, E., Zwiebach, B.: *Algebraic structures and differential geometry in two-dimensional string theory*. Nucl. Phys. B **377**, 55–112 (1992)

40. Wolf, M., Zwiebach, B.: *The plumbing of minimal area surfaces*. Preprint IASSNS-92/11. IAS, Princeton 1992, `hep-th/9202062`, Journal of Analysis and Geometry (to appear)

41. Zwiebach, B.: *Closed string field theory: Quantum action and the Batalin-Vilkovisky master equation*. Nucl. Phys. B **390**, 33–152 (1993)

DEPARTMENT OF MATHEMATICS, UNIVERSITY OF NORTH CAROLINA, CHAPEL HILL, NC 27599-3250

E-mail address: kimura@math.unc.edu

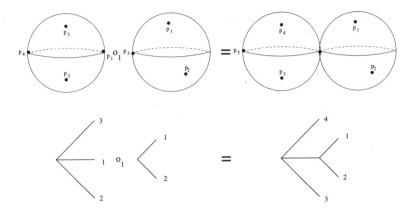

FIGURE 4. The composition map $\circ_1 : \overline{\mathcal{M}}(3) \times \overline{\mathcal{M}}(2) \to \overline{\mathcal{M}}(4)$ and the trees indexing the strata to which they belong

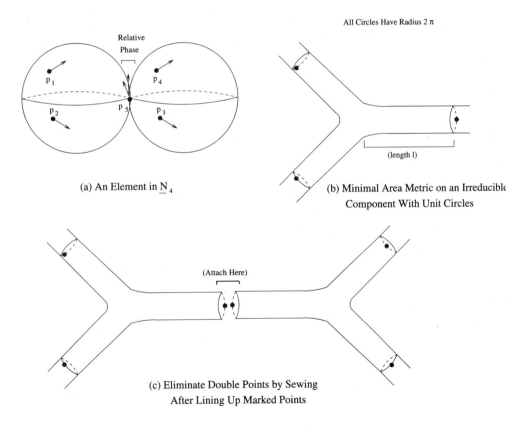

FIGURE 5. The morphism $\underline{\mathcal{N}}(4) \to \mathcal{P}(4)$

Contemporary Mathematics
Volume **193**, 1996

On Hopf algebras and the elimination theorem for free Lie algebras

J. LEPOWSKY AND R. L. WILSON

ABSTRACT. The elimination theorem for free Lie algebras, a general principle which describes the structure of a free Lie algebra in terms of free Lie subalgebras, has been recently used by E. Jurisich to prove that R. Borcherds' "Monster Lie algebra" has certain large free Lie subalgebras, illuminating part of Borcherds' proof that the moonshine module vertex operator algebra obeys the Conway-Norton conjectures. In the present expository note, we explain how the elimination theorem has a very simple and natural generalization to, and formulation in terms of, Hopf algebras. This fact already follows from general results contained in unpublished 1972 work, unknown to us when we wrote this note, of R. Block and P. Leroux.

1. Introduction

The elimination theorem for free Lie algebras (see [**10**] and [**3**]) describes how to "eliminate" generators; we state and discuss this theorem in detail below. It is of interest to us because it is an important step in E. Jurisich's proof of her theorem that R. Borcherds' "Monster Lie algebra" [**2**] is "almost free": This Lie algebra is the direct sum of the 4-dimensional Lie algebra $\mathfrak{gl}(2)$ and two $\mathfrak{gl}(2)$-submodules, each of which is a free Lie algebra over a specified infinite-dimensional $\mathfrak{gl}(2)$-module constructed naturally from the moonshine module ([**5**], [**6**]) for the Fischer-Griess Monster; see [**7**], [**8**]. This theorem has been used in [**7**], [**8**] and further, in [**9**], to illuminate and simplify Borcherds' proof [**2**] that the moonshine module vertex operator algebra satisfies the conditions of the Conway-Norton monstrous-moonshine conjectures [**4**]: The McKay-Thompson series for the the action of the Monster on this structure are modular functions which agree with the modular functions conjectured in [**4**] to be associated with a graded Monster-module. In fact, Jurisich's theorem, including its use of the

1991 *Mathematics Subject Classification.* Primary 16W30, 17B01; Secondary 08B20, 11F22, 16S10, 16S40.

Both authors were partially supported by NSF grant DMS-9401851.

elimination theorem, applies to the more general class of generalized Kac-Moody algebras, also called Borcherds algebras, which have no distinct orthogonal imaginary simple roots; again see [7], [8], and also [9] for further results.

The purpose of this note is to explain how the elimination theorem for free Lie algebras has a very natural and simple generalization to, and formulation in terms of, Hopf algebras (and in particular, quantum groups).

After completing this note we learned from Richard Block that our results follow from a still more general theorem contained in unpublished work [1] of R. Block and P. Leroux, which describes certain adjoint functors (cf. Remark 3.1 below). The present note, then, should be viewed as an exposition showing that the categorical considerations of Block and Leroux have concrete implications for the study of the Monster Lie algebra and monstrous moonshine.

Let us write $F(S)$ for the free Lie algebra over a given set S and $T(S)$ for the free associative algebra (the tensor algebra) over S. The elimination theorem states (see [3]) that the free Lie algebra over the disjoint union of two sets R and S is naturally isomorphic to the semidirect product of the free Lie algebra $F(S)$ with an ideal consisting of another free Lie algebra. This ideal is the ideal of $F(R \bigcup S)$ generated by R, and it is the free Lie algebra $F(T(S) \cdot R)$, where the dot denotes the natural adjoint action of $T(S)$, which is the universal enveloping algebra of $F(S)$, in $F(R \bigcup S)$:

$$F(R \bigcup S) = F(T(S) \cdot R) \rtimes F(S);$$

moreover, this $F(S)$-module $T(S) \cdot R$ is in fact the free $F(S)$-module generated by the set R. ("Module" means "left module" unless "right module" is specified.)

The main result of this note is a general structure theorem (which, we repeat, follows from the results in the unpublished work [1]) for the algebra freely generated by a given Hopf algebra and a given vector space, in a sense that we explain below. ("Algebra" means "associative algebra with unit.") The theorem expresses this freely generated algebra as the smash product of the given Hopf algebra with the tensor algebra over the module for the Hopf algebra freely generated by the given vector space. (We also point out that this freely generated algebra is a Hopf algebra itself in a natural way.) From the special case of this theorem when the given Hopf algebra is the universal enveloping algebra of some Lie algebra, we obtain as an immediate corollary a structure theorem expressing the Lie algebra freely generated by a given Lie algebra and a given vector space as the semidirect product of the given Lie algebra and the free Lie algebra over the module for the Lie algebra freely generated by the given vector space. Finally, from the further special case when this given Lie algebra is free, we recover the elimination theorem as an immediate corollary. (Actually, what we recover is the statement of the elimination theorem using free Lie algebras over vector spaces rather than free Lie algebras over sets; the distinction between the two equivalent formulations of the elimination theorem is simply a choice of basis for the underlying vector spaces, and such a choice is not necessary or natural for our purposes.) It is in this sense that we are "explaining" the elimination theorem as a very special manifestation of a general Hopf algebra principle.

The first of these two corollaries (the corollary involving a general Lie algebra) illuminates Jurisich's use of the elimination theorem in [7] and [8], and in fact makes her theorem exhibiting free Lie subalgebras of generalized Kac-Moody algebras, including the Monster Lie algebra, quite transparent. For the case of the Monster Lie algebra, we take the Lie algebra in our setting to be the Lie subalgebra $\mathfrak{gl}(2)$ of the Monster Lie algebra, and the vector space in our setting to be the moonshine module modulo its (one-dimensional) lowest weight space; then from the definition [2], [7], [8] of the Monster Lie algebra in terms of generators and relations, Jurisich's result follows immediately. This argument in fact really amounts to her argument in [8], with the difference that the Lie subalgebra $\mathfrak{gl}(2)$ is viewed as already given, rather than being presented in terms of generators and relations. Our main point is that what is really at work here is a very general, simple and natural Hopf-algebra principle. Moreover, our emphasis on tensor algebras (which are the universal enveloping algebras of free Lie algebras) illuminates the structure, presented in Section 4 of [9], of certain standard modules for the Monster Lie algebra as generalized Verma modules—as tensor algebras over certain modules for the Monster (see especially Theorem 4.5 of [9]).

This work is a continuation of some material in a talk of J. Lepowsky's at the June, 1994 Joint Summer Research Conference on Moonshine, the Monster and Related Topics at Mount Holyoke College, at which [9] was presented. J. L. wishes to thank the organizers, Chongying Dong, Geoff Mason and John McKay, for a stimulating conference. The authors are grateful to Elizabeth Jurisich, Wanglai Li and Siu-Hung Ng, and, as noted above, Richard Block, for helpful comments related to this work.

2. The results

In this section, we formulate the main theorem and deduce its consequences. We prove the theorem in the next section.

Throughout this note, we work over a field $\mathbb{F}$. Let H be a Hopf algebra, equipped with multiplication $M : H \otimes H \to H$, unit $u : \mathbb{F} \to H$, comultiplication (diagonal map) $\Delta : H \to H \otimes H$, counit $\epsilon : H \to \mathbb{F}$ and antipode $S : H \to H$, satisfying the usual axioms (cf. [11]). That is, H is an algebra with multiplication M (typically expressed as usual by juxtaposition of elements) and unit element $u(1)$ ($1 \in \mathbb{F}$), which we also write as 1_H, and is also a coalgebra such that the coalgebra operations Δ and ϵ are algebra homomorphisms (i.e., H is a bialgebra), and H is equipped with a linear endomorphism S such that for $h \in H$,

$$(1) \qquad \sum_i h_{1i} S(h_{2i}) = u\epsilon(h) = \epsilon(h)1_H = \sum_i S(h_{1i})h_{2i}.$$

Here and throughout this note, we use the following conventions for expressing the comultiplication (in any coalgebra) and its iterates: For $h \in H$, we write

$$\Delta(h) = \sum_i h_{1i} \otimes h_{2i}, \tag{2}$$

where h_{1i} and h_{2i} are suitable elements of H; we do not specify any index set in the summation. We similarly write

$$(\Delta \otimes 1)\Delta(h) = \sum_i h_{1i} \otimes h_{2i} \otimes h_{3i}, \tag{3}$$

where the index set and the elements h_{ji} in (2) and (3) are unrelated even though the notations are similar. (Note that by the coassociativity, the expression in (3) also equals $(1 \otimes \Delta)\Delta(h)$.) More generally, for $n > 0$ we write the result of applying the n-fold diagonal map to h as $\sum_i h_{1i} \otimes h_{2i} \otimes \cdots \otimes h_{n+1,i}$.

Note that the defining condition (1) for the antipode can also be written:

$$M \circ (1 \otimes S) \circ \Delta = u\epsilon = M \circ (S \otimes 1) \circ \Delta \tag{4}$$

as maps from H to H.

The antipode S is an algebra and coalgebra antimorphism (cf. [11]).

Our main object of interest is the algebra freely generated by H and a given vector space. First we define the algebra freely generated by a given algebra and a given vector space.

Given an algebra A and a vector space V, let $\mathcal{A}(A, V)$ denote the algebra freely generated by the algebra A and the vector space V, in the natural sense specified by the following universal property:

There is a given algebra homomorphism $A \to \mathcal{A}(A, V)$ and a given vector space homomorphism $V \to \mathcal{A}(A, V)$ such that for any algebra $\mathcal{B}$ and any algebra homomorphism $A \to \mathcal{B}$ and vector space homomorphism $V \to \mathcal{B}$, there is a unique algebra homomorphism $\mathcal{A}(A, V) \to \mathcal{B}$ such that the two obvious diagrams commute. Such a structure is of course unique up to unique isomorphism if it exists.

We shall now exhibit such a structure, confirming the existence. In the definition, we have not assumed that either A or V is embedded in $\mathcal{A}(A, V)$, but the construction which follows shows that both of these are in fact embedded in $\mathcal{A}(A, V)$.

Set

$$\mathcal{A}_n = (A \otimes V)^{\otimes n} \otimes A \text{ for } n \in \mathbb{N} \tag{5}$$

and

$$\mathcal{A}(A, V) = \coprod_{n \in \mathbb{N}} \mathcal{A}_n. \tag{6}$$

Then $\mathcal{A}(A, V)$ is an $\mathbb{N}$-graded algebra in a natural way; the product of an element of $\mathcal{A}_m$ with an element of $\mathcal{A}_n$ is obtained by juxtaposition followed multiplication of the adjacent A-factors. This structure clearly satisfies the universal property. Note that $\mathcal{A}(A, V)$ contains A as a subalgebra (in fact, $A = \mathcal{A}_0$) and V as a vector subspace ($V \subset \mathcal{A}_1$). It also contains the tensor algebras $T(A \otimes V)$ and $T(V \otimes A)$ as natural subalgebras, but $\mathcal{A}(A, V)$ is not the tensor product of either of these tensor algebras with the algebra A.

We shall take A to be the Hopf algebra H, and consider the algebra $\mathcal{A}(H, V)$, which contains H as a subalgebra.

Whenever the Hopf algebra H is a subalgebra of an algebra C, we have the following natural "adjoint" action of H on C: For $h \in H$ and $c \in C$,

$$(7) \qquad h \cdot c = \sum_i h_{1i} c S(h_{2i}).$$

It is easy to check from the definitions and the fact that S is an algebra anti-morphism that this action makes the algebra C an H-module algebra, where H is now viewed as a bialgebra rather than a Hopf algebra. This means (cf. [11]) that the linear action $H \otimes C \to C$ makes C an H-module (where H is viewed as an algebra and C as a vector space) and measures C to C (where H is viewed as a coalgebra and C as an algebra). This last condition means that for $h \in H$ and $c_1, c_2 \in C$,

$$(8) \qquad h \cdot c_1 c_2 = \sum_i (h_{1i} \cdot c_1)(h_{2i} \cdot c_2)$$

and

$$(9) \qquad h \cdot 1_C = u\epsilon(h) = \epsilon(h) 1_C.$$

Note that while the definition of "H-module algebra" requires only that H be a bialgebra and not a Hopf algebra, the definition of the adjoint action (7) of H on C uses the antipode. Note also (the case $C = H$) that under the action (7), any Hopf algebra is a module algebra for itself. From the definitions we see that the product in C and the action (7) are related as follows: For $h \in H$ and $c \in C$,

$$(10) \qquad hc = \sum_i (h_{1i} \cdot c) h_{2i}.$$

We also need the notion of smash product algebra $A \# B$ of an algebra A which is a B-module algebra, B a bialgebra, with the bialgebra B (cf. [11]): As a vector space, $A \# B$ is $A \otimes B$, and multiplication is defined by:

$$(11) \qquad (a_1 \otimes b^1)(a_2 \otimes b^2) = \sum_i a_1 (b_{1i}^1 \cdot a_2) \otimes b_{2i}^1 b^2$$

for $a_j \in A$, $b^j \in B$. Suppose that C is an algebra containing subalgebras A and H; that H is a Hopf algebra; that A is stable under the action (7); and that C is linearly the tensor product $A \otimes H$ in the sense that the natural linear map

$A \otimes H \to C$ induced by multiplication in C is a linear isomorphism. Then it is clear from (10) that (11) holds (for $B = H$) and hence that

(12) $$C = A \# H.$$

For a vector space V, we shall write $T(V)$ for the tensor algebra over V. (Recall that in the Introduction, we also used the notation $T(S)$ for the free associative algebra over a set S; then $T(V)$ is naturally isomorphic to $T(S)$ for any given basis S of V.)

We are now ready to state the main result. The proof is given in the next section. Note that statements (1), (2) and (3) in the Theorem include three different "freeness" assertions.

THEOREM 2.1. *Let H be a Hopf algebra and V a vector space, and consider the algebra $\mathcal{A} = \mathcal{A}(H, V)$ freely generated by H and V and the canonical adjoint action (7) of H on $\mathcal{A}$, making $\mathcal{A}$ an H-module algebra. Then:*

  (i) *The H-submodule $H \cdot V$ of $\mathcal{A}$ generated by V is naturally isomorphic to the free H-module over V.*

  (ii) *The subalgebra of $\mathcal{A}$ generated by $H \cdot V$ is naturally isomorphic to the tensor algebra $T(H \cdot V)$.*

  (iii) *As a vector space, the algebra $\mathcal{A}$ is naturally isomorphic to the tensor product*

$$\mathcal{A} = T(H \cdot V) \otimes H$$

under multiplication in $\mathcal{A}$; $T(H \cdot V)$ is an H-module subalgebra of $\mathcal{A}$; and with this structure, $\mathcal{A}$ is the smash product of $T(H \cdot V)$ and H:

$$\mathcal{A} = T(H \cdot V) \# H.$$

Here we observe several consequences. We describe first the special case in which H is the universal enveloping algebra $U(\mathfrak{g})$ of a Lie algebra $\mathfrak{g}$ and then the further special case in which $\mathfrak{g}$ is the free Lie algebra over a vector space. For a vector space W, we shall write $F(W)$ for the free Lie algebra over W, so that $F(W)$ is naturally isomorphic to the free Lie algebra $F(S)$ for any given basis S of W. (As with the notation $T(\cdot)$, we are using the notation $F(\cdot)$ for both sets and vector spaces.)

COROLLARY 2.2. *Let $\mathfrak{g}$ be a Lie algebra and V a vector space. Let $\mathcal{A}$ be the algebra freely generated by $\mathfrak{g}$ and V (defined by the obvious universal property), or equivalently, the algebra freely generated by $U(\mathfrak{g})$ and V, so that $\mathfrak{g} \subset U(\mathfrak{g}) \subset \mathcal{A}$. Let $\mathfrak{a}$ be the Lie subalgebra of $\mathcal{A}$ generated by $\mathfrak{g}$ and V. Then $\mathfrak{a}$ is naturally isomorphic to the Lie algebra freely generated by $\mathfrak{g}$ and V (in the obvious sense). Moreover, $\mathcal{A}$ is naturally isomorphic to the universal enveloping algebra of $\mathfrak{a}$. Consider the adjoint action of $\mathfrak{g}$ on $\mathcal{A}$; then the associated canonical action of $U(\mathfrak{g})$ on $\mathcal{A}$ is the adjoint action (7). We have:*

  (i) *The $\mathfrak{g}$-module $U(\mathfrak{g}) \cdot V \subset \mathfrak{a} \subset \mathcal{A}$ generated by V is naturally isomorphic to the free $\mathfrak{g}$-module (i.e., free $U(\mathfrak{g})$-module) over V.*

(ii) *The (associative) subalgebra of $\mathcal{A}$ generated by $U(\mathfrak{g}) \cdot V$ is naturally iso-morphic to the tensor algebra $T(U(\mathfrak{g}) \cdot V)$ and in particular, the Lie subalgebra of $\mathcal{A}$, and of $\mathfrak{a}$, generated by $U(\mathfrak{g}) \cdot V$ is naturally isomorphic to the free Lie algebra $F(U(\mathfrak{g}) \cdot V)$.*

(iii) *The algebra $\mathcal{A}$ is the smash product*

$$\mathcal{A} = T(U(\mathfrak{g}) \cdot V) \# U(\mathfrak{g})$$

and in particular,

$$\mathfrak{a} = F(U(\mathfrak{g}) \cdot V) \rtimes \mathfrak{g};$$

$F(U(\mathfrak{g}) \cdot V)$ *is the ideal of $\mathfrak{a}$ generated by V.*

Proof: All of this is clear; to verify the second part of (3), we note that $F(U(\mathfrak{g}) \cdot V) \oplus \mathfrak{g} \subset \mathfrak{a}$, and the adjoint action of $\mathfrak{g}$ on $\mathcal{A}$ preserves $F(U(\mathfrak{g}) \cdot V)$. Thus $F(U(\mathfrak{g}) \cdot V) \oplus \mathfrak{g}$ is a Lie subalgebra of $\mathfrak{a}$ containing $\mathfrak{g}$ and V, giving us the semidirect product. $\square$

Remark 2.3. The Poincaré-Birkhoff-Witt theorem has been used here in two places: It is used to show that $\mathfrak{g} \subset U(\mathfrak{g})$. It is also used to show that $\mathfrak{a}$ is naturally isomorphic to the Lie algebra freely generated by $\mathfrak{g}$ and V; the proof of the universal property for $\mathfrak{a}$—that for any Lie algebra $\mathfrak{b}$ and any Lie algebra map $\mathfrak{g} \to \mathfrak{b}$ and any vector space map $V \to \mathfrak{b}$, there exists exactly one Lie algebra map $\mathfrak{a} \to \mathfrak{b}$ making the two obvious diagrams commute—uses the fact that $\mathfrak{b} \subset U(\mathfrak{b})$. Note that this latter use of the Poincaré-Birkhoff-Witt theorem is a generalization of the similar use of this theorem (in what amounts to the case $\mathfrak{g} = 0$) in proving that the Lie subalgebra of $T(V)$ generated by V is the free Lie algebra over V and hence that $T(V) = U(F(V))$.

We restate the parts of this corollary involving only the Lie algebra structure:

COROLLARY 2.4. *Let $\mathfrak{g}$ be a Lie algebra and V a vector space, and let $\mathfrak{a}$ be the Lie algebra freely generated by $\mathfrak{g}$ and V, so that $\mathfrak{g} \subset \mathfrak{a}$ and $V \subset \mathfrak{a}$. Then:*

(i) *The $\mathfrak{g}$-module $U(\mathfrak{g}) \cdot V \subset \mathfrak{a}$ generated by V is naturally isomorphic to the free $\mathfrak{g}$-module over V.*

(ii) *The Lie subalgebra of $\mathfrak{a}$ generated by $U(\mathfrak{g}) \cdot V$ is naturally isomorphic to the free Lie algebra $F(U(\mathfrak{g}) \cdot V)$.*

(iii) *We have the semidirect product decomposition*

$$\mathfrak{a} = F(U(\mathfrak{g}) \cdot V) \rtimes \mathfrak{g},$$

and $F(U(\mathfrak{g}) \cdot V)$ is the ideal of $\mathfrak{a}$ generated by V. $\square$

Now we consider the further special case in which $\mathfrak{g} = F(W)$, the free Lie algebra over a given vector space W. We have immediately:

COROLLARY 2.5. *Let V and W be vector spaces, and consider $F(W), F(V \oplus W) \subset T(V \oplus W)$. Then:*

(i) *The $F(W)$-module*

$$U(F(W)) \cdot V = T(W) \cdot V \subset F(V \oplus W)$$

generated by V is naturally isomorphic to the free $F(W)$-module (i.e., free $T(W)$-module) over V.

(ii) *The Lie subalgebra of $F(V \oplus W)$ generated by $T(W) \cdot V$ is naturally isomorphic to the free Lie algebra $F(T(W) \cdot V)$ and the associative subalgebra of $T(V \oplus W)$ generated by $T(W) \cdot V$ is naturally isomorphic to the tensor algebra $T(T(W) \cdot V)$.*

(iii) *We have the semidirect product decomposition*

$$F(V \oplus W) = F(T(W) \cdot V) \rtimes F(W),$$

$F(T(W) \cdot V)$ is the ideal of $F(V \oplus W)$ generated by V, and

$$T(V \oplus W) = T(T(W) \cdot V) \# T(W). \quad \square$$

Remark 2.6. The elimination theorem for free Lie algebras over sets rather than over vector spaces, as recalled in the Introduction, is simply the obvious restatement of the last corollary for the free Lie algebra $F(R \bigcup S)$ over the disjoint union of sets R and S in place of the free Lie algebra $F(V \oplus W)$; we take R and S to be bases of V and W, respectively. Also, $T(V \oplus W)$ is replaced by $T(R \bigcup S)$.

Remark 2.7. Given H and V as in Theorem 2.1, we observe that the algebra $\mathcal{A} = \mathcal{A}(H, V)$ freely generated by H and V is naturally a Hopf algebra. In fact, define $\Delta : \mathcal{A} \to \mathcal{A} \otimes \mathcal{A}$ to be the unique algebra map which agrees with $\Delta : H \to H \otimes H$ on H and with the linear map $V \to V \otimes V$ taking $v \in V$ to $v \otimes 1 + 1 \otimes v$ on V. Define $\epsilon : \mathcal{A} \to \mathbb{F}$ to be the unique algebra map which agrees with $\epsilon : H \to \mathbb{F}$ and with the zero map on V. Then $\mathcal{A}$ is clearly a bialgebra since H and V generate $\mathcal{A}$. Define $S : \mathcal{A} \to \mathcal{A}$ to be the unique algebra antimorphism which agrees with the given antipode on H and with the map -1 on V. Then S is an antipode, since (4) holds on H and on V, and since the set of elements of a given bialgebra on which (4) holds, where S is a given algebra antimorphism, is closed under multiplication (cf. [**11**], p. 73).

3. Proof of Theorem 2.1

Now we prove Theorem 2.1.

Consider the free H-module $H \otimes V$ generated by V (viewing H as an algebra) and define the linear map

$$
\begin{aligned}
i : H \otimes V &\to H \otimes V \otimes H \\
h \otimes v &\mapsto \sum_i h_{1i} \otimes v \otimes S(h_{2i})
\end{aligned}
$$

(13)

$(h \in H, v \in V)$; that is,

(14) $$i = (1 \otimes \mathfrak{T}) \circ (1 \otimes S \otimes 1) \circ (\Delta \otimes 1),$$

where $\mathfrak{T}$ is the twist map:

$$\mathfrak{T} : K \otimes L \ \rightarrow \ L \otimes K$$
(15) $$k \otimes l \ \mapsto \ l \otimes k$$

$(k \in K, l \in L)$ for arbitrary vector spaces K and L. (There should be no confusion between the two uses of the notation "i.") Then i is clearly an H-module map, where $H \otimes V \otimes H$ is understood as the tensor product of the free H-module $H \otimes V$ and the H-module H equipped with the (left) action given by: $h \cdot k = kS(h)$ for $h, k \in H$. The image $i(H \otimes V)$ of i is the H-submodule of $H \otimes V \otimes H$ generated by V with respect to this action: $i(H \otimes V) = H \cdot V$. Furthermore, the H-module map

(16) $$i : H \otimes V \to H \cdot V$$

is an isomorphism (i.e., is an injection), since the linear map

$$1 \otimes 1 \otimes \epsilon : H \otimes V \otimes H \to H \otimes V$$

is a left inverse of i. In particular, $H \cdot V$ is a copy of the free H-module generated by V. Recalling (5), (6) and writing $\mathcal{A} = \mathcal{A}(H, V)$ as in the statement of Theorem 2.1, we observe that $H \cdot V \subset \mathcal{A}_1 \subset \mathcal{A}$, and that the first part of Theorem 2.1 is verified.

Now consider the linear map

$$i_1 : H \otimes V \otimes H \ \rightarrow \ H \otimes V \otimes H$$
(17) $$h^1 \otimes v \otimes h^2 \ \mapsto \ \sum_i h_{1i}^1 \otimes v \otimes S(h_{2i}^1)h^2$$

$(h^1, h^2 \in H, v \in V)$, which we may express as the map

(18) $$i_1 : \mathcal{A}_1 \to \mathcal{A}_1$$

given by the composition of $i \otimes 1 : \mathcal{A}_1 \to \mathcal{A}_1 \otimes H$ with the multiplication map in $\mathcal{A}$. We may also write:

(19) $i_1 = (1 \otimes \mathfrak{T}) \circ (1 \otimes M \otimes 1) \circ (1 \otimes S \otimes 1 \otimes 1) \circ (\Delta \otimes 1 \otimes 1) \circ (1 \otimes \mathfrak{T}).$

We now show that i_1 is a linear automorphism, and in fact that the map

$$j_1 : H \otimes V \otimes H \ \rightarrow \ H \otimes V \otimes H$$
(20) $$h^1 \otimes v \otimes h^2 \ \mapsto \ \sum_i h_{1i}^1 \otimes v \otimes h_{2i}^1 h^2$$

is a left and right inverse of i_1: For variety, in this argument we use the symbolism (4) and (19). Since

$$(21) \qquad j_1 = (1 \otimes \mathcal{T}) \circ (1 \otimes M \otimes 1) \circ (\Delta \otimes 1 \otimes 1) \circ (1 \otimes \mathcal{T}),$$

we have

$$
\begin{aligned}
j_1 \circ i_1 &= (1 \otimes \mathcal{T})(1 \otimes M \otimes 1)(\Delta \otimes 1 \otimes 1)(1 \otimes M \otimes 1) \cdot \\
&\qquad \cdot (1 \otimes S \otimes 1 \otimes 1)(\Delta \otimes 1 \otimes 1)(1 \otimes \mathcal{T}) \\
&= (1 \otimes \mathcal{T})(1 \otimes M \otimes 1)(1 \otimes 1 \otimes M \otimes 1) \cdot \\
&\qquad \cdot (\Delta \otimes 1 \otimes 1 \otimes 1)(1 \otimes S \otimes 1 \otimes 1)(\Delta \otimes 1 \otimes 1)(1 \otimes \mathcal{T}) \\
&= (1 \otimes \mathcal{T})(1 \otimes M \otimes 1)(1 \otimes 1 \otimes M \otimes 1)(1 \otimes 1 \otimes S \otimes 1 \otimes 1) \cdot \\
&\qquad \cdot (\Delta \otimes 1 \otimes 1 \otimes 1)(\Delta \otimes 1 \otimes 1)(1 \otimes \mathcal{T}) \\
&= (1 \otimes \mathcal{T})(1 \otimes M \otimes 1)(1 \otimes M \otimes 1 \otimes 1)(1 \otimes 1 \otimes S \otimes 1 \otimes 1) \cdot \\
&\qquad \cdot (1 \otimes \Delta \otimes 1 \otimes 1)(\Delta \otimes 1 \otimes 1)(1 \otimes \mathcal{T}) \\
&= (1 \otimes \mathcal{T})(1 \otimes M \otimes 1)(1 \otimes u\epsilon \otimes 1)(\Delta \otimes 1 \otimes 1)(1 \otimes \mathcal{T}) \\
&= (1 \otimes \mathcal{T})(1 \otimes \mathcal{T}) = 1,
\end{aligned}
$$

where we have used here essentially all the defining properties of a Hopf algebra: associativity, coassociativity, the definition (4) of the antipode, and the unit and counit properties. The fact the $i_1 \circ j_1 = 1$ is verified similarly. (The two parts of (4) are used in the two different arguments.)

For $n \geq 1$, define

$$(22) \qquad\qquad\qquad i_n, j_n : \mathcal{A}_n \to \mathcal{A}_n$$

by:

$$(23)$$
$$i_n = (i_1 \otimes 1 \otimes \cdots \otimes 1) \circ \cdots \circ (1 \otimes \cdots \otimes 1 \otimes i_1 \otimes 1 \otimes 1) \circ (1 \otimes \cdots \otimes 1 \otimes i_1),$$

$$(24)$$
$$j_n = (1 \otimes \cdots \otimes 1 \otimes j_1) \circ (1 \otimes \cdots \otimes 1 \otimes j_1 \otimes 1 \otimes 1) \circ \cdots \circ (j_1 \otimes 1 \otimes \cdots \otimes 1).$$

Using the facts that $j_1 \circ i_1 = 1$, $i_1 \circ j_1 = 1$, we obtain:

$$(25) \qquad\qquad\qquad j_n \circ i_n = 1$$
$$(26) \qquad\qquad\qquad i_n \circ j_n = 1.$$

Note that we may write $i_2 : \mathcal{A}_2 \to \mathcal{A}_2$, for example, explicitly as the map

$$(27)$$
$$h^1 \otimes v_1 \otimes h^2 \otimes v_2 \otimes h^3 \mapsto \sum_{i,j} \left(h_{1i}^1 \otimes v_1 \otimes S(h_{2i}^1) \right) \left(h_{1j}^2 \otimes v_2 \otimes S(h_{2j}^2) \right) h^3$$

$(h^1, h^2, h^3 \in H, v_1, v_2 \in V)$. Just as the map i_1 can be expressed using the map i and multiplication in $\mathcal{A}$ (recall (18)), the linear automorphism i_n is the composition of

$$i^{\otimes n} \otimes 1 : \mathcal{A}_n = (H \otimes V)^n \otimes H \to (H \otimes V \otimes H)^{\otimes n} \otimes H$$

with the multiplication map in $\mathcal{A}$, as in (27).

We combine the linear automorphisms i_n for $n \geq 0$, where we take i_0 to be the identity map on H, to form the graded linear automorphism

(28)
$$\mathfrak{I} = \coprod_{n \geq 0} i_n : \mathcal{A} \to \mathcal{A},$$

so that

(29)
$$\mathfrak{I}|_H = i_0 = 1 : H \to H,$$

the identity map, and

(30)
$$\mathfrak{I}|_{H \otimes V} = i : H \otimes V \xrightarrow{\sim} H \cdot V,$$

a linear isomorphism and in fact an H-module isomorphism (recall (16)). Moreover,

$$\mathfrak{I}|_{T(H \otimes V)} : T(H \otimes V) \to \mathcal{A}$$

is a linear isomorphism from the subalgebra $T(H \otimes V)$ of $\mathcal{A}$ to the subalgebra of $\mathcal{A}$ generated by $H \cdot V$, and this subalgebra is naturally isomorphic to the tensor algebra $T(H \cdot V)$, so that the second statement in the Theorem is proved:

(31)
$$\mathfrak{I}|_{T(H \otimes V)} : T(H \otimes V) \xrightarrow{\sim} T(H \cdot V);$$

in fact, this map is the isomorphism of tensor algebras induced by the linear isomorphism (30) of generating vector spaces of the two tensor algebras. Thus the linear automorphism $\mathfrak{I}$ restricts to algebra isomorphisms (29) and (31), which canonically extends (30), and the linear decomposition

$$\mathcal{A} = T(H \otimes V) \otimes H$$

of the domain of $\mathfrak{I}$ transports to a linear decomposition

(32)
$$\mathcal{A} = T(H \cdot V) \otimes H$$

of the codomain; the associated canonical linear map

(33)
$$T(H \cdot V) \otimes H \to \mathcal{A}$$

is the map induced by multiplication in $\mathcal{A}$. In particular, the natural linear map (33) induced by multiplication is a linear isomorphism. This proves the first part of the third statement in the Theorem.

Now we want to describe explicitly the multiplication operation in the algebra $\mathcal{A}$ in terms of the linear tensor product decomposition (32) of $\mathcal{A}$ into the two specified subalgebras of $\mathcal{A}$. But by the comments before (12), to prove the rest all we need to show is that $T(H \cdot V)$ is stable under the adjoint action of H. But this is clear from the iteration of (8) for a product of several elements of $H \cdot V$, together with (9). This completes the proof of Theorem 2.1. $\square$

Remark 3.1. The proof in [1] proceeds by viewing $T(H \otimes V)\#H$ as a universal object and by constructing mutually inverse canonical maps between this structure and $\mathcal{A}$. The argument above essentially carries this out "concretely."

References

1. R. E. Block and P. Leroux, Free Hopf module algebras and related functors, unpublished preprint, 1972; revised version, to appear.
2. R. Borcherds, Monstrous moonshine and monstrous Lie superalgebras, *Invent. Math.* **109** (1992), 405-444.
3. N. Bourbaki, *Lie Groups and Lie Algebras, Part 1*, Hermann, Paris, 1975.
4. J. H. Conway and S. P. Norton, Monstrous Moonshine, *Bull. London Math. Soc.* **11** (1979), 308-339.
5. I. Frenkel, J. Lepowsky and A. Meurman, A natural representation of the Fischer-Griess Monster with the modular function J as character, *Proc. Natl. Acad. Sci. USA* **81** (1984), 3256-3260.
6. I. Frenkel, J. Lepowsky and A. Meurman, *Vertex Operator Algebras and the Monster*, Academic Press, Boston, 1988.
7. E. Jurisich, *Generalized Kac-Moody algebras and their relation to free Lie algebras*, Ph.D. thesis, Rutgers University, May, 1994.
8. E. Jurisich, Generalized Kac-Moody Lie algebras, free Lie algebras and the structure of the Monster Lie algebra, *J. Pure and Applied Algebra*, to appear.
9. E. Jurisich, J. Lepowsky and R. L. Wilson, Realizations of the Monster Lie algebra, *Selecta Mathematica, New Series* **1** (1995), 129-161.
10. M. Lazard, *Groupes, anneaux de Lie et problème de Burnside*, Istituto Mat. dell' Università di Roma, 1960.
11. M. E. Sweedler, *Hopf Algebras*, W.A. Benjamin, New York, 1969.

DEPARTMENT OF MATHEMATICS, RUTGERS UNIVERSITY, NEW BRUNSWICK, NJ 08903
E-mail address: lepowsky@math.rutgers.edu, rwilson@math.rutgers.edu

Contemporary Mathematics
Volume **193**, 1996

LOCAL SYSTEMS OF TWISTED VERTEX OPERATORS, VERTEX OPERATOR SUPERALGEBRAS AND TWISTED MODULES

HAI-SHENG LI

ABSTRACT. We introduce the notion of "local system of $\mathbb{Z}_T$-twisted vertex operators" on a $\mathbb{Z}_2$-graded vector space M, generalizing the notion of local system of vertex operators [Li]. First, we prove that any local system of $\mathbb{Z}_T$-twisted vertex operators on M has a vertex superalgebra structure with an automorphism σ of order T with M as a σ-twisted module. Then we prove that for a vertex (operator) superalgebra V with an automorphism σ of order T, giving a σ-twisted V-module M is equivalent to giving a vertex (operator) superalgebra homomorphism from V to some local system of $\mathbb{Z}_T$-twisted vertex operators on M. As applications, we study the twisted modules for vertex operator (super)algebras associated to some well-known infinite-dimensional Lie (super)algebras and we prove the complete reducibility of $\mathbb{Z}_T$-twisted modules for vertex operator algebras associated to standard modules for an affine Lie algebra.

1. INTRODUCTION

In [Li], motivated by [G] we introduced the notion of local system of vertex operators and proved that any local system of vertex operators on a super vector space M has a vertex superalgebra structure with M as a module, so that if V is a vertex (operator) superalgebra, then giving a V-module M is equivalent to giving a vertex superalgebra homomorphism from V to some local system of vertex operators on M. Similar to the notion of local system of vertex operators, the notion of commutative quantum operator algebra was introduced in [LZ1-2]. For a vertex (operator) superalgebra V, in addition to having module theory, more generally we have twisted module theory ([D], [FFR], [FLM]). In this paper, generalizing the notion of local system of vertex operators ([G], [Li]), we introduce the notion of what we call "local system of $\mathbb{Z}_T$-twisted vertex operators" on a $\mathbb{Z}_2$-graded vector space M and we prove similar results.

Let M be a $\mathbb{Z}_2$-graded vector space and let T be a fixed positive integer. A *weak $\mathbb{Z}_T$-twisted vertex operator* on M is a formal series $a(z) \in (EndM)[[z^{\frac{1}{T}}, z^{-\frac{1}{T}}]]$ satisfying $a(z)u \in M((z^{\frac{1}{T}}))$ for any $u \in M$. Let $F(M,T)$ be the vector space of all weak $\mathbb{Z}_T$-twisted vertex operators on M. Set $\varepsilon = \exp\left(\frac{2\pi\sqrt{-1}}{T}\right)$. Let σ be the

---

1991 *Mathematics Subject Classification.* Primary 17B68; Secondary 17B69, 81T40.
This paper is in final form and no version of it will be submitted for publication elsewhere.

endomorphism of $(EndM)[[z^{\frac{1}{T}}, z^{-\frac{1}{T}}]]$ defined by: $\sigma f(z^{\frac{1}{T}}) = f(\varepsilon^{-1}z^{\frac{1}{T}})$. Then σ is an automorphism of $F(M,T)$ of order T. For any integer n, set

$$F(M,T)^n = \{f(z^{\frac{1}{T}}) \in F(M,T) | \sigma f(z^{\frac{1}{T}}) = \varepsilon^n f(z^{\frac{1}{T}})\}. \tag{1.1}$$

Then $z^{\frac{n}{T}}F(M,T)^n \subseteq (EndM)[[z, z^{-1}]]$ and $F(M,T)^n = F(M,T)^{T+n}$ for any $n \in \mathbb{Z}$. Thus

$$F(M,T) = F(M,T)^0 \oplus F(M,T)^1 \oplus \cdots \oplus F(M,T)^{T-1}, \tag{1.2}$$

so that $F(M,T)$ is a $(\mathbb{Z}_2 \times \mathbb{Z}_T)$-graded vector space.

We define *a local subspace* of $F(M,T)$ to be a $(\mathbb{Z}_2 \times \mathbb{Z}_T)$-graded subspace A such that for any two $\mathbb{Z}_2$-homogeneous elements $a(z)$ and $b(z)$ of A, there is a positive integer m such that

$$(z_1 - z_2)^m a(z_1)b(z_2) = (-1)^{|a(z)||b(z)|}(z_1 - z_2)^m b(z_2)a(z_1), \tag{1.3}$$

where $|a(z)| = 0$ if $a(z)$ is even, $|a(z)| = 1$ if $a(z)$ is odd. And we define *a local system of $\mathbb{Z}_T$-twisted vertex operators on M* to be a maximal local (graded) subspace of $F(M,T)$.

Let M be a $\mathbb{Z}_2$-graded vector space and let A be a local system of $\mathbb{Z}_T$-twisted vertex operators on M. Then σ is an automorphism of A such that $\sigma^T = Id_A$. Set $A^k = A \cap F(M,T)^k$ for any integer k. Similar to the untwisted case, the "multiplication" for twisted vertex operators comes from the twisted iterate formula. For any $a(z) \in A^k$, $b(z) \in A$ and any integer $n \in \mathbb{Z}$, $a(z)_n b(z)$ as an element of $F(M,T)$ is defined by:

$$Y(a(z), z_0)b(z)$$
$$=: \sum_{n \in \mathbb{Z}} (a(z)_n b(z))z_0^{-n-1}$$
$$= Res_{z_1} \left(\frac{z_1 - z_0}{z}\right)^{\frac{k}{T}} \left(z_0^{-1}\delta\left(\frac{z_1 - z}{z_0}\right)a(z_1)b(z) - z_0^{-1}\delta\left(\frac{z - z_1}{-z_0}\right)b(z)a(z_1)\right). \tag{1.4}$$

As in the untwisted case [Li], we first prove that A is closed under the "multiplication" (1.4) (Proposition 3.9). Then we prove that A under the "adjoint action" (1.4) is a local system of (untwisted) vertex operators on A (Proposition 3.13). By Proposition 2.5 (recalled from [Li]), A is a vertex superalgebra with a natural automorphism σ such that $\sigma^T = Id_A$. Furthermore, since the supercommutativity and the twisted iterate formula imply the twisted Jacobi identity (for a twisted module), M is a σ-twisted module for the vertex superalgebra (Theorem 3.14). Let V be a vertex superalgebra with an automorphism σ of order T. Then we prove that giving a σ-twisted V-module M is equivalent to giving a vertex superalgebra homomorphism from V to some local system of $\mathbb{Z}_T$-twisted vertex operators on M (Proposition 3.17).

Let M be any $\mathbb{Z}_2$-graded vector space and let S be any set of mutually local homogeneous $\mathbb{Z}_T$-twisted vertex operators on M. It follows from Zorn's lemma that there exists a local system A containing S. Therefore, S generates (inside A under the twisted "multiplication" (1.4)) a vertex superalgebra with σ as an automorphism such that M is a σ-twisted module (Corollary 3.15). Since the "multiplication"

(1.4) does not depend on the choice of the local system A, the vertex superalgebra $\langle S \rangle$ is canonical, so that we can speak about the vertex superalgebra generated by S.

To describe our results, for simplicity, we consider a special case. Let g be a finite-dimensional simple Lie algebra with a fixed Cartan subalgebra h. Let $\langle \cdot, \cdot \rangle$ be the normalized Killing form on g such that the square of the length of the longest root is 2. It has been proved ([DL], [FZ], [Li], [Lia]) that the generalized Verma $\tilde{g}$-module $M_g(\ell, C)$ has a natural vertex operator algebra structure for any $\ell \neq -\Omega$. Let σ be an automorphism of g of order T. Let M be any restricted module of level ℓ for the twisted affine Lie algebra $\tilde{g}[\sigma]$. By considering the generating function $a(z) = \sum_{n \in \mathbb{Z}} (t^{n+\frac{j}{T}}) z^{-n-1-\frac{j}{T}}$ for $a \in g^j$ as an element of $F(M, T)$, we obtain a local subspace $\{a(z) | a \in g\}$, so that $\{a(z) | a \in g\}$ generates a vertex algebra V with M as a σ-twisted module. Since V is a quotient vertex algebra of $M_g(\ell, C)$, M is a weak σ-twisted $M_g(\ell, C)$-module.

Let g be a finite-dimensional simple Lie algebra and let ℓ be a positive integer. It has been proved ([DL], [FZ], [Li]) that any lower truncated $\mathbb{Z}$-graded weak $L_g(\ell, 0)$-module is a direct sum of standard $\tilde{g}$-modules of level ℓ and that the set of equivalence classes of irreducible $L_g(\ell, 0)$-modules is exactly the set of equivalence classes of standard $\tilde{g}$-modules of level ℓ. Let μ be a Dynkin diagram automorphism of g. Then using a similar method used for ordinary V-modules ([DL], [Li]) we prove that any lower truncated $\frac{1}{T}\mathbb{Z}$-graded weak μ-twisted $L_g(\ell, 0)$-module is completely reducible and that the set of irreducible μ-twisted $L_g(\ell, 0)$-modules is exactly the set of the standard $\tilde{g}[\mu]$-modules of level ℓ (Theorem 5.8).

Let σ be any automorphism of order T of g. By [K], σ is conjugate to an automorphism $\mu\sigma_h$, where μ is a Dynkin diagram automorphism of g and σ_h is an inner automorphism of g associated with an element h of the Cartan subalgebra h such that $\sigma(h) = h$. If two automorphisms σ_1 and σ_2 of a vertex operator algebra V are conjugate each other, it is clear that the equivalence classes of σ_1-twisted V-modules one-to-one correspond to the equivalence classes of σ_2-twisted V-modules. In the Lie algebra level, it is well-known ([K], [H]) that the twisted affine Lie algebras $\tilde{g}[\mu]$ and $\tilde{g}[\mu\sigma_h]$ are isomorphic each other, so that the equivalence classes of $\tilde{g}[\mu]$-modules one-to-one correspond to the equivalence classes of $\tilde{g}[\mu\sigma_h]$-modules. But it is not obvious that we have a one-to-one correspondence between the equivalence classes of μ-twisted $L_g(\ell, 0)$-modules and the equivalence classes of $(\mu\sigma_h)$-twisted $L_g(\ell, 0)$-modules. In Section 5, such a one-to-one correspondence is established as a corollary of Proposition 5.4. Moreover, Proposition 5.4 has also its own interest. We will study certain applications of Proposition 5.4 in a coming paper. In this way, we show that for any automorphism σ of g, any lower truncated $(\frac{1}{T}\mathbb{Z})$-graded weak σ-twisted $L_g(\ell, 0)$-module is completely reducible and that there are only finitely many irreducible σ-twisted $L_g(\ell, 0)$-modules up to equivalence.

This paper is organized as follows: Section 2 is a preliminary section. In Section 3 we introduce local systems of $\mathbb{Z}_T$-twisted vertex operators. In Section 4 we study $\mathbb{Z}_T$-twisted modules for vertex operator superalgebras associated to the Neveu-Schwarz algebra and an affine Lie superalgebra. In Section 5 we give the semisimple twisted module theory for vertex operator algebras associated to standard modules for an affine Lie algebra.

Acknowledgment We would like to thank Professors Chongying Dong, James

Lepowsky and Robert Wilson for many useful discussions.

2. VERTEX SUPERALGEBRAS AND TWISTED MODULES

In this section, we review the definitions of vertex (operator) superalgebra and twisted module and we also present some elementary results.

Let $z, z_0, z_1, \cdots$ be commuting formal variables. For a vector space V, we recall the following notations and formal variable calculus from [FLM]:

$$V\{z\} = \{\sum_{n\in C} v_n z^n | v_n \in V\}, \ V[[z, z^{-1}]] = \{\sum_{n\in\mathbb{Z}} v_n z^n | v_n \in V\}, \tag{2.1}$$

$$V[z] = \{\sum_{n\in N} v_n z^n | v_n \in V, v_n = 0 \ \text{for } n \text{ sufficiently large}\}, \tag{2.2}$$

$$V[z, z^{-1}] = \{\sum_{n\in\mathbb{Z}} v_n z^n | v_n \in V, v_n = 0 \ \text{for all but finitely many } n\}, \tag{2.3}$$

$$V[[z]] = \{\sum_{n\in N} v_n z^n | v_n \in V\}, \tag{2.4}$$

$$V((z)) = \{\sum_{n\in\mathbb{Z}} v_n z^n | v_n \in V, v_n = 0 \ \text{for } n \text{ sufficiently small}\}. \tag{2.5}$$

For $f(z) = \sum_{n\in\mathbb{Z}} v_n z^n \in V[[z, z^{-1}]]$, the formal derivative is defined to be

$$\frac{d}{dz} f(z) = f'(z) = \sum_{n\in\mathbb{Z}} n v_n z^{n-1}, \tag{2.6}$$

and the formal residue is defined as follows:

$$Res_z f(z) = v_{-1} \quad \text{(the coefficient of } z^{-1} \text{ in } f(z)). \tag{2.7}$$

If $f(z) \in C((z)), g(z) \in V((z))$, we have:

$$Res_z(f'(z)g(z)) = -Res_z(f(z)g'(z)). \tag{2.8}$$

For $\alpha \in \mathbb{C}$, as a formal power series, $(z_1 + z_2)^\alpha$ is defined to be

$$(z_1 + z_2)^\alpha = \sum_{k=0}^{\infty} \binom{\alpha}{k} z_1^{\alpha-k} z_2^k, \tag{2.9}$$

where $\binom{\alpha}{k} = \dfrac{\alpha(\alpha-1)\cdots(\alpha-k+1)}{k!}$. If $f(z) \in V\{z\}$, then we have the following Taylor formula:

$$e^{z_0 \frac{\partial}{\partial z}} f(z) = f(z + z_0). \tag{2.10}$$

The formal δ-function is defined to be

$$\delta(z) = \sum_{n\in\mathbb{Z}} z^n \in \mathbb{C}[[z, z^{-1}]]. \tag{2.11}$$

Thus

$$\delta\left(\frac{z_1 - z_2}{z_0}\right) = \sum_{n\in\mathbb{Z}} z_0^{-n}(z_1 - z_2)^n = \sum_{n\in\mathbb{Z}} \sum_{k\in\mathbb{N}} (-1)^k \binom{n}{k} z_0^{-n} z_1^{n-k} z_2^k. \tag{2.12}$$

Furthermore, for $\alpha \in \mathbb{C}$, we have:

$$z_0^{-1}\delta\left(\frac{z_1 - z_2}{z_0}\right)\left(\frac{z_1 - z_2}{z_0}\right)^\alpha = z_1^{-1}\delta\left(\frac{z_0 + z_2}{z_1}\right)\left(\frac{z_0 + z_2}{z_1}\right)^{-\alpha}. \quad (2.13)$$

Lemma 2.1. *[FLM] If $f(z_1, z_2) \in V[[z_1, z_1^{-1}]]((z_2))$, then*

$$\delta\left(\frac{z_0 + z_2}{z_1}\right)f(z_1, z_2) = \delta\left(\frac{z_0 + z_2}{z_1}\right)f(z_0 + z_2, z_2). \quad (2.14)$$

Lemma 2.2. *Let $m, n \in \mathbb{Z}_+$ such that $m > n$. Then*

$$(z_1 - z_2)^m \delta^{(n)}\left(\frac{z_1}{z_2}\right) = 0. \quad (2.15)$$

Proof. It easily follows from Lemma 2.1 and an induction on n. $\square$

Lemma 2.3. *Let V be any vector space, let α and β be rational numbers and let $f_j(z_2)$ $(j = 0, \cdots, n)$ be elements of $V((z_2^\beta))$. Let*

$$g(z_1, z_2) = z_2^{-1}\delta\left(\frac{z_1}{z_2}\right)\left(\frac{z_1}{z_2}\right)^\alpha. \quad (2.16)$$

Then

$$g(z_1, z_2)f_0(z_2) + \left(\frac{\partial}{\partial z_2}g(z_1, z_2)\right)f_1(z_2) + \cdots + \left(\frac{\partial^n}{\partial z_2^n}g(z_1, z_2)\right)f_n(z_2) = 0 \quad (2.17)$$

if and only if $f_j(z_2) = 0$ for all j.

Proof. First, we prove that for any nonnegative integer k we have:

$$\left(\frac{\partial}{\partial z_2}\right)^k g(z_1, z_2) = (-1)^k \left(\frac{\partial}{\partial z_1}\right)^k g(z_1, z_2). \quad (2.18)$$

Using definition, we have:

$$\frac{\partial}{\partial z_2}g(z_1, z_2) = -\frac{\partial}{\partial z_1}g(z_1, z_2). \quad (2.19)$$

Then (2.18) easily follows from an induction on k. Suppose $f_j(z_2) \neq 0$ for some j. Without losing generality, we may assume $f_n(z_2) \neq 0$. Applying $Res_{z_1} z_1^{-\alpha}(z_1 - z_2)^n$ to (2.17), then using Leibniz rule and Lemma 2.2 we obtain:

$$\begin{aligned}
0 &= Res_{z_1} z_1^{-\alpha}(z_1 - z_2)^n \left(\frac{\partial^n}{\partial z_2^n}g(z_1, z_2)\right)f_n(z_2) \\
&= (-1)^n Res_{z_1} z_1^{-\alpha}(z_1 - z_2)^n \left(\frac{\partial^n}{\partial z_1^n}g(z_1, z_2)\right)f_n(z_2) \\
&= Res_{z_1} \left(\frac{\partial^n}{\partial z_1^n}z_1^{-\alpha}(z_1 - z_2)^n\right)g(z_1, z_2)f_n(z_2) \\
&= n! z_2^{-\alpha} f_n(z_2). \quad (2.20)
\end{aligned}$$

It is a contradiction. $\square$

Let $M = M^{(0)} \oplus M^{(1)}$ be any $\mathbb{Z}_2$-graded vector space. Then any element u in $M^{(0)}$ (resp. $M^{(1)}$) is said to be *even* (resp. *odd*). For any homogeneous element u, we define $|u| = 0$ if u is even, $|u| = 1$ if u is odd. If M and W are any two

$\mathbb{Z}_2$-graded vector spaces, we define $\varepsilon_{u,v} = (-1)^{|u||v|}$ for any homogeneous elements $u \in M, v \in W$.

Definition 2.4. *A vertex superalgebra is a quadruple* $(V, 1, D, Y)$, *where* $V = V^{(0)} \oplus V^{(1)}$ *is a* $\mathbb{Z}_2$-*graded vector space,* D *is a* $\mathbb{Z}_2$-*endomorphism of* V, 1 *is a specified vector called the vacuum of* V, *and* Y *is a linear map*

$$Y(\cdot, z): \quad V \to (EndV)[[z, z^{-1}]];$$
$$a \mapsto Y(a, z) = \sum_{n \in \mathbb{Z}} a_n z^{-n-1} \quad (where \ a_n \in EndV) \qquad (2.21)$$

such that

(V1)   *For any* $a, b \in V, a_n b = 0$ *for* n *sufficiently large;*

(V2) $[D, Y(a, z)] = Y(D(a), z) = \dfrac{d}{dz} Y(a, z)$ *for any* $a \in V$;

(V3) $Y(1, z) = Id_V$ *(the identity operator of* V*)*;

(V4) $Y(a, z)1 \in (EndV)[[z]]$ *and* $\lim_{z \to 0} Y(a, z)1 = a$ *for any* $a \in V$;

(V5)   *For* $\mathbb{Z}_2$ *-homogeneous* $a, b \in V$, *the following Jacobi identity holds:*

$$z_0^{-1}\delta\left(\frac{z_1 - z_2}{z_0}\right) Y(a, z_1)Y(b, z_2) - \varepsilon_{a,b} z_0^{-1}\delta\left(\frac{z_2 - z_1}{-z_0}\right) Y(b, z_2)Y(a, z_1)$$
$$= z_2^{-1}\delta\left(\frac{z_1 - z_0}{z_2}\right) Y(Y(a, z_0)b, z_2). \qquad (2.22)$$

This completes the definition of vertex superalgebra. For any $\mathbb{Z}_2$-homogeneous elements $a, b \in V$, there is a positive integer m such that (cf. [DL], [Li])

$$(z_1 - z_2)^m Y(a, z_1)Y(b, z_2) = \varepsilon_{a,b}(z_1 - z_2)^m Y(b, z_2)Y(a, z_1). \qquad (2.23)$$

A vertex superalgebra V is called a *vertex operator superalgebra* (cf. [T], [FFR], [DL]) if there is another distinguished vector ω of V such that

(V6) $[L(m), L(n)] = (m - n)L(m + n) + \dfrac{m^3 - m}{12}\delta_{m+n,0}(rankV)$

   for $m, n \in \mathbb{Z}$, where $Y(\omega, z) = \sum_{n \in \mathbb{Z}} L(n)z^{-n-2}$, $rankV \in \mathbb{C}$;

(V7) $L(-1) = D$, *i.e.,* $Y(L(-1)a, z) = \dfrac{d}{dz} Y(a, z)$ *for any* $a \in V$;

(V8) $V = \oplus_{n \in \frac{1}{2}\mathbb{Z}} V_{(n)}$ is $\dfrac{1}{2}\mathbb{Z}$-graded such that $L(0)|_{V_{(n)}} = nId_{V_{(n)}}$, $\dim V_{(n)} < \infty$,

   and $V_{(n)} = 0$ for n sufficiently small.

Recall the following proposition from [FLM], [FHL], [DL] and [Li]:

Proposition 2.5. *The Jacobi identity for vertex superalgebra can be equivalently replaced by the supercommutativity (2.23).*

Let V_1 and V_2 be vertex superalgebras. A vertex superalgebra *homomorphism* from V_1 to V_2 is a linear map σ such that

$$\sigma(1) = 1, \ \sigma(Y(a,z)b) = Y(\sigma(a),z)\sigma(b) \quad \text{for any } a, b \in V_1. \tag{2.24}$$

If both V_1 and V_2 are vertex operator superalgebras, a homomorphism is required to map the Virasoro element of V_1 to the Virasoro element of V_2. An *automorphism* of a vertex (operator) superalgebra V is 1-1 onto homomorphism from V to V. Let σ be an automorphism of order T of a vertex superalgebra V. Then σ acts semisimply on V. Therefore

$$V = V^0 \oplus V^1 \oplus \cdots \oplus V^{T-1} \tag{2.25}$$

where V^k is the eigenspace of V for σ with eigenvalue $exp\left(\frac{2k\pi\sqrt{-1}}{T}\right)$. It is clear that V^0 is a vertex subsuperalgebra of V and all V^k $(k = 0, 1, \cdots, T-1)$ are V^0-modules.

Definition 2.6. *Let $(V, 1, D, Y)$ be a vertex superalgebra with an automorphism σ of order T. A σ-twisted V-module is a triple (M, d, Y_M) consisting of a super vector space M, a $\mathbb{Z}_2$-endomorphism d of M and a linear map $Y_M(\cdot, z)$ from V to $(EndM)[[z^{\frac{1}{T}}, z^{-\frac{1}{T}}]]$ satisfying the following conditions:*

(M1)   *For any $a \in V, u \in M, a_n u = 0$ for $n \in \frac{1}{T}\mathbb{Z}$ sufficiently large;*

(M2) $Y_M(1, z) = Id_M$;

(M3) $[d, Y_M(a,z)] = Y_M(D(a), z) = \dfrac{d}{dz}Y_M(a,z)$ *for any $a \in V$;*

(M4)   *For any $\mathbb{Z}_2$-homogeneous $a, b \in V$, the following σ-twisted Jacobi identity holds:*

$$z_0^{-1}\delta\left(\frac{z_1 - z_2}{z_0}\right)Y_M(a, z_1)Y_M(b, z_2) - \varepsilon_{a,b}z_0^{-1}\delta\left(\frac{z_2 - z_1}{-z_0}\right)Y_M(b, z_2)Y_M(a, z_1)$$

$$= z_2^{-1}\sum_{j=0}^{T-1}\frac{1}{T}\delta\left(\left(\frac{z_1 - z_0}{z_2}\right)^{\frac{1}{T}}\right)Y_M(Y(\sigma^j a, z_0)b, z_2). \tag{2.26}$$

If V is a vertex operator superalgebra, a σ-twisted V-module for V as a vertex superalgebra is called a σ-twisted *weak* module for V as a vertex operator superalgebra. A σ-twisted weak V-module M is said to be $\frac{1}{2T}\mathbb{Z}$-*graded* if $M = \oplus_{n\in\frac{1}{2T}\mathbb{Z}}M(n)$ such that

(M5) $a_n M(r) \subseteq M(r + m - n - 1)$ for $a \in V_{(m)}, m \in \mathbb{Z}, n \in \frac{1}{T}\mathbb{Z}, r \in \frac{1}{2T}\mathbb{Z}$.

A σ-twisted module M for V as a vertex superalgebra is called a σ-*twisted module for V as a vertex operator superalgebra* if $M = \oplus_{\alpha\in\mathbb{C}}M_{(\alpha)}$ such that

(M6) $L(0)u = \alpha u$ for $\alpha \in \mathbb{C}, u \in M_{(\alpha)}$;

(M7)   For any fixed α, $M_{(\alpha+n)} = 0$ for $n \in \frac{1}{2T}\mathbb{Z}$ sufficiently small;

(M8) $\dim M_{(\alpha)} < \infty$ for any $\alpha \in \mathbb{C}$.

It is clear that any σ-twisted module is a direct sum of $\frac{1}{2T}\mathbb{Z}$-graded σ-twisted modules.

Remark 2.7. *As in the untwisted case, the Virasoro algebra relation follows from (M2)-(M4).*

If $a \in V^k$ is even or odd, the σ-twisted Jacobi identity (2.26) becomes:

$$z_0^{-1}\delta\left(\frac{z_1 - z_2}{z_0}\right)Y_M(a, z_1)Y_M(b, z_2) - \varepsilon_{a,b}z_0^{-1}\delta\left(\frac{z_2 - z_1}{-z_0}\right)Y_M(b, z_2)Y_M(a, z_1)$$

$$= z_2^{-1}\delta\left(\frac{z_1 - z_0}{z_2}\right)\left(\frac{z_1 - z_0}{z_2}\right)^{-\frac{k}{T}}Y_M(Y(a, z_0)b, z_2). \tag{2.27}$$

Setting $b = 1$, we get

$$z_2^{-1}\delta\left(\frac{z_1 - z_0}{z_2}\right)Y_M(a, z_1) = z_2^{-1}\delta\left(\frac{z_1 - z_0}{z_2}\right)\left(\frac{z_1 - z_0}{z_2}\right)^{-\frac{k}{T}}Y_M(Y(a, z_0)1, z_2). \tag{2.28}$$

By taking $Res_{z_0}z_0^{-1}$, we obtain

$$z_2^{-1}\delta\left(\frac{z_1}{z_2}\right)Y_M(a, z_1) = z_2^{-1}\delta\left(\frac{z_1}{z_2}\right)\left(\frac{z_1}{z_2}\right)^{-\frac{k}{T}}Y_M(a, z_2). \tag{2.29}$$

Therefore

$$z^{\frac{k}{T}}Y_M(a, z) \in (EndM)[[z, z^{-1}]] \quad \text{for any } a \in V^k. \tag{2.30}$$

Taking Res_{z_0} of (2.27), we obtain the *twisted supercommutator formula*:

$$\begin{aligned}
&[Y_M(a, z_1), Y_M(b, z_2)]_{\pm} \\
=:\ &Y_M(a, z_1)Y_M(b, z_2) - \varepsilon_{a,b}Y_M(b, z_2)Y_M(a, z_1) \\
=\ &Res_{z_0}z_2^{-1}\delta\left(\frac{z_1 - z_0}{z_2}\right)\left(\frac{z_1 - z_0}{z_2}\right)^{-\frac{k}{T}}Y_M(Y(a, z_0)b, z_2) \\
=\ &\sum_{j=0}^{\infty}\frac{1}{j!}\left(\left(\frac{\partial}{\partial z_2}\right)^j z_1^{-1}\delta\left(\frac{z_2}{z_1}\right)\left(\frac{z_2}{z_1}\right)^{\frac{k}{T}}\right)Y_M(a_j b, z_2).
\end{aligned} \tag{2.31}$$

Then we get the same supercommutativity as the one in the untwisted case. Multiplying (2.27) by $\left(\frac{z_1 - z_0}{z_2}\right)^{\frac{k}{T}}$, then taking Res_{z_1}, we obtain the *twisted iterate formula*:

$$Y_M(Y(a, z_0)b, z_2) = Res_{z_1}\left(\frac{z_1 - z_0}{z_2}\right)^{\frac{k}{T}} \cdot X \tag{2.32}$$

where

$$X = z_0^{-1}\delta\left(\frac{z_1 - z_2}{z_0}\right)Y_M(a, z_1)Y_M(b, z_2) - \varepsilon_{a,b}z_0^{-1}\delta\left(\frac{z_2 - z_1}{-z_0}\right)Y_M(b, z_2)Y_M(a, z_1).$$

Let r be a positive integer such that the supercommutativity (2.23) holds. Then

$$\begin{aligned}
z_0^r X &= (z_1 - z_2)^r X \\
&= z_2^{-1}\delta\left(\frac{z_1 - z_0}{z_2}\right)\left((z_1 - z_2)^r Y_M(a, z_1)Y_M(b, z_2)\right).
\end{aligned} \tag{2.33}$$

Thus

$$z_0^r(z_2 + z_0)^{\frac{k}{T}} Y_M(Y(a, z_0)b, z_2)$$

$$= Res_{z_1} z_2^{-1} \delta\left(\frac{z_1 - z_0}{z_2}\right) \left(\frac{z_1 - z_0}{z_2}\right)^{\frac{k}{T}} (z_2 + z_0)^{\frac{k}{T}} ((z_1 - z_2)^r Y_M(a, z_1) Y_M(b, z_2))$$

$$= Res_{z_1} z_1^{-1} \delta\left(\frac{z_2 + z_0}{z_1}\right) \left(\frac{z_2 + z_0}{z_1}\right)^{-\frac{k}{T}} (z_2 + z_0)^{\frac{k}{T}} ((z_1 - z_2)^r Y_M(a, z_1) Y_M(b, z_2))$$

$$= Res_{z_1} z_1^{-1} \delta\left(\frac{z_2 + z_0}{z_1}\right) ((z_1 - z_2)^r Y_M^o(a, z_1) Y_M(b, z_2))$$

$$= (z_0^r Y_M^o(a, z_2 + z_0) Y_M(b, z_2)), \tag{2.34}$$

where $Y_M^o(a, z) = z^{\frac{k}{T}} Y_M(a, z) \in (EndM)[[z, z^{-1}]]$. For any $u \in M$, let k be a positive integer such that $z^k Y_M^o(a, z)u \in M[[z]]$. Then we obtain

$$(z_2 + z_0)^{\frac{k}{T}} (z_0 + z_2)^k z_0^r Y_M(Y(a, z_0)b, z_2)u$$

$$= z_0^r(z_0 + z_2)^k Y_M^o(a, z_0 + z_2) Y_M(b, z_2)u. \tag{2.35}$$

Multiplying both sides by z_0^{-r}, we obtain the following twisted associativity:

$$(z_0 + z_2)^k (z_2 + z_0)^{\frac{k}{T}} Y_M(Y(a, z_0)b, z_2)u$$

$$= (z_0 + z_2)^k Y_M^o(a, z_0 + z_2) Y_M(b, z_2)u. \tag{2.36}$$

Lemma 2.8. *Let σ be an automorphism of order T of a vertex superalgebra V. Then the σ-twisted Jacobi identity (2.26) can be equivalently replaced by the supercommutativity and the σ-twisted associativity (2.36).*

Proof. For any homogeneous $a, b \in V, u \in M$, let m be a positive integer such that both the supercommutativity and the twisted associativity hold and that

$z^m Y_M^o(a, z)u$ involves only positive powers of z. Then

$$z_0^m z_1^m z_0^{-1} \delta\left(\frac{z_1 - z_2}{z_0}\right) Y_M(a, z_1) Y_M(b, z_2) u$$

$$-z_0^m z_1^m \varepsilon_{a,b} z_0^{-1} \delta\left(\frac{z_2 - z_1}{-z_0}\right) Y_M(b, z_2) Y_M(a, z_1) u$$

$$= z_0^{-1} \delta\left(\frac{z_1 - z_2}{z_0}\right) \left(z_1^m (z_1 - z_2)^m Y_M(a, z_1) Y_M(b, z_2) u\right)$$

$$- \varepsilon_{a,b} z_0^{-1} \delta\left(\frac{-z_2 + z_1}{z_0}\right) \left(z_1^m (z_1 - z_2)^m Y_M(b, z_2) Y_M(a, z_1) u\right)$$

$$= z_2^{-1} \delta\left(\frac{z_1 - z_0}{z_2}\right) \varepsilon_{a,b} \left(z_1^m (z_1 - z_2)^m Y_M(b, z_2) Y_M(a, z_1) u\right)$$

$$= z_1^{-1} \delta\left(\frac{z_2 + z_0}{z_1}\right) \varepsilon_{a,b} \left(z_1^{-\frac{k}{T}} (z_2 + z_0)^m z_0^m Y_M(b, z_2) Y_M^o(a, z_2 + z_0) u\right)$$

$$= z_1^{-1} \delta\left(\frac{z_2 + z_0}{z_1}\right) \varepsilon_{a,b} \left(z_1^{-\frac{k}{T}} z_0^m (z_0 + z_2)^m Y_M(b, z_2) Y_M^o(a, z_0 + z_2) u\right)$$

$$= z_2^{-1} \delta\left(\frac{z_1 - z_0}{z_2}\right) \left(z_1^{-\frac{k}{T}} z_0^m (z_0 + z_2)^m Y_M^o(a, z_0 + z_2) Y_M(b, z_2) u\right)$$

$$= z_2^{-1} \delta\left(\frac{z_1 - z_0}{z_2}\right) z_1^{-\frac{k}{T}} z_0^m (z_0 + z_2)^m (z_2 + z_0)^{\frac{k}{T}} Y_M(Y(a, z_0)b, z_2) u$$

$$= z_0^m z_1^m z_2^{-1} \delta\left(\frac{z_1 - z_0}{z_2}\right) \left(\frac{z_2 + z_0}{z_1}\right)^{\frac{k}{T}} Y_M(Y(a, z_0)b, z_2) u.$$

Multiplying by $z_0^{-m} z_1^{-m}$, we obtain the σ-twisted Jacobi identity. □

Remark 2.9. *Since the twisted associativity follows from the supercommutativity and the iterate formula (2.32), the supercommutativity and the iterate formula (2.32) imply the σ-twisted Jacobi identity.*

Let $a, b \in V$ be $\mathbb{Z}_2$-homogeneous such that $[Y_M(a, z_1), Y_M(b, z_2)]_\pm = 0$. Then

$$Y_M(Y(a, z_0)b, z_2)$$

$$= \text{Res}_{z_1} \left(\frac{z_2 + z_0}{z_1}\right)^{-\frac{k}{T}} z_1^{-1} \delta\left(\frac{z_2 + z_0}{z_1}\right) (Y_M(a, z_1) Y_M(b, z_2))$$

$$= (z_2 + z_0)^{-\frac{k}{T}} Y_M^o(a, z_2 + z_0) Y_M(b, z_2)$$

$$= Y_M(a, z_2 + z_0) Y_M(b, z_2). \tag{2.37}$$

Therefore

$$Y_M(a_{-1}b, z_2) = Y_M(a, z_2) Y_M(b, z_2). \tag{2.38}$$

If $a \in V$ is $\mathbb{Z}_2$-homogeneous such that $[Y_M(a, z_1), Y_M(a, z_2)]_\pm = 0$, then

$$Y_M((a_{-1})^N 1, z) = Y_M(a, z)^N \quad \text{for any } N \in \mathbb{Z}_+. \tag{2.39}$$

This is Dong and Lepowsky's formula (13.70) in [DL] in the twisted case. Then we have the following generalization of Proposition 13.16 in [DL].

Proposition 2.10. *Let V be a vertex superalgebra, let M be a σ-twisted V-module and let $a \in V$ be a $\mathbb{Z}_2$-homogeneous element such that $[Y_M(a, z_1), Y_M(a, z_2)]_\pm = 0$, so that $Y(a, z)^N$ is well defined for $N \in \mathbb{Z}_+$. Then $Y_V(a, z)^N = 0$ implies $Y_M(a, z)^N = 0$ on M. On the other hand, the converse is true if M is faithful.*

Let σ be an automorphism of V and let M be a faithful σ-twisted V-module. Let $a, b, u^0, u^1, \cdots, u^n \in V$. It follows from Lemma 2.3 and the σ-twisted commutator formula (2.31) that

$$[Y_M(a, z_1), Y_M(b, z_2)]_\pm = \sum_{j=0}^{\infty} \frac{1}{j!} \left(\left(\frac{\partial}{\partial z_2} \right)^j z_1^{-1} \delta \left(\frac{z_2}{z_1} \right) \left(\frac{z_2}{z_1} \right)^{\frac{k}{T}} \right) Y_M(u^j, z_2) \quad (2.40)$$

if and only if $a_i b = u^j$ for $0 \le j \le n$, $a_i b = 0$ for $j > n$. Since V is always a faithful V-module, we have:

Lemma 2.11. *Let V be a vertex superalgebra with an automorphism of order T and let (M, Y_M) be a faithful σ-twisted V-module. Let $a, b, u^0, \cdots, u^n$ be $\mathbb{Z}_T$-homogeneous elements of V. Then*

$$[Y_V(a, z_1), Y_V(b, z_2)]_\pm = \sum_{j=0}^{\infty} \frac{1}{j!} \left(\left(\frac{\partial}{\partial z_2} \right)^j z_1^{-1} \delta \left(\frac{z_2}{z_1} \right) \right) Y_V(u^j, z_2) \quad (2.41)$$

if and only if (2.40) holds.

3. Local Systems of Twisted Vertex Operators

Let us first define three basic categories C, C^o and C_ℓ. The set of objects of the category C is the set of $\mathbb{Z}_2$-graded vector spaces and the set of morphisms for any two objects is the set of all $\mathbb{Z}_2$-homomorphisms. The set of objects of the category C^o consists of all pairs (M, d) where M is an object of the category C and d is an endomorphism of the object M. If (M_i, d_i) $(i = 1, 2)$ are two objects of C^o, then a morphism is a morphism f from M_1 to M_2 such that $d_2 f = f d_1$. Finally, C_ℓ is the category of $\mathbb{Z}_2$-graded vector spaces which are also restricted Vir-modules with ℓ as its central charge and with the even subspace and the odd subspace as submodules.

Throughout this section, T will be a fixed positive integer.

Definition 3.1. *Let M be any $\mathbb{Z}_2$-graded vector space. A $\mathbb{Z}_T$-twisted weak vertex operator on M is an element $a(z) = \sum_{n \in \frac{1}{T}\mathbb{Z}} a_n z^{-n-1} \in (End\, M)[[z^{\frac{1}{T}}, z^{-\frac{1}{T}}]]$ such that for any $u \in M, a_n u = 0$ for $n \in \frac{1}{T}\mathbb{Z}$ sufficiently large. Let (M, d) be an object of the category C^o. A $\mathbb{Z}_T$-twisted weak vertex operator $a(z)$ on M is called a $\mathbb{Z}_T$-twisted weak vertex operator on (M, d) if it satisfies the following condition:*

$$[d, a(z)] = a'(z) \left(= \frac{d}{dz} a(z) \right). \quad (3.1)$$

Let M be an object of C_ℓ. A $\mathbb{Z}_T$-twisted weak vertex operator $a(z)$ on $(M, L(-1))$ is said to be of weight $\lambda \in \mathbb{C}$ if it satisfies the following condition:

$$[L(0), a(z)] = \lambda a(z) + za'(z). \quad (3.2)$$

Denote by $F(M,T)$ (resp. $F(M,d,T)$) the space of all $\mathbb{Z}_T$-twisted weak vertex operators on M (resp. (M,d)). Set $\varepsilon = \exp\left(\frac{2\pi\sqrt{-1}}{T}\right)$. Let σ be the endomorphism of $(EndM)[[z^{\frac{1}{T}}, z^{-\frac{1}{T}}]]$ defined by: $\sigma f(z^{\frac{1}{T}}) = f(\varepsilon^{-1}z^{\frac{1}{T}})$. It is clear that σ restricted to $F(M,T)$ is an automorphism of order T. Then

$$F(M,T) = F(M,T)^0 \oplus F(M,T)^1 \oplus \cdots \oplus F(M,T)^{T-1}, \qquad (3.3)$$

where $F(M,T)^k = \{f(z) \in F(M,T)|\sigma f(z) = \varepsilon^k f(z)\}$ for $0 \le k \le T-1$. Therefore, $F(M,T)$ is $\mathbb{Z}_2 \times \mathbb{Z}_T$-graded. If $a(z)$ is a $\mathbb{Z}_T$-twisted weak vertex operator on M (resp. (M,d)), then $a'(z) = \frac{d}{dz}a(z)$ is a $\mathbb{Z}_T$-twisted weak vertex operator on M (resp. (M,d)). Then we have an endomorphism $D = \frac{d}{dz}$:

$$D : F(M,T) \to F(M,T); \; D \cdot a(z) = \frac{d}{dz}a(z) = a'(z). \qquad (3.4)$$

It is clear that $D\sigma = \sigma D$.

Definition 3.2. *Two $\mathbb{Z}_2$-homogeneous $\mathbb{Z}_T$-twisted weak vertex operators $a(z)$ and $b(z)$ are said to be mutually local if there is a positive integer n such that*

$$(z_1 - z_2)^n a(z_1)b(z_2) = (-1)^{|a(z)||b(z)|}(z_1 - z_2)^n b(z_2)a(z_1). \qquad (3.5)$$

A $\mathbb{Z}_T$-twisted weak vertex operator $a(z)$ is called a $\mathbb{Z}_T$-twisted vertex operator if $a(z)$ is local with itself. A $\mathbb{Z}_2$-graded subspace A of $F(M,T)$ is said to be local if any two $\mathbb{Z}_2$-homogeneous elements of A are mutually local. We define a local system of $\mathbb{Z}_T$-twisted vertex operators on M is a maximal $\mathbb{Z}_2$-graded local space of $\mathbb{Z}_T$-twisted vertex operators on M.

It is clear that any local system A of $F(M,T)$ has a natural automorphism σ such that $\sigma^T = Id_A$.

Replacing n with $n+1$ in (3.5), then differentiating (3.5) with respect to z_2 we have:

Lemma 3.3. *[Li] Let M be a $\mathbb{Z}_2$-graded vector space and let $a(z)$ and $b(z)$ be mutually local $\mathbb{Z}_T$-twisted vertex operators on M. Then $a(z)$ is local with $b'(z)$.*

Remark 3.4. *Let M be any $\mathbb{Z}_2$-graded vector space. Then the identity operator $I(z) = Id_M$ is a $\mathbb{Z}_T$-twisted vertex operator which is mutually local with any $\mathbb{Z}_T$-twisted weak vertex operator on M. Therefore, any local system of $\mathbb{Z}_T$-twisted vertex operators on M contains $I(z)$. It follows from Lemma 3.3 that any local system is closed under the derivative operation $D = \frac{d}{dz}$.*

Using the same proof for the untwisted case in [Li] we get:

Lemma 3.5. *[Li] Let M be a restricted Vir-module of central charge ℓ. Then $L(z) = \sum_{n\in\mathbb{Z}} L(n)z^{-n-2}$ is a (local) $\mathbb{Z}_T$-twisted vertex operator (of weight two) on M.*

Remark 3.6. *Let V be a vertex superalgebra with an automorphism σ of order T and let (M, Y_M) be a σ-twisted V-module. Then the image of V under the linear map $Y_M(\cdot, z)$ is a local subspace of $F(M,T)$.*

Definition 3.7. *Let M be a $\mathbb{Z}_2$-graded vector space and let $a(z)$ and $b(z)$ be mutually local $\mathbb{Z}_T$-twisted vertex operators on M such that $a(z) \in F(M,T)^k$. Then for any integer n we define $a(z)_n b(z)$ as an element of $F(M,T)$ as follows:*

$$a(z)_n b(z) = \text{Res}_{z_1} \text{Res}_{z_0} \left(\frac{z_1 - z_0}{z} \right)^{\frac{k}{T}} z_0^n \cdot X \tag{3.6}$$

where

$$X = z_0^{-1} \delta \left(\frac{z_1 - z}{z_0} \right) a(z_1) b(z) - \varepsilon_{a,b} z_0^{-1} \delta \left(\frac{z - z_1}{-z_0} \right) b(z) a(z_1),$$

or, equivalently, $a(z)_n b(z)$ is defined by:

$$\sum_{n \in \mathbb{Z}} (a(z)_n b(z)) z_0^{-n-1} = \text{Res}_{z_1} \left(\frac{z_1 - z_0}{z} \right)^{\frac{k}{T}} \cdot X. \tag{3.7}$$

Remark 3.8. *For any $u \in M$, we have:*

$$(a(z)_n b(z)) u$$
$$= \sum_{i=0}^{\infty} \binom{\frac{k}{T}}{i} (-1)^i z_1^{\frac{k}{T} - i} z^{\frac{k}{T}} \cdot$$
$$\cdot \left((z_1 - z)^{n+i} a(z_1) b(z) u - \varepsilon_{a,b} (-z + z_1)^{n+i} b(z) a(z_1) u \right)$$
$$= \sum_{i=0}^{N} \binom{\frac{k}{T}}{i} (-1)^i z_1^{\frac{k}{T} - i} z^{\frac{k}{T}} \cdot$$
$$\cdot \left((z_1 - z)^{n+i} a(z_1) b(z) u - \varepsilon_{a,b} (-z + z_1)^{n+i} b(z) a(z_1) u \right). \tag{3.8}$$

Then it is clear that $(a(z)_n b(z)) u \in M((z))$. Thus $a(z)_n b(z) \in F(M,T)$. It follows from the locality that $a(z)_n b(z) = 0$ for n sufficiently large. Furthermore, for any nonnegative integer n, we have:

$$z_0^{-1} \delta \left(\frac{z_1 - z_2}{z_0} \right) \left(\frac{z_1 - z_0}{z_2} \right)^n = z_0^{-1} \delta \left(\frac{z_1 - z_2}{z_0} \right), \tag{3.9}$$

$$z_0^{-1} \delta \left(\frac{-z_2 + z_1}{z_0} \right) \left(\frac{z_1 - z_0}{z_2} \right)^n = z_0^{-1} \delta \left(\frac{-z_2 + z_1}{z_0} \right). \tag{3.10}$$

Then

$$\left(\frac{z_1 - z_0}{z_2} \right)^n X = X. \tag{3.11}$$

Since $\left(\frac{z_1 - z_0}{z_2} \right)^n X$ exists for any integer n, (3.11) is true for any integer n. Then the right hand side of (3.7) depends only on k mod $T\mathbb{Z}$.

The proof of the following proposition is similar to the proof in the ordinary case [Li], which was given by Professor Chongying Dong.

Proposition 3.9. *Let $a(z)$, $b(z)$ and $c(z)$ be $\mathbb{Z}_2$-graded $\mathbb{Z}_T$-twisted weak vertex operators on M. If both $a(z)$ and $b(z)$ are local with $c(z)$, then $a(z)_n b(z)$ is local with $c(z)$ for all $n \in \mathbb{Z}$.*

Proof. Let r be a positive integer such that $r + n > 0$ and that the following identities hold:

$$(z_1 - z_2)^r a(z_1)b(z_2) = \varepsilon_{a,b}(z_1 - z_2)^r b(z_2)a(z_1),$$
$$(z_1 - z_2)^r a(z_1)c(z_2) = \varepsilon_{a,c}(z_1 - z_2)^r c(z_2)a(z_1),$$
$$(z_1 - z_2)^r b(z_1)c(z_2) = \varepsilon_{b,c}(z_1 - z_2)^r c(z_2)b(z_1).$$

Assume $a(z) \in F(M, T)^i$. Then we have

$$
\begin{aligned}
a(z)_n b(z) &= Res_{z_1} \sum_{k=0}^{\infty} (-1)^k \begin{pmatrix} \frac{i}{T} \\ k \end{pmatrix} z_1^{\frac{i}{T}-k} z_2^{-\frac{i}{T}} A \\
&= Res_{z_1} \sum_{k=0}^{2r} (-1)^k \begin{pmatrix} \frac{i}{T} \\ k \end{pmatrix} z_1^{\frac{i}{T}-k} z^{-\frac{i}{T}} A,
\end{aligned}
\tag{3.12}
$$

where $A = (z_1 - z)^{n+k} a(z_1)b(z) - \varepsilon_{a,b}(-z + z_1)^{n+k} b(z)a(z_1)$. Since

$$
\begin{aligned}
&(z - z_3)^{4r} \left((z_1 - z)^{n+k} a(z_1)b(z)c(z_3) - \varepsilon_{a,b}(-z + z_1)^{n+k} b(z)a(z_1)c(z_3) \right) \\
&= \sum_{s=0}^{3r} \begin{pmatrix} 3r \\ s \end{pmatrix} (z - z_1)^{3r-s}(z_1 - z_3)^s (z - z_3)^r \cdot \\
&\quad \cdot \left((z_1 - z)^{n+k} a(z_1)b(z)c(z_3) - \varepsilon_{a,b}(-z + z_1)^{n+k} b(z)a(z_1)c(z_3) \right) \\
&= \sum_{s=r+1}^{3r} \begin{pmatrix} 3r \\ s \end{pmatrix} (z - z_1)^{3r-s}(z_1 - z_3)^s (z - z_3)^r \cdot \\
&\quad \cdot \left((z_1 - z)^{n+k} a(z_1)b(z)c(z_3) - \varepsilon_{a,b}(-z + z_1)^{n+k} b(z)a(z_1)c(z_3) \right) \cdot \\
&= \sum_{s=r+1}^{3r} \begin{pmatrix} 3r \\ s \end{pmatrix} (z - z_1)^{3r-s}(z_1 - z_3)^s (z - z_3)^r \cdot \\
&\quad \cdot \left((z_1 - z)^{n+k} c(z_3)a(z_1)b(z) - \varepsilon_{a,b}(-z + z_1)^{n+k} c(z_3)b(z)a(z_1) \right) \\
&= (z - z_3)^{4r} \left((z_1 - z)^{n+k} c(z_3)a(z_1)b(z) - \varepsilon_{a,b}(-z + z_1)^{n+k} c(z_3)b(z)a(z_1) \right),
\end{aligned}
$$

we have

$$(z - z_3)^{4r}(a(z)_n b(z))c(z_3) = (z - z_3)^{4r} c(z_3)(a(z)_n b(z)). \quad \square$$

Let A be any local systems of $\mathbb{Z}_T$-twisted vertex operators on M. Then it follows from Remark 3.8 and Proposition 3.9 that $a(z)_n b(z) \in A$ for $a(z) \in A^k, b(z) \in A$. Then for any $a(z) \in A^k, b(z) \in A$ we define

$$Y_A(a(z), z_0)b(z) = \sum_{n \in \mathbb{Z}} (a(z)_n b(z)) z_0^{-n-1}. \tag{3.13}$$

Linearly extending the definition of $Y_A(\cdot, z_0)$ to any element of A, we obtain a linear map:

$$Y_A(\cdot, z_0) : A \to (EndA)[[z_0, z_0^{-1}]]; a(z) \mapsto Y_A(a(z), z_0). \tag{3.14}$$

Also notice that A contains the identity operator $I(z)$ and that A is closed under the derivative operation. For the rest of this section, A will be a fixed local system of $\mathbb{Z}_T$-twisted vertex operators on M.

Lemma 3.10. *For any $a(z) \in A$, we have*

$$Y_A(I(z), z_0)a(z) = a(z),$$

(3.15)

$$Y_A(a(z), z_0)I(z) = e^{z_0 \frac{\partial}{\partial z}} a(z) = a(z + z_0).$$

Proof. The proof of the first identity is the same as that in the untwisted case. For the second one, without losing generality, we may assume that $a(z) \in A^k$. Then

$$Y_A(a(z), z_0)I(z)$$

$$= Res_{z_1} \left(\frac{z_1 - z_0}{z} \right)^{\frac{k}{T}} \left(z_0^{-1} \delta \left(\frac{z_1 - z}{z_0} \right) a(z_1)I(z) - z_0^{-1}\delta \left(\frac{z - z_1}{-z_0} \right) I(z)a(z_1) \right)$$

$$= Res_{z_1} \left(\frac{z_1 - z_0}{z} \right)^{\frac{k}{T}} z^{-1}\delta \left(\frac{z_1 - z_0}{z} \right) a(z_1)$$

$$= Res_{z_1} \left(\frac{z + z_0}{z_1} \right)^{-\frac{k}{T}} z_1^{-1}\delta \left(\frac{z + z_0}{z_1} \right) a(z_1)$$

$$= Res_{z_1} (z + z_0)^{-\frac{k}{T}} z_1^{-1}\delta \left(\frac{z + z_0}{z_1} \right) \left(z_1^{\frac{k}{T}} a(z_1) \right)$$

$$= Res_{z_1} (z + z_0)^{-\frac{k}{T}} z_1^{-1}\delta \left(\frac{z + z_0}{z_1} \right) \left((z + z_0)^{\frac{k}{T}} a(z + z_0) \right)$$

$$= a(z + z_0)$$

$$= e^{z_0 \frac{\partial}{\partial z}} a(z). \quad \square$$

(3.16)

Lemma 3.11. *For any $a(z), b(z) \in A$, we have:*

$$\frac{\partial}{\partial z_0} Y_A(a(z), z_0)b(z) = Y_A(a'(z), z_0)b(z) = Y_A(D(a(z)), z_0)b(z).$$

(3.17)

Proof. By linearity, we may assume that $a(z) \in F(M, T)^k$. Then we have

$$
\frac{\partial}{\partial z_0} Y_A(a(z), z_0) b(z)
$$

$$
= \frac{\partial}{\partial z_0} Res_{z_1} \left(\frac{z_1 - z_0}{z} \right)^{\frac{k}{T}} \cdot
$$

$$
\cdot \left(z_0^{-1} \delta \left(\frac{z_1 - z}{z_0} \right) a(z_1) b(z) - \varepsilon_{a,b} z_0^{-1} \delta \left(\frac{z - z_1}{-z_0} \right) b(z) a(z_1) \right)
$$

$$
= Res_{z_1} \left(\frac{\partial}{\partial z_0} \left(\frac{z_1 - z_0}{z} \right)^{\frac{k}{T}} \right) \cdot
$$

$$
\cdot \left(z_0^{-1} \delta \left(\frac{z_1 - z}{z_0} \right) a(z_1) b(z) - \varepsilon_{a,b} z_0^{-1} \delta \left(\frac{z - z_1}{-z_0} \right) b(z) a(z_1) \right)
$$

$$
+ Res_{z_1} \left(\frac{z_1 - z_0}{z} \right)^{\frac{k}{T}} \cdot
$$

$$
\cdot \frac{\partial}{\partial z_0} \left(z_0^{-1} \delta \left(\frac{z_1 - z}{z_0} \right) a(z_1) b(z) + \varepsilon_{a,b} z_0^{-2} \delta \left(\frac{z - z_1}{-z_0} \right) b(z) a(z_1) \right)
$$

$$
= - Res_{z_1} \left(\frac{\partial}{\partial z_1} \left(\frac{z_1 - z_0}{z} \right)^{\frac{k}{T}} \right) \cdot
$$

$$
\cdot \left(z_0^{-1} \delta \left(\frac{z_1 - z}{z_0} \right) a(z_1) b(z) - \varepsilon_{a,b} z_0^{-1} \delta \left(\frac{z - z_1}{-z_0} \right) b(z) a(z_1) \right)
$$

$$
- Res_{z_1} \left(\frac{z_1 - z_0}{z} \right)^{\frac{k}{T}} \cdot
$$

$$
\cdot \left(\left(\frac{\partial}{\partial z_1} z_0^{-1} \delta \left(\frac{z_1 - z}{z_0} \right) \right) a(z_1) b(z) - \varepsilon_{a,b} \left(\frac{\partial}{\partial z_1} z_0^{-1} \delta \left(\frac{z - z_1}{-z_0} \right) \right) b(z) a(z_1) \right)
$$

$$
= Res_{z_1} \left(\frac{z_1 - z_0}{z} \right)^{\frac{k}{T}} \left(z_0^{-1} \delta \left(\frac{z_1 - z}{z_0} \right) a'(z_1) b(z) - \varepsilon_{a,b} z_0^{-1} \delta \left(\frac{z - z_1}{-z_0} \right) b(z) a'(z_1) \right)
$$

$$
= Y_A(a'(z), z_0) b(z). \quad \square \tag{3.18}
$$

Proposition 3.12. *Let M be an object of category C_ℓ and let A be a local system of $\mathbb{Z}_T$-twisted vertex operators on $(M, L(-1))$. Let $a(z) \in A^j, b(z) \in A^k$ of weights α and β, respectively. Then for any integer n, $a(z)_n b(z) \in A^{j+k}$ has weight $(\alpha + \beta - n - 1)$.*

Proof. It is enough to prove the following identities:

$$
[L(-1), Y_A(a(z), z_0) b(z)] = \frac{\partial}{\partial z} (Y_A(a(z), z_0) b(z)), \tag{3.19}
$$

$$
[L(0), Y_A(a(z), z_0) b(z)]
$$

$$
= (\alpha + \beta) Y_A(a(z), z_0) b(z) + z_0 \frac{\partial}{\partial z_0} (Y_A(a(z), z_0) b(z)) + z \frac{\partial}{\partial z} (Y_A(a(z), z_0) b(z)). \tag{3.20}
$$

By definition we have:

$$\frac{\partial}{\partial z}(Y_A(a(z), z_0)b(z))$$

$$= \frac{\partial}{\partial z} Res_{z_1} \left(\frac{z_1 - z_0}{z}\right)^{\frac{j}{T}} \cdot$$

$$\cdot \left(z_0^{-1}\delta\left(\frac{z_1 - z}{z_0}\right) a(z_1)b(z) - \varepsilon_{a,b}z_0^{-1}\delta\left(\frac{z - z_1}{-z_0}\right) b(z)a(z_1)\right)$$

$$= Res_{z_1} \left(-\frac{j}{T}\right) z^{-1} \left(\frac{z_1 - z_0}{z}\right)^{\frac{j}{T}} \cdot$$

$$\cdot \left(z_0^{-1}\delta\left(\frac{z_1 - z}{z_0}\right) a(z_1)b(z) - \varepsilon_{a,b}z_0^{-1}\delta\left(\frac{z - z_1}{-z_0}\right) b(z)a(z_1)\right)$$

$$+ Res_{z_1} \left(\frac{z_1 - z_0}{z}\right)^{\frac{j}{T}} \cdot$$

$$\cdot \left(-z_0^{-2}\delta'\left(\frac{z_1 - z}{z_0}\right) a(z_1)b(z) + \varepsilon_{a,b}z_0^{-2}\delta'\left(\frac{z - z_1}{-z_0}\right) b(z)a(z_1)\right)$$

$$+ Res_{z_1} \left(\frac{z_1 - z_0}{z}\right)^{\frac{j}{T}} \cdot$$

$$\cdot \left(\delta\left(\frac{z_1 - z}{z_0}\right) a(z_1)b'(z) - \varepsilon_{a,b}z_0^{-1}\delta\left(\frac{z - z_1}{-z_0}\right) b'(z)a(z_1)\right)$$

$$= -Res_{z_1} \left(\frac{\partial}{\partial z_1}\left(\frac{z_1 - z_0}{z}\right)^{\frac{j}{T}}\right)\left(\frac{z_1 - z_0}{z}\right) \cdot$$

$$\cdot \left(z_0^{-1}\delta\left(\frac{z_1 - z}{z_0}\right) a(z_1)b(z) - \varepsilon_{a,b}z_0^{-1}\delta\left(\frac{z - z_1}{-z_0}\right) b(z)a(z_1)\right)$$

$$- Res_{z_1} \left(\frac{z_1 - z_0}{z}\right)^{\frac{j}{T}} \cdot$$

$$\cdot \left(\left(\frac{\partial}{\partial z_1}z_0^{-1}\delta\left(\frac{z_1 - z}{z_0}\right)\right) a(z_1)b(z) - \left(\frac{\partial}{\partial z_1}z_0^{-1}\delta\left(\frac{z_1 - z}{z_0}\right)\right) b(z)a(z_1)\right)$$

$$+ Res_{z_1} \left(\frac{z_1 - z_0}{z}\right)^{\frac{j}{T}} \cdot$$

$$\cdot \left(z_0^{-1}\delta\left(\frac{z_1 - z}{z_0}\right) a(z_1)b'(z) - \varepsilon_{a,b}z_0^{-1}\delta\left(\frac{z - z_1}{-z_0}\right) b'(z)a(z_1)\right)$$

$$= Res_{z_1} \left(\frac{z_1 - z_0}{z}\right)^{\frac{j}{T}} \cdot$$

$$\cdot \left(z_0^{-1}\delta\left(\frac{z_1 - z}{z_0}\right) a'(z_1)b(z) - \varepsilon_{a,b}z_0^{-1}\delta\left(\frac{z - z_1}{-z_0}\right) b(z)a'(z_1)\right)$$

$$+ Res_{z_1} \left(\frac{z_1 - z_0}{z}\right)^{\frac{j}{T}} \cdot$$

$$\cdot \left(z_0^{-1}\delta\left(\frac{z_1 - z}{z_0}\right) a(z_1)b'(z) - \varepsilon_{a,b}z_0^{-1}\delta\left(\frac{z - z_1}{-z_0}\right) b'(z)a(z_1)\right)$$

$$= [L(-1), Y_A(a(z), z_0)b(z)]$$

and

$$[L(0), Y_A(a(z), z_0)b(z)]$$

$$= Res_{z_1} \left(\frac{z_1 - z_0}{z} \right)^{\frac{j}{T}} \cdot$$

$$\cdot \left(z_0^{-1} \delta \left(\frac{z_1 - z_2}{z_0} \right) [L(0), a(z_1)b(z)] - \varepsilon_{a,b} z_0^{-1} \delta \left(\frac{z - z_1}{-z_0} \right) [L(0), b(z)a(z_1)] \right)$$

$$= Res_{z_1} \left(\frac{z_1 - z_0}{z} \right)^{\frac{j}{T}} \cdot B$$

where

$$B = z_0^{-1} \delta \left(\frac{z_1 - z}{z_0} \right) (a(z_1)[L(0), b(z)] + [L(0), a(z_1)]b(z))$$

$$- \varepsilon_{a,b} z_0^{-1} \delta \left(\frac{z_1 - z}{z_0} \right) (b(z)[L(0), a(z_1)] + [L(0), b(z)]a(z_1))$$

$$= z_0^{-1} \delta \left(\frac{z_1 - z}{z_0} \right) (\beta a(z_1)b(z) + z a(z_1)b'(z) + \alpha a(z_1)b(z) + z_1 a'(z_1)b(z))$$

$$- z_0^{-1} \delta \left(\frac{z_1 - z}{z_0} \right) (\alpha b(z)a(z_1) + z_1 b(z)a'(z_1) + \beta b(z)a(z_1) + z b'(z)a(z_1))$$

$$= z_0^{-1} \delta \left(\frac{z_1 - z}{z_0} \right) (\beta a(z_1)b(z) + z a(z_1)b'(z) + \alpha a(z_1)b(z) + (z_0 + z)a'(z_1)b(z))$$

$$- z_0^{-1} \delta \left(\frac{z_1 - z}{z_0} \right) (\alpha b(z)a(z_1) + (z_0 + z)b(z)a'(z_1) + \beta b(z)a(z_1) + z b'(z)a(z_1)).$$

Using Lemma 3.10 we obtain

$$[L(0), Y_A(a(z), z_0)b(z)]$$

$$= (\alpha + \beta) Y_A(a(z), z_0)b(z) + z \frac{\partial}{\partial z} (Y_A(a(z), z_0)b(z)) + z_0 \frac{\partial}{\partial z_0} (Y_A(a(z), z_0)b(z)).$$

$\square$

Proposition 3.13. *Let M be a $\mathbb{Z}_2$-graded vector space and let A be any local system of $\mathbb{Z}_T$-twisted vertex operators on M. Then $Y_A(a(z), z_1)$ and $Y_A(b(z), z_2)$ are mutually local on V for $a(z), b(z) \in A$.*

Proof. By linearity, we may assume that $a(z) \in A^j$ and $b(z) \in A^k$. Let $c(z) \in A$. Then we have:

$$Y_A(a(z), z_3) Y_A(b(z), z_0)c(z)$$

$$= Res_{z_1} \left(\frac{z_1 - z_3}{z} \right)^{\frac{j}{T}} z_3^{-1} \delta \left(\frac{z_1 - z}{z_3} \right) a(z_1)(Y_A(b(z), z_0)c(z))$$

$$- \varepsilon_{a,b} \varepsilon_{a,c} z_3^{-1} \delta \left(\frac{-z + z_1}{z_3} \right) (Y_A(b(z), z_0)c(z))a(z_1)$$

$$= Res_{z_1} Res_{z_4} \left(\frac{z_1 - z_3}{z} \right)^{\frac{j}{T}} \left(\frac{z_4 - z_0}{z} \right)^{\frac{k}{T}} \cdot P$$

where

$$
\begin{aligned}
P &= z_3^{-1}\delta\left(\frac{z_1 - z}{z_3}\right) z_0^{-1}\delta\left(\frac{z_4 - z}{z_0}\right) a(z_1)b(z_4)c(z) \\
&\quad -\varepsilon_{b,c}z_3^{-1}\delta\left(\frac{z_1 - z}{z_3}\right) z_0^{-1}\delta\left(\frac{-z + z_4}{z_0}\right) a(z_1)c(z)b(z_4) \\
&\quad -\varepsilon_{a,b}\varepsilon_{a,c}z_3^{-1}\delta\left(\frac{-z + z_1}{z_3}\right) z_0^{-1}\delta\left(\frac{z_4 - z}{z_0}\right) b(z_4)c(z)a(z_1) \\
&\quad +\varepsilon_{a,b}\varepsilon_{a,c}\varepsilon_{b,c}z_3^{-1}\delta\left(\frac{-z + z_1}{z_3}\right) z_0^{-1}\delta\left(\frac{-z + z_4}{z_0}\right) c(z)b(z_4)a(z_1).
\end{aligned}
$$

Similarly, we have

$$
\begin{aligned}
&Y_A(b(z), z_0)Y_A(a(z), z_3)c(z) \\
&= \operatorname{Res}_{z_1}\operatorname{Res}_{z_4}\left(\frac{z_1 - z_3}{z}\right)^{\frac{j}{T}}\left(\frac{z_4 - z_0}{z}\right)^{\frac{k}{T}} \cdot Q
\end{aligned}
$$

where

$$
\begin{aligned}
Q &= z_3^{-1}\delta\left(\frac{z_1 - z}{z_3}\right) z_0^{-1}\delta\left(\frac{z_4 - z}{z_0}\right) b(z_4)a(z_1)c(z) \\
&\quad -\varepsilon_{b,c}z_3^{-1}\delta\left(\frac{-z + z_1}{z_3}\right) z_0^{-1}\delta\left(\frac{z_4 - z}{z_0}\right) b(z_4)c(z)a(z_1) \\
&\quad -\varepsilon_{a,b}\varepsilon_{a,c}z_3^{-1}\delta\left(\frac{z_1 - z}{z_3}\right) z_0^{-1}\delta\left(\frac{-z + z_4}{z_0}\right) a(z_1)c(z)b(z_4) \\
&\quad +\varepsilon_{a,b}\varepsilon_{a,c}\varepsilon_{b,c}z_3^{-1}\delta\left(\frac{-z + z_1}{z_3}\right) z_0^{-1}\delta\left(\frac{-z + z_4}{z_0}\right) c(z)a(z_1)b(z_4).
\end{aligned}
$$

Let k be any positive integer such that

$$
(z_1 - z_4)^k a(z_1)b(z_4) = \varepsilon_{a,b}(z_1 - z_4)^k b(z_4)a(z_1).
$$

Since

$$
\begin{aligned}
&(z_3 - z_0)^k z_3^{-1}\delta\left(\frac{z_1 - z_2}{z_3}\right) z_0^{-1}\delta\left(\frac{z_4 - z_2}{z_0}\right) \\
&= (z_1 - z_4)^k z_3^{-1}\delta\left(\frac{z_1 - z_2}{z_3}\right) z_0^{-1}\delta\left(\frac{z_4 - z_2}{z_0}\right),
\end{aligned} \tag{3.21}
$$

it is clear that locality of $a(z)$ with $b(z)$ implies the locality of $Y_A(a(z), z_1)$ with $Y_A(b(z), z_2)$. $\quad\square$

Theorem 3.14. *Let M be a $\mathbb{Z}_2$-graded vector space and let A be any local system of $\mathbb{Z}_T$-twisted vertex operators on M. Then V is a vertex superalgebra with an automorphism σ of order T and M is a natural σ-twisted A-module in the sense $Y_M(a(z), z_1) = a(z_1)$ for $a(z) \in A$.*

Proof. It follows from Proposition 3.9, Lemmas 3.10 and 3.11, and Proposition 2.5 that A is a vertex superalgebra. It is clear that σ is an automorphism of A. If we define $Y_M(a(z), z_0) = a(z_0)$ for $a(z) \in A$, the twisted vertex operator multiplication formula (3.6) becomes the twisted iterate formula for $Y_M(\cdot, z_0)$. Since the supercommutativity (3.5) and iterate formula imply the twisted Jacobi identity (Lemma 2.10), $Y_M(\cdot, z_0)$ satisfies the twisted Jacobi identity. Therefore, (M, Y_M) is a σ-twisted A-module. $\square$

Let M be any $\mathbb{Z}_2$-graded vector space and let S be a set of $\mathbb{Z}_2$-homogeneous mutually local $\mathbb{Z}_T$-twisted vertex operators on M. It follows from Zorn's lemma that there exists a local system A of $\mathbb{Z}_T$-twisted vertex operators on M containing S. Let $\langle S \rangle$ be the vertex superalgebra generated by S inside A (which is a vertex superalgebra). Since the "multiplication" (3.6) is canonical (which is independent of the choice of the local system A), the vertex superalgebra $\langle S \rangle$ is canonical, so that we may speak about the vertex superalgebra generated by S without any confusion. Summarizing the previous argument, we have:

Corollary 3.15. *Let M be any $\mathbb{Z}_2$-graded vector space. Then any set of mutually local $\mathbb{Z}_T$-twisted vertex operators on M generates a (canonical) vertex superalgebra with an automorphism σ of order T such that M is a σ-twisted module for this vertex superalgebra.*

Proposition 3.16. *[Li] Let M be an object of the category C_ℓ and let V be a local system of $\mathbb{Z}_T$-twisted vertex operators on $(M, L(-1))$, which contains $L(z)$. Then the vertex operator $L(z)$ is a Virasoro element of the vertex superalgebra V.*

Proof. The proof is the same as the proof for untwisted case in [Li]. $\square$

Proposition 3.17. *Let V be a vertex superalgebra with an automorphism σ of order T. Then giving a σ-twisted V-module (M, d) is equivalent to giving a vertex superalgebra homomorphism from V to some local system of $\mathbb{Z}_T$-twisted vertex operators on (M, d).*

Proof. Let (M, d, Y_M) be a σ-twisted V-module. Then $\bar{V} = \{Y_M(a, z) | a \in V\}$ is a set of mutually local $\mathbb{Z}_T$-twisted vertex operators on (M, d). By Zorn's lemma, there exist local systems which contain $\bar{V}$. Let A be any one of those. Then by definition $Y_M(\cdot, z)$ is a linear map from V to A. For homogeneous $a \in V^j, b \in V$ we have:

$$Y_M(\cdot, z)(Y_V(a, z_0)b)$$
$$= Y_M(Y_V(a, z_0)b, z)$$
$$= Res_{z_1} \left(\frac{z_1 - z_0}{z_2} \right)^{\frac{j}{T}} \cdot$$
$$\left(z_0^{-1} \delta \left(\frac{z_1 - z}{z_0} \right) Y_M(a, z_1) Y_M(b, z) - \varepsilon_{a,b} z_0^{-1} \delta \left(\frac{z - z_1}{-z_0} \right) Y_M(b, z) Y_M(a, z_1) \right)$$
$$= Y_A(Y_M(a, z), z_0) Y_M(b, z)$$
$$= Y_A(Y_M(\cdot, z)(a), z_0) Y_M(\cdot, z)(b). \tag{3.22}$$

By definition we have:

$$Y_M(\cdot, z)(1_V) = Y_M(1, z) = Id_M = I(z). \tag{3.23}$$

Thus $Y_M(\cdot, z)$ is a vertex superalgebra homomorphism from V to A. Conversely, let A be a local system of $\mathbb{Z}_T$-twisted vertex operators on M and let $Y_M(\cdot, z)$ be a vertex superalgebra homomorphism from V to A. Then

$$Y_M(1, z) = Y_M(\cdot, z)(1) = I(z) = Id_M. \tag{3.24}$$

Since A is a local system, for any $a, b \in V$, $Y_M(a, z)$ and $Y_M(b, z)$ satisfy the supercommutativity (2.23). By reversing (3.22) we obtain the iterate formula for $Y_M(\cdot, z)$. Since the twisted Jacobi identity follows from the supercommutativity and the iterate formula, $Y_M(\cdot, z)$ satisfies the twisted Jacobi identity. Therefore, (M, Y_M) is a σ-twisted V-module. $\quad \square$

4. Twisted modules for vertex operator superalgebras associated to some infinite-dimensional Lie superalgebras

In this section, we shall use the machinery we built in Section 3 to study twisted modules for vertex operator (super)algebras associated to the Neveu-Schwarz algebra and an affine Lie superalgebra.

4.1. Twisted modules for the Neveu-Schwarz vertex operator superalgebra. Let us first recall the Neveu-Schwarz algebra (cf. [FFR], [KW], [T]). The Neveu-Schwarz algebra is the Lie superalgebra

$$NS = \oplus_{m \in \mathbb{Z}} CL(m) \oplus \oplus_{n \in \mathbb{Z}} CG(n + \frac{1}{2}) \oplus \mathbb{C}c \tag{4.1}$$

with the following commutation relations:

$$[L(m), L(n)] = (m - n)L(m + n) + \frac{m^3 - m}{12} \delta_{m+n,0} c, \tag{4.2}$$

$$[L(m), G(n + \frac{1}{2})] = \left(\frac{m}{2} - n - \frac{1}{2}\right) G(m + n + \frac{1}{2}), \tag{4.3}$$

$$[G(m + \frac{1}{2}), G(n - \frac{1}{2})]_+ = 2L(m + n) + \frac{1}{3} m(m + 1) \delta_{m+n,0} c, \tag{4.4}$$

$$[L(m), c] = 0, \quad [G(n + \frac{1}{2}), c] = 0. \tag{4.5}$$

Or, equivalently

$$[L(z_1), L(z_2)] = z_1^{-1} \delta\left(\frac{z_2}{z_1}\right) L'(z_2) + 2z_1^{-2} \delta'\left(\frac{z_2}{z_1}\right) L(z_2) + \frac{\ell}{12} z_1^{-4} \delta^{(3)}\left(\frac{z_2}{z_1}\right), \tag{4.6}$$

$$[L(z_1), G(z_2)] = z_1^{-1} \delta\left(\frac{z_2}{z_1}\right) \frac{\partial}{\partial z_2} G(z_2) + \frac{3}{2}\left(\frac{\partial}{\partial z_2} z_1^{-1} \delta\left(\frac{z_2}{z_1}\right)\right) G(z_2), \tag{4.7}$$

and

$$[G(z_1), G(z_2)]_+ = 2z_1^{-1} \delta\left(\frac{z_2}{z_1}\right) L(z_2) + \frac{1}{3} c \left(\frac{\partial}{\partial z_2}\right)^2 z_1^{-1} \delta\left(\frac{z_2}{z_1}\right) \tag{4.8}$$

where $L(z) = \sum_{m \in \mathbb{Z}} L(m) z^{-m-2}$ and $G(z) = \sum_{n \in \mathbb{Z}} G(n + \frac{1}{2}) z^{-n-2}$.

By definition, $NS^{(0)} = \oplus_{m \in \mathbb{Z}} \mathbb{C}L(m) \oplus \mathbb{C}c$ and $NS^{(1)} = \oplus_{n \in \mathbb{Z}} \mathbb{C}G(n + \frac{1}{2})$. $NS = \oplus_{n \in \frac{1}{2}\mathbb{Z}} NS_n$ is a $\frac{1}{2}\mathbb{Z}$-graded Lie superalgebra by defining

$$\deg L(m) = m, \ \deg c = 0, \ \deg G(n) = n \quad \text{for } m \in \mathbb{Z}, n \in \frac{1}{2} + \mathbb{Z}. \tag{4.9}$$

We have a triangular decomposition $NS = NS_+ \oplus NS_0 \oplus NS_+$ where

$$NS_\pm = \sum_{n=1}^{\infty} (\mathbb{C}L(\pm n) + \mathbb{C}G(\pm n \mp \frac{1}{2})), \ NS_0 = \mathbb{C}L(0) \oplus \mathbb{C}c. \tag{4.10}$$

Let $\bar{M}(c,0) = M(c,0)/\langle G(-\frac{1}{2})1 \rangle$, where $M(c,0)$ is the Verma module over NS of central charge c and $\langle G(-\frac{1}{2})1 \rangle$ is the submodule generated by $G(-\frac{1}{2})1$. It has been proved ([KW], [Li]) that $\bar{M}(c,0)$ has a natural vertex operator superalgebra structure and any restricted NS-module of charge c is a weak $\bar{M}(c,0)$-module.

Let V be any vertex superalgebra. Then the linear map $\sigma \colon V \to V; \sigma(a + b) = a - b$ for $a \in V^{(0)}, b \in V^{(1)}$, is an automorphism of V, which was called the *canonical automorphism* [FFR]. It is easy to see that $Aut\bar{M}(c,0) = \mathbb{Z}_2 = \langle \sigma \rangle$.

The Ramond algebra R [GSW] is the Lie superalgebra with a basis $L(m), F(n), c$ for $m, n \in \mathbb{Z}$ and with the following defining relations:

$$[L(m), L(n)] = (m - n)L(m + n) + \frac{m^3 - m}{12}\delta_{m+n,0}c, \tag{4.11}$$

$$[L(m), F(n)] = (\frac{m}{2} - n)F(m + n), \tag{4.12}$$

$$[F(m), F(n)]_+ = 2L(m + n) + \frac{1}{3}(m^2 - \frac{1}{4})\delta_{m+n,0}c, \tag{4.13}$$

$$[L(m), c] = [F(m), c] = 0 \tag{4.14}$$

for $m, n \in \mathbb{Z}$. Set

$$L(z) = \sum_{m \in \mathbb{Z}} L(m)z^{-m-2}, \ F(z) = \sum_{n \in \mathbb{Z}} F(n)z^{-n-\frac{3}{2}}. \tag{4.15}$$

Then the defining relations (4.12) and (4.13) are equivalent to the following identities:

$$[L(z_1), F(z_2)] = z_1^{-1}\delta\left(\frac{z_2}{z_1}\right)F(z_2) + \frac{3}{2}\left(\frac{\partial}{\partial z_2}z_1^{-1}\delta\left(\frac{z_2}{z_1}\right)\left(\frac{z_2}{z_1}\right)^{\frac{1}{2}}\right)F(z_2), \tag{4.16}$$

$$[F(z_1), F(z_2)]_+ = 2\delta\left(\frac{z_2}{z_1}\right)\left(\frac{z_2}{z_1}\right)^{\frac{1}{2}}L(z_1) + \frac{2}{3}\left(\frac{\partial}{\partial z_2}\right)^2\left(\delta\left(\frac{z_2}{z_1}\right)\left(\frac{z_2}{z_1}\right)^{\frac{1}{2}}\right). \tag{4.17}$$

Let M be any restricted module for the Ramond algebra or the twisted Neveu-Schwarz algebra $\mathbb{R}$ with central charge c. Then $\{L(z), F(z), I(z)\}$ is a set of mutually local homogeneous $\mathbb{Z}_2$-twisted vertex operators on M. Let V be the vertex superalgebra generated by $\{L(z), F(z), I(z)\}$. Then M is a faithful σ-twisted V-module. By Lemma 2.12, $Y_V(L(z), z_1)$ and $Y_V(F(z), z_1)$ satisfy the Neveu-Schwarz relations (4.6) and (4.7). Then V is a lowest weight module for NS under the linear map: $L(m) \mapsto L(z)_{m+1}, G(n + \frac{1}{2}) \mapsto F(z)_n, c \mapsto c$. Then V is a vertex operator superalgebra, which is a quotient algebra of $\bar{M}(c,0)$. Consequently, M is a weak σ-twisted module for $\bar{M}(c,0)$. Therefore, we have proved:

Proposition 4.1. *Let c be any complex number. Then any restricted module for the Ramond algebra R of central charge c is a weak σ-twisted $\bar{M}(c, 0)$-module.*

4.2. Twisted modules for vertex operator superalgebras associated to affine Lie superalgebras.

Let (g, B) be a pair consisting of a finite-dimensional Lie superalgebra $g = g^{(0)} \oplus g^{(1)}$ such that $[g^{(1)}, g^{(1)}]_+ = 0$ and a nondegenerate symmetric bilinear form B such that

$$B(g^{(0)}, g^{(1)}) = 0, \ B([a, u], v) = -B(u, [a, v]) \quad \text{for } a \in g^{(0)}, u, v \in g. \tag{4.18}$$

This amounts to having a finite-dimensional Lie algebra $g^{(0)}$ with a nondegenerate symmetric invariant bilinear form $B_0(\cdot, \cdot)$ and a finite-dimensional $g^{(0)}$-module $g^{(1)}$ with a nondegenerate symmetric bilinear form $B_1(\cdot, \cdot)$ such that

$$B_1(au, v) = -B_1(u, av) \quad \text{for any } a \in g^{(0)}, u, v \in g^{(1)}. \tag{4.19}$$

The affine Lie superalgebra $\tilde{g}$ is defined to be $\mathbb{C}[t, t^{-1}] \otimes g \oplus \mathbb{C}c$ with the following defining relations:

$$[a_m, b_n] = [a, b]_{m+n} + m\delta_{m+n,0}B(a, b)c, \tag{4.20}$$

$$[a_m, u_n] = -[u_n, a_m] = (au)_{m+n}, \tag{4.21}$$

$$[u_m, v_n]_+ = \delta_{m+n+1,0}B(u, v)c, \tag{4.22}$$

$$[c, \tilde{g}] = 0 \tag{4.23}$$

for $a, b \in g^{(0)}, u, v \in g^{(1)}$ and $m, n \in \mathbb{Z}$, where x_m stands for $t^m \otimes x$. Or, equivalently

$$[a(z_1), b(z_2)] = z_1^{-1}\delta\left(\frac{z_2}{z_1}\right)[a, b](z_2) + z_1^{-2}\delta'\left(\frac{z_2}{z_1}\right)B(a, b)c, \tag{4.24}$$

$$[a(z_1), u(z_2)] = z_1^{-1}\delta\left(\frac{z_2}{z_1}\right)(au)(z_2), \tag{4.25}$$

$$[u(z_1), v(z_2)]_+ = B(u, v)z_1^{-1}\delta\left(\frac{z_2}{z_1}\right)c, \tag{4.26}$$

$$[x(z), c] = 0 \tag{4.27}$$

for any $a, b \in g^{(0)}, u, v \in g^{(1)}, x \in g, m, n \in \mathbb{Z}$, where $x(z) = \sum_{n \in \mathbb{Z}} x_n z^{-n-1}$ for $x \in g$. The even subspace and the odd subspace are:

$$\tilde{g}^{(0)} = \mathbb{C}[t, t^{-1}] \otimes g^{(0)} \oplus \mathbb{C}, \ \tilde{g}^{(1)} = C[t, t^{-1}] \otimes g^{(1)}. \tag{4.28}$$

Furthermore, $\tilde{g}$ is a $\frac{1}{2}\mathbb{Z}$-graded Lie superalgebra with the degree defined as follows:

$$\deg a_m = -m, \ \deg u_n = -n + \frac{1}{2}, \ \deg c = 0 \tag{4.29}$$

for any $a \in g^{(0)}, u \in g^{(1)}, m, n \in \mathbb{Z}$. We have a triangular decomposition $\tilde{g} = N_+ \oplus N_0 \oplus N_-$, where

$$N_- = t\mathbb{C}[t] \otimes g, \ N_+ = t^{-1}\mathbb{C}[t^{-1}] \otimes g^{(0)} \oplus \mathbb{C}[t^{-1}] \otimes g^{(1)}, \ N_0 = g^{(0)} \oplus \mathbb{C}c. \tag{4.30}$$

Let $P = N_- + N_0$ be a parabolic subalgebra. For any $g^{(0)}$-module U and any complex number ℓ, denote by $M_{(g,B)}(\ell, U)$ the generalized Verma module or Weyl module for $\tilde{g}$ with c acting as scalar ℓ. Namely, $M_{(g,B)}(\ell, U) = U(\tilde{g}) \otimes_{U(P)} U$.

It has been proved ([KW],[Li]) that for any complex number ℓ, $M_{(g,B)}(\ell, \mathbb{C})$ has a natural vertex superalgebra structure and that any restricted $\tilde{g}$-module M of level ℓ is a module for the vertex superalgebra $M_{(g,B)}(\ell, \mathbb{C})$. Furthermore, if $g^{(0)}$ is simple and ℓ is not zero, then $M_{(g,B)}(\ell, \mathbb{C})$ is a vertex operator superalgebra. For any $a \in g$, we may identify a with $a_{-1}1 \in M_{(g,B)}(\ell, \mathbb{C})$, so that we may consider g as a subspace of $M_{(g,B)}(\ell, \mathbb{C})$.

An automorphism of (g, B) is an automorphism σ of the Lie superalgebra g, which preserves the even and the odd subspaces and the bilinear form B. Let σ be an automorphism of (g, B) of order T. Then we have:

$$g = g_0 \oplus g_1 \oplus \cdots \oplus g_{T-1} \quad \text{where } g_j = \{u \in g | \sigma u = \varepsilon^j u\}. \tag{4.31}$$

The twisted Lie superalgebra $\tilde{g}[\sigma]$ is defined to be $\tilde{g}[\sigma] = \oplus_{j=0}^{T-1} g_j \otimes t^{\frac{j}{T}} \mathbb{C}[t, t^{-1}] \oplus \mathbb{C}c$ with the following defining relations:

$$[t^{m+\frac{j}{T}} \otimes a, t^{n+\frac{k}{T}} \otimes b] = t^{m+n+\frac{j+k}{T}} \otimes [a, b] + B(a, b)\left(m + \frac{j}{T}\right)\delta_{m+n+\frac{j+k}{T},0}c, \tag{4.32}$$

$$[t^{m+\frac{j}{T}} \otimes u, t^{n+\frac{k}{T}} \otimes v] = B(a, b)\delta_{m+n+1+\frac{j+k}{T},0}c, \tag{4.33}$$

$$[c, \tilde{g}] = 0 \tag{4.34}$$

for $a \in (g^{(0)})_j, b \in g_k, u \in (g^{(1)})_j, v \in (g^{(1)})_k$. For $a \in g_j$, set

$$a(z) = \sum_{n \in \mathbb{Z}}(t^{n+\frac{j}{T}} \otimes a)z^{-n-1-\frac{j}{T}} \in (EndM)[[z^{\frac{1}{T}}, z^{-\frac{1}{T}}]]. \tag{4.35}$$

Then for $a \in (g^{(0)})_j, b \in g_k, u \in (g^{(1)})_j, v \in g^{(1)}$, we have:

$$\begin{aligned}
&[a(z_1), b(z_2)] \\
=& \sum_{m,n \in \mathbb{Z}} t^{m+n+\frac{j}{T}+\frac{k}{T}} \otimes [a, b]z_1^{-m-1-\frac{j}{T}}z_2^{-n-1-\frac{k}{T}} \\
&+ c(m + \frac{j}{T})B(a, b)\delta_{m,-n}\delta_{j,-k}z_1^{-m-1-\frac{j}{T}}z_2^{-n-1-\frac{k}{T}} \\
=& z_1^{-1}\left(\frac{z_2}{z_1}\right)^{\frac{j}{T}}\delta\left(\frac{z_2}{z_1}\right)[a, b](z_2) + B(a, b)\frac{\partial}{\partial z_2}\left(z_1^{-1}\delta\left(\frac{z_2}{z_1}\right)\left(\frac{z_2}{z_1}\right)^{\frac{j}{T}}\right)c, \tag{4.36}
\end{aligned}$$

$$[u(z_1), v(z_2)]_+ = B(u, v)z_1^{-1}\delta\left(\frac{z_2}{z_1}\right)\left(\frac{z_2}{z_1}\right)^{\frac{j}{T}}c. \tag{4.37}$$

For any $\tilde{g}$-module M, we may consider $a(z)$ as an element of $(EndM)[[z^{\frac{1}{T}}, z^{-\frac{1}{T}}]]$ for $a \in g$. A $\tilde{g}[\sigma]$-module M is said to be *restricted* if for any $u \in M$, $\left(t^{n+\frac{j}{T}} \otimes g_j\right)u = 0$ for n sufficiently large. Then a $\tilde{g}[\sigma]$-module M is restricted if and only if every $a(z)$ for $a \in g$ is a weak $\mathbb{Z}_T$-twisted vertex operators on M. Let M be a restricted $\tilde{g}[\sigma]$-module of level ℓ. It follows from (4.36), (4.37) and Lemma 2.2 that $\{a(z)|a \in g\}$ is a local subspace of $F(M, T)$. Let V be the vertex superalgebra generated by $\{a(z)|a \in g\}$ from the identity operator $I(z)$. Then M is a faithful σ-twisted V module. It follows from (4.36), (4.37) and Lemma 2.13 that V is a $\tilde{g}$-module where a_m for $a \in g, m \in \mathbb{Z}$ is represented by $a(z)_m$. It is easy to see that V is a lowest weight $\tilde{g}$-module with a lowest weight vector $I(z)$, so that V is a quotient vertex superalgebra of $M_{(g,B)}(\ell, \mathbb{C})$. Through the natural vertex superalgebra homomorphism, M becomes a σ-twisted $M_{(g,B)}(\ell, \mathbb{C})$-module. Thus we have proved:

Theorem 4.2. *Let (g, B) be a pair given as before, let σ be any automorphism of order T of $M_{(g,B)}(\ell, \mathbb{C})$ or (g, B) and let ℓ be any complex number. Then any restricted $\tilde{g}[\sigma]$-module of level ℓ is a σ-twisted $M_{(g,B)}(\ell, \mathbb{C})$-module.*

Next we shall consider two special cases.

Case 1. Let $g^{(0)} = 0$ and let $g^{(1)} = A$ be an n-dimensional vector space with a nondegenerate symmetric bilinear form B. It has been proved ([KW], [Li]) that if $\ell \neq 0$, $M_{(g_1,B)}(\ell, C)$ ($= \mathbb{F}^n$ [KW], [Li], or $= CM(\mathbb{Z} + \frac{1}{2})$ [DM1], [FFR]) is a rational vertex operator superalgebra which has only one irreducible module up to equivalence.

Let σ be an automorphism of (A, B) of order T. Then

$$A = A_0 \oplus A_2 \oplus \cdots \oplus A_{T-1} \tag{4.38}$$

where $A_j = \{u \in A | \sigma(u) = \varepsilon^j u\}$. Set

$$\tilde{A}[\sigma] = \oplus_{j=0}^{T-1} t^{\frac{j}{T}} C[t, t^{-1}] \otimes A_j. \tag{4.39}$$

We define a symmetric bilinear form on $\tilde{A}[\sigma]$ as follows:

$$\langle t^{m+\frac{j}{T}} \otimes a, t^{n+\frac{k}{T}} \otimes b \rangle = B(a, b) \delta_{m+n+1+\frac{j+k}{T}, 0} \tag{4.40}$$

for $a \in A_j, b \in A_k, m, n \in \mathbb{Z}$. Let $C(\tilde{A}[\sigma])$ be the Clifford algebra of $\tilde{A}[\sigma]$ with respect to this defined symmetric bilinear form.

Let M be any restricted $C(\tilde{A}[\sigma])$-module, i.e., for any $u \in M$, $(t^{n+\frac{j}{T}} \otimes A_j)u = 0$ for sufficiently large n. Then $\{u(z) | u \in A\}$ is a local subspace of $F(M, T)$. Let V be the vertex superalgebra generated by $\{u(z) | u \in A\}$. Then M is a faithful σ-twisted V-module. By repeating our technique used in the first subsection, we see that V is a module for $C(\tilde{A})$. Since V contains $I(z)$, V is a lowest weight $C(\tilde{A})$-module so that $V = \mathbb{F}^n$. Thus M is a σ-twisted F^n-module.

Next, we follow [DM1] or [FFR] to find the irreducible $C(\tilde{A}[\sigma])$-modules. For $u \in A_j, n \in \mathbb{Z}$, we define

$$\deg(t^{n+\frac{j}{T}} \otimes u) = -n - \frac{1}{2} - \frac{j}{T}. \tag{4.41}$$

Then $\tilde{A}[\sigma]$ becomes a $\frac{1}{T}\mathbb{Z}$-graded (resp. $\frac{1}{2T}\mathbb{Z}$-graded) space if T is even (resp. odd). Let $\tilde{A}[\sigma]_\pm$ be the sum of homogeneous subspaces of positive or negative degrees and let $\tilde{A}[\sigma]_0$ be the degree-zero homogeneous subspace. Then $\tilde{A}[\sigma]_0 = 0$ if T is odd, $\tilde{A}[\sigma]_0 = t^{-\frac{1}{2}} \otimes A_{\frac{T}{2}}$ if T is even. Then $\tilde{A}[\sigma]_\pm$ are isotropic subspaces. Let $\Lambda(\tilde{A}[\sigma]_+)$ be the exterior algebra of $\tilde{A}[\sigma]_+$. If T is odd, there is no nonzero element of degree-zero in the twisted Clifford algebra. Then it follows from the Stone-Von Neumann theorem that $\Lambda(\tilde{A}[\sigma]_+)$ is the unique irreducible $C(\tilde{A}[\sigma])$-module and that if M is a $C(\tilde{A}[\sigma])$-module such that for any $u \in M$, there is a positive integer N such that $\tilde{A}[\sigma]_-^N u = 0$, then M is a direct sum of some copies of $\Lambda(\tilde{A}[\sigma]_+)$.

If $T = 2k$ is even, the situation is a little bit complicated. If $\dim A_k = 2m$ is even, we can decompose $A_k = (A_k)_+ \oplus (A_k)_-$ where $(A_k)_\pm$ are isotropic subspaces of dimension m. Then it follows from the Stone-Von Neumann theorem that $\Lambda(\tilde{A}[\sigma]_+ \oplus (A_k)_+)$ is the unique irreducible $C(\tilde{A}[\sigma])$-module up to equivalence, so that $\Lambda(\tilde{A}[\sigma]_+ \oplus (A_k)_+)$ is the unique irreducible σ-twisted $\mathbb{F}^n$-module up to equivalence.

If $\dim A_k = 2m + 1$ $(m \in \mathbb{Z}_+)$ is odd, let $e \in A_k$ such that $\langle e, e \rangle = 2$. Then we have:

$$A_k = Ce \oplus (A_k)_+ \oplus (A_k)_- \qquad (4.42)$$

where $(A_k)_\pm$ are isotropic subspaces of dimension m such that $\langle e, (A_k)_\pm \rangle = 0$. Let

$$\Lambda(\tilde{A}[\sigma]_+ \oplus (A_k)_+) = \Lambda^{even}(\tilde{A}[\sigma]_+ \oplus (A_k)_+) \oplus \Lambda^{odd}(\tilde{A}[\sigma]_+ \oplus (A_k)_+). \qquad (4.43)$$

Let e act as $(\pm 1, \mp 1)$. Then we obtain two irreducible $C(\tilde{A}[\sigma])$-modules, so that we get two irreducible σ-twisted $\mathbb{F}^n$-modules. It is easy to prove that these are the only irreducible modules, up to equivalence. Summarizing all the arguments, we have:

Proposition 4.3. *Any restricted $C(\tilde{A}[\sigma])$-module M is a weak σ-twisted $\mathbb{F}^n$-module. Furthermore, if T is odd, $\Lambda(\tilde{A}[\sigma]_+)$ is the unique irreducible σ-twisted module for the vertex operator algebra $\mathbb{F}^n$; if T is even and $\dim A_{\frac{T}{2}}$ is even, $\Lambda(\tilde{A}[\sigma]_+ \oplus (A_k)_+)$ is the unique irreducible σ-twisted $\mathbb{F}^n$-module up to equivalence; if T is even and $\dim A_{\frac{T}{2}}$ is odd, $\mathbb{F}^n$ has two irreducible σ-twisted modules.*

Proposition 4.4. *Let M be any σ-twisted weak $\mathbb{F}^n$-module such that for any $u \in M$, there is a positive integer N such that $\tilde{A}[\sigma]^N_- u = 0$. Then M is a direct sum of irreducible σ-twisted $\mathbb{F}^n$-modules $\Lambda(\tilde{A}[\sigma]_+)$.*

Case 2. Let g be a finite-dimensional simple Lie algebra with a fixed Cartan subalgebra h. Let Δ be the set of all roots, let Π be a set of simple roots, and let θ be the highest root. Let $\langle \cdot, \cdot \rangle$ be the normalized Killing form such that $\langle \theta, \theta \rangle = 2$. Then the Killing form is $2\Omega \langle \cdot, \cdot \rangle$, where Ω is the dual Coxeter number of g. Let σ be an automorphism of g such that $\sigma^T = Id$. Then σ preserves the invariant bilinear form. Denote by $M_{(g,\sigma)}(\ell, \lambda)$ the Verma module for $\tilde{g}[\sigma]$ with lowest weight λ of level ℓ. Denote the irreducible quotient module by $L_{(g,\sigma)}(\ell, \lambda)$.

Corollary 4.5. *Let σ be any automorphism of order T of $M_g(\ell, \mathbb{C})$ or g and let ℓ be any complex number such that $\ell \neq -\Omega$. Then any restricted $\tilde{g}[\sigma]$-module of level ℓ is a weak σ-twisted $M_g(\ell, \mathbb{C})$-module. In particular, each irreducible $\tilde{g}[\sigma]$-module $L_{(g,\mu)}(\ell, \lambda)$ is a σ-twisted $M_g(\ell, \mathbb{C})$-module.*

In the next section we will study the σ-twisted module theory for the vertex operator algebra $L_g(\ell, 0)$.

5. THE SEMISIMPLE TWISTED MODULE THEORY FOR $L_g(\ell, 0)$

Let g be a finite-dimensional simple Lie algebra with a fixed Cartan subalgebra h and let Δ be the set of roots. Fix a set Π of simple roots and let θ be the highest root. Let $\langle \cdot, \cdot \rangle$ be the normalized Killing form such that $\langle \theta, \theta \rangle = 2$.

Let σ be an automorphism of g such that $\sigma^T = 1$. Then we may consider σ as an automorphism of the vertex operator algebra $L_g(\ell, 0)$ for any $\ell \neq -\Omega$. In this section, we first prove that there is a Dynkin diagram automorphism μ of g such that the equivalence classes of σ-twisted $L_g(\ell, 0)$-modules one-to-one correspond to the equivalence classes of μ-twisted $L_g(\ell, 0)$-modules. Then we prove that if ℓ is a positive integer, then any lower truncated $\frac{1}{T}\mathbb{Z}$-graded weak μ-twisted $L_g(\ell, 0)$ module is completely reducible and that the set of equivalence classes of irreducible

μ-twisted $L_g(\ell, 0)$-modules is just the set of equivalence classes of standard $\tilde{g}[\mu]$-modules of level ℓ.

Throughout this section, ℓ will be fixed a positive integer.

Recall the following proposition from [DL], [Li] or [MP] about untwisted representation theory.

Proposition 5.1. *Let e be any root vector of g with root α. If $L_g(\ell, \lambda)$ is an integrable $\tilde{g}$-module, then $e(z)^{t\ell+1} = 0$ acting on $L_g(\ell, \lambda)$ where $t = 1$ if α is a long root, $t = 2$ if α is a short root and $g \neq g_2$, $t = 3$ if α is a short root and $g = g_2$.*

The following proposition was proved in [DL], [FZ] and [Li].

Proposition 5.2. *Let $M = \oplus_{n \in \mathbb{Z}} M(n)$ be any lower truncated $\mathbb{Z}$-graded weak module for $L_g(\ell, 0)$ as a vertex operator algebra. Then M is a direct sum of standard $\tilde{g}$-modules of level ℓ.*

Recall the following theorem of Kac from [K]:

Proposition 5.3. *Let g be a finite-dimensional simple Lie algebra, let h be a Cartan subalgebra and let $\Pi = \{\alpha_1, \cdots, \alpha_n\}$ be a set of simple roots. Let $\sigma \in Autg$ be such that $\sigma^T = 1$. Then σ is conjugate to an automorphism of g in the form*

$$\mu exp\left(ad\left(\frac{2\pi\sqrt{-1}}{T}h\right)\right), \quad h \in h_{\bar{0}}, \tag{5.1}$$

where μ is a diagram automorphism preserving h and Π, $h_{\bar{0}}$ is the fixed-point set of μ in h, and $\langle \alpha_i, h \rangle \in \mathbb{Z}$ $(i = 1, \cdots, n)$.

If σ_1 and σ_2 are conjugate automorphisms of a vertex operator algebra V, then it is clear that the equivalence classes of σ_1-twisted V-modules one-to-one correspond to the equivalence classes of σ_2-twisted V-modules (see also [DM1]). By Proposition 5.3, we only need to study the twisted module theory for the automorphisms in form (5.1). Let $G(g)$ be the subgroup of $Autg$ generated by all Dynkin diagram automorphisms of g. Then $G(g) = \mathbb{Z}_k$ where $k = 1$ for B_n, C_n, G_n, $k = 2$ for A_n, D_n, E_n $(n \neq 4)$; $k = 3$ for D_4 [K].

Let V be a vertex (operator) superalgebra. A *derivation* ([B], [Lia]) of V is an endomorphism f of V such that

$$f(Y(u, z)v) = Y(f(u), z)v + Y(u, z)f(v) \quad \text{for any } u, v \in V. \tag{5.2}$$

Let f be a derivation of V. Then

$$e^{z_0 f} Y(u, z)v = Y(e^{z_0 f}u, z)e^{z_0 f}v \quad \text{for any } u, v \in V. \tag{5.3}$$

If a derivation f of V is locally finite on V, i.e., for any $u \in M$, $\{f^j u | j \in \mathbb{Z}_+\}$ linearly spans a finite-dimensional subspace of M, then e^f is an automorphism of V.

Let a be an even element of V. Then a_0 (the zero mode of the vertex operator $Y(a, z)$) is a derivation of V. Let V be a vertex operator superalgebra and let a be an even element of V with weight $k \leq 1$. Since $wta_0 = k - 1 \leq 0$, by the two grading-restriction assumptions on V, a_0 is a locally finite derivation on V. Then

e^{a_0} is an automorphism of V except that e^{a_0} may change the Virasoro element ω. Furthermore, let a be a primary element. Then

$$a_0\omega = Res_z Y(a,z)\omega = Res_z e^{zL(-1)}Y(\omega,-z)a = (k-1)L(-1)a. \qquad (5.4)$$

Therefore, if a is an even weight-one primary element of a vertex operator superalgebra V, then $e^{a_0}\omega = \omega$, so that e^{a_0} is an automorphism of V. Furthermore, suppose that a_0 is semisimple on V. Then e^{a_0} is of finite-order if and only if there is a positive integer T such that $Spec\,(a_0) \subseteq \dfrac{2\pi i}{T}\mathbb{Z}$.

Let V be a vertex operator superalgebra, let σ be an automorphism of order S of V and let $h \in V$ be an even element such that

$$L(n)h = \delta_{n,0}h,\ \sigma(h) = h,\ [h_m, h_n] = 0 \text{ for } m, n \in \mathbb{Z}_+. \qquad (5.5)$$

Furthermore, we assume that $h(0)$ is semisimple on V such that $Spec\,h(0) \subseteq \frac{1}{T}\mathbb{Z}$ for some positive integer T. Then $\sigma_h = \exp(2\pi\sqrt{-1}h(0))$ is an automorphism of V such that $\sigma_h^T = 1$. Since $\sigma(h) = h$, we get $\sigma h(0) = h(0)\sigma$. Let (M, Y_M) be a σ-twisted V-module. For any $a \in V$, we define

$$\bar{Y}_M(a,z) = Y_M\left(z^{h(0)}\exp\left(\sum_{n=1}^\infty \frac{h(n)}{-n}(-z)^{-n}\right)a, z\right). \qquad (5.6)$$

Proposition 5.4. $(M, \bar{Y}_M(\cdot, z))$ is a weak $(\sigma\sigma_h)$-twisted V-module.

Proof. For any $a \in V$, it follows from the grading-restriction assumptions on V that $\Delta(z)a$ is a finite sum of rational powers of z, so that $\bar{Y}_M(a, z)$ satisfies the truncation condition. Next, we check the $L(-1)$-derivative property. Set

$$\Delta(z) = z^{h(0)}\exp\left(\sum_{n=1}^\infty \frac{h(n)}{-n}(-z)^{-n}\right). \qquad (5.7)$$

Notice that $\Delta(z)$ is invertible. By the definition of $\Delta(z)$, we get

$$\frac{d}{dz}\Delta(z) = \left(-\sum_{n=0}^\infty h(n)(-z)^{-n-1}\right)\Delta(z). \qquad (5.8)$$

Since $[L(-1), h(0)] = 0$ and

$$[L(-1), \sum_{n=1}^\infty \frac{h(n)}{-n}(-z)^{-n}] = \sum_{n=1}^\infty h(n-1)(-z)^{-n} = \sum_{n=0}^\infty h(n)(-z)^{-n-1}, \qquad (5.9)$$

we get

$$[L(-1), \Delta(z)] = \left(\sum_{n=1}^\infty h(n-1)(-z)^{-n}\right)\Delta(z) = \left(\sum_{n=0}^\infty h(n)(-z)^{-n-1}\right)\Delta(z). \qquad (5.10)$$

Therefore

$$
\begin{aligned}
\frac{d}{dz}\bar{Y}_M(a,z) &= \frac{d}{dz}Y_M(\Delta(z)a,z) \\
&= Y_M\left(\frac{d}{dz}\Delta(z)a,z\right) + \frac{\partial}{\partial z_0}Y_M(\Delta(z)a,z_0)|_{z_0=z} \\
&= Y_M\left(\frac{d}{dz}\Delta(z)a,z\right) + Y_M(L(-1)\Delta(z)a,z) \\
&= Y_M\left(\left(\frac{d}{dz}\Delta(z) + [L(-1),\Delta(z)]\right)a,z\right) + \bar{Y}_M(L(-1)a,z) \\
&= \bar{Y}_M(L(-1)a,z).
\end{aligned}
\tag{5.11}
$$

For the $(\sigma\sigma_h)$-twisted Jacobi identity, let us assume that $\sigma a = \varepsilon_S^j a$ and $h(0)a = \frac{k}{T}a$. Then

$$
\begin{aligned}
&z_0^{-1}\delta\left(\frac{z_1-z_2}{z_0}\right)\bar{Y}_M(a,z_1)\bar{Y}_M(b,z_2) - \varepsilon_{a,b}z_0^{-1}\delta\left(\frac{z_2-z_1}{-z_0}\right)\bar{Y}_M(b,z_2)\bar{Y}_M(a,z_1) \\
&= z_0^{-1}\delta\left(\frac{z_1-z_2}{z_0}\right)Y_M(\Delta(z_1)a,z_1)Y_M(\Delta(z_2)b,z_2) \\
&\quad -\varepsilon_{a,b}z_0^{-1}\delta\left(\frac{z_2-z_1}{-z_0}\right)Y_M(\Delta(z_2)b,z_2)Y_M(\Delta(z_1)a,z_1) \\
&= z_2^{-1}\delta\left(\frac{z_1-z_0}{z_2}\right)\left(\frac{z_1-z_0}{z_2}\right)^{-\frac{j}{S}}Y_M(Y(\Delta(z_1)a,z_0)\Delta(z_2)b,z_2).
\end{aligned}
\tag{5.12}
$$

Our desired Jacobi identity is:

$$
\begin{aligned}
&z_0^{-1}\delta\left(\frac{z_1-z_2}{z_0}\right)\bar{Y}_M(a,z_1)\bar{Y}_M(b,z_2) - \varepsilon_{a,b}z_0^{-1}\delta\left(\frac{z_2-z_1}{-z_0}\right)\bar{Y}_M(b,z_2)\bar{Y}_M(a,z_1) \\
&= z_2^{-1}\delta\left(\frac{z_1-z_0}{z_2}\right)\left(\frac{z_1-z_0}{z_2}\right)^{-\frac{j}{S}-\frac{k}{T}}\bar{Y}_M(Y(a,z_0)b,z_2) \\
&= z_2^{-1}\delta\left(\frac{z_1-z_0}{z_2}\right)\left(\frac{z_1-z_0}{z_2}\right)^{-\frac{j}{S}-\frac{k}{T}}Y_M(\Delta(z_2)Y(a,z_0)b,z_2).
\end{aligned}
\tag{5.13}
$$

Therefore, it is equivalent to prove:

$$
\begin{aligned}
&z_2^{-1}\delta\left(\frac{z_1-z_0}{z_2}\right)\left(\frac{z_1-z_0}{z_2}\right)^{-\frac{j}{S}-\frac{k}{T}}\Delta(z_2)Y(a,z_0) \\
&= z_2^{-1}\delta\left(\frac{z_1-z_0}{z_2}\right)\left(\frac{z_1-z_0}{z_2}\right)^{-\frac{j}{S}}Y(\Delta(z_1)a,z_0)\Delta(z_2),
\end{aligned}
\tag{5.14}
$$

or

$$
\begin{aligned}
&z_2^{-1}\delta\left(\frac{z_1-z_0}{z_2}\right)\left(\frac{z_1-z_0}{z_2}\right)^{-\frac{j}{S}-\frac{k}{T}}\Delta(z_2)Y(a,z_0)\Delta(z_2)^{-1} \\
&= z_2^{-1}\delta\left(\frac{z_1-z_0}{z_2}\right)\left(\frac{z_1-z_0}{z_2}\right)^{-\frac{j}{S}}Y(\Delta(z_1)a,z_0).
\end{aligned}
\tag{5.15}
$$

From the commutator formula, we have

$$[h(n), Y(a, z_0)] = \sum_{i=0}^{\infty} \binom{n}{i} z_0^{n-i} Y(h(i)a, z_0). \qquad (5.16)$$

Then

$$[\sum_{n=1}^{\infty} \frac{h(n)}{-n}(-z_2)^{-n}, Y(a, z_0)]$$

$$= -\sum_{n=1}^{\infty}\sum_{i=0}^{\infty} \frac{1}{n}\binom{n}{i}(-z_2)^{-n} z_0^{n-i} Y(h(i)a, z_0)$$

$$= -\sum_{n=1}^{\infty} \frac{1}{n}(-z_2)^{-n} z_0^n Y(h(0)a, z_0)$$

$$\quad - \sum_{i=1}^{\infty}\sum_{n=1}^{\infty} \frac{1}{i}\binom{n-1}{i-1}(-z_2)^{-n} z_0^{n-i} Y(h(i)a, z_0)$$

$$= \log\left(1 + \frac{z_0}{z_2}\right)^{-\frac{k}{T}} Y(a, z_0) + \sum_{i=1}^{\infty} \frac{(-z_2 - z_0)^{-i}}{-i} Y(h(i)a, z_0). \quad (5.17)$$

Thus

$$\exp\left(\sum_{n=1}^{\infty} \frac{h(n)}{-n}(-z_2)^{-n}\right) Y(a, z_0) \exp\left(-\sum_{n=1}^{\infty} \frac{h(n)}{-n}(-z_2)^{-n}\right)$$

$$= Y\left(\left(\exp\log\left(1 + \frac{z_0}{z_2}\right)^{\frac{k}{T}} \exp\sum_{i=1}^{\infty} \frac{(-z_2 - z_0)^{-i}}{-i} h(i)\right) a, z_0\right)$$

$$= \left(1 + \frac{z_0}{z_2}\right)^{\frac{k}{T}} Y\left((z_2 + z_0)^{-h(0)}\Delta(z_2 + z_0)a, z_0\right)$$

$$= z_2^{-\frac{k}{T}} Y\left(\Delta(z_2 + z_0)a, z_0\right). \qquad (5.18)$$

Since $[h(0), Y(a, z_0)] = Y(h(0)a, z_0)$, we get:

$$z^{h(0)} Y(a, z_0) z^{-h(0)} = Y(z^{h(0)}a, z_0). \qquad (5.19)$$

Therefore

$$\Delta(z_2) Y(a, z_0)\Delta(z_2)^{-1} = z_2^{-\frac{k}{T}} Y\left(z_2^{h(0)}\Delta(z_2 + z_0)a, z_0\right)$$

$$= Y\left(\Delta(z_2 + z_0)a, z_0\right). \qquad (5.20)$$

Then

$$z_2^{-1}\delta\left(\frac{z_1 - z_0}{z_2}\right)\left(\frac{z_1 - z_0}{z_2}\right)^{-\frac{j}{S} - \frac{k}{T}} \Delta(z_2) Y(a, z_0)\Delta(z_2)^{-1}$$

$$= z_1^{-1}\delta\left(\frac{z_2 + z_0}{z_1}\right)\left(\frac{z_2 + z_0}{z_1}\right)^{\frac{j}{S} + \frac{k}{T}} Y\left(\Delta(z_2 + z_0)a, z_0\right)$$

$$= z_1^{-1}\delta\left(\frac{z_2 + z_0}{z_1}\right)\left(\frac{z_2 + z_0}{z_1}\right)^{\frac{j}{S}} Y\left(\Delta(z_1)a, z_0\right). \qquad (5.21)$$

Therefore, the $(\sigma\sigma_h)$-twisted Jacobi identity holds. Thus $(M, \bar{Y})$ is a weak $(\sigma\sigma_h)$-twisted V-module. $\square$

Remark 5.5. *Since $\Delta(z)$ is invertible, $(M, \bar{Y})$ is irreducible if (M, Y) is irreducible. In particular, if $\sigma = Id_V$, any σ_h-twisted V-module can be constructed from a V-module. Proposition 5.4 also gives an isomorphism between the twisted affine Lie algebras $\tilde{g}[\mu]$ and $\tilde{g}[\mu\sigma_h]$. Although Proposition 5.4 is only about an inner automorphism, we believe that it has more interesting applications. We will study this subject in a coming paper.*

From Propositions 5.3 and 5.4, it is enough for us to study the μ-twisted module theory for $L_g(\ell, 0)$. Notice that g_0 is a simple subalgebra of g [K]. For any linear functional λ on h_0, denote by $L_{(g,\mu)}(\ell, \lambda)$ the irreducible lowest weight $\tilde{g}[\mu]$-module of level ℓ with lowest weight λ. Recall from [K] that θ_0 is the highest weight of g_1 with respect to the Cartan subalgebra of g_0 and that e_{θ_0} is a root vector.

Proposition 5.6. *An irreducible $\tilde{g}[\mu]$-module $L_{(g,\mu)}(\ell, \lambda)$ is a μ-twisted module for the vertex operator algebra $L_g(\ell, 0)$ if and only if it is a standard $\tilde{g}[\mu]$-module.*

Proof. For any $a \in g_j$, by considering $a(z) = \sum_{n\in\mathbb{Z}}(t^{n+\frac{j}{T}} \otimes a)z^{-n-1-\frac{j}{T}}$ as an element of $F(L_{g,\mu}(\ell, \lambda), T)$, we obtain a local subspace $\{a(z)|a \in g\}$ of $F(L_{(g,\mu)}(\ell, \lambda), T)$. Let V be the vertex algebra generated by $\{a(z)|a \in g\}$. Being a quotient vertex algebra of the vertex operator algebra $M_g(\ell, \mathbb{C})$, V is a vertex operator algebra. Then $L_{(g,\mu)}(\ell, \lambda)$ is a μ-twisted $L_g(\ell, 0)$-module if and only if $V = L_g(\ell, 0)$. It follows from [Li] that $V = L_g(\ell, 0)$ if and only if $Y_V(e_\theta, z)^{\ell+1} = 0$ on V. Since $L_{(g,\mu)}(\ell, \lambda)$ is a faithful μ-twisted V-module, it follows from Proposition 2.12 that $Y_V(e_\theta, z)^{\ell+1} = 0$ on V if and only if $Y_M(e_\theta, z)^{\ell+1} = 0$ for $M = L_{(g,\mu)}(\ell, \lambda)$. Then we only need to prove that $Y_M(e_\theta, z)^{\ell+1} = 0$ on $L_{(g,\mu)}(\ell, \lambda)$ if and only if $L_{(g,\mu)}(\ell, \lambda)$ is an integrable $\tilde{g}[\mu]$-module.

If $L_{(g,\mu)}(\ell, \lambda)$ is an integrable $\tilde{g}[\mu]$-module, then it is an integrable $\tilde{g}^\theta$-module. By Proposition 5.1, we have: $Y_M(e_\theta, z)^{\ell+1} = 0$ for $M = L_{(g,\mu)}(\ell, \lambda)$. Then $L_{(g,\mu)}(\ell, \lambda)$ is a μ-twisted module for the vertex operator algebra $L_g(\ell, 0)$. Conversely, suppose that $L_{(g,\mu)}(\ell, \lambda)$ is a μ-twisted $L_g(\ell, 0)$-module. From the untwisted theory [Li] $L_{(g,\mu)}(\ell, \lambda)$ is an integrable $\tilde{g}_0$-module. By Proposition 5.1, $Y(e_{\theta_0}, z)$ is nilpotent on $L_g(\ell, 0)$. By Proposition 2.12, $Y(e_{\theta_0}, z)$ is nilpotent on $L_{(g,\mu)}(\ell, \lambda)$. Therefore, $L_{(g,\mu)}(\ell, \lambda)$ is an integrable $\tilde{g}[\mu]$-module. $\square$

Let us recall a complete reducibility theorem of Kac's (Theorem 10.7 [K]).

Proposition 5.7. *Let $g(A)$ be a Kac-Moody algebra associated to a symmetrizable generalized Cartan matrix A of rank n, let e_i, f_i, h_i $(i = 1, \cdots, n)$ be the Chevalley generators for the derived subalgebra $g'(A)$ and let $g(A) = g(A)_+ \oplus H \oplus g(A)_-$ be the triangular decomposition. Let M be a $g'(A)$-module such that for any $u \in M$ there is a positive integer k such that*

$$g(A)_+^k u = 0, \quad f_i^k u = 0 \quad for \ i = 1, \cdots, n. \tag{5.22}$$

Then M is a direct sum of irreducible highest weight integrable $g'(A)$-modules.

For our purpose, we are only concerned about (both twisted and untwisted) affine Lie algebras. Let g be a finite-dimensional simple Lie algebra, let μ be a Dynkin diagram automorphism of order T of g and let $\tilde{g}[\mu]$ be the corresponding twisted affine Lie algebra as before. Then we have the following modified complete reducibility theorem.

Proposition 5.8. *Let $\tilde{g}[\mu]$ be the twisted affine Lie algebra as above and let M be a $\tilde{g}[\mu]$-module such that for any $u \in M$, there are positive integers k and r such that*

$$\tilde{g}[\mu]_+^k u = 0, \; e_\alpha(z)^r M = 0 \tag{5.23}$$

where α is a root of g_0, e_α is a root vector of root α. Then M is a direct sum of irreducible highest weight integrable $\tilde{g}[\mu]$-modules (of level less than r).

Proof. For such a $\tilde{g}[\mu]$-module M, we set

$$\Omega(M) = \{u \in M | \tilde{g}[\mu]_+ u = 0\}. \tag{5.24}$$

Then $\Omega(M) \neq 0$ if $M \neq 0$. It is also clear that $\Omega(M)$ is a g_0-module. Since $e_\alpha(z)^r \Omega(M) = 0$, by considering the coefficient of z^{-r} we obtain $e_\alpha(0)^r \Omega(M) = 0$. It follows from Proposition 5.1.5 [Li] that $\Omega(M)$ is a direct sum of finite-dimensional irreducible g_0-modules. Let u be a highest weight vector of $\Omega(M)$ as a g_0-module. Then by considering the constant term of $e_\alpha(z)^r u = 0$ for a positive root α we obtain $e_\alpha(-1)^r u = 0$. Thus the submodule M^o of M generated by $\tilde{g}[\mu]$ from $\Omega(M)$ satisfies the conditions assumed in Proposition 5.7, so that M^o is a direct sum of irreducible highest weight integrable $\tilde{g}[\mu]$-modules.

Next we shall prove that $M = M^o$. Suppose $M \neq M^o$. Then $\Omega(M/M^o) \neq 0$, so that (M/M^o) has a (nonzero) irreducible highest weight integrable submodule M^1/M^o. It is easy to see that M^1 satisfies the conditions assumed in Proposition 5.7. Thus M^1 is completely reducible, so that $M^1 = M^o \oplus W$ where W is a standard module. But by definition the highest weight vectors of W are contained in M^o. This is a contradiction. $\square$

Theorem 5.9. *Let M be a μ-twisted weak $L_g(\ell, 0)$-module such that for any $u \in M$ there is a positive integer k such that $\tilde{g}[\mu]_+^k u = 0$. Then M is a direct sum of irreducible μ-twisted $L_g(\ell, 0)$-modules, which are standard $\tilde{g}[\mu]$-modules of level ℓ.*

Proof. It follows directly from Propositions 5.1, 5.6 and 5.8. $\square$

Since any lower truncated $\frac{1}{T}\mathbb{Z}$-graded weak σ-twisted $L_g(\ell, 0)$-module M satisfies the conditions of Theorem 5.9, we have:

Corollary 5.10. *Any lower truncated $\frac{1}{T}\mathbb{Z}$-graded weak σ-twisted $L_g(\ell, 0)$-module M is a direct sum of standard $\tilde{g}[\sigma]$-modules of level ℓ.*

By Proposition 5.4 and Remark 5.5 we have:

Theorem 5.11. *Let σ be any automorphism of finite order of g and let M be any weak σ-twisted $L_g(\ell, 0)$-module such that for any $u \in M$ there is a positive integer k such that $\tilde{g}[\sigma]_+^k u = 0$. Then M is completely reducible and there are only finite many irreducible σ-twisted $L_g(\ell, 0)$-modules up to equivalence.*

For the rest of this section, we apply Proposition 5.4 to the study of certain twisted $\mathbb{F}^{2n}$-modules. Following [DM1], let $A = A_+ \oplus A_-$ and define $G = GL(A^+)$. For any $\sigma \in G$, we define an action of σ on A^- by:

$$\langle u, \sigma v \rangle = \langle \sigma^{-1}u, v \rangle \quad \text{for any } u \in A^+, v \in A^-. \tag{5.25}$$

Then G acts on A such that G preserves the bilinear form on A. For $\sigma \in G$ of order T, let $\{a_{1,\sigma}, \cdots, a_{n,\sigma}\}$ be a basis of A^+ such that

$$\sigma a_{i,\sigma} = \varepsilon^{n_{i,\sigma}} a_{i,\sigma} \tag{5.26}$$

where $0 \le n_{i,\sigma} < T$ for each $1 \le i \le n$. Let $\{a_{1,\sigma}^*, \cdots, a_{n,\sigma}^*\}$ be a dual basis of A^- so that $\langle a_{i,\sigma}, a_{j,\sigma}^* \rangle = \delta_{i,j}$. Then

$$\sigma a_{i,\sigma}^* = \varepsilon^{-n_{i,\sigma}} a_{i,\sigma}^*. \tag{5.27}$$

Set

$$h_\sigma = \frac{1}{T} \sum_{j=1}^n n_{j,\sigma} (a_{j,\sigma})_{-1} a_{j,\sigma}^*. \tag{5.28}$$

Then h_σ is an even weight-one element of $\mathbb{F}^{2n}$. For any $a, b, c \in A$, we have:

$$
\begin{aligned}
&(a_{-1}b)_0 c \\
=\ & Res_{z_1} Res_{z_2} \left((z_1 - z_2)^{-1} Y(a, z_1) Y(b, z_2) + (-z_2 + z_1)^{-1} Y(b, z_2) Y(a, z_1) \right) c \\
=\ & \sum_{r=0}^{\infty} (a_{-r-1}b_r c - b_{-r-1}a_r c) \\
=\ & a_{-1}b_0 c - b_{-1}a_0 c \\
=\ & \langle b, c \rangle a - \langle a, c \rangle b.
\end{aligned} \tag{5.29}
$$

Then

$$h_\sigma(0)a_{i,\sigma} = \frac{1}{T}n_{i,\sigma}a_{i,\sigma}, \ h_\sigma(0)a_{i,\sigma}^* = -\frac{1}{T}n_{i,\sigma}a_{i,\sigma}^*. \tag{5.30}$$

Since A generates the vertex operator superalgebra $\mathbb{F}^{2n}$, we have:

$$\sigma = \exp\left(2\pi\sqrt{-1}h_\sigma(0)\right). \tag{5.31}$$

(See [DM1].) Then as a corollary of Proposition 5.4, we have the following proposition of Dong and Mason [DM1]:

Proposition 5.12. *For any $\sigma \in G$, up to equivalence there is only one irreducible σ-twisted $\mathbb{F}^{2n}$-module. Furthermore, it can be constructed by using Proposition 5.4 from the adjoint $\mathbb{F}^{2n}$-module.*

References

[B]   R. E. Borcherds, Vertex algebras, Kac-Moody algebras, and the Monster, *Proc. Natl. Acad. Sci. USA* **83** (1986), 3068-3071.

[D]   C. Dong, Twisted modules for vertex algebras associated with even lattices, *J. of Algebra* **165** (1994), 91-112.

[DL]   C. Dong and J. Lepowsky, *Generalized Vertex Algebras and Relative Vertex Operators*, Progress in Math. **Vol.** 112, Birkhäuser, Boston, 1993.

[DM1]   C. Dong and G. Mason, Nonabelian orbifolds and the boson-fermion correspondence, *Comm. Math. Phys.* **163** (1994), 523-529.

[DM2]   C. Dong and G. Mason, An orbifold theory of genus zero associated to the sporadic group M_{24}, *Comm. Math. Phys.* **164** (1994), 87-104.

[FFR]   A. J. Feingold, I. B. Frenkel, J. F. X. Ries, Spinor construction of vertex operator algebras, triality and $E_8^{(1)}$, *Contemporary Math.* **Vol.** 121, Amer. Math. Soc., Providence, 1991.

[FHL]   I. B. Frenkel, Y.-Z. Huang and J. Lepowsky, On axiomatic approaches to vertex operator algebras and modules, Preprint, 1989; *Memoirs American Math. Soc.* **104**, 1993.

[FLM]   I. B. Frenkel, J. Lepowsky and A. Meurman, *Vertex Operator Algebras and the Monster*, *Pure and Applied Math.* **Vol.** 134, Academic Press, New York, 1988.

[FZ]   I. B. Frenkel and Y.-C. Zhu, Vertex operator algebras associated to representations of affine and Virasoro algebras, *Duke Math. J.* **66** (1992), 123-168.

[GSW]   Michael B. Green, John H. Schwarz and Edward Witten, *Superstring Theory*, **Volume 1**, Cambridge University Press, 1987.

[G]   P. Goddard, Meromorphic conformal field theory, in *Infinite Dimensional Lie Algebras and Groups*, Proceedings of the conference held at CIRM, Luminy, edited by Victor G. Kac (1988).

[H]   S. Helgason, *Differential Geometry, Lie Groups and Symmetric spaces*, Academic Press, 1978.

[K]   V. G. Kac, *Infinite-dimensional Lie Algebras*, 3rd ed., Cambridge Univ. Press, Cambridge, 1990.

[KW]   V. G. Kac and W.-Q. Wang, Vertex operator superalgebras and representations, *Contemp. Math.* **Vol.** 175 (1994), 161-191.

[Le]   J. Lepowsky, Calculus of twisted vertex operators, *Proc. Natl. Acad. Sci. USA* **82** (1985), 8295-8299.

[Li]   H.-S. Li, Local systems of vertex operators, vertex superalgebras and modules, hep-th/9406185, *J. of Pure and Appl. Alg.*, to appear.

[Lia]   B.-H. Lian, On the classification of simple vertex operator algebras, *Comm. Math. Phys.* **163** (1994), 307-357.

[LZ1]   B. H. Lian and G. J. Zuckerman, Some classical and quantum algebras, hep-th/9404010.

[LZ1]   B. H. Lian and G. J. Zuckerman, Commutative quantum operator algebras, q-alg/9501014, to appear in Journ. Pure Appl. Alg. (1995).

[MP]   A. Meurman and M. Primc, Annihilating fields of standard modules of $\tilde{sl}(2, C)$ and combinatorial identities, preprint, 1994.

[T1]   H. Tsukada, Vertex operator superalgebras, *Comm. in Alg.*, **18** (1990), 2249-2274.

[T2]   H. Tsukada, Shifted vertex operator algebras and G-elliptic systems, *J. Math. Phys.* **33** (1992), 2546-2556.

[X]   X. Xu, Twisted modules of colored lattice vertex operator superalgebras, preprint (1992).

[Z]   Y.-C. Zhu, Vertex operator algebras, elliptic functions and modular forms, Ph.D. thesis, Yale University, 1990.

Department of Mathematics, Rutgers University, New Brunswick, NJ 08903
Current address: Department of Mathematics, University of California, Santa Cruz, CA 95064
E-mail address: hli@math.ucsc.edu

Contemporary Mathematics
Volume **193**, 1996

MODULAR FORMS AND TOPOLOGY

KEFENG LIU

We want to discuss various applications of modular forms in topology. The starting point is elliptic genus and its generalizations. The main techniques are the Atiyah-Singer index theorem, the Atiyah-Bott-Segal-Singer Lefschetz fixed point formula, Kac-Moody Lie algebras, modular forms and theta-functions. Just as the representations theory of classical Lie groups has close connections with the Atiyah-Singer index formula as exposed in [A1], the representation theory of loop groups plays very important role in our study. One of the most important new features of loop group representations is the modular invariance of the Kac-Weyl character formula, which allows us to derive many interesting new results and to unify many important old results in topology. In this paper we will develope along this line. We hope that the other features of loop group representations, such as fusion rules and tensor category structure may also be applied to topology. See the discussions in §3.

The contents of this paper is organized in the following way. In §1 we introduce elliptic genus by combining index theory and the representation theory of loop groups. The relation of the classical index theory with representation theory of classical Lie groups was discussed in [A1]. Here one finds that just replacing the classical Lie groups by their corresponding loop groups, we get the complete theory of elliptic genus, or more generally the index theory on loop space. Especially the Dirac operator and the Witten genus of loop space are derived more convincingly in this way. Then all of the other well-known properties of elliptic genus, such as functional equations, characterizations by rigidity and fibrations, can be easily obtained. Here we only pick some less well-known results to discuss, for example the expressions of the parameters in elliptic genera in terms of theta-functions.

Different from the classical Lie group case, a new feature in our situation appears, the modular invariance which, by combining with index theory, is applied to obtain many new topological results. This is the content of §2. Most results in this section are special cases of more general results. For simplicity we only give the main ideas of the proofs.

We make some geometric constructions in §3 to understand elliptic cohomology. The construction in §3.3 is motivated by the vertex operator algebra

1991 *Mathematics Subject Classification*. 55P35, 55N91, 58G10.

construction of the monstrous moonshine module. In §3.1 we introduce vector bundles with infinite dimensional structure groups. We study the corresponding Grothendieck groups and the Riemann-Roch properties in §3.2. An easy corollary is that elliptic genus can be realized as the difference of two infinite dimensional vector bundles on a sphere with the action of Virasoro algebra on each of them. For each modular subgroup, a ring of graded bundles with modularity is introduced in §3.3, to which there is a natural homomorphism from elliptic cohomology. Several simple theorems are stated without proof. The detail of this section will appear in a forthcoming paper.

This survey article is basically an expanded version of my lectures given in the topology seminars at MIT, Harvard, and the AMS conference on the monster and the moonshine module. Many ideas are certainly already well-known to experts. The reader may find related discussions in the references. I have also benefited from discussions with many people. I would like to thank the organizers and the audience of the seminars and conference, especially R. Bott, J.-L. Brylinski, C. Dong, M. Hopkins, Y. Huang, V. Kac, G. Katz, H. Miller, A. Radul, W. Wang, E. Weinstein, S.-T. Yau, Y. Zhu.

CONTENTS

1. Index Theory, Elliptic Curves and Loop Groups

One can look at elliptic genus from several different points of view; from index theory, from representation theory of Kac-Moody affine Lie algebras or from the theory of elliptic functions and modular forms. Each of them shows us some quite different interesting features of ellitic genus. On the other hand we can also combine the forces of these three different mathematical fields to derive many interesting results in topology such as rigidity, divisibility and vanishing of topological invariants.

In this section, we first introduce elliptic genus by combing the representation theory of affine Lie algebras and the Atiyah-Singer index theory from which we derive all the other properties of elliptic genus such as functional equations and logrithms, etc. This section is a 'loop' analogue of [A1].

1.1. Atiyah-Singer Index theorem. Let M be a smooth compact spin manifold of dimension $2k$. We then have the following principal $\mathrm{Spin}(2k)$ bundle

$$\mathrm{Spin}(2k) \to Q \to M$$

which is the double cover of the frame bundle of TM, the tangent bundle of M. From the two half spinor representations of $\mathrm{Spin}(2k)$, $\{\triangle^+, \triangle^-\}$ we get two associated bundles on M which we still denote by $\triangle^\pm$ respectively. The Dirac operator on M is a basic elliptic operator between the section space $\Gamma(\triangle^+)$ and $\Gamma(\triangle^-)$. Given another principal bundle P on M with structure group G, and any (real) representation E of G, we can construct the corresponding associated vector bundle, still denoted by E. The Atiyah-Singer index theorem in this case is given by

$$\mathrm{Ind}\, D \otimes E = \int_M \hat{A}(M) \mathrm{ch}\, E$$

where recall that $D \otimes E$ is the twisted Dirac operator

$$D \otimes E:\ \Gamma(\triangle^+ \otimes E) \to \Gamma(\triangle^- \otimes E)$$

and the index is defined to be

$$\mathrm{Ind} D \otimes E = \dim \mathrm{Ker}\, D \otimes E - \dim \mathrm{Coker}\, D \otimes E.$$

If $\{\pm x_j\}$ and $\{\pm y_j\}$ are the formal Chern roots of $TM \otimes C$ and $E \otimes C$ respectively, then

$$\hat{A}(M) = \prod_{j=1}^{k} \frac{x_j}{2\sinh x_j/2}, \ \mathrm{ch}\, E = \sum_{j=1}^{l} e^{y_j} + e^{-y_j}$$

are respectively the $\hat{A}$-class of M and the Chern character of E.

Another important elliptic operator, the signature operator d_s is obtained by taking E to be $\triangle = \triangle^+ \oplus \triangle^-$. We obviously have

$$\operatorname{Ind} D = \int_M \hat{A}(M), \text{ and } \operatorname{Ind} d_s = \int_M L(M)$$

where

$$L(M) = \prod_{j=1}^{k} \frac{x_j}{\tanh x_j/2}$$

is the Hirzebruch L-class of M.

As one can see, the starting point of the index formula is the representation theory of spin groups. More precisely, the Chern character of a bundle is induced from the character of the corresponding representation through transgression as in [BH]. Since

$$\operatorname{ch}(\triangle^+ - \triangle^-) = \prod_{j=1}^{k} (e^{x_j/2} - e^{-x_j/2}) = 2 \prod_{j=1}^{k} \sinh x_j/2,$$

the $\hat{A}$-class is essentially the ratio of the Euler class and the Chern character of the basic element $\triangle^+ - \triangle^-$, i.e.

$$\hat{A}(M) = \frac{e(M)}{\operatorname{ch}(\triangle^+ - \triangle^-)}.$$

Most interestingly the denominator of the $\hat{A}$-class can also be viewed as induced from the Weyl denominator in the representation theory of compact Lie groups.

1.2. **Loop groups and index theory.** In the following, we will show that, if we replace the representations of the classical Lie groups by their correponding loop groups, we exactly recover the complete theory of elliptic genus. For this we need the following simple construction.

Given a principal G-bundle P on M, and a positive energy representation E of $\tilde{L}G$ which is the central extension of the loop group of G, we decompose E according to the rotation action of the loop to get $E = \sum_{\geq 0} E_n$ where each E_n is a finite dimensional representation of G. First assume that E is irreducible. Construct associated bundles to P from each E_n, which we still denote by E_n, we get an element

$$\psi(P, E) = q^{m_\Lambda} \sum_n E_n q^n$$

where $q = e^{2\pi i \tau}$ with τ in the upper half plane and m_Λ is a rational number which is the so-called modular anomaly of the representation E, see [K]. This construction extends linearly. If $E = \oplus_j E^j$, then $\psi(P, E) = \sum_j \psi(P, E^j)$.

For any positive integer l, the loop group $\tilde{L}\mathrm{Spin}(2l)$ has four irreducible level 1 positive energy representations. Let us denote them by S^+, S^- and S_+, S_- respectively. Let $\{\pm\alpha_j\}$ be the roots of $\mathrm{Spin}(2l)$. Then the normalized Kac-Weyl characters of these four representations can be expressed in terms of the four Jacobi theta-functions and the Dedekind eta-function [Ch] as follows,

$$\chi_{S^+-S^-} = \prod_{j=1}^{l} \frac{\theta(\alpha_j,\tau)}{\eta(\tau)} \quad , \quad \chi_{S_+-S_-} = \prod_{j=1}^{l} \frac{\theta_2(\alpha_j,\tau)}{\eta(\tau)},$$

$$\chi_{S^++S^-} = \prod_{j=1}^{l} \frac{\theta_1(\alpha_j,\tau)}{\eta(\tau)} \quad , \quad \chi_{S_++S_-} = \prod_{j=1}^{l} \frac{\theta_3(\alpha_j,\tau)}{\eta(\tau)}.$$

Note that here we view $\pi i v$ in the theta-functions in [Ch] as one variable v. That is, for example

$$\theta(v,\tau) = -q^{1/8}2i\sinh v/2 \prod_{n=1}^{\infty}(1-q^n)(1-e^vq^n)(1-e^{-v}q^n)$$

It turns out that $S^+ \pm S^-$ are the loop group analogues of the finite dimensional spinor representations $\triangle^+ \pm \triangle^-$. Recall that each representation of loop group $\tilde{L}G$ is induced from a representation of G. We note that $S^{\pm}$ are induced from $\triangle^{\pm}$ respectively. Let Q be the principal spin bundle of M as in §1.1. For a vector bundle V on M, let

$$\Lambda_t V = 1 + t\Lambda^1 V + t^2\Lambda^2 V + \cdots,$$

$$S_t V = 1 + tS^1 V + t^2 S^2 V + \cdots$$

be the two operations in $K(M)[[t]]$. It is easy to get the following

$$\psi(Q, S^- - S^-) = q^{-\frac{k}{12}}(\triangle^+ - \triangle^-) \otimes_{j=1}^{\infty} \Lambda_{-q^j}(TM),$$

$$\psi(Q, S^+ + S^-) = q^{-\frac{k}{12}}(\triangle^+ + \triangle^-) \otimes_{j=1}^{\infty} \Lambda_{q^j}(TM);$$

$$\psi(Q, S_+ - S_-) = q^{-\frac{k}{24}} \otimes_{j=1}^{\infty} \Lambda_{-q^{j-1/2}}(TM)$$

$$\psi(Q, S_+ + S_-) = q^{-\frac{k}{24}} \otimes_{j=1}^{\infty} \Lambda_{q^{j-1/2}}(TM).$$

Just as $\triangle^+ - \triangle^-$ induces the Dirac operator on M, $S^+ - S^-$ also induces the Dirac operator on LM. Similar to the above $\hat{A}$-calss of M, introduce the loop space $\hat{A}$-class $\hat{\Theta}(M)$ as

$$\hat{\Theta}(M) = \frac{e(M)}{\mathrm{ch}\left(\psi(Q,S^+) - \psi(Q,S^-)\right)} = \eta(\tau)^k \cdot \prod_{j=1}^{k} \frac{x_j}{\theta(x_j,\tau)}.$$

Then we have similarly the Dirac operator for loop space

$$D^L = q^{-\frac{k}{12}} D \otimes \otimes_{j=1}^{\infty} S_{q^j}(TM)$$

and the corresponding index formula

$$\text{Ind}\, D^L = \int_M \hat{\Theta}(M).$$

A general index theorem in this loop group setting is

$$\text{Ind}\, D^L \otimes \psi(P, E) = \int_M \hat{\Theta}(M)\text{ch}\,\psi(P, E).$$

Especially take $P = Q$ as the principal Spin(2k)-bundle of M and $E = S^+ + S^-$, we get the signature of the loop space LM

$$\text{Ind}\, D^L \otimes \psi(Q, S^+ + S^-) = \int_M \prod_{j=1}^k x_j \frac{\theta_1(x_j, \tau)}{\theta(x_j, \tau)}.$$

We call

$$\prod_{j=1}^k x_j \frac{\theta_1(x_j, \tau)}{\theta(x_j, \tau)}$$

the elliptic L-class.

Associated to $S_+ \pm S_-$ are the other two elliptic operators which do not have finite dimensional analogues. Their indice are

$$\text{Ind}\, D^L \otimes \psi(Q, S_- - S_-) = \int_M \prod_{j=1}^k x_j \frac{\theta_2(x_j, \tau)}{\theta(x_j, \tau)},$$

$$\text{Ind}\, D^L \otimes \psi(Q, S_- + S_-) = \int_M \prod_{j=1}^k x_j \frac{\theta_3(x_j, \tau)}{\theta(x_j, \tau)}.$$

It is just because of these two new elliptic operators that make the proof of the famous Witten rigidity theorems very simple. This is the magic of infinite dimensional geometry and topology.

1.3. Elliptic genera. The indices of the above three elliptic operators are actually modular functions. To get modular forms instead, we consider their virtual versions. This means that we replace TM by its virtual bundle $\bar{T}M = TM - 2k$ in the above symmetric and wedge products. Denote by

$$\mathfrak{D}^L = D \otimes \otimes_{j=1}^\infty S_{q^j}(\bar{T}M),$$

we get

$$\text{Ind}\, \mathfrak{D}^L = \int_M \prod_{j=1}^k x_j \frac{\theta'(0, \tau)}{\theta(x_j, \tau)}$$

which is called the Witten genus; and

$$\operatorname{Ind} \mathfrak{D}^L \otimes \triangle \otimes_{m=1}^{\infty} \Lambda_{q^m}(\bar{T}M) = \int_M \prod_{j=1}^{k} x_j \frac{\theta'(0,\tau)\theta_1(x_j,\tau)}{\theta(x_j,\tau)\theta_1(0,\tau)},$$

$$\operatorname{Ind} \mathfrak{D}^L \otimes_{m=1}^{\infty} \Lambda_{-q^{m-1/2}}(\bar{T}M) = \int_M \prod_{j=1}^{k} x_j \frac{\theta'(0,\tau)\theta_2(x_j,\tau)}{\theta(x_j,\tau)\theta_2(0,\tau)},$$

$$\operatorname{Ind} \mathfrak{D}^L \otimes_{m=1}^{\infty} \Lambda_{q^{m-1/2}}(\bar{T}M) = \int_M \prod_{j=1}^{k} x_j \frac{\theta'(0,\tau)\theta_3(x_j,\tau)}{\theta(x_j,\tau)\theta_3(0,\tau)}$$

which are usually called universal elliptic genera. The three functions, suitably normalized, appeared in the above index formulas,

$$F_j(x) = \frac{1}{2i} \frac{\theta_j(x,\tau)\theta'(0,\tau)}{\theta(x,\tau)\theta_j(0,\tau)}$$

with $j = 1, 2, 3$ are called their generating series correspondingly. Also we simply call

$$\prod_{j=1}^{k} x_j \frac{\theta'(0,\tau)}{\theta(x_j,\tau)}$$

the Witten class and write it as $W(M)$. Obviously

$$W(M) = \eta(\tau)^{2k}\hat{\Theta}(M).$$

Similarly

$$\prod_{j=1}^{k} x_j \frac{\theta'(0,\tau)\theta_1(x_j,\tau)}{\theta(x_j,\tau)\theta_1(0,\tau)}$$

is called the normalized elliptic L-class of M.

It is worthwhile to record the following remarks. Assume there exists a torus action on a manifold M with fixed point set F and normal bundle N. Then according to the action, N can be decomposed into sums of complex line bundles $\{L_j\}$ and the torus acts on L_j by g^{m_j}, where g is a generator of the torus and m_j is called the weight of L_j.

The equivariant Euler class of N is by definition $\prod_j (m_j + x_j)$ where x_j is the first Chern class of L_j. We define the normalized equivariant Euler class of N to be $\prod_j (1 + x_j/m_j)$.

Let LM denote the loop space of M and LLM the double loop space of M. The rotation of the circles induces a natural circle action on LM and an $S^1 \times S^1$ action on LM with fixed points M. The normal bundle N_L of M in LM is $\oplus_{n \neq 0} TM_n$ where each TM_n is the same as TM with weight n. The normal bundle N_{LL} of M in LLM is $\oplus_{n,m\neq 0}TM_{n,m}$ with $TM_{n,m}$ the same as TM with weight $n + m\tau$.

As a simple consequence of Eisenstein's product formulas for sine function and theta-function [We], we have the following

    1) the normalized equivariant Euler class of N_L is the inverse of the $\hat{A}$-class of M.

    2) the normalized equivariant Euler calss of N_{LL} is the inverse of the $\hat{\Theta}$-class of M.

  In certain sense we can say that the classical Hirzebruch genera, such as $\hat{A}$-genus and L-genus, are 1-periodic genera, since they are associated to triognometric functions; while the elliptic genera, associated to elliptic functions, are 2-periodic genera. Note that the first observation above is due to Witten. From a pure functional theoretic point of view, make the following replacement in the Hirzebruch $\hat{A}$-class and L-class,

$$\sinh x/2 \to \frac{\theta(x,\tau)}{\eta(\tau)}$$

$$\cosh x/2 \to \frac{\theta_1(x,\tau)}{\eta(\tau)},$$

one exactly recovers the $\hat{\Theta}$-class and the elliptic L-calss. On the other hand, make the following replacement

$$\sinh x/2 \to \frac{\theta(x,\tau)}{\theta'(0,\tau)}$$

$$\cosh x/2 \to \frac{\theta_1(x,\tau)}{\theta_1(0,\tau)},$$

we get the Witten class and the normalized elliptic L-class.

1.4. Elliptic genera and theta-functions.
Let $\mathfrak{P}(x)$ be the Weierstrass elliptic function associated to the lattice $\{2m\pi + 2n\pi\tau\}$. Then one has the following parametrization of elliptic curves

$$\mathfrak{P}'(x) = 4(\mathfrak{P}(x) - e_1)(\mathfrak{P}(x) - e_2)(\mathfrak{P}(x) - e_3)$$

with $e_j = \mathfrak{P}(\omega_j)$ where $\omega_1 = \pi, \omega_2 = \pi\tau, \omega_3 = \pi(1 + \tau)$.

The interesting connection of elliptic genera with elliptic curves are manifested by the following relations, for $j = 1, 2, 3$

$$2iF_j(x) = \sqrt{\mathfrak{P}(x) - e_j}$$

which is easily seen by comparing poles of the functions of both sides.

For $j = 1, 2, 3$, let us denote $\theta_j(0, \tau)$ by θ_j, we then have relations

$$e_3 - e_2 = \theta_1^4, \ e_1 - e_3 = \theta_2^4, \ e_1 - e_2 = \theta_3^4.$$

Here note that we use v, instead of πv as the variable in the theta-functions in [Ch]. This is also slightly different from §1.2. This change is only for convenience and does not matter much. Since $\mathfrak{P}'(x) = -8F_j(x)F_j'(x)$, plug into the Weierstrass equation, we get the functional equations of the three elliptic genera.

$$F_1(x)^2 = (F_1(x)^2 - 1/4\theta_3^4)(F_1(x)^2 - 1/4\theta_2^4),$$

$$F_2(x)^2 = (F_2(x)^2 + 1/4\theta_3^4)(F_2(x)^2 + 1/4\theta_1^4),$$

$$F_3(x)^2 = (F_3(x)^2 + 1/4\theta_2^4)(F_3(x)^2 - 1/4\theta_1^4).$$

From these equations, we easily get the logrithms of the three elliptic genera

$$g_1(x) = \int_0^x \frac{du}{\sqrt{(1 - 1/4\theta_3^4 u^2)(1 - 1/4\theta_2^4 u^2)}},$$

$$g_2(x) = \int_0^x \frac{du}{\sqrt{(1 + 1/4\theta_3^4 u^2)(1 + 1/4\theta_1^4 u^2)}},$$

$$g_3(x) = \int_0^x \frac{du}{\sqrt{(1 + 1/4\theta_2^4 u^2)(1 - 1/4\theta_1^4 u^2)}}.$$

In fact, let z denote the first Chern class of the universal line bundle on CP^n. By definition,

$$g_j(x) = \sum_{n=0}^{\infty} \varphi_j(CP^{2n})\frac{x^{2n+1}}{2n+1}, \ j = 1, 2, 3,$$

where φ_j denotes the corresponding elliptic genus associated to $xF_j(x)$. Since

$$\varphi_j(CP^{2n}) = \int_{CP^{2n}} (zF_j(z))^{2n+1} = \frac{1}{n!}\frac{d^n}{dz^n}[zF_j(z)]^{2n+1},$$

it is easy to show, by Lagrange theorem [Co], that

$$g_j^{-1}(x) = \frac{1}{F_j(x)}.$$

Compare with the Ochanine's standard equation for elliptic genus,

$$y^2 = 1 - 2\delta x^2 + \varepsilon x^4,$$

we have the following expressions,

$$\delta_1 = 1/8(\theta_2^4 + \theta_3^4), \varepsilon_1 = 1/16\theta_2^4\theta_3^4,$$

$$\delta_2 = -1/8(\theta_1^4 + \theta_3^4), \varepsilon_2 = 1/16\theta_1^4\theta_3^4,$$

$$\delta_3 = 1/8(\theta_2^4 - \theta_2^4), \varepsilon_3 = -1/16\theta_1^4\theta_2^4.$$

The δ_j's are modular forms of level 2 and the ε_j's are modular forms of level 4, as easily follow from the properties of theta-functions. Note that these formulas give us the infinite product expansions of the ε's. They are very useful in following discussions. These three elliptic genera are uniquely characterized by their functional equations, as well as by their topological properties, such as rigidity under group action and multiplicativaty under spin fibrations. See [HBJ], [O], [S] and below for further discussions.

2. APPLICATIONS IN TOPOLOGY

The combining strength of index theory, Kac-Moody algebras, elliptic functions and modular forms in the above discussion can help us get many interesting topological results. Here for simplicity, I will only explain several special examples. For more general discussion see [Liu1]. The following two lists may be helpful in understanding the general picture of applications. First many results on a compact smooth spin manifold M have analogues on its loop space LM

$$M \quad - - - \quad LM$$

$\hat{A}$ − vanishing theorem	−−	$\hat{\Theta}$ − vanishing theorem
d_s is rigid	−−	the Witten rigidity theorem
signature $\equiv 0(\text{mod } 16)$	−−	signature $\equiv 0(\text{mod } 16)$.

But the results on LM are much more stronger and many of them have no finite dimensional analogues. More precisely we have the following list

$$M \quad - - - \quad LM$$

?	−−	higher level rigidity theorems
?	−−	higher level vanishing theorems
?	−−	general miraculous cancellation.

Here higher level means higher level loop group representation which appears natrually in our study. Of course one can get results on M by specializing the results on LM. We hope quantum group, which is the finite dimensional counterpart of affine Lie algebra, may eventually help explain the above ?'s.

2.1. Miraculous cancellation formula.

In [AW], a gravitational anomaly cancellation formula was derived from direct computations. This is a formula relating the L-class to the $\hat{A}$-class and a twisted $\hat{A}$-class of a 12-dimensional manifold. More precisely, let M be a smooth manifold of dimension 12, then the miraculous cancellation formula is

$$L(M) = 8\hat{A}(M,T) - 32\hat{A}(M)$$

where $T = TM$ denotes the tangent bundle of M and the equality holds at the top degree of each *differential form*. Here

$$\hat{A}(M,T) = \hat{A}(M)\mathrm{ch}\,T$$

is the $\hat{A}$-class twisted by tangent bundle.

Using elliptic genera, we can easily derive a much general formula for manifolds of any dimension. We can even get a formula with a general vector bundle involved. For simplicity, let us just take a manifold M of dimension $8k + 4$ and only consider the tangent bundle case. Write

$$\Theta_1(M) = \otimes_{j=1}^{\infty} S_{q^j}(\bar{T}M) \otimes \otimes_{m=1}^{\infty} \Lambda_{q^m}(\bar{T}M),$$

$$\Theta_2(M) = \otimes_{j=1}^{\infty} S_{q^j}(\bar{T}M) \otimes \otimes_{m=1}^{\infty} \Lambda_{-q^{m-1/2}}(\bar{T}M).$$

They have the following expansions in q with coefficients in $K(M)$

$$\Theta_1(M) = A_0 + A_1 q + \cdots,$$

$$\Theta_2(M) = B_0 + B_1 q^{\frac{1}{2}} + \cdots.$$

Denote the top degree terms in

$$L(M)\mathrm{ch}\,\Theta_1(M) \text{ and } \hat{A}(M)\mathrm{ch}\,\Theta_2(M)$$

by $P_1(\tau)$ and $P_2(\tau)$ respectively.

Let $\Gamma_0(2), \Gamma^0(2)$ be the modular subgroups,

$$\Gamma_0(2) = \{ \begin{pmatrix} a & b \\ c & d \end{pmatrix} \in SL_2(Z) | c \equiv 0 \ (\mathrm{mod}\ 2) \},$$

$$\Gamma^0(2) = \{ \begin{pmatrix} a & b \\ c & d \end{pmatrix} \in SL_2(Z) | b \equiv 0 \ (\mathrm{mod}\ 2) \}.$$

Recall that a modular form over a modular subgroup Γ is a holomorphic function $f(\tau)$ defined on the upper half plane H such that for any $g = \begin{pmatrix} a & b \\ c & d \end{pmatrix} \in \Gamma$, one has

$$f(\frac{a\tau + b}{c\tau + d}) = \chi(g)(c\tau + d)^k f(\tau)$$

where $\chi : \Gamma \to C^*$ is a character on Γ and k is called the weight of f. We also assume that f is holomorphic at $\tau = i\infty$. The following lemma can be proved by using the theta-function expressions of $P_1(\tau)$ and $P_2(\tau)$ as given in §1.

Lemma 2.1.1. $P_1(\tau)$ *is a modular form of weight* $4k + 2$ *over* $\Gamma_0(2)$; $P_2(\tau)$ *is a modular form of weight* $4k + 2$ *over* $\Gamma^0(2)$.

Let

$$\delta_1(\tau) = \frac{1}{8}(\theta_2^4 + \theta_3^4) \quad , \quad \varepsilon_1(\tau) = \frac{1}{16}\theta_2^4\theta_3^4,$$
$$\delta_2(\tau) = -\frac{1}{8}(\theta_1^4 + \theta_3^4) \quad , \quad \varepsilon_2(\tau) = \frac{1}{16}\theta_1^4\theta_3^4$$

be as in §1.4. They have the following Fourier expansions in q

$$\delta_1(\tau) = \frac{1}{4} + 6q + \cdots \quad , \quad \varepsilon_1(\tau) = \frac{1}{16} - q + \cdots,$$
$$\delta_2(\tau) = -\frac{1}{8} - 3q^{\frac{1}{2}} + \cdots \quad , \quad \varepsilon_2(\tau) = q^{\frac{1}{2}} + \cdots,$$

where "$\cdots$" are the higher degree terms all of which are of integral coefficients.

Let $M(\Gamma)$ denote the ring of modular forms over Γ with integral Fourier coefficients.

Using the transformation formulas of the Jacobi theta-functions [Ch], we get

Lemma 2.1.2. δ_1, δ_2 *are modular forms of weight* 2 *and* $\varepsilon_1, \varepsilon_2$ *are modular forms of weight* 4, *and furthermore* $M(\Gamma^0(2)) = Z[8\delta_2(\tau), \varepsilon_2(\tau)]$.

In view of these two Lemmas we can write

$$P_2(\tau) = b_0(8\delta_2)^{2k+1} + b_1(8\delta_2)^{2k-1}\varepsilon_2 + \cdots + b_k(8\delta_2)\varepsilon_2^k,$$

where the b_j's are integral linear combinations of the top degree terms of the $\{\hat{A}(M)\mathrm{ch}B_j\}$'s.

Apply the modular transformation $S : \tau \to -\frac{1}{\tau}$, we have

$$\delta_2(-\frac{1}{\tau}) = \tau^2 \delta_1(\tau) \quad , \quad \varepsilon_2(-\frac{1}{\tau}) = \tau^4 \varepsilon_1(\tau),$$

$$P_2(-\frac{1}{\tau}) = 2^{-(4k+2)} \tau^{4k+2} P_1(\tau).$$

Therefore,

$$P_1(\tau) = 2^{4k+2}[b_0(8\delta_1)^{2k+1} + b_1(8\delta_1)^{2k-1}\varepsilon_1 + \cdots + b_k(8\delta_1)\varepsilon_1^k].$$

At $q = 0$, $8\delta_1 = 2$ and $\varepsilon_1 = 2^{-4}$. We get a special case of the general miraculus cancellation formula,

Theorem 2.1. *At top degree, the following identity holds,*

$$L(M) = 2^3 \sum_{j=0}^{k} 2^{6k-6j} b_j.$$

For a more general formula, see [Liu1]. Denote by $b_j(TM)$ the integral linear combinations of the $B_i(TM)$'s which gives the polynomials b_j in the theorem. It is rather clear that these $b_j(TM)$'s can be determined canonically.

We have, for a compact oriented smooth manifold M,

Corollary 2.1. *The follwoing identity holds,*

$$Sign(M) = 2^3 \sum_{j=0}^{k} 2^{6k-6j} \int_M \hat{A}(M) ch\, b_j(TM).$$

The left hand side denotes the signature of M. In particular, if M is spin, then each characteristic number on the right hand side is an even integer. One thus recovers the Ochanine theorem [O1],

Corollary 2.2. *The signature of an $8k+4$ dimensional compact spin manifold is divisible by* 16.

Actually the proof shows that $P_1(\tau)$, the signature of the corresponding loop space LM is divisible by 16. Combining with the Atiyah-Patodi-Singer index formula, the above miraculous cancellation formula gives us interesting analytic expressions of some topological invariants. It can also be used to express the holonomy of certain determinant line bundles in terms of η-invariants. See [Liu1] and [LZ] for more details. Note that the main idea of the above proof are due to Hirzebruch [H1] and Landweber [La].

2.2. Rigidity. Elliptic genus originated from many people's trying to find the generating series of rigid elliptic operators. Given a smooth compact manifold with an action of a group G. Let P be an elliptic operator on M which commutes with the action. Then both the kernal and the cokernal of P are finite dimensional representations of G. The Lefschetz number of P at $g \in G$ is defined to be

$$L_P(g) = Tr_g \text{Ker}\, P - Tr_g \text{Coker}\, P$$

which is a character of G.

We say P is rigid with respect to G, if $L(g)$ is independent of g, that is a constant character of G. To prove the rigidity of an elliptic operator with respect to a general compact connected Lie group action, we obviously only need to study its rigidity with respect to S^1-action which will be our only concern in the following.

The rigidity of the signature operator d_s with respect to S^1-action on M is a trivial fact, since its kernal and cokernal are contained in the deRham cohomology. If M is spin, the rigidity of the Dirac operator with respect to the S^1-action is the famous $\hat{A}$-vanishing theorem of [AH]. As will be seen in the following, the rigidity of the signature operator with respect to S^1-action on loop space was conjectured by Witten and is highly non-trivial. The correponding $\hat{A}$-vanishing theorem for loop space will be discussed in next section.

In [W] Witten first conjectured the rigidity of $D \otimes TM$ and proved it for compact homogeneous spin manifolds. In trying to prove it in general and to find other rigid elliptic operators, Landweber-Stong and Ochanine discovered elliptic genus [LS], [O]. Witten, motivated by quantum field theory, conjectured that all of the three elliptic operators associted to the three elliptic genera, i.e.

$$\mathfrak{D}^L \otimes \triangle \otimes_{j=1}^{\infty} \Lambda_{q^j}(\bar{T}M)$$

$$\mathfrak{D}^L \otimes \otimes_{j=1}^{\infty} \Lambda_{-q^{j-1/2}}(\bar{T}M)$$

$$\mathfrak{D}^L \otimes \otimes_{j=1}^{\infty} \Lambda_{q^{j-1/2}}(\bar{T}M)$$

are rigid. Note that, so far no body has been able to give a direct proof of the rigidity of $D \otimes TM$ for a general spin manifold M which is only known through the following theorem.

Theorem 2.2. *The above three elliptic operators are rigid.*

This means that, if we expand the above elliptic operators into formal power series in q, for example

$$\mathfrak{D}^L \otimes \otimes_{m=1}^{\infty} \Lambda_{q^{m-1/2}}(\bar{T}M) = \sum_{j=1}^{\infty} D \otimes E_j q^{j-1/2},$$

then each $D \otimes E_j$ is rigid. Note that $D \otimes TM$ is the second term.

This theorem was first proved in [T], [BT]. Our proof uses the key properties of the Jacobi theta-functions and the Atiyah-Bott-Segal-Singer fixed point formula. For simplicity we only consider the isolated fixed point situation. Let $\{p\} \subset M$ be the fixed points of a generator $g = e^{2\pi i t} \in S^1$. Let $\{m_j\}$ be the exponents of TM at the fixed point p. That is, we have orientation-compatible decompositions

$$TM|_p = E_1 \oplus \cdots \oplus E_k, \ k = \frac{1}{2}\mathrm{dim}M,$$

and g acts on E_j by g^{m_j}.

let us denote the Lefschetz numbers of the above three elliptic operators by

$$F_1(t,\tau), \ F_2(t,\tau) \text{ and } F_3(t,\tau)$$

respectively. Apply the Lefschetz fixed point formula [AB], we have

$$F_1(t,\tau) = (2\pi i)^{-k} \sum_p \prod_{j=1}^{k} \frac{\theta_1(m_j t, \tau)\theta'(0,\tau)}{\theta(m_j t, \tau)\theta_1(0,\tau)},$$

$$F_2(t,\tau) = (2\pi i)^{-k} \sum_p \prod_{j=1}^{k} \frac{\theta_2(m_j t, \tau)\theta'(0,\tau)}{\theta(m_j t, \tau)\theta_2(0,\tau)},$$

$$F_3(t,\tau) = (2\pi i)^{-k} \sum_p \prod_{j=1}^{k} \frac{\theta_3(m_j t, \tau)\theta'(0,\tau)}{\theta(m_j t, \tau)\theta_3(0,\tau)}.$$

Here the theta-functions are the same as in [Ch]. The rigidity is equivalent to the fact that the above F's are independent of t. First we can obviously extend these F's to well-defined meromorphic functions on $(t,\tau) \in C \times H$. The key point is to show that they are actually holomorphic. The proof of the theorem is divided into three steps.

Lemma 2.2.1. *The F's are doubly periodic on the lattice $\{n + m\tau\}$.*

That is we have $F_j(t + n + m\tau) = F_j(t,\tau)$. For $g = \begin{pmatrix} a & b \\ c & d \end{pmatrix} \in SL_2(Z)$,

define modular transformation of g on the F's as

$$F(g(t,\tau)) = (c\tau + d)^{-k} F(\frac{t}{c\tau + d}, \frac{a\tau + b}{c\tau + d}).$$

Then we have

Lemma 2.2.2. *The integral span of the three F's are invariant under the action of $SL_2(Z)$.*

These two lemmas can be proved by using the transformation formulas of the four Jacobi theta-functions [Ch]. It depends essentially on the 'elliptic property' of these elliptic genera.

Lemma 2.2.3. *The three F's are holomorphic for t real and τ in upper half plane.*

Note that t real implies that $e^{2\pi i t}$ lies in S^1, and that the definition of Lefschetz number imples that it is the character of the action group S^1, therefore it is holomorphic as a function of $g \in S^1$. In our case, these F's can be expanded as power series in q with coefficients the combinations of the characters of S^1 representations. The Lefschetz fixed point forumla comes into play crucially at this point.

Then a key observation is that these three lemmas implies the rigidity. In fact, modular transformation, together with lemma 2, transforms the regularity of the F's on S^1 to everywhere in the complex plane; the first lemma immediately implies that they are constant. See [Liu0] for the detail.

2.3. A vanishing theorem of the Witten genus. Let M be a compact smooth spin manifold with an S^1-action, in [AH], Atiyah-Hirzebruch proved the following theorem,

   Theorem [AH]: *The $\hat{A}$-genus of M is zero.*

   As a corollary, we know that a compact smooth spin manifold with non-zero $\hat{A}$-genus does not admit any compact connected Lie group action.

   Let M be compact smooth and spin, with an S^1-action, and $MS^1 = ES^1 \times_{S^1} M$ be the Borel model of M. Let u be the generator of $H^*(BS^1, Z)$. Recall that $BS^1 = CP^\infty$ and $ES^1 = S^\infty$. The equivariant characteristic classes of M are defined to be the usual characteristic classes of MS^1. Let $p_1(M)_{S^1}$ denote the first equivariant Pontrjagin class with respect to the S^1-action. Let

$$\pi : \quad MS^1 \to BS^1$$

be the canonical projection and

$$\pi^* : \quad H^*(BS^1, Z) \to H^*(MS^1, Z)$$

be the pull-back in cohomology. Corresponding to the above theorem of Atiyah-Hirzebruch, we have the following theorem for LM,

Theorem 2.3. *If $p_1(M)_{S^1} = n \cdot \pi^* u^2$ for some integer n, then the Witten genus of M is zero.*

Note that the Witten genus is the $\hat{A}$-genus of LM which is the index of

$$D \otimes \otimes_{m=1}^{\infty} S_{q^m}(\bar{T}M).$$

The Witten genus is the virtual version of the $\hat{\Theta}$-genus which is the index of

$$D \otimes \otimes_{m=1}^{\infty} S_{q^m}(TM).$$

They differ by a factor $\eta(\tau)^{2k}$.

From quantum field theory, we know that, given a compact smooth manifold M, the loop space LM is orientable, if and only if M is spin; LM is spin, if and only if $p_1(M)$, the first Pontrjagin class of M is zero. The condition on the first equivariant Pontrjagin class in Theorem 2.3 is equivalent to that $p_1(M) = 0$ and the S^1-action preserves this condition, i.e. LM is spin and the S^1-action preserves this spin structure. Note that in Atiyah-Hirzebruch's situation, one can always lift the action to a double cover to make the action preserve the spin structure of M. The reason is that the second Stieffel-Whitney class whose vanishing is equivalent to the existence of spin structure on M is 2-torsion.

If there is an non-abelian compact connected Lie group action, the condition on $p_1(M)_{S^1}$ in Theorem 2.3 is equivalent to $p_1(M) = 0$ which gives us the following corollary due to Dessai [D],

Corollary 2.3. *Let M be a compact smooth spin manifold with a non-abelian compact connected Lie group action. If $p_1(M) = 0$, then the Witten genus of M is zero.*

Especially the Witten genus of any compact homogeneous spin manifold with $p_1 = 0$ vanishes. This is the first concrete evidence of the following conjecture of Hoehn and Stolz:

Conjecture: *Any compact smooth spin manifold with $p_1 = 0$ and positive Ricci curvature has vanishing Witten genus.*

The proof of our loop space $\hat{A}$-vanishing theorem is a refinement of the proof of the rigidity theorem in last section. Still let us only consider the isolated fixed point case and use the same notation as in last section. The Lefschetz number $H(t, \tau)$ of

$$D \otimes \otimes_{m=1}^{\infty} S_{q^m}(\bar{T}M)$$

is given by

$$H(t, \tau) = (2\pi i)^{-k} \sum_p \prod_{j=1}^{k} \frac{\theta'(0, \tau)}{\theta(m_j t, \tau)}.$$

Obviously $H(t, \tau)$ can be extended to $C \times H$ as a meromorphic function.

Lemma 2.3.1. *For $m, n \in Z$,*

$$H(t + n + m\tau, \tau) = e^{|n|\pi i(2t+\tau)} H(t, \tau).$$

This tells us the behaviour of $H(t, \tau)$ under the translation of lattice. Its behaviour under modular transformation is given by the following

Lemma 2.3.2. *For* $g = \begin{pmatrix} a & b \\ c & d \end{pmatrix} \in SL_2(Z)$, *we have*

$$H(\frac{t}{c\tau + d}, \frac{a\tau + b}{c\tau + d}) = (c\tau + d)^k e^{-|n|ct^2/(c\tau+d)} H(t, \tau).$$

These two lemmas imply that $H(t, \tau)$ is a (meromorphic) Jacobi form of index $-|n|/2$ and weight k. On the other hand, as in Lemma 2.2.3, the Atiyah-Bott-Segal-Singer Lefschetz fixed point formula tells us that $H(t, \tau)$ is holomorphic for $t \in R$ and τ in the upper half plane which together with the above two lemmas gives us the regularity of $H(t, \tau)$ in t on the whole complex plane. Therefore $H(t, \tau)$ is a holomorphic Jacobi form of negative index which is impossible, except it is zero. Actually it is a simple exercise to prove that the number of zeroes in a fundamental domain of the lattice of a holomorphic Jacobi form is equal to its index [EZ].

2.4. Kac-Moody algebras in topolgy. Let G be a simple and simply connected Lie group, and $\tilde{L}G$ be the central extension of its loop group. Let E be a positive energy representation of $\tilde{L}G$ of level m and P be a principal G-bundle on a compact smooth spin manifold M. Let $\psi(P, E)$ be the element constructed in §1 from P and E, and assume that there exists an S^1-action on M which lifts to P.

Theorem 2.4. *If* $p_1(M)_{S^1} = m \cdot p_1(P)_{S^1}$, *then*

$$D^L \otimes \psi(P, E)$$

is rigid.

Here $p_1(P)_{S^1}$ is the equivariant first Pontrjagin class of P. The proof of this theorem follows the same idea of the above modular invariance argument. The modular invariance of Kac character formula for the corresponding affine Lie algebras, instead of the classical Jacobi theta-functions comes into play.

Similarly we have the following

Theorem 2.5. *If* $m \cdot p_1(P)_{S^1} - p_1(M)_{S^1} = n \cdot \pi^* u^2$ *with* n *a negative integer,* *then the Lefschetz number, especially the index of*

$$D^L \otimes \psi(P, E)$$

vanishes.

We refer to [Liu] and [GL] for the proofs of the above results. It should be interesting to generalize them from affine Lie algebras to vertex operator algebras.

3. VECTOR BUNDLES AND MODULAR FORMS

In this section we first introduce infinite dimensional vector bundles on finite dimensional manifolds, and then describe the corresponding K-group and Riemann-Roch type theorems. We then construct rings from families of vector bundles with certain modularity restriction. Modulo torsion, these rings give construction of elliptic cohomology in terms of families of vector bundles. Part of the work in this section is still under progress.

3.1. Infinite dimensional vector bundles. Let M be a manifold. By using transition functions [L] we can define an infinite dimensional vector bundle E with Hilbert space as fiber just in the same way as define a finite dimensional vector bundle. Let H be the structure group of E which may be infinite dimensional. First assume that the fiber of E is an irreducible representation of H. We say E is a vector bundle of positive energy, if

1. There exists a fiberwise S^1-action on E, such that under this action E is decomposed into direct sum of finite dimensional vector bundles, $\oplus_{n \in Z} E_n$;

2. The character $\chi(\tau) = \sum_n dim E_n q^{n+a}$ for some rational number a is a modular form.

This definition is motivated by the definition of positive energy representations of infinite dimensional Lie algebras [K] and Lie groups [PS]. In general, we also call the direct sum and tensor product of two positive energy bundles of positive energy.

The examples of positive energy bundles include the case that $H = \tilde{L}G$ with G a simple and simply connected Lie group. Each positive energy representation of loop group admits an action of $\mathrm{Diff}^+ S^1$ which, in our case extends to fiberwise action on the corresponding infinite dimensional vector bundle on M. Therefore in loop group case, one has very rich structures on the infinite dimensional vector bundles. On the other hand, the Virasoro equivariant bundles on LM in [Br], when restricted to $M \subset LM$, gives infinite dimensional vector bundles of positive energy on M.

Corresponding to each positive energy vector bundle E, we introduce a formal power series with genuine vector bundle coefficients, $\psi(E) = q^a \sum_n E_n q^n$ and consider the set $\{\psi(E)\}$ for all E's. The sum and product in this set are induced by the sum and the product of the corresponding formal power series. In the standard way, we can make it into an abelian group with respect to the sum. Together with product, it becomes a ring which we denote by $FK(M)$. Especially interesting is the subring generated by those elements coming from

positive energy vector bundles with loop group as structure group, we denote
this subring by $LK(X)$. Obviously the equivalence relation in $FK(X)$ and
$LK(X)$ is not the one induced from the K-group of X.

3.2. Riemann-Roch and Virasoro algebra. Let X, Y be two compact
smooth spin manifolds with an embedding $f : Y \to X$. Assume that $\dim Y =
2l$ and $\dim X = 2k$. Denote by $FH^*(X)$ the image of the Chern character

$$\text{ch} : \ FK(X) \to H^*(X, R)$$

where R is the ring

$$R = \lim_{N \to \infty} Q[[q^{-1/N}, q^{1/N}]].$$

Let f_* and f^* denote the push-forward and pull-back map in cohomology
respectively.

The following theorem is an analogue of the Atiyah-Hirzebruch Riemann-
Roch theorem [AH1].

Theorem 3.1. *Assume $f^* p_1(X) = p_1(Y)$, then for any element U in $FK(Y)$,
there exists an element $f_! U$ in $FK(X)$, such that*

$$f_*(\hat{\Theta}(Y) \, ch \, U) = \hat{\Theta}(X) \, ch \, f_! U$$

holds in $FH^*(X)$.

We can also consider the subring $LK(X)$ in $FK(X)$ generated by those elements coming from positive energy vector bundles with loop group as structure
group. We can replace $FK(X)$ by $LK(X)$ in which case the above theorem
still holds. More precisely we have

Theorem 3.2. *Assume $f^* p_1(X) = p_1(Y)$, then for any element U in $LK(Y)$,
there exists an element $f_! U$ in $LK(X)$, such that*

$$f_*(\hat{\Theta}(Y) \, ch \, U) = \hat{\Theta}(X) \, ch \, f_! U$$

holds in $LH^*(X)$.

Here $LH^*(X)$ denotes the image of the Chern character of $LK(X)$ in $H^*(X, R)$.
Note that every element in $LK(X)$ admits a fiberwise action of $\text{Diff}^+ S^1$. Especially one can take $X = S^{2L}$ for a big integer L and Y an $MO < 8 >$-manifold
to get the

Corollary 3.1. *Every elliptic genus of Y can be realized as a (virtual) infinite
dimensional vector bundle on S^{2L} on which there exists a fiberwise action of
$\text{Diff}^+ S^1$.*

Here virtual means the difference of two bundles and the $Diff^+ S^1$ acts on each of them. The proof of Theorems 3.1 and 3.2 are basically the same as the proof of the Atiyah-Hirzebruch Riemann-Roch. We use the existence of the level 1 positive energy representations of $\tilde{L}\text{Spin}(2l)$, S^+, S^- such that the character of $S^+ - S^-$ is given by

$$\chi_{S^+ - S^-} = \prod_{j=1}^{l} \frac{\theta(\alpha_j, \tau)}{\eta(\tau)}$$

as in §1.2. For a different proof of Corollary 3.1, see [Ta].

Tensor category may enter this picture in the following way. Take a principal bundle on a manifold M with structure group $\tilde{L}G$ with G a semisimple Lie group. Let $P_m(G)$ be the set of highest weight representations of of $\tilde{L}G$ of level m. Each element in $P_m(G)$ induces one element in $LK(X)$ by the above construction. We denote the set consisting of these elements by $LK_m(M, G)$.

There exists an interesting exotic tensor structure on $P_m(G)$ which is given by fusion rule [KL]. This naturally induces an exotic tensor structure on $LK_m(M, G)$. For complex Grassmanniann, this is in turn isomorphic to the quantum cohomology. According to [KL], as tensor category $P_m(G)$ is isomorphic to the tensor category of representations of the quantum group $U_q G$ with q a root of unity. $U_q G$ is the deformation of the universal eveloping algebra of G. Hopefully the three very interesting subjects, loop groups, quantum groups and tensor category may bring some new light to topology.

3.3. Vector bundles and Jacobi forms. This section is motivated by the vertex operator algebra construction of the monstrous moonshine module. We will discuss a construction from Jacobi forms and relate it to elliptic cohomology.

Let $R(G)$ be the Grothendieck ring of representations of a Lie group G. Let $q = e^{2\pi i \tau}$ with τ in the upper half plane be a parameter. Let $\{V_n\}$ be a family of elements in $R(G)$ and $\chi(V_n)$ denote their characters.

A Jacobi form of level m and weight k with l variables over $L \rtimes \Gamma$, where L is an integral lattice in the complex plane C preserved by the modular subgroup Γ is a holomorphic function $f(t, \tau)$ with $t = (t_1, \cdots, t_l)$ on $C^l \times H$ such that

$$(1) \ f(\frac{t}{c\tau + b}, \frac{a\tau + b}{c\tau + d}) = (c\tau + d)^k e^{2\pi i m c \sum_j t_j^2/(c\tau + d)} f(t, \tau),$$

$$(2) \ f(t + \lambda\tau + \mu, \tau) = e^{-2\pi i m(l\lambda^2 \tau + 2\lambda \sum_j t_j)} f(t, \tau)$$

where $(\lambda, \mu) \in L$ and $g = \begin{pmatrix} a & b \\ c & d \end{pmatrix} \in \Gamma$. Here

$$\frac{t}{c\tau + d} = (\frac{t_1}{c\tau + d}, \cdots, \frac{t_l}{c\tau + d})$$

and

$$t + \lambda\tau + \mu = (t_1 + \lambda\tau + \mu, \cdots, t_l + \lambda\tau + \mu).$$

Let $(x_1, \cdots, x_l)$ denote the standard root basis of G. Consider the formal power series $\sum_n V_n \cdot q^n$.

Definition: *A family $\{V_n\}$ in $R(G)$ is called a Jacobian family of level n and weight k over $L \rtimes \Gamma$, if there exists a rational number a, such that*

$$q^a \sum_n \chi(V_n) q^n$$

is a Jacobi form in $(x_1, \cdots, x_l; \tau)$ of level n and weight k over $L \rtimes \Gamma$.

Let M be a smooth compact manifold. Let P be a principal G bundle on M. From each V_n we get a vector bundle on M by associating V_n to P. For convenience we still denote the corresponding bundle by V_n. This induces an element in the ring of formal power series in q with vector bundle coefficients. Let us denote the Grothendieck ring generated by these induced Jacobian elements over $L \rtimes \Gamma$ for some L by $JK(M)_\Gamma$.

The examples of Jacobian elements include the elements induced from loop group representations.

1) Bundles induced from loop group representations. Let E be an irreducible positive energy represenation of $\tilde{L}G$. According to the rotation of the circle, E has decomosition $\oplus_j E_j$ and there exists a rational number a such that

$$q^a \sum_j E_j q^j$$

is Jacobian. Its weight and level depend on G and E.

Actually the category of Jacobi elements may be bigger than that of loop group representations.

2) For any Jacobi form over $SL_2(Z)$, one can get a Jacobian element. Let $J_{k,m}$ be the ring of Jacobi forms over $L \rtimes SL_2(Z)$ for some lattice L [EZ]. For any $f(z, \tau) \in J_{k,m}$, under the action of $-I$ in $SL_2(Z)$ we have by definition

$$f(-z, \tau) = (-1)^k f(z, \tau)$$

which implies that

$$k \text{ even}, \quad f(z, \tau) = \sum_n P_n(\xi + \xi^{-1}) q^n$$

$$k \text{ odd}, \quad f(z, \tau) = \sum_n (\xi - \xi^{-1}) P_n(\xi + \xi^{-1}) q^n$$

where $\xi = e^{2\pi i z}$ and P_n is a polynomial.

Let E be a complex line bundle on M, then

$$F(E, \tau) = \sum_n P_n(E + E^{-1})q^n \text{ or } F(E, \tau) = \sum_n (E - E^{-1})P_n(E + E^{-1})q^n$$

is a Jacobian element over $L \rtimes SL_2(Z)$. Apply splitting principle, we can get a Jacobian element from any vector bundle on M.

3) The following interesting element from the definition of elliptic genus is a Jacobian element over $L \ltimes \Gamma_0(2)$ with $L = Z + Z\tau$,

$$[(\triangle_V^+ - \triangle_V^-) \otimes_{n=1}^{\infty} \Lambda_{-q^n} V_C] \cdot [(\triangle_V^+ + \triangle_V^-) \otimes_{m=1}^{\infty} \Lambda_{q^m} V_C]^{-1}.$$

where V_C is the complexification of V and $\triangle_V^{\pm}$ are the spinor bundles of V.

3.4. Relation with elliptic cohomology.

The Jacobian elements defined in last section can be easily related to elliptic cohomology. We take the modular subgroup $\Gamma = \Gamma_0(2)$ and denote the corresponding subring of level 0 elements by $JK(M)_\Gamma^0$ and denote $JK(M)_\Gamma \otimes Z[\frac{1}{2}]$ simply by $JK(M)$. The reason to invert 2, which will be clear from the following discussions, is to get a homomorphism from the Landweber-Ravenel-Stong elliptic cohomology [La]. Note that on a $MO < 8 >$ manifold, every Jacobian element has level 0. Let M_* be the ring of modular forms over $\Gamma_0(2)$ with Fourier coefficients in $Z[\frac{1}{2}]$, and B be the Bott periodicity element in $K(S^8)$. First it is easy to see the following

Lemma 3.4.1.
$$JK(S^8) = M_* \oplus M_*[B].$$

Define the reduced JK-group to be the kernal of the induced map by the inclusion of a point x_0 in M,

$$i^* : JK(M) \to JK(x_0).$$

It is clear that $JK(x_0) = M_*$. Let us denote the reduced JK-group by $\tilde{JK}(M)$. For a manifold $N \subset M$, one can accordingly define the relative JK-group by $JK(M, N) = \tilde{JK}(M/N)$ which is a subring of formal power series in q with coeffcents in $K(M, N)$ and with modularity. It can also be described by the standard difference construction in topological K-theory.

As in topological K-theory, for $n \geq 1$, using suspension, we define

$$\tilde{JK}^{-n}(M) = \tilde{JK}(S^n M),$$

$$JK^{-n}(M, N) = \tilde{JK}(S^n(M/N)),$$

$$JK^{-n}(M) = JK^{-n}(M, \phi).$$

Let $\varepsilon = \varepsilon_1 = 1/16\,\theta_2^4\theta_3^4$ be as in §1.4, we then have the following simple lemma:

Lemma 3.4.2. *The multiplication by $B\varepsilon$ induces an isomorphism*

$$JK^{-n}(M,N) \to JK^{-n-8}(M,N).$$

Consider

$$JK''(M,N) = \sum_n JK^{-n}(M,N).$$

Let

$$JK^*(M,N) = JK''(M,N)/(1 - B\varepsilon).$$

It is clear that

$$JK^*(pt) = M_*[\varepsilon^{-1}] = Z[\frac{1}{2}, \delta, \varepsilon, \varepsilon^{-1}].$$

The following obvious theorem establishes the relation of this ring with elliptic cohomology

Theorem 3.3. *There is a natural homomorphism Φ from $\Omega^*_{Spin}(M)[\frac{1}{2}]$ to $JK^*(M)$ which factorizes through elliptic cohomology.*

Φ is constructed in a standard way as in [CF]. Let

$$\pi : E \to BSpin(8k)$$

be the universal spin bundle, P be the corresponding principal bundle and $S^\pm$ be the two level 1 positive energy representations as given in §1.2. Denote by $S = S^+ + S^-$. Let $\psi(P, S^\pm)$ and $\psi(P, S)$ be the corresponding power series constructed from $S^\pm$ and respectively S as in §1.2. Then

$$\beta = (\pi^*(\psi(P, S^+)\psi(P, S)^{-1}), \pi^*(\psi(P, S^-)\psi(P, S)^{-1}); \mu)$$

where μ is the Clifford multiplication by e at $(x, e) \in E$, is an element in $JK^*(MSpin(8k))$. Note that $\psi(P, S)$ is invertable in $JK^*(BSpin(8k))$.

Let

$$f : S^{8k-n} \wedge M \to MSpin(8k)$$

be the map inducing an element $[X]_f$ in $\Omega^*_{Spin}(M)$, then

$$\Phi([X]_f) = f^*\beta \in JK(S^{8k-n} \wedge M).$$

This theorem implies the following diagram of natural homomorphisms,

$$
\begin{array}{ccc}
 & \mathrm{Ell}^*(M) & \\
{\scriptstyle p} \nearrow & & \searrow \bar{\Phi} \\
\Omega^*_{Spin}(M)[[\frac{1}{2}]] & \xrightarrow{\ \Phi\ } & JK^*(M)
\end{array}
$$

where p is the natural transformation and $\bar{\Phi}$ is the induced map by Φ. This diagram is also compatible with push-forward map [M]. More precisely for $\pi : M \to pt$, we have the K-theory push-forward

$$\pi_*^K : JK^*(M) \to M_*[\varepsilon^{-1}],$$

as well as the elliptic cohomology push-forward

$$\pi_*^E : \text{Ell}^*(M) \to M_*[\varepsilon^{-1}].$$

For an element $\alpha \in \text{Ell}^*(M)$, one has

$$\pi_*^E \alpha = \pi_*^K \bar{\Phi} \alpha$$

which should be compared with [M].

It is very easy to see that

$$\text{Ell}^*(M) \otimes Q \simeq JK^*(M) \otimes Q.$$

Actually both of them are isomorphic to $H^*(M, M_*[\varepsilon^{-1}] \otimes Q)$.

Obviously the $JK^*(M)$ is slightly 'bigger' than $\text{Ell}^*(M)$. At this point, we do not know how to put an equivalence relation in $JK^*(M)$ to make it a cohomology theory. If this can be done properly, then the quotient theory should be very close to a vector bundle description of $\text{Ell}(M)$.

REFERENCES

[A] Atiyah, M. F.: *Collected Works.* Oxford Science Publication 1989

[A1] Atiyah, M. F.: Classical Groups and Classical Differential Operators. In [A1], Vol. 4, 341-386

[AB] Atiyah, M. F. and Bott, R.: The Lefschetz Fixed Point Theorems for Elliptic Complexes I, II. in [A] Volume 3 91-170

[AH] Atiyah, M. F. and Hirzebruch, F.: Spin Manifolds and Group Actions. In [A] Vol. 3 417-429

[AH1] Atiyah, M. F. and Hirzebruch, F.: Riemann-Roch Theorems for Differentiable Manifolds. Bull. Amer. Math. Soc. 65 (1959), 276-281

[AS] Atiyah, M., Singer, I.: The Index of the Dirac Operator. In [A], Vol3 239-300

[BH] Borel, A., Hirzebruch, F.: Characteristic Classes and Homogeneous Spaces I. Amer. J. Math. 80 (1958) 458-538

[BT] Bott, R. and Taubes, C.: On the Rigidity Theorems of Witten. J. of AMS. No.2 (1989) 137-186

[Br] Brylinski, J-L.: Representations of Loop Groups, Dirac Operators on Loop Spaces and Modular Forms. Top. Vol. 29 No. 4 (1990) 461-480

[Co] Copson, E. T.: *Theory of Functions of a Complex Variable.* Oxford Univ. Press 1935

[D] Dessai, A.: S^3-action and the Witten Genus. (Preprint 1993)

[EZ] Eichler, M. and Zagier, D.: *The Theorey of Jacobi Forms.* Birkhauser 1985

[GL] Gong, D., Liu, K.: On the Rigidity of Higher Elliptic Genera. (Preprint 1994)

[H] Hirzebruch, F.: A Riemann-Roch Theorem for Differentiable Manifolds. Sem. Bourbaki, No. 177 (1959)

[H1] Hirzebruch, F.: Mannigfaltigkeiten und Modulformen. In Jahresber. Deutsch Math.-Verein. Jubilaumstagung, Teubner, Stuttgart, 1990, 20-38.

[HBJ] Hirzebruch, F., Berger, T., Jung, R.: *Manifolds and Modular Forms.* Vieweg 1991

[K]   Kac, V.: *Infinite Dimensional Lie Groups.* Cambridge Press (1991)

[KL] Kazhdan, D., Luztig, G.: Affine Lie Algebras and Quantum Groups. Int. Math. Res. Notices 21-29 (1991)

[L]   Lang, S.: *Differential Manifolds.* Addison Wesley 1972

[La] Landweber, P. S.: *Elliptic Curves and Modular Forms in Algebraic Topology.* Lecture Notes in Math. 1326

[LS] Landweber, P., Stong, R.: Circle Actions on Spin Manifolds and Characteristic Numbers. Topology 27 (1988) p 145-161

[Liu] Liu, K.: On Modular Invariance and Rigidity Theorems. J. of Diff. Geom. 1994, Vol. 4

[Liu0] Liu, K.: On Theta-functions and Elliptic Genera. Toplogy, to appear

[Liu1] Liu, K.: Modular Invariance and Characteristic Numbers. Commun. in Math. Physics, to appear

[Liu2] Liu, K.: On $SL_2(Z)$ and Topology. Math. Res. Letter Vol.1 No.1 53-64

[LZ] Liu, K., Zhang, W.: Elliptic Genus and η-invariant. Int. Math. Res. Notice, No. 8 (1994) 319-327.

[M]   Miller, H.: The Elliptic Character and the Witten Genus. Contemprory Math. Vol. 96 (1989) 281-289

[O]   Ochanine, S.: Sur les Genres Multiplicatifs Definis par Integrales Elliptiques. Topology 26 (1987) 143-151

[S]   Segal, G.: Elliptic Cohomology. Seminaire Bourbaki 40eannee, 1987-88 No.695

[Ta] Tamanoi, H.: Tamanoi, H.: Elliptic Genera and Vertex Operator Super Algebras. (Preprint 1994).

[T]   Taubes, C.: S^1-Actions and Elliptic Genera. Comm. in Math. Physics. Vol. 122 No. 3 (1989) 455-526

[We] Weil, A.: *Elliptic Functions According to Eisenstein and Kronecker.* Springer 1976

[W]   Witten, E.: The Index of the Dirac Operator in Loop Space. In [La] 161-186

Department of Mathematics
MIT
Cambridge, MA 02139

Contemporary Mathematics
Volume **193**, 1996

On Modular Invariance of Completely Replicable Functions

Yves Martin

1 Introduction

Let $J(q) = q^{-1}+196884q+21493760q^2+\ldots$ be the unimodular invariant function. Among its many properties we recall the following; its image under the action of the n-th Hecke operator is equal to a degree n polynomial in $J(q)$. John H. Conway and Simon Norton observed that a similar property is satisfied by the modular functions associated to the Monster simple group in [4]. That is, the coefficients in the Fourier series of these modular functions verify an infinite set of polynomial identities, analogous to the algebraic relations among the coefficients of $J(q)$ (see [4], [8]). These polynomial equations, called replication formulae in [4], are the objects studied in this paper.

Our starting point is the concept of replicable function as defined by Alexander, Cummins, McKay and Simons in [2].

Definition: *Let $f(q)$ be the formal power series with integers coefficients*

$$f(q) = q^{-1} + \sum_{m=1}^{\infty} a_m q^m. \tag{1}$$

For any positive integer n there is a unique monic polynomial with coefficients in $\mathbb{Z}$, say $P_{n,f}(t)$, such that $P_{n,f}(f(q)) - q^{-n}$ is a series in positive powers of q.

We say that $f(q)$ is a replicable function if for each positive integer n there is a power series $f^{(n)}(q) = q^{-1} + \sum_{m=1}^{\infty} a_m^{(n)} q^m$, called the n-th replicate of $f(q)$, such that

$$\frac{1}{n} \sum_{\substack{ad=n \\ 0 \le b < d}} f^{(a)}(\exp{(2\pi i \frac{b}{d})} q^{\frac{a}{d}}) = \frac{1}{n} P_{n,f}(f(q)). \tag{2}$$

1991 Mathematics Subject Classification: Primary 11F22; Secondary 11F11.

The n-th generalized Hecke operator T_n is defined to be the transformation on formal power series in q given by the left hand side of equation (2).

Observe that whenever $f^{(a)}(q) = f(q)$ for all divisors a of n, the operator T_n is just a classical Hecke operator. In the following we often drop the word generalized when we refer to T_n.

S. Norton has studied these replicable functions in several papers ([10], [11], [12]) and he has generalized this concept to power series in q with no restrictions on the integrality of its coefficients. In particular, Norton gave in [10] a different, equivalent definition of replicable function to the one presented here (see [2]).

In [5] Cummins and Norton show that a finite power series (1) is a replicable function if, and only if, $f(q)$ is of the form $q^{-1} + a_1 q$. This statement is part of a remarkable conjecture due to S. Norton [10], which characterize any replicable function as follows:

Conjecture: *Let $f(q)$ be the power series (1). Then $f(q)$ is a replicable function if, and only if, either $f(q) = q^{-1} + a_1 q$ or the series (1) has infinitely many nonzero coefficients and there is some group G with $\Gamma_0(N) \subseteq G \subseteq PSL(2, \mathbb{R})$ for a positive integer N such that:*

i) The group G contains a translation $\tau \mapsto \tau + k$ only if k is an integer.

ii) The series $f(q)$ represent a modular function invariant under G.

iii) The compactified Riemann surface X, defined by the action of G on the complex upper half plane $\mathcal{H}$, has genus zero and the field of meromorphic functions on X is the field of rational functions on $f(q)$, $\mathbb{C}(f(q))$.

We say that $f(q)$ is a Hauptmodul if $f(q)$ is a modular function on some subgroup G of $PSL(2, \mathbb{R})$, it has a Fourier expansion at infinity as in (1), and it satisfies condition (iii) above.

In [5] Cummins and Norton prove another part of this conjecture by showing that any Hauptmodul $f(q)$ is a replicable function.

In the work that we are now presenting we look at the converse implication for $f(q)$ in a proper subclass of replicable functions.

Definition: *A replicable function (1) is said to be completely replicable if each one of its replicates $f^{(n)}(q)$ is itself replicable, and*

for each positive integer s the s-replicate of $f^{(n)}(q)$ is $f^{(ns)}(q)$, i.e.
$f^{(n)(s)}(q) = f^{(ns)}(q)$.

Indeed, we consider only completely replicable functions $f(q)$ with an infinite series (1) and a finite number of distinct replicates. This last property implies the existence of only finitely many primes l for which $f^{(l)}(q) \neq f(q)$, and is equivalent to the existence of a positive integer N for which $f^{(Nn)}(q) = f^{(N)}(q)$ for all n. It follows from a result of John McKay and Harvey Cohn [3] that in such a case the power series $f(q)$ is J-final, i.e. $f^{(N)}(q) = J(q)$.

The number of completely replicable functions of this type is finite, and all modular functions associated to the Monster group in [4] are of this form. The complete list of these functions can be obtained from [6]. Since these completely replicable functions are known explicitly, it has been verified, case by case, that each one of them is the Hauptmodul of some genus zero subgroup of $PSL(2, \mathbb{R})$. This fact is used by R. Borcherds in the proof of the Monster Moonshine conjecture [1].

In an effort to find a theoretical explanation of the Hauptmodul property for these completely replicable functions we were able to prove the following theorem, the main result in this paper.

Theorem 1 *Let $f(q) = q^{-1} + \sum_{m=1}^{\infty} a_m q^m$ be a completely replicable function. Assume that infinitely many coefficients a_m are non-zero and that the number of distinct replicates of $f(q)$ is finite.*

Then $f(q)$ defines a single-valued holomorphic function on $\mathcal{H} - B$, where $\mathcal{H}$ is the complex upper half plane and $B \subseteq \mathcal{H}$ is some discrete set.

Let p be a prime such that $f^{(p)}(q)$, the p-th replicate of $f(q)$, is a modular function on some group G. Assume that $f^{(p)}(q)$ is holomorphic on $\mathcal{H}$, and $\Gamma_0(N) \subseteq G$ for some positive integer N.

Then $f(q)$ is a modular function on some group G_f contained in the intersection $G_p = G \cap \begin{pmatrix} \frac{1}{p} & 0 \\ 0 & 1 \end{pmatrix} G \begin{pmatrix} p & 0 \\ 0 & 1 \end{pmatrix}$. The index $[G_p : G_f]$ is less or equal to $p^2 + p$, and $\Gamma_0(p^3 N) \subseteq G_f$.

Furthermore, the function $f(q)$ is holomorphic on the whole upper half plane $\mathcal{H}$.

An immediate consequence of theorem 1 is the next

Corollary 2 *Let $f(q)$ be as above, and N a positive integer such that $f^{(N)}(q) = J(q)$.*

Then $f(q)$ is a modular function on some group G_f which contains $\Gamma_0(N^3)$.

The basic idea in the proof of theorem 1 is to show that a completely replicable function $f(q)$ is related to both $f^{(p)}(q)$ and $f^{(p)}(q^p)$ by a polynomial expression. Then one uses that $f^{(p)}(q)$ is a modular function on the group $\Gamma_0(N)$ in order to prove that $f(q)$ is a modular function on $\Gamma_0(p^3 N)$.

This result is part of an attempt to obtain the Hauptmodul property for $f(q)$ from the function $J(q)$ using some iterative argument. As one can see from the statement of theorem 1, the remaining obstacle for achieving this goal is that in general the group G_f is not large enough for the function $f : \mathcal{H}/G_f \mapsto \mathbb{C} \cup \{\infty\}$ to be injective, so that one needs to find more modular transformations that leave $f(q)$ invariant.

The rest of this paper is devoted to the proof of theorem 1, and it has been organized as follows:

In the next section we set up a formal and ad-hoc structure reflecting the main property of completely replicable functions. We begin by considering a ring $\mathcal{R}_1$ defined by infinitely many generators $\{x_m^{(n)}\}_{m,n\geq 1} \cup \{q\}$, and the set of algebraic relations in $\{x_m^{(n)}\}_{m,n\geq 1}$ that are satisfied by the coefficients of any completely replicable function and its replicates. Then any algebraic equation in $\mathcal{R}_1$ is still valid after the substitution of $x_m^{(n)}$ by $a_m^{(n)}$, where $\{a_m^{(n)}\}_{m\geq 1}$ is the set of coefficients of $f^{(n)}(q)$. Once $\mathcal{R}_1$ and similar rings $\mathcal{R}_M$ have been defined we show how a semigroup of upper triangular matrices in $GL^+(2, \mathbb{Q})$ induces ring homomorphisms among them. This allow us to introduce the generalized Hecke operators T_k and the Adams operators Ψ_k, $k \geq 1$, as endomorphisms of $\mathcal{R}_1$.

We spend the second half of section 2 studying the algebra of operators generated by $\{T_k, \Psi_k\}_{k\geq 1}$. We show that these endomorphisms satisfy algebraic relations completely analogous to those found in a more classical Hecke algebra, and in particular we prove that they define a commutative algebra.

In section 3 we use these properties of T_k and Ψ_k in the construction of an explicit polynomial in four variables $\Phi_l(W, X, Y, Z)$ with coefficients in $\mathcal{R}_1$ for every prime l. These polynomials satisfy the following; if $W_0 = X_0 = q^{-l} + \sum_{m=1}^{\infty} x_m^{(l)} q^{lm}$, $Y_0 = q^{-1} + \sum_{m=1}^{\infty} x_m^{(l)} q^m$

and $Z_0 = q^{-1} + \sum_{m=1}^{\infty} x_m^{(1)} q^m$ then $\Phi_l(W_0, X_0, Y_0, Z_0) = 0$. After the evaluation $x_m^{(n)} = a_m^{(n)}$ this gives an explicit algebraic relation among $f(q)$, $f^{(l)}(q)$ and $f^{(l)}(q^l)$. (For our purposes it is enough to consider these polynomials, but it will be clear that Φ_l can be defined similarly for composite integers l).

We should observe that $\Phi_l(X, X, Y, Z)$ is by no means the only polynomial that has the triple $f^{(l)}(q^l), f^{(l)}(q), f(q)$ as a root. Nevertheless, it is convenient since one can easily see from its construction that for each prime l such that $f^{(l)}(q) = f(q)$, $\Phi_l(X, X, Y, Y)$ gives the polynomial associated by Mahler to the power series $f(q)$ in [8]. In particular, if one substitutes $x_m^{(n)}$ by the coefficient of q^m in $J(q)$, then $\Phi_l(X, X, Y, Y)$ becomes the classical modular equation relating $J(q)$ with $J(q^l)$.

At this point in the paper the preliminary work has been done, and we proceed to prove theorem 1.

We fix a power series $f(q)$ satisfying the hypothesis of the theorem and finish section 3 defining two polynomials derived of $\Phi_l(W, X, Y, Z)$; $\Phi_p(Z) = \Phi_p(f^{(p)}(q^p), f^{(p)}(q^p), f^{(p)}(q), Z)$ and, for those primes l with $f^{(l)}(q) = f(q)$, the polynomial $\Phi_l(W, Z) = \Phi_l(W, W, Z, Z)$.

Section 4 contains the proof that the series $f(q)$ defines a holomorphic function on the complex upper half plane with the possible exception of a countable set. The argument here requires the existence of two distinct primes l_1, l_2 such that $f^{(l_i)}(q) = f(q)$, $i = 1, 2$, and uses the polynomials $\Phi_{l_i}(W, Z)$ defined before.

Section 5 contains our proof of the invariance of $f(q)$ under some congruence subgroup. Firstly we consider the polynomial $\Phi_p(Z) = \Phi_p(f^{(p)}(q^p), f^{(p)}(q^p), f^{(p)}(q), Z)$. Since $f(q)$ is a root of $\Phi_p(Z)$ and the coefficients of $\Phi_p(Z)$ are modular functions we get a precise characterization of the power series representing $f(q)|\gamma$ for any $\gamma \in SL(2, \mathbb{Z})$. From this, one shows the invariance of $f(q)$ under any element in $\Gamma(m_p pN)$ (N as in theorem 1 and m_p equal to the least common multiple of $\{1, 2, \ldots, p^2 + p\}$). As an easy consequence of his fact we can show that $f(q)$ must be holomorphic on the whole upper half plane and the groups G_p and G_f in theorem 1 have index at most $p^2 + p$.

In section 6 we use the definition of replication and the existence of infinitely many primes l with $f^{(l)}(q) = f(q)$ for the proof of the $\Gamma_0(m_p pN)$-invariance of $f(q)$. The generalization of these arguments show the invariance of $f(q)$ under $\Gamma_0(p^3 N)$.

2   Generalized Hecke Operators

For every positive integer r consider the formal power series in q

$$h^{(r)}(q) = q^{-1} + \sum_{m=1}^{\infty} x_m^{(r)} q^m \tag{3}$$

whose coefficients are the indeterminates $x_m^{(r)}$.

To any pair $r, k \geq 1$ we associate the polynomial $P_{k,r}(t)$ defined inductively by

$$P_{1,r}(t) = t, \qquad P_{k-1,r}(t) = k x_{k-1}^{(r)} + P_{k,r}(t) + \sum_{s=1}^{k-2} x_s^{(r)} P_{k-s-1,r}(t) \tag{4}$$

(see [2]). This is a polynomial in $\mathbb{Z}[x_1^{(r)}, x_2^{(r)}, \ldots, x_{k-1}^{(r)}][t]$ characterized by the fact that $P_{k,r}(h^{(r)}(q)) - q^{-k}$ is a power series in positive powers of q.

Let $\mathcal{L}$ be a field extension of $\mathbb{Q}$ containing all roots of unity and put $K = \mathcal{L}[\ldots, x_m^{(n)}, \ldots]$ where $m, n \geq 1$.

Then fix $r \geq 1$ and consider the set of equations indexed by $k \geq 1$

$$\frac{1}{k} \sum_{\substack{ad=k \\ 0 \leq b < d}} h^{(ra)}(\exp{(2\pi i \frac{b}{d})} q^{\frac{a}{d}}) = \frac{1}{k} P_{k,r}(h^{(r)}(q)) \tag{5}$$

These equations yield an infinite set of identities in K by considering the coefficients of the powers of q from both sides of (5) at each k. We denote by $I^{(r)}$ the ideal in K generated by them, and write I for the ideal in K generated by $\cup_{r=1}^{\infty} I^{(r)}$.

Observe that the unimodular invariant function $J(q) = q^{-1} + \sum_{m \geq 1} c_m$ satisfy the equations in (5) if we take $h^{(r)}(q) = J(q)$ for all r. Hence the map $x_m^{(r)} \mapsto c_m$ defines a non-trivial homomorphism $E : K \to \mathbb{Z}$ whose kernel contains I. Thus, I is properly contained in the ideal of K generated by $\{\ldots, x_m^{(n)}, \ldots\}$.

If u is a positive integer we define a $\mathbb{Z}$-algebra endomorphism ψ_u of K as follows; ψ_u fixes every element in $\mathcal{L}$ and maps each $x_m^{(n)}$ onto $x_m^{(nu)}$. Since the equations defining $I^{(r)}$ and $I^{(ru)}$ have the same form, it is clear that $\psi_u(I^{(r)}) = I^{(ru)}$ for any $r \geq 1$. Consequently $\psi_u(I) \subseteq I$, and we can think of ψ_u as a $\mathcal{L}$-algebra endomorphism of the quotient ring K/I.

Definition: For each $M \geq 1$ let $\mathcal{R}_M = \frac{K}{I}[[q^{\frac{1}{M}}]]$.

Definition: *Let Δ be the semigroup of $GL^+(2,\mathbb{Q})$ given by*

$$\Delta = \{ \begin{pmatrix} a & b \\ 0 & d \end{pmatrix} ; a,\ b,\ d \in \mathbb{Z},\ a > 0 \}.$$

For $M \geq 1$ and $\alpha = \begin{pmatrix} u & v \\ 0 & y \end{pmatrix} \in \Delta$ we set

$$e\|\alpha = e, \quad x_m^{(r)}\|\alpha = \psi_u(x_m^{(r)}) = x_m^{(ru)}, \quad q^{\frac{1}{M}}\|\alpha = \exp\left(2\pi i \frac{v}{My}\right) q^{\frac{u}{My}}$$

for all $e \in \mathcal{L}$, $m, r \geq 1$. Then we extend the mapping $\|\alpha$ to an $\mathcal{L}$-algebra homomorphism from $K[[q^{\frac{1}{M}}]]$ into $K[[q^{\frac{1}{My}}]]$.

Remarks:

i) The ideal I of K is stable under $\|\alpha$ for every $\alpha \in \Delta$, hence each $\|\alpha$ is an $\mathcal{L}$-algebra homomorphism from $\mathcal{R}_M$ into $\mathcal{R}_{My}$, where y is the lowest right entry of α.

ii) For every $\alpha, \beta \in \Delta$ and $g(q) \in \mathcal{R}_M$,

$$(g(q)\|\alpha)\|\beta = g(q)\|\alpha\beta.$$

iii) If $\alpha = \begin{pmatrix} u & v \\ 0 & y \end{pmatrix} \in \Delta$ then

$$h^{(r)}(q)\|\alpha = h^{(ru)}\left(\exp\left(2\pi i \frac{v}{y}\right) q^{\frac{u}{y}}\right)$$

iv) For every positive integer j, $(h^{(r)}(q))^j\|\alpha = (h^{(r)}(q)\|\alpha)^j$ as $\|\alpha$ is a ring homomorphism.

Definition: *Let n be a positive integer and $g(q) \in \mathcal{R}_1$. We define T_n by*

$$T_n g(q) = \frac{1}{n} \sum_{\substack{uy=n \\ 0 \leq v < y}} g(q)\| \begin{pmatrix} u & v \\ 0 & y \end{pmatrix} \tag{6}$$

and call T_n a (generalized) Hecke operator.

Definition: *Let n be a positive integer and $g(q) \in \mathcal{R}_1$. We denote by Ψ_n the mapping*

$$\Psi_n g(q) = g(q)\| \begin{pmatrix} n & 0 \\ 0 & n \end{pmatrix}. \tag{7}$$

Each Ψ_n is called an Adams operator, and obviously it extends the function ψ_n introduced at the beginning of this section.

It is immediate that both T_n and Ψ_n are linear endomorphisms of $\mathcal{R}_1$, and that the ideal I of K was defined so that

$$T_n h^{(r)}(q) = \frac{1}{n} P_{n,r}(h^{(r)}(q)) \tag{8}$$

in $\mathcal{R}_1$. Furthermore, for all $n \geq 1$, $\alpha \in \Delta$ and $g(q) \in \mathcal{R}_M$ we have $\Psi_n(g(q)\|\alpha) = (\Psi_n g(q))\|\alpha$. Therefore $(\Psi_m \circ T_n)h^{(r)}(q) = (T_n \circ \Psi_m)h^{(r)}(q)$.

In the following we prove some others relations verified by the operators $\{T_u, \Psi_u\}_{u \geq 1}$ of $\mathcal{R}_1$.

Using the same arguments as in the classical case one can show

Proposition 3 *Let l_1 and l_2 be two relatively prime positive integers, and $g(q) \in \mathcal{R}_1$. Then*

$$T_{l_1} T_{l_2} g(q) = T_{l_1 l_2} g(q). \tag{9}$$

In particular $T_{l_1} T_{l_2} g(q) = T_{l_2} T_{l_1} g(q)$.

Proposition 4 *If l is a prime, s a positive integer and $g(q) \in \mathcal{R}_1$, then*

$$T_{l^s} T_l g(q) = T_{l^{s+1}} g(q) + \frac{1}{l} T_{l^{s-1}} \Psi_l g(q) \tag{10}$$

where T_0 denotes the identity operator.

Proof. Using the definition of T_l we have that $T_{l^s} T_l g(q)$ is

$$\frac{1}{l^s} \sum_{\substack{0 \leq i \leq s \\ b_i}} \left(\frac{1}{l} \sum_b g(q) \middle\| \begin{pmatrix} 1 & b \\ 0 & l \end{pmatrix} + \frac{1}{l} g(q) \middle\| \begin{pmatrix} l & 0 \\ 0 & 1 \end{pmatrix} \right) \middle\| \begin{pmatrix} l^{s-i} & b_i \\ 0 & l^i \end{pmatrix}$$

where b_i in the exterior sum runs modulo l^i, and b in the interior sum runs modulo l. Hence $b_i + l^i b$ runs through a complete system modulo l^{i+1}. Therefore $T_{l^s} T_l g(q)$ can be written as

$$T_{l^{s+1}} g(q) - \frac{1}{l^{s+1}} g(q) \middle\| \begin{pmatrix} l^{s+1} & 0 \\ 0 & 1 \end{pmatrix} + \frac{1}{l^{s+1}} \sum_{\substack{0 \leq i \leq s \\ b_i}} g(q) \middle\| \begin{pmatrix} l^{s+1-i} & l b_i \\ 0 & l^i \end{pmatrix}$$

where again the sum on b_i is over the integers modulo l^i. On the other hand

$$\sum_{\substack{0 \leq i \leq s \\ b_i \pmod{l^i}}} g(q) \| \begin{pmatrix} l^{s+1-i} & lb_i \\ 0 & l^i \end{pmatrix} = g(q) \| \begin{pmatrix} l^{s+1} & 0 \\ 0 & 1 \end{pmatrix} +$$

$$\sum_{i=1}^{s} \left(\sum_{\substack{b_{i-1} \pmod{l^i} \\ b \pmod{l}}} g(q) \| \begin{pmatrix} l & 0 \\ 0 & l \end{pmatrix} \begin{pmatrix} l^{s-i} & b_{i-1} + l^{i-1}b \\ 0 & l^{i-1} \end{pmatrix} \right).$$

Thus $T_{l^s} T_l g(q)$ is equal to

$$T_{l^{s+1}} g(q) + \frac{1}{l^{s+1}} \sum_{i=0}^{s-1} \left(\sum_{\substack{b_i \pmod{l^i} \\ b \pmod{l}}} \Psi_l(g(q)) \| \begin{pmatrix} 1 & b \\ 0 & 1 \end{pmatrix} \begin{pmatrix} l^{s-1-i} & b_i \\ 0 & l \end{pmatrix} \right)$$

$$= T_{l^{s+1}} g(q) + \frac{1}{l} T_{l^{s-1}} \Psi_l(g(q))$$

using that $\Psi_l(g(q))$ is invariant under translations $\begin{pmatrix} 1 & b \\ 0 & 1 \end{pmatrix}$ with b in $\mathbb{Z}$. $\square$

Corollary 5 *Let l be a prime, r and s any positive integers and $g(q) \in \mathcal{R}_1$. Then*

$$T_{l^r} T_{l^s} g(q) = T_{l^s} T_{l^r} g(q).$$

Proof. By induction on r and s. $\square$

Corollary 6 *Let m and n be any two positive integers, $g(q) \in \mathcal{R}_1$. Then*

$$T_m T_n g(q) = T_n T_m g(q). \tag{11}$$

From our next result we get information about the action of the Hecke operators on powers of completely replicable functions. This proposition and its corollary will be important in showing that the series $f(q)$ in theorem 1 satisfies a polynomial in $\mathbb{Z}[f^{(p)}(q^p), f^{(p)}(q)][Z]$.

Let l be a fixed prime. We write Q_j for $lT_l(h(q)^j)$, $j \geq 1$, and $b^{(l)}$ for $b \| \begin{pmatrix} l & * \\ 0 & * \end{pmatrix}$ whenever $b \in K/I$. Moreover, we put $Q_0 = l + 1$ and define $T_{\frac{j}{l}}$ to be the operator that maps $\mathcal{R}_1$ to zero whenever $\frac{j}{l}$ is not an integer. With some abuse of notation we refer to the obvious image of $P_{j,r}(t) \in K[t]$ in the ring $\frac{K}{l}[t]$ by the same symbol.

Proposition 7 *Let $j \in \mathbb{Z}$, $j > 0$ and $P_{j,1}(t) = t^j + \sum_{i=1}^l b_{j,i} t^{j-i} \in \frac{K}{l}[t]$. Then*

$$Q_j + \sum_{i=1}^j b_{j,i} Q_{j-i} + \sum_{i=1}^j (b_{j,i}^{(l)} - b_{j,i}) h^{(l)}(q^l)^{j-i} =$$

$$P_{jl,1}(h(q)) + jT_{\frac{j}{l}} \Psi_l(h(q)). \qquad (12)$$

(here we further simplify the notation writing $h(q)$ instead of $h^{(1)}(q)$).

Proof. By equation (8)

$$T_j h(q) = \frac{1}{j}\left(h(q)^j + \sum_{i=1}^j b_{j,i} h(q)^{j-i} \right).$$

On the other hand $T_l(b_{j,i} h(q)^{j-i})$ is equal to

$$b_{j,i} \left(\frac{1}{l} \sum_{0 \leq b \leq l-1} h(q)^{j-i} \| \begin{pmatrix} 1 & b \\ 0 & l \end{pmatrix} + \frac{1}{l} h(q)^{j-i} \| \begin{pmatrix} l & 0 \\ 0 & 1 \end{pmatrix} \right)$$

$$-\frac{1}{l} b_{j,i} h(q)^{j-i} \| \begin{pmatrix} l & 0 \\ 0 & 1 \end{pmatrix} + \frac{1}{l} b_{j,i}^{(l)} \left(h(q)^{j-i} \| \begin{pmatrix} l & 0 \\ 0 & 1 \end{pmatrix} \right) =$$

$$b_{j,i} T_l(h(q)^{j-i}) + \frac{1}{l}(b_{j,i}^{(l)} - b_{j,i}) h^{(l)}(q^l)^{j-i}$$

(notice that one uses $T_l(1) = \frac{l+1}{l}$ for the case $i = j$).

Hence, by linearity and commutativity of the operators T_l and T_j we have

$$T_l(h(q)^j) = T_l(jT_j h(q) - \sum_{i=1}^j b_{j,i} h(q)^{j-i})$$

$$= jT_j T_l h(q) - \sum_{i=1}^j b_{j,i} T_l(h(q)^{j-i}) - \frac{1}{l}\sum_{i=1}^j (b_{j,i}^{(l)} - b_{j,i}) h^{(l)}(q^l)^{j-i}.$$

i) Suppose $\gcd(j,l) = 1$. Then $T_j T_l h(q) = \frac{1}{jl} P_{jl,1}(h(q))$ by proposition 3 and equation (8).

ii) Suppose $\gcd(j,l) \neq 1$, say l^s is the l-part of j with $s \geq 1$. Then

$$
\begin{aligned}
T_j T_l h(q) &= T_{\frac{j}{l^s}}\left(T_{l^{s+1}}h(q) + \frac{1}{l}T_{l^{s-1}}\Psi_l h(q)\right) \\
&= T_{jl}h(q) + \frac{1}{l}T_{\frac{j}{l}}\Psi_l h(q)
\end{aligned}
$$

and therefore

$$
ljT_j T_l h(q) = P_{jl}(h(q)) + jT_{\frac{j}{l}}\Psi_l h(q).\square
$$

From now on and for the rest of this article we fix a power series $f(q)$ as in (1) satisfying the hypothesis of theorem 1. The first step in the proof of our main result is to show that $f(q)$ verifies equation (12).

Let $E : K \to \mathcal{L}$ be the ring homomorphism given by $E(e) = e$ for all $e \in \mathcal{L}$ and $E(x_m^{(r)}) = a_m^{(r)}$ for all $m, r \geq 1$. Since $f(q)$ is a completely replicable function, $f(q)$ and its replicates verify equations (2), hence $I \subseteq \mathrm{Ker}E$. This shows that E can be viewed as a ring homomorphism from $\mathcal{R}_1$ into $\mathcal{L}$. In particular, any polynomial relation in $\mathcal{R}_1$ also holds in $\mathcal{L}$ for the corresponding images under E. Put $E(b_{j,i}) = \tilde{b}_{j,i}$, $E(Q_j) = Q_{j,f}$. Then $E(h^{(r)}(q)) = f^{(r)}(q)$ implies $Q_{j,f} = lT_l(f(q)^j)$. Moreover $E(P_{j,1}(h(q))) = P_{j,f}(f(q))$, and from proposition 7 we get

Corollary 8 *Let j be a positive integer. Then*

$$
Q_{j,f} + \sum_{i=1}^{j}\tilde{b}_{j,i}Q_{j-i,f} + \sum_{i=1}^{j}(\tilde{b}_{j,i}^{(l)} - \tilde{b}_{j,i})f^{(l)}(q^l)^{j-i} =
$$

$$
P_{jl,f}(f(q)) + jT_{\frac{j}{l}}f^{(l)}(q). \qquad (13)
$$

3   Generalized modular equations

The goal of this section is to present an explicit polynomial in Z with coefficients in $\mathcal{L}[f^{(l)}(q^l), f^{(l)}(q)]$ that has $f(q)$ as a root. From the same construction we get polynomials in $\mathcal{L}[W, Z]$ that have the pair $(f(q^l), f(q))$ as a solution whenever l is prime and $f^{(l)}(q) = f(q)$.

First we fix an arbitrary prime l, and for $j = 1, 2, \ldots, l+1$ recall the polynomials $P_{j,1}(t) = t^j + \sum_{i=1}^{l} b_{j,i}t^{j-i} \in K/I[t]$.

Definition: *Define $\tilde{Q}_j$ in $\frac{K}{I}[X, Y, Z]$ for $j = 0, 1, \ldots, l+1$ by*

$$
\tilde{Q}_0 = l+1 \quad and
$$

$$
\tilde{Q}_j + \sum_{i=1}^{j}b_{j,i}\tilde{Q}_{j-i} + \sum_{i=1}^{j}(b_{j,i}^{(l)} - b_{j,i})X^{j-i} = P_{jl,1}(Z) + j\delta_{j,l}Y \qquad (14)
$$

where X, Y and Z are variables, $b_{j,i}^{(l)}$ stand, as above, for the image of $b_{j,i}$ under $\| \begin{pmatrix} l & * \\ 0 & * \end{pmatrix}$ in K/I, and $\delta_{j,l}$ is the map that takes the value 1 at $j = l$ and is 0 everywhere else.

Definition: Let $\tilde{S}_j \in \frac{K}{I}[X, Y, Z]$, $j = 1, 2, \ldots, l+1$, be the polynomial given by

$$\tilde{Q}_j + \tilde{S}_1 \tilde{Q}_{j-1} + \tilde{S}_2 \tilde{Q}_{j-2} + \ldots + \tilde{S}_{j-1} \tilde{Q}_1 + j \tilde{S}_j = 0, \qquad (15)$$

and let $\Phi_l(W, X, Y, Z)$ be the polynomial in $\frac{K}{I}[W, X, Y, Z]$

$$\Phi_l(W, X, Y, Z) = W^{l+1} + \sum_{j=1}^{l+1} \tilde{S}_j W^{l+1-j}. \qquad (16)$$

Lemma 9 *The degree of $\Phi_l(W, X, Y, Z)$ in the variable Z is at most $l(l+1)$.*

Proof. It follows from the recursive definitions of $\tilde{S}_j$ and $\tilde{Q}_j$.$\square$

Observe that for the homomorphism $E : \mathcal{R}_1 \to \mathcal{L}$ given in the previous section one has $E(b_{j,i}) = \tilde{b}_{j,i} \in \mathbb{Z}$. Thus the image of $\Phi_l(W, X, Y, Z)$ under E defines a polynomial, that we also denote by $\Phi_l(W, X, Y, Z)$, in $\mathbb{Z}[W, X, Y, Z]$.

Proposition 10 *For any prime l*

$$\Phi_l(W, f^{(l)}(q^l), f^{(l)}(q), f(q)) =$$
$$(W - f^{(l)}(q^l)) \prod_{b=0}^{l-1} (W - f(\exp(2\pi i \frac{b}{l}) q^{\frac{1}{l}})). \qquad (17)$$

Proof. The polynomial $\Phi_l(W, f^{(l)}(q^l), f^{(l)}(q), f(q))$ is an element of $\mathbb{Z}[f^{(l)}(q^l), f^{(l)}(q), f(q)][W]$ and is equal to the image under the homomorphism E of $\Phi_l(W, X, Y, Z) \in \frac{K}{I}[W, X, Y, Z]$ after the substitutions $X = h^{(l)}(q^l)$, $Y = h^{(l)}(q)$ and $Z = h(q)$.

We put $\Phi_l(W, f^{(l)}(q^l), f^{(l)}(q), f(q)) = W^{l+1} + \sum_{j=1}^{l+1} E(\tilde{S}_j) W^{l+1-j}$.

Also, we set $(W - f^{(l)}(q^l)) \prod_{b=0}^{l-1} (W - f(\exp(2\pi i \frac{b}{l}) q^{\frac{1}{l}})) = W^{l+1} + \sum_{j=1}^{l+1} S_{j,f} W^{l+1-j}$.

Then one has to show $E(\tilde{S}_j) = S_{j,f}$ for all $j = 1, 2, \ldots, l+1$.

Since E is a ring homomorphism, all $E(\tilde{S}_j)$ are completely determined by the polynomials $\tilde{Q}_i$, $i = 1, 2, \ldots, j + 1$, the substitutions $X = h^{(l)}(q^l)$, $Y = h^{(l)}(q)$, $Z = h(q)$, and the mapping E (see equation (15)).

On the other hand the image under E of every $\tilde{Q}_j$ with $X = h^{(l)}(q^l)$, $Y = h^{(l)}(q)$, $Z = h(q)$, say $E(\tilde{Q}_j)$, satisfy the recursive relation (13) (see equation (14)). Thus $E(\tilde{Q}_0) = \tilde{Q}_0 = Q_{0,f}$ implies $E(\tilde{Q}_j) = Q_{j,f} = lT_l(f(q)^j)$ for all $j = 1, 2, \ldots, l + 1$, and the equality $E(\tilde{S}_j) = S_{j,f}$ follows.$\square$

Corollary 11

$$\Phi_l(f^{(l)}(q^l), f^{(l)}(q^l), f^{(l)}(q), f(q)) = 0. \tag{18}$$

Corollary 12 *Consider the polynomial* $\Phi_l(W, X, Y, Z)$ *given by equation (16) and let* $W = X = h^{(l)}(q^l)$, $Y = h^{(l)}(q)$. *Then its image under the ring homomorphism* E,

$$\Phi_l(Z) =: \Phi_l(f^{(l)}(q^l), f^{(l)}(q^l), f^{(l)}(q), Z), \tag{19}$$

is a polynomial in $\mathbb{Z}[f^{(l)}(q^l), f^{(l)}(q)][Z]$ *with the series* $f(q)$ *as a root.*

Corollary 13 *Assume that the prime* l *is such that* $f^{(l)}(q) = f(q)$. *Then*

$$\Phi_l(W, W, f(q), f(q)) = (W - f(q^l)) \prod_{b=0}^{l-1} (W - f(\exp{(2\pi i \frac{b}{l})} q^{\frac{1}{l}})). \tag{20}$$

In particular $\Phi_l(W, Z) =: \Phi_l(W, W, Z, Z) \in \mathbb{Z}[W, Z]$ *is the* l-*th polynomial associated to* $f(q)$ *by Mahler [8].*

Proof. It follows from equation (14) and [8], p. 70.$\square$

The polynomials $\Phi_l(Z)$ and $\Phi_l(W, Z)$ defined in the previous corollaries will be used in two ways. In the next section we take two distinct primes p_1, p_2 with $f^{(p_i)}(q) = f(q)$ and get from $\Phi_{p_i}(W, Z)$ $i = 1, 2$ that $f(q)$ is a holomorphic function on the complex upper half plane minus a discrete subset. Latter we consider the polynomial $\Phi_p(Z)$, where p is the prime in theorem 1, and prove the invariance of $f(q)$ under some congruence subgroup.

Before ending this section we show that $\Phi_p(Z)$ is a non-zero polynomial. From equation (16) it follows that the constant term of $\Phi_p(Z)$

is equal to a polynomial in $\mathbb{Z}[f^{(p)}(q^p), f^{(p)}(q)]$ which is monic of degree $p+1$ in $f^{(p)}(q^p)$, and at most of degree 1 in $f^{(p)}(q)$ (see equation (14)). If this polynomial were zero we should have either $f^{(p)}(q) \in \mathbb{C}(f^{(p)}(q^p))$ or $f^{(p)}(q^p)$ is algebraic over $\mathbb{C}$. Both cases are impossible as $f^{(p)}(q)$ is a non-constant modular function on $\Gamma_0(N)$.

4   The holomorphicity of f(q)

We prove here that the power series (1) represents a singled-valued, holomorphic function on $\mathcal{H} - B$, where B is some discrete subset of $\mathcal{H}$.

Let p_1 and p_2 be two different primes for which the replicates $f^{(p_1)}(q)$ and $f^{(p_2)}(q)$ are equal to $f(q)$. Consider the polynomials $\Phi_{p_i}(W, Z)$, $i = 1, 2$. These are two symmetric polynomials of degree $p_i + 1$ in $\mathbb{C}[W, Z]$ (see [8]). As $\Phi_{p_i}(f(q^{p_i}), f(q)) = 0$, we get from [8] that $f(q)$ converges and defines a singled-valued holomorphic function on a punctured disc $0 < |q| \leq R'$ for some R' in $\mathbb{R}$. Obviously, this function has a simple pole at the origin $q = 0$.

For any positive real number r let $D(r) = \{q \in \mathbb{C}; \ 0 < |q| < r\}$. Denote by $B(R')$ the subset in $D(R')$ of all those q_0 for which either polynomial $\Phi_{p_1}(W, f(q_0))$ or $\Phi_{p_2}(W, f(q_0))$ has a multiple root. Since $f(q)$ has a simple pole at $q = 0$ we can choose R' such that $B(R')$ is finite.

Suppose $R' < 1$. Then there exist $R > R'$ such that $R^{p_1} < R'$ and $R^{p_2} < R'$. We now prove that $f(q)$ can be continued analytically to a singled-valued function on $D(R)$ minus some finite subset.

Let q_1 be in $D(R)$ such that $\Phi_{p_1}(f(q_1^{p_1}), Y)$ has no repeated roots. We choose a path $c : [0, 1] \to D(R)$ from any $c(0) = q_0 \in D(R')$ to $c(1) = q_1$ such that $c(t)^{p_1} \notin B(R')$ for $0 \leq t \leq 1$. Then there exist an analytic continuation of $f(q)$ along the curve $c(t)$ which determines a holomorphic germ ϕ at q_1 satisfying the equation $\Phi_{p_1}(f(q_1^{p_1}), \phi(q_1)) = 0$.

Now let $\tilde{c} : [0, 1] \to D(R)$ be any closed path such that $\tilde{c}(0) = \tilde{c}(1) \in D(R')$. Let $\{(O_i, f_i); \ 1 \leq i \leq k\}$ be the set of germs of holomorphic functions defining some analytic continuation of $f(q)$ along $\tilde{c}(t)$. Without lost of generality we take both O_1 and O_k inside $D(R')$ and each open set $O_k^{p_1}$ and $O_k^{p_2}$ away from $B(R')$. Since $\Phi_{p_1}(f(q), f(q^{p_1})) = 0$ for $q \in O_1$, the holomorphicity of each f_i yields $\Phi_{p_1}(f_k(q), f(q^{p_1})) = 0$ on O_k. Similarly, $\Phi_{p_2}(f_k(q), f(q^{p_2})) = 0$ on O_k.

The equation $\Phi_{p_1}(W, f(q^{p_1})) = 0$ (resp. $\Phi_{p_2}(W, f(q^{p_2})) = 0$) defines a $(p_1 + 1)$-sheeted (resp. $(p_2 + 1)$-sheeted) unbranched covering of O_k. The $(p_1 + 1)$ sheets of this covering are

$$\{f(q), f(\zeta_1 q), f(\zeta_1^2 q), \ldots, f(\zeta_1^{p_1-1} q), f(q^{p_1^2})\}$$

where ζ_1 is a primitive p_1-root of unity. The $(p_2 + 1)$ sheets of the covering defined by $\Phi_{p_2}(X, f(q^{p_2})) = 0$ on O_k are

$$\{f(q), f(\zeta_2 q), f(\zeta_2^2 q), \ldots, f(\zeta_2^{p_2-1} q), f(q^{p_2^2})\}$$

where ζ_2 is a primitive p_2-root of unity. By the previous paragraph we know that $f_k(q)$ is in the intersection of these two sets.

Suppose $f_k(q) \neq f(q)$. Then $f(\zeta_1 q) = f(\zeta_2 q)$ or $f(\zeta_1 q) = f(q^{p_2^2})$ or $f(q^{p_1^2}) = f(q^{p_2^2})$ on O_k after a suitable choice of ζ_1, ζ_2. Since the mappings $q \mapsto f(\zeta_i q)$ and $q \mapsto f(q^{p_i^2})$ are holomorphic on $D(R')$ one of these identities must hold on the whole $D(R')$. But then, for any $\xi > 0$ we can find $q' \neq q''$ in $D(\xi)$ such that $f(q') = f(q'')$. This is a contradiction as $q = 0$ is a simple pole of $f(q)$. Consequently $f_k(q) = f(q)$.

If we put $B(R) = \{q \in D(R); \ q^{p_1} \in B(R')\}$ then $B(R)$ is finite. Hence the analytic continuation of $f(q)$ from $D(R')$ to $D(R) - B(R)$ gives a well-defined single-valued function on the latter set. This shows

Proposition 14 *The replicable function $f(q)$ with $q = \exp(2\pi i\tau)$, τ in $\mathcal{H}$, defines a singled-valued holomorphic function on $\mathcal{H} - B$, where B is some discrete subset of the complex upper half plane $\mathcal{H}$.*

It is not hard to see that the set B in proposition 14 can be assumed to be $SL(2, \mathbb{Z})$-invariant.

5   Invariance of $f(q)$ under some modular transformations

We just showed that the power series $f(q)$ represents a holomorphic function defined on $\mathcal{H} - B$ after the substitution $q = \exp(2\pi i\tau)$, $\tau \in \mathcal{H}$. From now on we denote this function by $f(\tau)$. Next we prove that $f(\tau)$ is invariant under the action of some parabolic and hyperbolic elements in $\Gamma_0(pN)$, and deduce from it the invariance of $f(\tau)$ under some congruence subgroup.

Let p to be the prime in the hypothesis of theorem 1, and adopt the more convenient notation $g(\tau) = f^{(p)}(q)$, $g(p\tau) = f^{(p)}(q^p)$. Both $g(\tau)$ and $g(p\tau)$ are modular functions invariant under the group

$$G_p = G \cap \begin{pmatrix} p & 0 \\ 0 & 1 \end{pmatrix}^{-1} G \begin{pmatrix} p & 0 \\ 0 & 1 \end{pmatrix}. \qquad (21)$$

We denote by X_p the compact Riemann surface defined by the quotient $\mathcal{H}/G_p$, and let $\mathcal{M}(X_p)$ be the field of meromorphic functions on X_p. Then $\Phi_p(Z)$ is a non-zero polynomial in $\mathcal{M}(X_p)[Z]$.

Lemma 15 *Let* $\gamma = \begin{pmatrix} a & b \\ c & d \end{pmatrix} \in SL(2, \mathbb{Z})$ *and* $h_c = \frac{N}{\gcd(c^2, N)}$. *Then there exist* $n_0 \in \mathbb{Z}$ *and* $1 \leq n' \leq p^2 + p$ *such that*

$$f \circ \gamma(\tau) = \sum_{n \geq n_0} d_n \exp\left(2\pi i n\tau / n' p h_c\right) \qquad (22)$$

on some open neighborhood of $i\infty$. *If* $g(\tau)$ *and* $g(p\tau)$ *are holomorphic at* $\gamma(i\infty)$ *then* $n_0 \geq 0$.

Proof. Let U be a subset of $\mathcal{H} - B$ such that $U \cup \{i\infty\}$ is an open neighborhood of $i\infty$. Since B is $SL(2, \mathbb{Z})$-invariant, the open set $\gamma'(U)$ is also contained in $\mathcal{H} - B$ for any γ' in $SL(2, \mathbb{Z})$. Therefore the functions $f \circ \gamma'(\tau)$ are holomorphic on U. As $f(\tau)$ is a root of the polynomial $\Phi_p(Z) = \Phi_p(g(p\tau), g(p\tau), g(\tau), Z)$ given in (19) it follows that $f \circ \gamma'(\tau) = f(\tau)|\gamma'$ is a root of

$$\Phi_p\left(g(\tau)\Big|\begin{pmatrix} p & 0 \\ 0 & 1 \end{pmatrix}\gamma', g(\tau)\Big|\begin{pmatrix} p & 0 \\ 0 & 1 \end{pmatrix}\gamma', g(\tau)|\gamma', Z\right). \qquad (23)$$

Hence for every $\gamma' \in G_p$, $f \circ \gamma'(\tau)$ is a root of $\Phi_p(Z)$. In particular, the set $\{f \circ \gamma'(\tau) : \gamma' \in G_p\}$ have at most $p^2 + p$ elements by lemma 9. Clearly, the group G_p acts transitively on this set.

Consider the group $H_f = \{\gamma' \in \Gamma_0(pN) : f \circ \gamma'(\tau) = f(\tau) \text{ on } \mathcal{H} - B\}$. This is the stabilizer of $f(\tau)$ in $\Gamma_0(pN)$, hence its index in $\Gamma_0(pN)$ is $[\Gamma_0(pN) : H_f] = n$ for some n with $1 \leq n \leq p^2 + p$.

For fixed $\gamma = \begin{pmatrix} a & b \\ c & d \end{pmatrix} \in SL(2, \mathbb{Z})$ the integer $h'_c = \frac{pN}{\gcd(c^2, pN)}$ is the smallest positive integer such that $\begin{pmatrix} 1 & h'_c \\ 0 & 1 \end{pmatrix} \in \gamma^{-1}\Gamma_0(pN)\gamma$. Now

$[\gamma^{-1}\Gamma_0(pN)\gamma : \gamma^{-1}H_f\gamma] = n$ implies that for some n' with $1 \leq n' \leq n$ the matrix $\begin{pmatrix} 1 & n'h'_c \\ 0 & 1 \end{pmatrix}$ is in $\gamma^{-1}H_f\gamma$. Thus the function $f \circ \gamma(\tau)$ can be represented on U by a series in powers of $q^{\frac{1}{n'h'_c}}$.

Both $g(\tau)$ and $g(p\tau)$ are modular functions on $\Gamma_0(pN)$, and so they are meromorphic at $\gamma(i\infty)$. We recall that $f \circ \gamma(\tau)$ is a root of the polynomial (23) with $\gamma' = \gamma$, and deduce from it that $f \circ \gamma(\tau)$ is meromorphic at $i\infty$ (see proposition 2.7 in [13] p. 31). By the same argument, the holomorphicity of $g(\tau)$ and $g(p\tau)$ at $\gamma(i\infty)$ imply that $f \circ \gamma(\tau)$ is holomorphic at $i\infty$.

Finally we only have to check that $\frac{h'_c}{h_c} \in \{1, p\}$.$\Box$

Observe that the proof above also yields the inequality $[G_p : G_f] \leq p^2 + p$, where G_f denotes the stabilizer of f in G_p.

Let m_p be the least common multiple of $\{1, 2, \ldots, p^2 + p\}$.

Proposition 16 *Let α be an element in the principal congruence subgroup $\Gamma(m_p pN)$.*

Then $f \circ \alpha(\tau) = f(\tau)$ for all $\tau \in \mathcal{H} - B$.

Proof. If $\alpha(i\infty) = i\infty$ then α is an upper triangular matrix, and the invariance of $f(\tau)$ is clear.

Otherwise there exist $s \in \mathbb{Q}$ which is fixed by α, since α is either a parabolic or hyperbolic element. Let $\gamma \in \text{SL}(2, \mathbb{Z})$ such that $\gamma(i\infty) = s$. Then $\gamma^{-1}\alpha\gamma = \begin{pmatrix} 1 & t \\ 0 & 1 \end{pmatrix}$ for some $t \in \mathbb{Z}$. As α is in $\Gamma(m_p pN)$, t must be divisible by $m_p pN$.

The Fourier expansion of $f \circ \gamma(\tau)$ on some neighborhood of $i\infty$ is given by equation (22), thus

$$f \circ \alpha \circ \gamma(\tau) = f \circ \gamma \circ (\gamma^{-1}\alpha\gamma)(\tau) = f \circ \gamma(\tau + t) = f \circ \gamma(\tau).$$

Hence $f(\alpha(\tau')) = f(\tau')$ on an open neighborhood of s. Since $f(\tau)$ is holomorphic on $\mathcal{H} - B$ the proposition follows.$\Box$

By the last result we may consider $f(\tau)$ as a function on the Riemann surface defined by the action of $\Gamma(m_p pN)$ on $\mathcal{H} - B$. Around any $b \in B$ there is a neighborhood on which the function $f(\tau)$ is bounded (since $g(\tau)$ is holomorphic on $\mathcal{H}$ and $\Phi_p(g(p\tau), g(p\tau), g(\tau), f(\tau)) = 0$). Hence Riemann's removable singularities theorem implies that $f(\tau)$

can be extended uniquely to a holomorphic function on $\mathcal{H}$, invariant under $\Gamma(m_p pN)$. In particular $f(\tau)$ is a modular function on $\Gamma(m_p pN)$, and therefore on $\Gamma_1(m_p pN)$.

6   Invariance of $f(\tau)$ under $\Gamma_0(p^3 N)$

In this final section we complete the proof of theorem 1 by showing that the function $f(\tau)$ is invariant under $\Gamma_0(p^3 N)$. Firstly we prove that $f(\tau)$ is invariant under $\Gamma_0(m_p pN)$. Latter, the same arguments will be generalized for the case of the group $\Gamma_0(p^3 N)$.

Lemma 17 *Let $\gamma \in \Gamma_0(m_p pN)$ and p_0 a prime which does not divide $m_p pN$. Then there exist $\delta \in \Gamma_1(m_p pN)$ such that $\delta\gamma \in \Gamma_0(p_0 m_p pN)$.*

Proof. Let $\gamma = \begin{pmatrix} a & b \\ c & d \end{pmatrix}$. Since $\gcd(d, m_p pN) = 1$ we can choose $a' \in \mathbb{Z}$ such that $a'd \equiv 1 \pmod{m_p pN}$ and $\gcd(a', p_0) = 1$. Then there exists some $\gamma' = \begin{pmatrix} a' & b' \\ p_0 m_p pN & d' \end{pmatrix}$ in $\Gamma_0(p_0 m_p pN)$. Take $\delta = \gamma'\gamma^{-1}$.□

Lemma 18 *Let p_0 be an arbitrary prime and $\gamma \in \Gamma_0(p_0 m_p pN)$. Then for each l, $0 \le l \le p_0 - 1$, there is a unique l' with $0 \le l' \le p_0 - 1$ such that*

$$\gamma \begin{pmatrix} 1 & l \\ 0 & p_0 \end{pmatrix} \gamma^{-1} \begin{pmatrix} 1 & l' \\ 0 & p_0 \end{pmatrix}^{-1} \in \Gamma_1(m_p pN). \qquad (24)$$

Moreover

$$\gamma \begin{pmatrix} p_0 & 0 \\ 0 & 1 \end{pmatrix} \gamma^{-1} \begin{pmatrix} p_0 & 0 \\ 0 & 1 \end{pmatrix}^{-1} \in \Gamma_1(m_p pN). \qquad (25)$$

Proof. Let $\gamma = \begin{pmatrix} a & b \\ c & d \end{pmatrix}$. Then, for any given l choose l' between 0 and $p_0 - 1$ such that $al - b - dl' \equiv 0 \pmod{p_0}$.□

Notice that equation (24) in the previous lemma also holds for $\gamma \in \Gamma_0(m_p p\acute{N})$ and $p_0 = p$.

Lemma 19 *Let* $\gamma \in \Gamma_0(m_p pN)$. *If* p_0 *is a prime,* $\gcd(p_0, m_p pN) = 1$ *and* $f^{(p_0)}(q) = f(q)$ *then*

$$f(\tau)|\gamma \begin{pmatrix} p_0 & 0 \\ 0 & 1 \end{pmatrix} + \sum_{l=0}^{p_0-1} f(\tau)|\gamma \begin{pmatrix} 1 & l \\ 0 & p_0 \end{pmatrix} = P_{p_0,f}(f(\tau)|\gamma). \qquad (26)$$

Furthermore

$$g(\tau)|\gamma \begin{pmatrix} p & 0 \\ 0 & 1 \end{pmatrix} + \sum_{l=0}^{p-1} f(\tau)|\gamma \begin{pmatrix} 1 & l \\ 0 & p \end{pmatrix} = P_{p,f}(f(\tau)|\gamma). \qquad (27)$$

Proof. It follows from lemmas 17, 18, and the invariance of $f(\tau)$ (resp. $g(\tau)$ and $g(p\tau)$) under $\Gamma_1(m_p pN)$ (resp. $\Gamma_0(m_p pN)$).□

Lemma 20 *Let* $\gamma \in \Gamma_0(m_p pN)$. *Then* $f(\tau)|\gamma = f(s\tau)$ *for some positive integer* s.

Proof. Since $f(\tau)$ is a modular function on $\Gamma_1(m_p pN)$ there exist some $s \in \mathbb{Z}$ such that $f(\tau)|\gamma = d_{-s}q^{-s} + d_{-s+1}q^{-s+1} + \ldots + d_m q^m + \ldots$ with $d_{-s} \neq 0$. Indeed, $s > 0$ as $\Phi_p(f(\tau)|\gamma)$ is both zero and equal to a polynomial in $\mathbb{C}[g(p\tau), g(\tau), f(\tau)|\gamma]$ of degree $p+1$ in $g(p\tau)$ and degree at most 1 in $g(\tau)$ (see equations (14), (15), (16)).

Let p_0 be any prime such that $\gcd(p_0, m_p pN) = 1$ and $f^{(p_0)}(q) = f(q)$. Then equation (26) yields

$$\sum_{m=-s}^{\infty} d_m q^{mp_0} + \sum_{\substack{m=-s \\ p_0|m}}^{\infty} p_0 d_m q^{\frac{m}{p_0}} =$$

$$d_{-s}^{p_0} q^{-sp_0} + p_0 d_{-s}^{p_0-1} d_{-s+1} q^{-sp_0+1} + \ldots - p_0 a_1 d_{-s}^{p_0-2} q^{-s(p_0-2)} + \ldots \qquad (28)$$

as $P_{p_0,f}(t) = t^{p_0} - p_0 a_1 t^{p_0-2} + \ldots$.

In particular one obtains $d_{-s}^{p_0-1} = 1$ and $d_{-s+1} = 0$. As the set of primes p_1 for which $f^{(p_1)}(q) \neq f(q)$ is finite, a suitable choice of distinct primes p_0 in the argument above implies $d_{-s} = 1$.

Suppose now that we have $i_0 > 0$ so that

$$f(\tau)|\gamma = q^{-s} + \sum_{i=1}^{i_0} a_i q^{is} + d_{i_0s+j} q^{i_0s+j} + d_{i_0s+j+1} q^{i_0s+j+1} + \ldots$$

for some $1 \leq j \leq s$.

It is easy to see that $q^{-s(p_0-1)+i_0s+j}$ is the smallest power of q in the right hand side of equation (28) whose coefficient is not an element of $\mathbb{Z}[a_1, a_2, \ldots, a_{i_0}]$. In fact, this coefficient is of the form $p_0 d_{i_0s+j} + A$ for some A in $\mathbb{Z}[a_1, a_2, \ldots, a_{i_0}]$.

On the other hand, if we choose p_0 so that $p_0 \geq i_0 + 3$ and $p_0 > i_0 s + s + j$, we have $-sp_0^2 + (i_0+1)sp_0 + jp_0 < -s$ and $\gcd(p_0, i_0 s + s + j) = 1$. Hence the power $q^{-sp_0+(i_0+1)s+j}$ must have a zero coefficient in both series

$$\sum_{m=-s,\,p_0|m}^{\infty} p_0 d_m q^{\frac{m}{p_0}} \quad \text{and} \quad \sum_{m=-s}^{\infty} d_m q^{mp_0}.$$

Consequently the coefficient of $q^{-sp_0+(i_0+1)s+j}$ in the left hand side of (28) is zero. Thus $p_0 d_{i_0s+j} + A = 0$.

If $j \neq s$ then $A = 0$ because A is a polynomial expression in the coefficients of $\{q^{is}\}_{i=-1}^{i_0}$. Hence $d_{i_0s+j} = 0$. Otherwise we have $d_{(i_0+1)s} = \frac{1}{p_0}A$. But one knows $a_{i_0+1} = \frac{1}{p_0}A$ since $f(\tau)$ is p_0-th self-replicable, i.e. the identity $\sum_{m=-s}^{\infty} a_m q^{mp_0} + \sum_{m=-s,\,p_0|m}^{\infty} p_0 a_m q^{\frac{m}{p_0}} = P_{p_0,f}(f(\tau))$ holds. Thus $d_{(i_0+1)s} = a_{i_0+1}$ and

$$f(\tau)|\gamma = q^{-s} + \sum_{i=1}^{i_0+1} a_i q^{is} + d_{(i_0+1)s+1} q^{(i_0+1)s+1} + \ldots$$

By induction the lemma follows.□

Proposition 21 *The function $f(\tau)$ is invariant under $\Gamma_0(m_p p N)$.*

Proof. Let $\gamma \in \Gamma_0(m_p p N)$. By the previous lemma there is a positive integer s such that $f(\tau)|\gamma = f(s\tau)$, so we only need to show $s = 1$.

Let $g(\tau) = q^{-1} + \sum_{m=1}^{\infty} e_m q^m$. If we consider equation (27) then

$$q^{-p} + \sum_{m=1}^{\infty} e_m q^{pm} + \sum_{\substack{m=-1 \\ p|ms}}^{\infty} pa_m q^{\frac{ms}{p}} =$$

$$q^{-sp} + pa_1 q^{-sp+2s} + \ldots - pa_1 q^{-s(p-2)} + \ldots$$

Since $-sp < 0$ and $-sp \neq \frac{-s}{p}$ we must have $-sp = -p$. Therefore $s = 1$.□

Next, in order to prove that $f(\tau)$ is indeed invariant under $\Gamma_0(p^3 N)$ we use an inductive argument. We assume that $f(\tau)$ is a modular function on $\Gamma_0(p'M)$, where $p'M$ is a factor of $m_p pN$ divisible by pN, and p' is a prime factor of m_p distinct to p. From results analogous to lemmas 17, 18, 19, 20 in this more general context we deduce the invariance under $\Gamma_0(M)$. This process and the fact that p^2 is the p-part of m_p yield that $f(\tau)$ is a modular function on $\Gamma_0(p^3N)$.

Lemma 22 *Let $\gamma \in \Gamma_0(M)$ and p_0, p' any primes with $\gcd(p_0, p'M) = 1$. Then there exist $\delta \in \Gamma_1(p'M)$ such that $\delta\gamma \in \Gamma_0(p_0 M) \cap \Gamma^0(p_0)$.*

Proof. Let $\gamma = \begin{pmatrix} a & b \\ c & d \end{pmatrix}$. If $\gcd(p_0, c) = 1$ there is v in $\mathbb{Z}$ such that $ap'M + cv \equiv 0 \pmod{p_0}$ and $v \equiv 1 \pmod{p'M}$. Hence there exist $\gamma_1 = \begin{pmatrix} x & y \\ p'M & v \end{pmatrix}$ in $\Gamma_1(p'M)$. By our choice of v, $\gamma_1\gamma \in \Gamma_0(p_0 M)$.

By the argument above we assume $\gcd(p_0, c) = p_0$. If $\gcd(p_0, b) = 1$ there is x' in $\mathbb{Z}$ such that $bx' + d \equiv 0 \pmod{p_0}$ and $x' \equiv 1 \pmod{p'M}$. Consequently there is $\gamma_2 = \begin{pmatrix} x' & 1 \\ p_0 p'My' & v' \end{pmatrix} \in \Gamma_1(p'M)$, and $\gamma_2\gamma \in \Gamma_0(p_0 M) \cap \Gamma^0(p_0)$. $\square$

Lemma 23 *Let p_0 and p' be any primes such that $\gcd(p_0, p'M) = 1$ and $p_0 \equiv 1 \pmod{p'}$.*

Let $\gamma \in \Gamma_0(p_0 M) \cap \Gamma^0(p_0)$ and $\{l_0, l_1, \ldots, l_{p_0-1}\}$ a set of integers such that $l_i \equiv i \pmod{p_0}$, $l_i \equiv 0 \pmod{p'}$.

Then, there are integers $l'_0, l'_1, \ldots, l'_{p_0-1}$ such that $l'_i \not\equiv l'_j \pmod{p_0}$ if $i \neq j$, and

$$\gamma \begin{pmatrix} 1 & l \\ 0 & p_0 \end{pmatrix} \gamma^{-1} \begin{pmatrix} 1 & l' \\ 0 & p_0 \end{pmatrix}^{-1} \in \Gamma_0(p'M). \qquad (29)$$

Moreover

$$\gamma \begin{pmatrix} p_0 & 0 \\ 0 & 1 \end{pmatrix} \gamma^{-1} \begin{pmatrix} p_0 & 0 \\ 0 & 1 \end{pmatrix}^{-1} \in \Gamma_0(p'M). \qquad (30)$$

Lemma 24 *Let $f(\tau)$ be invariant under $\Gamma_0(p'M)$ and let p_0 be any prime such that $\gcd(p_0, p'M) = 1$, $p_0 \equiv 1 \pmod{p'}$ and $f^{(p_0)}(q) = f(q)$.*

Then for any $\gamma \in \Gamma_0(M)$ and integers $l_0, l_1, \ldots, l_{p_0-1}$ with $l_i \equiv i \pmod{p_0}$ and $l_i \equiv 0 \pmod{p'}$, the following holds

$$f(\tau)|\gamma \begin{pmatrix} p_0 & 0 \\ 0 & 1 \end{pmatrix} + \sum_{i=0}^{p_0-1} f(\tau)|\gamma \begin{pmatrix} 1 & l_i \\ 0 & p_0 \end{pmatrix} = P_{p_0, f}(f(\tau)|\gamma). \qquad (31)$$

The proofs of lemmas 23 and 24 are completely analogous to the proofs of lemmas 18 and 19 respectively.

Lemma 25 *Let $f(\tau)$ be invariant under $\Gamma_0(p'M)$, and let $\gamma \in \Gamma_0(M)$.*

Then $f(\tau)|\gamma = f(\frac{s\tau+r}{p'})$ for some positive integer s and some r in $\{0, 1, \ldots, p'-1\}$.

Proof. Let $\zeta = q^{\frac{1}{p'}}$. By hypothesis $f(\tau)$ is a modular function on $\Gamma_0(p'M)$, hence there is $s \in \mathbb{Z}$ so that $f(\tau)|\gamma = d_{-s}\zeta^{-s} + d_{-s+1}\zeta^{-s+1} + \ldots + d_m\zeta^m + \ldots$ with $d_{-s} \neq 0$.

If we use equation (31) instead of equation (26) in the proof of the lemma 20, the same arguments yield $s > 0$ and

$$f(\tau)|\gamma = d_{-s}\zeta^{-s} + a_1 d_{-s}^{-1}\zeta^s + \ldots + a_m d_{-s}^{-m}\zeta^{ms} + \ldots$$

where d_{-s} is some p'-th root of unity.

Take $r \in \{0, 1, \ldots, p'-1\}$ such that $d_{-s}^{-1} = \exp\left(\frac{2\pi i r}{p'}\right)$ and the lemma follows. $\square$

Lemma 26 *Let $f(\tau)$ be invariant under $\Gamma_1(p'M)$ and $\gamma \in \Gamma_0(M)$.*

Then $f(\tau)|\gamma = f(\tau)$.

Proof. By the previous result we have $f(\tau)|\gamma = f(\frac{s\tau+r}{p'})$ for some $s > 0$, $0 \leq r \leq p'-1$.

Let $0 \leq r' \leq p'-1$ such that $rp - r' \equiv 0 \pmod{p'}$, and consider a set of integers $\{l_0, l_1, \ldots, l_{p-1}\}$ so that $l_i \equiv i \pmod{p}$ and $l_i \equiv 0 \pmod{p'}$. Then, by solving the equations $r + sl_i + p'l_i' \equiv 0 \pmod{p}$ we get integers $l_0', l_1', \ldots, l_{p-1}'$ such that $l_i' \not\equiv l_j' \pmod{p}$ if $i \neq j$ and

$$\begin{pmatrix} s & r \\ 0 & p' \end{pmatrix}\begin{pmatrix} 1 & l \\ 0 & p \end{pmatrix}\begin{pmatrix} s & r' \\ 0 & p' \end{pmatrix}^{-1}\begin{pmatrix} 1 & l' \\ 0 & p \end{pmatrix}^{-1} \in \Gamma_1(p'M).$$

Thus

$$g(\tau)|\gamma\begin{pmatrix} p & 0 \\ 0 & 1 \end{pmatrix} + \sum_{i=0}^{p-1} f(\tau)|\gamma\begin{pmatrix} 1 & l_i \\ 0 & p \end{pmatrix} = P_{p,f}(f(\tau)|\begin{pmatrix} s & r' \\ 0 & p' \end{pmatrix}) \quad (32)$$

since both $g(\tau)$ and $g(p\tau)$ are invariant under γ.

If we write $g(\tau) = q^{-1} + \sum_{m=1}^{\infty} e_m q^m$ as before, equation (32) is equal to

$$q^{-p} + \sum_{m=1}^{\infty} e_m q^{mp} + \sum_{\substack{m=-1 \\ p|ms}}^{\infty} p a_m \exp\left(2\pi i \frac{r}{p'} m\right) \zeta^{\frac{sm}{p}} =$$

$$\exp\left(-2\pi i \frac{r'}{p'} p\right) \zeta^{-sp} + p a_1 \exp\left(-2\pi i \frac{r'}{p'}(p-2)\right) \zeta^{-sp+2s} + \cdots$$

where $\zeta = q^{\frac{1}{p'}}$.

Hence $\frac{-sp}{p'} = -p$ and $\exp\left(-2\pi i \frac{r'}{p'} p\right) = 1$. Therefore $s = p'$ and $r' = 0$, which implies $r = 0$ as desired. $\square$

Recall that the arguments above are valid whenever pN divides M and p' is a prime distinct to p. Thus, if we apply them to $f(\tau)$ we get

Proposition 27 $f(\tau)$ *is a modular function on* $\Gamma_0(p^3 N)$.

Acknowledgment. I would like to thank Prof. John McKay for both introducing me to the topic of replicable functions and for suggesting me the approach to the subject used in this work, (the study of generalized Hecke operators). Also, I want to thank the referee for his/her valuable remarks.

References

[1] R. E. Borcherds, Monstrous Moonshine and monstrous Lie super-algebras, *Invent. Math.*, **109** (1992), 405-444.

[2] D. Alexander, C. Cummins, J. McKay, C. Simons, Completely replicable functions, Groups, Combinatorics and Geometry. Ed. M. W. Liebeck, J. Saxl, Cambridge University Press 1992, 87-95.

[3] H. Cohn, J. McKay, Spontaneous generation of modular invariants, to appear in Math. Comp.

[4] J. H. Conway, S. P. Norton, Monstrous Moonshine, *Bull. London Math. Soc.*, **11** (1979), 308-339.

[5] C. J. Cummins, S. P. Norton, Rational Hauptmoduls are replicable, to appear in the *Canadian J. Math.*

[6] D. Ford, J. McKay, S. Norton, More on replicable functions, *Communications in Algebra*, **22** (1994), 5175-5193.

[7] O. Forster, Lectures on Riemann surfaces, Springer-Verlag, New York, 1981.

[8] K. Mahler, On a class of non-linear functional equations connected with modular functions, *J. Austral. Math. Soc.*, **22A** (1976), 65-118.

[9] G. Mason, Finite groups and modular functions, *Proc. Symp. Pure Math.*, **47** (1987), 181-210.

[10] S. P. Norton, More on Moonshine, *Computational Group Theory*. Ed M. D. Atkinson, Academic Press, London (1984), 185-193.

[11] S. P. Norton, Generalized Moonshine, *Proc. Symp. Pure Math.*, **47** (1987), 208-209.

[12] S. P. Norton, Non-monstrous Moonshine, preprint.

[13] G. Shimura, Introduction to the arithmetic theory of automorphic functions, Princeton University Press, Princeton, 1971.

Centre de Recherches Mathématiques and
Department of Mathematics and Statistics, Concordia University
1455 de Maisonneuve W., Montréal, Québec, H3G 1M8

e-mail: ymartin@abacus.concordia.ca

Contemporary Mathematics
Volume **193**, 1996

ON EMBEDDING OF INTEGRABLE HIGHEST
WEIGHT MODULES OF AFFINE LIE ALGEBRAS

KAILASH C. MISRA[*]

Introduction

The integrable highest weight modules of affine Lie algebras are the infinite-dimensional analogs of irreducible highest weight modules of finite-dimensional semisimple Lie algebras (see [Kac]). The level of an integrable highest weight module is a scalar by which the canonical central element of the affine Lie algebra acts. The existence of explicit construction of level one integrable highest weight modules in terms of vertex operators has led to many applications of affine Lie algebra representation theory. One way to give explicit constructions of higher level integrable highest weight modules is to realize it in the tensor product of known level one modules. Recently, it has been proved that any integrable highest weight module can be embedded in the tensor product of level one integrable highest weight modules ([Ag], [BM], [X]). In particular, using the path realizations of crystal bases [KMN 1,2] for the corresponding quantum affine Lie algebras it has been shown in [BM] that for the affine lie algebras $B_n^{(1)}$, $D_n^{(1)}$ and $D_{n+1}^{(2)}$, up to degree shift the level two fundamental modules $L(\wedge_k)$ occur as a direct summand in the tensor product of the level one modules $L(\wedge_0) \otimes L(\wedge_i)$, where $k \equiv i \pmod 2$ for $B_n^{(1)}$ or $D_n^{(1)}$ and $i = 0$ for $D_{n+1}^{(2)}$. However, $L(\wedge_n)$ for $B_n^{(1)}$; $L(\wedge_n)$, $L(\wedge_{n-1})$ for $D_n^{(1)}$ and $L(\wedge_n)$ for $D_{n+1}^{(2)}$ are also level one fundamental modules. In this paper, using the crystal base techniques of [BM], we show that any level two fundamental module $L(\wedge_{n-k})$ (without any degree shift) occurs as a direct summand in $L(\wedge_n) \otimes L(\wedge_n)$ for $B_n^{(1)}$; $L(\wedge_{n-j}) \otimes L(\wedge_n)$ ($k \equiv j \pmod 2$) for $D_n^{(1)}$ and $L(\wedge_n) \otimes L(\wedge_n)$ for $D_{n+1}^{(2)}$. As a consequence, it follows that any integrable highest weight module for $\mathfrak{g} = B_n^{(1)}$, $D_n^{(1)}$ or $D_{n+1}^{(2)}$, can be embedded in the tensor product of level one fundamental modules.

1. Affine Lie algebra representations and associated crystals:

Let $\mathfrak{g} = \mathfrak{g}(A)$ denote the affine Lie algebra over a field F of characteristic zero, associated with the affine generalized Cartan matrix $A = (a_{ij})_{i,j=0}^n$ (see [Kac]). Let $\mathfrak{g}' = [\mathfrak{g}, \mathfrak{g}]$ be the derived algebra of $\mathfrak{g}$. Then $\mathfrak{g} = \mathfrak{g}' \oplus Fd$, where d denotes the scaling element of $\mathfrak{g}$. Let $\{e_i, f_i, h_i | i \in I\}$, $I = \{0, 1, \dots, n\}$, denote the set of Chevalley

1991 Mathematics Subject Classification: 17B67, 17B10, 17B37
*Supported in part by NSA/MSP grant #MDA92-H-3076.
This paper is in final form and no version of it will be submitted for publication elsewhere.

generators for $\mathfrak{g}'$. Then $\mathfrak{h} = \text{span}\{h_0, h_1, \ldots, h_n, d\}$ is a Cartan subalgebra of $\mathfrak{g}$ and $\dim(\mathfrak{h}) = n + 2$. Let $\{\alpha_0, \alpha_1, \ldots, \alpha_n\} \subset \mathfrak{h}^*$ denote the simple roots with $\alpha_0(d) = 1$, and $\alpha_i(d) = 0$ for $i \neq 0$. There exists positive integers $a_0, a_1, \ldots, a_n$ and $\check{a}_0, \check{a}_1, \ldots, \check{a}_n$ with $gcd(a_0, a_1, \ldots, a_n) = 1$ and $gcd(\check{a}_0, \check{a}_1, \ldots, \check{a}_n) = 1$, such

that $A \begin{pmatrix} a_0 \\ a_1 \\ \vdots \\ a_n \end{pmatrix} = \begin{pmatrix} 0 \\ 0 \\ \vdots \\ 0 \end{pmatrix}$ and $(\check{a}_0, \check{a}_1, \ldots \check{a}_n)A = (0, 0, \ldots, 0)$. Then $c = \sum_{i=0}^{n} \check{a}_i h_i \in \mathfrak{h}$

and $\delta = \sum_{i=0}^{n} a_i h_i \in \mathfrak{h}^*$ are the canonical central element and null root respectively.

Let $P \subset \mathfrak{h}^*$ denote the weight lattice. Then $P^+ = \{\lambda \in P | \lambda(h_i) \in \mathbb{Z}_{\geq 0}, 0 \leq i \leq n\}$ is the set of dominant integral weights. It is known (see [Kac]) that for $\lambda \in P^+$ there exists a unique (up to isomorphism) highest weight $\mathfrak{g}$-module $L(\lambda)$ called the *integrable highest weight* (or *standard*) *module*. The scalar $\lambda(c)$ is called the *level* of $L(\lambda)$. It has been recently proved ([Ag], [BM], [X]) that any integrable highest weight $\mathfrak{g}$-module can be embedded in the tensor product of level one integrable highest weight $\mathfrak{g}$-modules.

The fundamental weights are the dominant integral weights $\wedge_i \in P^+$ defined by $\wedge_i(h_j) = \delta_{ij}$ and $\wedge_i(d) = 0$. The corresponding integrable highest weight modules $L(\wedge_i)$ are called the fundamental modules. Any dominant integral weight $\lambda \in P^+$ can be written uniquely as

$$(1.1) \qquad \lambda = \sum_{i=0}^{n} k_i \wedge_i + k_\delta \delta, \; k_i \in \mathbb{Z}_{\geq 0}, \; k_\delta \in F.$$

Hence we have the $\mathfrak{g}$-module embedding

$$(1.2) \qquad L(\lambda) \subset \bigotimes_{i=0}^{n} \left(\otimes^{k_i} L(\wedge_i) \right) \otimes L(k_\delta \delta),$$

where $\otimes^{k_i} L(\wedge_i)$ denotes the k_i-fold tensor product of $L(\wedge_i)$. Note that $\dim(L(k_\delta \delta)) = 1$. Thus to show that $L(\lambda)$ is contained in the tensor product of level one integrable highest weight modules, it suffices to show that the fundamental modules $L(\wedge_i)$ have this property. Thus, if all fundamental modules $L(\wedge_i)$ are of level one (as in case of $A_n^{(1)}$ and $C_n^{(1)}$) then the result follows trivially. For the affine Lie algebras $\mathfrak{g} = B_n^{(1)}, D_n^{(1)}, A_{2n-1}^{(2)}, D_{n+1}^{(2)}$ and $A_{2n}^{(2)}$ the following Theorem have been proved in [BM].

Theorem 1.3. (a) *For the affine Lie algebra* $\mathfrak{g} = B_n^{(1)}, D_n^{(1)}$ *or* $A_{2n-1}^{(2)}$, *the level two fundamental modules* $L(\wedge_k)$ *occur as direct summands in* $L(\wedge_i) \otimes L(\wedge_0)$, *where* $k \equiv i(\text{mod } 2)$, *up to degree shifts.*

(b) *For the affine Lie algebra* $\mathfrak{g} = A_{2n}^{(2)}$ *or* $D_{n+1}^{(2)}$, *the level two fundamental modules* $L(\wedge_k)$ *occur as direct summands in* $L(\wedge_0) \otimes L(\wedge_0)$ *up to degree shifts.*

However, for the affine Lie algebras $\mathfrak{g} = B_n^{(1)}, D_n^{(1)}$ and $D_{n+1}^{(2)}$, the fundamental modules $L(\wedge_n); L(\wedge_{n-1})$ and $L(\wedge_n);$ and $L(\wedge_n)$, respectively are also of level one. In this paper, we prove the following theorems.

Theorem 1.4. *For the affine Lie algebra $\mathfrak{g} = B_n^{(1)}$ $(n \geq 3)$, the fundamental modules $L(\wedge_{n-k})$, $k = 1, 2, \ldots, n-2$, occur as direct summands in $L(\wedge_n) \otimes L(\wedge_n)$.*

Theorem 1.5. *For the affine Lie algebra $\mathfrak{g} = D_n^{(1)}(n \geq 4)$, the level two fundamental modules $L(\wedge_{n-k})$, $k = 2, 3, \ldots, n-2$, occur as direct summands in $L(\wedge_{n-i}) \otimes L(\wedge_n)$, where $k \equiv i \pmod 2$.*

Theorem 1.6. *For the affine Lie algebra $\mathfrak{g} = D_{n+1}^{(2)}(n \geq 2)$, the level two fundamental modules $L(\wedge_{n-k})$, $k = 1, 2, 3, \ldots, n-1$, occur as direct summands in $L(\wedge_n) \otimes L(\wedge_n)$.*

Note that unlike in Theorem 1.3 we do not require any degree shift for the embeddings in Theorems 1.4, 1.5 and 1.6. As in [BM], we will use the crystal base theory of Kashiwara [K1, K2] to prove Theorems 1.4, 1.5 and 1.6. By the crystal base theory an integrable $\mathfrak{g}$-module V is uniquely associated with a *P-weighted crystal* (see [KMN1], [KKM]) which is defined as follows.

A *P-weighted crystal B* is a set B equipped with maps

$$wt : B \longrightarrow P,$$
$$\epsilon_i, \varphi_i : B \longrightarrow \mathbb{Z} \sqcup \{-\infty\}, \quad i \in I,$$
$$\tilde{e}_i, \tilde{f}_i : B \longrightarrow B \sqcup \{0\}, \quad i \in I,$$

satisfying the following conditions:

(1.7) $$\varphi_i(b) = \epsilon_i(b) + <h_i, wt(b)>, \quad b \in B,$$

(1.8)   if $b \in B$ and $\tilde{e}_i b \in B$ (resp. $\tilde{f}_i b \in B$), then

$$wt(\tilde{e}_i b) = wt(b) + \alpha_i \text{ (resp. } wt(\tilde{f}_i b) = wt(b) - \alpha_i),$$
$$\epsilon_i(\tilde{e}_i b) = \epsilon_i(b) - 1 \text{ (resp. } \epsilon_i(\tilde{f}_i b) = \epsilon_i(b) + 1),$$
and $$\varphi_i(\tilde{e}_i b) = \varphi_i(b) + 1 \text{ (resp. } \varphi_i(\tilde{f}_i b) = \varphi_i(b) - 1),$$

(1.9)   for $b, b' \in B$, $i \in I$, $b' = \tilde{f}_i b$ if and only if $b = \tilde{e}_i b'$,

(1.10)   for $b \in B$, if $\varphi_i(b) = -\infty$, then $\tilde{e}_i b = 0 = \tilde{f}_i b$.

Note that if B is a P-weighted crystal then $B = \sqcup_{\lambda \in P} B_\lambda$, where $B_\lambda = \{b \in B | wt(b) = \lambda\}$. Also note that, for $b \in B$, $wt(b) = \varphi(b) - \epsilon(b)$, where $\varphi(b) = \sum_{i \in I} \varphi_i(b) \wedge_i$ and $\epsilon(b) = \sum_{i \in I} \epsilon_i(b) \wedge_i$. Furthermore, a crystal B can be viewed as a oriented colored (by $i \in I$) graph with the set B as the set of vertices and for $b, b' \in B$, we draw i-arrows $b \xrightarrow{i} b'$ if $b' = \tilde{f}_i b$. It is known [K1, K2] that the integrable $\mathfrak{g}$-module V is irreducible if and only if the associated crystal B is connected. A P-weighted crystal B is said to be a *crystal with highest weight* $\lambda \in P^+$ if there is an element $b_\lambda \in B$ (called the highest weight element), such that $wt(b_\lambda) = \lambda$ and $\tilde{e}_i b_\lambda = 0$ for all $i \in I$. Thus an integrable highest weight or standard $\mathfrak{g}$-module $L(\lambda)$ is uniquely associated with a connected crystal $B(\lambda)$ with highest weight λ.

The set of P-weighted crystals form a tensor category (see [KKM]). In particular, for two P-weighted crystals B_1 and B_2 their tensor product $B_1 \otimes B_2$ is defined as follows.

The underlying set in $B_1 \otimes B_2$ is the set $B_1 \times B_2$. For $b_1 \in B_1$, $b_2 \in B_2$, write $b_1 \otimes b_2$ for (b_1, b_2). We understand $b_1 \otimes 0 = 0 = 0 \otimes b_2$ for all $b_1 \in B_1$, $b_2 \in B_2$.

Then $B_1 \otimes B_2$ equipped with the maps wt, ϵ_i, φ_i, $\tilde{e}_i$, $\tilde{f}_i$, $(i \in I)$ defined below is a P-weighted crystal. For $b_1 \in B_1$, $b_2 \in B_2$, and $i \in I$, define

$$(1.11) \qquad wt(b_1 \otimes b_2) = wt(b_1) + wt(b_2),$$

$$(1.12) \qquad \epsilon_i(b_1 \otimes b_2) = \max(\epsilon_i(b_1), \epsilon_i(b_2) - < h_i, \, wt(b_1) >),$$

$$(1.13) \qquad \varphi_i(b_1 \otimes b_2) = \max(\varphi_i(b_2), \varphi_i(b_1) + < h_i, \, wt(b_2) >),$$

$$(1.14) \qquad \tilde{e}_i(b_1 \otimes b_2) = \begin{cases} \tilde{e}_i b_1 \otimes b_2 & \text{if} \quad \varphi_i(b_1) \geq \epsilon_i(b_2), \\ b_1 \otimes \tilde{e}_i b_2 & \text{if} \quad \varphi_i(b_1) < \epsilon_i(b_2), \end{cases}$$

$$(1.15) \qquad \tilde{f}_i(b_1 \otimes b_2) = \begin{cases} \tilde{f}_i b_1 \otimes b_2 & \text{if} \quad \varphi_i(b_1) > \epsilon_i(b_2), \\ b_1 \otimes \tilde{f}_i b_2 & \text{if} \quad \varphi_i(b_1) \leq \epsilon_i(b_2). \end{cases}$$

Let $B(\lambda)$ and $B(\mu)$ denote the crystals associated with the integrable highest weight $\mathfrak{g}$-modules $L(\lambda)$ and $L(\mu)$ $(\lambda, \mu \in P^+)$ respectively. To decompose the tensor product $L(\lambda) \otimes L(\mu)$ it suffices to determine the connected components of the crystal $B(\lambda) \otimes B(\mu)$. The following Lemma is very useful in determining the highest weight elements in $B(\lambda) \otimes B(\mu)$ which in turn can be used in decomposing $L(\lambda) \otimes L(\mu)$.

Lemma 1.16. *(see [K1, K2]). For $\lambda, \mu \in P^+$, the element $b_1 \otimes b_2 \in B(\lambda) \otimes B(\mu)$ is a highest weight element if and only if $b_1 = b_\lambda$, $\tilde{e}_i^{\lambda(h_i)+1} b_2 = 0$ for all $i \in I$, where b_λ denotes the highest weight element in $B(\lambda)$ of weight λ.*

In the next three sections we give the proofs of Theorems 1.4, 1.5 and 1.6.

As in [BM], we will use the path realizations of the crystals $B(\lambda)$ associated with integrable highest weight $\mathfrak{g}$-modules $L(\lambda)$ given in [KMN2].

2. Proof of Theorem 1.4:

First we recall the path realization of the crystal $B(\wedge_n)$ associated with the integrable highest weight $B_n^{(1)}$-module $L(\wedge_n)$. For more details see [KMN2].

Consider the set $B = \{b_j = (\delta_{ij})_{i=1}^{2n+1} | 1 \leq j \leq 2n+1\} \subseteq \mathbb{Z}^{2n+1}$, with the maps $\tilde{e}_i$, $\tilde{f}_i$, φ_i, ϵ_i, and $\overline{wt}$ for $i \in I$, given by:

$$(2.1) \quad \begin{cases} \tilde{e}_i b_{i+1} = b_i, \; \tilde{e}_i b_{2n+2-i} = b_{2n+1-i}, \; i = 1, 2, \dots, n, \\ \tilde{e}_0 b_2 = b_{2n+1}, \; \tilde{e}_0 b_1 = b_{2n}, \; \tilde{e}_i b_j = 0, \; \text{otherwise}, \\ \tilde{f}_i b = b' \text{ if and only if } \tilde{e}_i b' = b \text{ for } b, \, b' \in B, \text{ and } \tilde{f}_i b_j = 0, \; \text{otherwise}, \end{cases}$$

$$(2.2) \quad \begin{cases} \varphi_i(b_j) = \delta_{i,j} + \delta_{2n+1-i,j} + \delta_{i,n}\delta_{j,n} + \delta_{i,0}\delta_{j,2n}, \\ \epsilon_i(b_j) = \delta_{i+1,j} + \delta_{2n+2-i,j} + \delta_{i,n}\delta_{j,n+2} + \delta_{i,0}\delta_{j,2}, \\ 0 \le i \le n, \quad 1 \le j \le 2n+1, \end{cases}$$

and $\overline{wt}(b_j) = \sum_{i=0}^{n}(\varphi_i(b_j) - \epsilon_i(b_j))\wedge_i$, for $j = 1, 2, \ldots, 2n+1$. In particular, we have

$$(2.3) \quad \begin{cases} \overline{wt}(b_1) = \wedge_1 - \wedge_0, \ \overline{wt}(b_2) = \wedge_2 - \wedge_1 - \wedge_0, \\ \overline{wt}(b_j) = \wedge_j - \wedge_{j-1}, \ j = 3, 4, \ldots, n-1, \\ \overline{wt}(b_n) = 2\wedge_n - \wedge_{n-1}, \ \overline{wt}(b_{n+1}) = 0, \ \overline{wt}(b_{n+2}) = \wedge_{n-1} - 2\wedge_n, \\ \overline{wt}(b_{2n+1-j}) = \wedge_j - \wedge_{j+1}, \ j = 2, 3, \ldots, n-2, \\ \overline{wt}(b_{2n}) = \wedge_0 + \wedge_1 - \wedge_2, \ \text{and} \ \overline{wt}(b_{2n+1}) = \wedge_0 - \wedge_1. \end{cases}$$

Note that the set B equipped with the maps $\tilde{e}_i, \tilde{f}_i, \varphi_i, \epsilon_i, \overline{wt}$, for $i \in I \backslash \{0\}$, is the crystal associated with the irreducible B_n-module with highest weight $\wedge_1$ (see [KN]). The sequence $\eta = (\eta_k)_{k \ge 1} = (\ldots \eta_3\eta_2\eta_1)$ in B, where $\eta_k = b_{n+1}$ for all $k \ge 1$, is called the *ground-state path* of weight $\wedge_n$. A $\wedge_n$-*path*, by definition, is a sequence $p = (p_k)_{k \ge 1} = (\ldots p_3p_2p_1)$ such that $p_k \in B$ and $p_k = \eta_k$ for $k \gg 0$. As shown in [KMN2], the set $\mathcal{P}(\wedge_n) \subset \ldots \otimes B \otimes B \otimes B$, of $\wedge_n$-paths, as a crystal, is isomorphic to the crystal $B(\wedge_n)$ associated with the $B_n^{(1)}$-module $L(\wedge_n)$. For instance, the actions of $\tilde{e}_i$ and $\tilde{f}_i$ on $\mathcal{P}(\wedge_n)$ are given via (1.14) and (1.15). For a purely combinatorial description of the actions of $\tilde{e}_i$ and $\tilde{f}_i$ on $\mathcal{P}(\wedge_n)$ see Section 2 of [BM]. We also recall that (see [KMN1]) the weight of a $\wedge_n$-path $p = (p_k)_{k \ge 1} \in \mathcal{P}(\wedge_n)$ is given by the formula:

$$(2.4) \quad wt(p) = \wedge_n + \sum_{k=1}^{\infty}(\overline{wt}(p_k) - \overline{wt}(\eta_k)) - \left(\sum_{k=1}^{\infty} k(H(p_{k+1} \otimes p_k) - H(\eta_{k+1} \otimes \eta_k))\right)\delta,$$

where $H : B \otimes B \longrightarrow \mathbb{Z}$ is a $\mathbb{Z}$-valued function on $B \otimes B$, called the *energy function on B*, which has the following properties: for $i \in I$ and $b \otimes b' \in B \otimes B$ such that $\tilde{e}_i(b \otimes b') \ne 0$,

$$(2.5) \quad H(\tilde{e}_i(b \otimes b')) = \begin{cases} H(b \otimes b') & \text{if} \ i \ne 0, \\ H(b \otimes b') + 1 & \text{if} \ i = 0 \ \text{and} \ \varphi_0(b) \ge \epsilon_0(b'), \\ H(b \otimes b') - 1 & \text{if} \ i = 0 \ \text{and} \ \varphi_0(b) < \epsilon_0(b'). \end{cases}$$

Now for $k = 1, 2, \ldots, n-2$, consider the paths

$$(2.6) \quad p = (\ldots b_{n+1}b_{n+2}\ldots b_{n+k+1}) \in \mathcal{P}(\wedge_n).$$

Since it can be easily checked using (2.5) that $H(b_{n+1} \otimes b_{n+1}) = H(b_{n+i-1} \otimes b_{n+i})$ for $i = 1, 2, \ldots, n-1$, (also see [KKM, Section 5.3]), it follows from (2.3) and (2.4) that

$$wt(p) = \wedge_n + \sum_{i=2}^{k+1}(\overline{wt}(b_{n+i}) - \overline{wt}(b_{n+1}))$$

$$= \wedge_n + (\wedge_{n-1} - 2\wedge_n) + (\wedge_{n-2} - \wedge_{n-1}) + \ldots + (\wedge_{n-k} - \wedge_{n-k+1})$$

$$= \wedge_{n-k} - \wedge_n.$$

Hence $\eta \otimes p \in \mathcal{P}(\wedge_n) \otimes \mathcal{P}(\wedge_n)$ has weight $wt(\eta) + wt(p) = \wedge_{n-k}$, $k = 1, 2, \ldots, n-2$. Furthermore, $\tilde{e}_i p = 0$ for $i = 0, 1, \ldots, n-1$, $\tilde{e}_n p = (\ldots b_{n+1} b_{n+3} b_{n+4} \ldots b_{n+k+1})$ and $\tilde{e}_n^2 p = \tilde{e}_n^{\wedge_n(h_n)+1} p = 0$. Hence it follows from Lemma 1.16 that $L(\wedge_{n-k})$ occurs as a direct summand in the decomposition of $L(\wedge_n) \otimes L(\wedge_n)$ for $k = 1, 2, \ldots, n-2$. This completes the proof of Theorem 1.4.

3. Proof of Theorem 1.5:

First we recall the path realizations of the crystals $B(\wedge_{n-1})$ and $B(\wedge_n)$ associated with the integrable highest weight $D_n^{(1)}$-modules $L(\wedge_{n-1})$ and $L(\wedge_n)$ respectively. For more details see [KMN2].

Consider the set $B = \{b_j = (\delta_{ij})_{i=1}^{2n} \mid j = 1, 2, \ldots, 2n\} \subset \mathbb{Z}^{2n}$, with maps $\tilde{e}_i, \tilde{f}_i, \epsilon_i, \varphi_i$ and $\overline{wt}$ for $i \in I$, given by:

$$(3.1) \quad \begin{cases} \tilde{e}_i b_{i+1} = b_i, \ \tilde{e}_i b_{2n+1-i} = b_{2n-i}, \ i = 1, 2, \ldots, n-1, \\ \tilde{e}_n b_{n+1} = b_{n-1}, \ \tilde{e}_n b_{n+2} = b_n, \ \tilde{e}_0 b_2 = b_{2n}, \\ \tilde{e}_0 b_1 = b_{2n-1} \text{ and } \tilde{e}_i b_j = 0, \text{ otherwise,} \\ \tilde{f}_i b = b' \text{ if and only if } \tilde{e}_i b' = b \text{ for } b, b' \in B, \\ \text{and } \tilde{f}_i b_j = 0, \text{ otherwise,} \end{cases}$$

$$(3.2) \quad \begin{cases} \varphi_i(b_j) = \delta_{i,j} + \delta_{2n-i,j} - \delta_{i,n}\delta_{j,n} + \delta_{i,n}\delta_{j,n-1} + \delta_{i,0}\delta_{j,2n-1}, \\ \epsilon_i(b_j) = \delta_{i+1,j} + \delta_{2n+1-i,j} - \delta_{i,n}\delta_{j,n+1} + \delta_{i,n}\delta_{j,n+2} + \delta_{i,0}\delta_{j,2}, \\ \text{for } 0 \leq i \leq n, \ 1 \leq j \leq 2n, \end{cases}$$

and $\overline{wt}(b_j) = \sum_{i=0}^{n} (\varphi_i(b_j) - \epsilon_i(b_j))\wedge_i$, for $j = 1, 2, \ldots, 2n$. In particular, we have

$$(3.3) \quad \begin{cases} \overline{wt}(b_1) = \wedge_1 - \wedge_0, \ \overline{wt}(b_2) = \wedge_2 - \wedge_1 - \wedge_0, \\ \overline{wt}(b_j) = \wedge_j - \wedge_{j-1}, \ j = 3, 4, \ldots, n-2, \\ \overline{wt}(b_{n-1}) = \wedge_n + \wedge_{n-1} - \wedge_{n-2}, \ \overline{wt}(b_n) = \wedge_n - \wedge_{n-1}, \\ \overline{wt}(b_{n+1}) = \wedge_{n-1} - \wedge_n, \ \overline{wt}(b_{n+2}) = \wedge_{n-2} - \wedge_{n-1} - \wedge_n, \\ \overline{wt}(b_{2n-j}) = \wedge_j - \wedge_{j+1}, \ j = 2, 3, \ldots, n-3, \\ \overline{wt}(b_{2n-1}) = \wedge_0 + \wedge_1 - \wedge_2, \text{ and } \overline{wt}(b_{2n}) = \wedge_0 - \wedge_1. \end{cases}$$

Note that the set B equipped with the maps $\tilde{e}, \tilde{f}_i, \varphi_i, \epsilon_i, \overline{wt}$, for $i \in I \setminus \{0\}$, is the crystal associated with the irreducible D_n-module with highest weight $\wedge_1$ (see [KN]).

For $i = 0, 1$, let $\eta^{(n-i)} = (\eta_k^{(n-i)})_{k \geq 1} = (\ldots \eta_3^{(n-i)} \eta_2^{(n-i)} \eta_1^{(n-i)})$ denote the sequence (with period two) in B defined by

$$(3.4) \quad \eta_k^{(n-i)} = \begin{cases} b_n, & \text{if } k+i \text{ is odd,} \\ b_{n+1}, & \text{if } k+i \text{ is even,} \end{cases}$$

The sequence $\eta^{(n-i)}$ $(i = 0, 1)$ is called a ground-state path of weight $\wedge_{n-i}$. A $\wedge_{n-i}$-path, by definition, is a sequence $p = (p_k)_{k \geq 1} = (\ldots p_3 p_2 p_1)$ such that $p_k \in B$ and $p_k = \eta_k^{(n-i)}$ for $k \gg 0$. The set $\mathcal{P}(\wedge_{n-i}) \subseteq \ldots \otimes B \otimes B \otimes B$, $(i = 0, 1)$ of $\wedge_{n-i}$-paths, as a crystal is isomorphic to the crystal $B(\wedge_{n-i})$ associated with the $D_n^{(1)}$-module $L(\wedge_{n-i})$.

For instance, the actions of $\tilde{e}_j$ and $\tilde{f}_j$ on $\mathcal{P}(\wedge_{n-i})$ are given via (1.14) and (1.15), which in turn can be interpreted purely combinatorially (see Section 2 of [BM]).

Now for $k = 2, 3, \ldots, n - 2$, consider the paths

$$p = \begin{cases} (\ldots b_{n+1} b_n b_{n+1} b_{n+2} b_{n+3} \ldots b_{n+k}), & \text{if } k \text{ is even}, \\ (\ldots b_{n+1} b_n b_{n+2} b_{n+3} \ldots b_{n+k}), & \text{if } k \text{ is odd}, \end{cases}$$

in $\mathcal{P}(\wedge_n)$. Using the properties (2.5) of the energy function H, it can be easily checked in this case that $H(b_{n+1} \otimes b_{n+2}) = H(b_n \otimes b_{n+2}) = H(b_{n+i-1} \otimes b_{n+i})$ for $i = 2, 3, \ldots, n - 1$. (In fact, $b_{n+1} \otimes b_{n+2}$, $b_n \otimes b_{n+2}$, and $b_{n+i-1} \otimes b_{n+i}$, $i = 2, 3, \ldots, n - 1$, lie in the $\wedge_2$-connected component of the D_n-crystal $B \otimes B$). Therefore, it follows from (2.4) and (3.3) that

$$(3.6) \qquad wt(p) = \begin{cases} \wedge_{n-k} - \wedge_n, & \text{if } k \text{ even}, \\ \wedge_{n-k} - \wedge_{n-1}, & \text{if } k \text{ odd}, \end{cases}$$

Hence $\eta^{(n-i)} \otimes p \in \mathcal{P}(\wedge_{n-i}) \otimes \mathcal{P}(\wedge_n)$ has weight $wt(\eta^{(n-i)}) + wt(p) = \wedge_{n-k}$, $k = 2, 3, \ldots, n - 2$, where $k \equiv i \pmod{2}$. Furthermore, observe that for $p = (\ldots b_{n+1} b_n b_{n+1} b_{n+2} b_{n+3} \ldots b_{n+k}) \in \mathcal{P}(\wedge_n)$,

$$(3.7) \qquad \tilde{e}_i(p) = \begin{cases} 0 & \text{if } i \neq n, \\ (\ldots b_{n+1} b_n b_{n+1} b_n b_{n+3} \ldots b_{n+k}), & \text{if } i = n, \end{cases}$$

and $\tilde{e}_n^2(p) = 0$. Also for $p = (\ldots b_{n+1} b_n b_{n+2} b_{n+3} \ldots b_{n+k})$,

$$(3.8) \qquad \tilde{e}_i(p) = \begin{cases} 0, & \text{if } i \neq n - 1, \\ (\ldots b_{n+1} b_n b_{n+1} b_{n+3} \ldots b_{n+k}), & \text{if } i = n - 1. \end{cases}$$

Hence we have $\tilde{e}_j^{\wedge_{n-i}(h_j)+1}(p) = 0$, for $j = 0, 1, \ldots, n$, $k = 2, 3, \ldots, n - 2$ and $k \equiv i \pmod{2}$. Therefore, it now follows from Lemma 1.16 that $L(\wedge_{n-k})$ occurs as a direct summand in the decomposition of $L(\wedge_{n-i}) \otimes L(\wedge_n)$, for $k = 2, 3, \ldots, n - 2$, where $k \equiv i \pmod{2}$. This completes the proof of Theorem 1.5.

4. Proof of Theorem 1.6:

First we recall the path realization of the crystal $B(\wedge_n)$ associated with the integrable highest weight $D_{n+1}^{(2)}$-module $L(\wedge_n)$. For more details see [KMN2].

Consider the set $B = \{b_0, b_1, \ldots, b_{2n+1}\} \subset \mathbb{Z}^{2n+1}$, where $b_0 = (0, 0, \ldots, 0)$ and $b_j = (\delta_{ij})_{i=1}^{2n+1}$, for $j = 1, 2, \ldots, 2n + 1$, with maps $\tilde{e}_i, \tilde{f}_i, \epsilon_i, \varphi_i$, and $\overline{wt}$, for $i \in I$, given by:

$$(4.1) \qquad \begin{cases} \tilde{e}_i b_{i+1} = b_i, \ \tilde{e}_i b_{2n+2-i} = b_{2n+1-i}, \ i = 1, 2, \ldots, n, \\ \tilde{e}_0 b_1 = b_0, \ \tilde{e}_0 b_0 = b_{2n+1}, \ \tilde{e}_i b_j = 0, \ \text{otherwise}, \\ \tilde{f}_i b = b' \text{ if and only if } \tilde{e}_i b' = b \text{ for } b, b' \in B, \\ \text{and } \tilde{f}_i b_j = 0, \ \text{otherwise}, \end{cases}$$

(4.2)
$$\begin{cases} \varphi_i(b_j) = \delta_{i,j} + \delta_{2n+1-i,j} + \delta_{i,0}\delta_{j,2n+1} + \delta_{i,n}\delta_{j,n+1}, \\ \epsilon_i(b_j) = \delta_{i+1,j} + \delta_{2n+2-i,j} + \delta_{i,n}\delta_{j,n+2} + \delta_{i,0}\delta_{j,1} + \delta_{i,0}\delta_{j,0}, \\ \text{for} \quad 0 \le i \le n, \ 0 \le j \le 2n+1, \end{cases}$$

and $\overline{wt}(b_j) = \sum_{i=0}^{n}(\varphi_i(b_j) - \epsilon_i(b_j))\wedge_i$, for $j = 0, 1, 2, \ldots, 2n+1$. In particular, we have

(4.3)
$$\begin{cases} \overline{wt}(b_0) = 0, \ \overline{wt}(b_1) = \wedge_1 - 2\wedge_0, \\ \overline{wt}(b_j) = \wedge_j - \wedge_{j-1}, \ j = 2, 3, \ldots, n-1, \\ \overline{wt}(b_n) = 2\wedge_n - \wedge_{n-1}, \ \overline{wt}(b_{n+1}) = 0, \\ \overline{wt}(b_{n+2}) = \wedge_{n-1} - 2\wedge_n, \ \overline{wt}(b_{2n+1}) = 2\wedge_0 - \wedge_1, \\ \text{and } \overline{wt}(b_{2n+1-j}) = \wedge_j - \wedge_{j+1}, \ j = 1, 2, \ldots, n-2. \end{cases}$$

Note that the set B equipped with the maps $\tilde{e}$, $\tilde{f}_i$, φ_i, ϵ_i, $\overline{wt}$, for $i \in I \setminus \{0\}$, is the crystal associated with the B_n-module $V(\wedge_1) \oplus V(0)$ (see [KN]).

The sequence $\eta = (\eta_k)_{k \ge 1} = (\ldots \eta_3 \eta_2 \eta_1)$ in B, where $\eta_k = b_{n+1}$, for all $k \ge 1$, is called the *ground-state path* of weight $\wedge_n$. A $\wedge_n$-*path*, by definition is a sequence $p = (p_k)_{k \ge 1} = (\ldots p_3 p_2 p_1)$ such that $p_k \in B$ and $p_k = \eta_k$ for $k \gg 0$. As shown in [KMN2], the set $\mathcal{P}(\wedge_n) \subseteq \ldots \otimes B \otimes B \otimes B$, of $\wedge_n$-paths, as a crystal is isomorphic to the crystal $B(\wedge_n)$ associated with the $D_{n+1}^{(2)}$-module $L(\wedge_n)$. In particular, the actions of $\tilde{e}_i$ and $\tilde{f}_i$ on $\mathcal{P}(\wedge_n)$ are given via (1.14) and (1.15) (also see Section 2 of [BM]).

Now for $k = 1, 2, \ldots, n-1$, consider the path $p = (\ldots b_{n+1} b_{n+2} b_{n+3} \ldots b_{n+k+1}) \in \mathcal{P}(\wedge_n)$. It can be easily seen using property (2.5) of the energy function H that in this case

(4.4)
$$H(b_{n+1} \otimes b_{n+1}) = H(b_{n+i} \otimes b_{n+i+1}),$$

for $i = 1, 2, \ldots, k$ (also see [KKM, Section 5.6]). Hence, it follows from (2.3), (4.3) and (4.4) that $wt(p) = \wedge_{n-k} - \wedge_n$. Hence $\eta \otimes p \in \mathcal{P}(\wedge_n) \otimes \mathcal{P}(\wedge_n)$ has weight $wt(\eta) + wt(p) = \wedge_{n-k}$ for $k = 1, 2, \ldots, n-1$. Furthermore, $\tilde{e}_i p = 0$, for $i = 0, 1, \ldots, n-1$, $\tilde{e}_n p = (\ldots b_{n+1} b_{n+3} b_{n+4} \ldots b_{n+k+1})$ and $\tilde{e}_n^2 p = \tilde{e}_n^{\wedge_n(h_n)+1} p = 0$. Hence it follows from Lemma 1.16 that $L(\wedge_{n-k})$ occurs as a direct summand in the decomposition of $L(\wedge_n) \otimes L(\wedge_n)$ for $k = 1, 2, \ldots, n-1$. This completes the proof of Theorem 1.6.

References

[Ag] Agrebaoui, B.: Standard modules and standard modules of level one, (to appear).

[BM] Bos, M. K., Misra, K. C.: An application of crystal bases to representations of affine Lie algebras. J. Algebra (to appear).

[Kac] Kac, V. G.: Infinite dimensional Lie algebras. Cambridge Univ. Press. (1990).

[KMN1] Kang, S., Kashiwara, M., Misra, K., Miwa, T., Nakashima, T., Nakayashiki, A.: Affine crystals and vertex models. Int. J. Mod. Phys. A., **7** (1A), 449-484 (1992).

[KMN2] Kang, S., Kashiwara, M., Misra, K., Miwa, T., Nakashima, T., Nakayashiki, A.: Perfect crystals of quantum affine Lie algebras. Duke Math. J., **68** (3), 499-607 (1992).

[KKM] Kang, S., Kashiwara, M., Misra, K.: Crystal bases of Verma modules for quantum affine Lie algebras. Compositio Math. **92**, 299-325 (1994).

[K1] Kashiwara, M.: Crystalizing the q-analogue of universal enveloping algebras. Commun. Math. Phys **133**, 249-260 (1990).

[K2] Kashiwara, M.: On crystal bases of the q-analogue of universal enveloping algebras. Duke Math. J. **63**, 465-516 (1991).

[KN] Kashiwara, M., Nakashima. T.: Crystal graphs for representations of the q-analogue of classical Lie algebras. J. Algebra **165**, 295-345 (1994).

[X] Xie, C.: Standard modules embedding in tensor products of level one fundamental modules. Commun. Algebra (to appear).

North Carolina State University
Mathematics Department
Raleigh, NC 27695-8205

Contemporary Mathematics
Volume **193**, 1996

THE MONSTER ALGEBRA: SOME NEW FORMULAE

S. NORTON

ABSTRACT. We prove some formulae linking the group and algebra structures of the Monster. Some of these were stated without proof in the Atlas of Finite Groups; others are new. We also speculate on relations with Moonshine.

1. INTRODUCTION

In [4] Griess proved the existence of the Fischer-Griess Monster group M by constructing an algebra on its 196883-dimensional representation. This algebra is normally known as the Griess algebra. In [1] Conway considerably simplified the construction by considering a related algebra on the 196884-dimensional space obtained by adjoining the trivial representation to the 196883-space. We call his algebra, which is also described in the Atlas [2], the *Griess-Conway algebra*.

The Monster is a 6-transposition group in the sense that the product of any two involutions of class $2A$, in the notation of [2] (which, with one proviso which we describe shortly, is used throughout this paper), has order at most 6. We therefore call such involutions *transpositions*.

On page 230 of [2] there is a table, reproduced from [1], which completely describes the various algebras generated by pairs of elements corresponding to transpositions. However, to eliminate fractions and for other reasons that will appear in due course, it is convenient to double all inner products from the ones shown in this table. The relevant formulae, adjusted to this new convention, are shown in Table 1 (see page 2). The notation used in this table is as follows.

All group elements are in M. We denote the algebra product by $*$ and call the algebra elements *vectors*. There is an invariant correspondence between group elements of conjugacy class $2A$, $3A$, $4A$ or $5A$ up to inversion, and certain vectors; we reserve for such vectors the letters t, u, v and w, respectively, possibly with primes and/or suffices, according to the order of the corresponding group element. If the vector x corresponds to the group element y, we write $x \leftrightarrow y$. We may also write $y = |x|$ when it doesn't matter whether we choose y or y^{-1}, e.g. when y is a transposition. In Table 1, a and b are transpositions, $t_i \leftrightarrow a(ab)^i$, and t, u, v and $\pm w$ denote vectors corresponding to powers of ab having the right order.

1991 *Mathematics Subject Classification*. Primary: 17D99 Secondary: 20B25, 20C34.
Key words and phrases. Monster, algebra, transposition.

Table 1

$1A$	$t_0 * t_0 = 64t_0$, $(t_0, t_0) = 256$, $(t_0, 1) = 4$, $(1, 1) = 3$
$2A, 1A$	$t_0 * t_1 = 8(t_0 + t_1 - t)$, $(t_0, t_1) = 32$, $t \leftrightarrow ab$
$2B, 1A$	$t_0 * t_1 = 0 = (t_0, t_1)$
$3A, 1A$	$t_0 * t_1 = 4t_0 + 4t_1 + 2t_2 - 3u$, $(t_0, t_1) = 13$, $u \leftrightarrow ab$, $(u, 1) = 9$
	$t_0 * u = 10(2t_0 - t_1 - t_2 + u)$, $(t_0, u) = 90$
	$u * u = 90u$, $(u, u) = 810$
$3C, 1A$	$t_0 * t_1 = t_0 + t_1 - t_2$, $(t_0, t_1) = 4$
$4A, 2B, 1A$	$t_0 * t_1 = 3t_0 + 3t_1 + t_2 + t_3 - v$, $(t_0, t_1) = 8$, $v \leftrightarrow ab$, $(v, 1) = 24$
	$t_0 * v = 12(5t_0 - 2t_1 - t_2 - 2t_3 + v)$, $(t_0, v) = 288$
	$v * v = 192v$, $(v, v) = 4608$
$4B, 2A, 1A$	$t_0 * t_1 = t_0 + t_1 - t_2 - t_3 + t$, $(t_0, t_1) = 4$, $t \leftrightarrow (ab)^2$
$5A, 1A$	$t_0 * t_1 = \frac{1}{2}(3t_0 + 3t_1 - t_2 - t_3 - t_4 + w)$, $(t_0, t_1) = 6$, $w \leftrightarrow ab$
	$t_0 * t_2 = \frac{1}{2}(3t_0 + 3t_2 - t_4 - t_1 - t_3 - w)$, $(t_0, t_2) = 6$, $-w \leftrightarrow (ab)^2$
	$t_0 * w = 14(t_1 - t_2 - t_3 + t_4 + w)$, $(t_0, w) = 0$, $(w, 1) = 0$
	$w * w = 350(t_0 + t_1 + t_2 + t_3 + t_4)$, $(w, w) = 7000$
$6A, 3A, 2A, 1A$	$t_0 * t_1 = t_0 + t_1 - t_2 - t_3 - t_4 - t_5 + t + u$, $(t_0, t_1) = 5$
	$t * u = 0 = (t, u)$, $t \leftrightarrow (ab)^3$, $u \leftrightarrow (ab)^2$

$$x^{|t|} = x - \frac{1}{14}[12(x, t)t + 16x * t - (x * t) * t]$$

2. Proof of the $3C$ and $5A$ cases of Table 1

In [1] it was stated that while the cases apart from $3C$ and $5A$ were computed explicitly from Conway's construction of the Griess-Conway algebra, these two, and $5A$ in particular, needed a more theoretical calculation. Here is how one can deal with those two cases. We use ad_x or $\mathrm{ad}(x)$ to denote the operation of algebra multiplication by x, and start with

Lemma 1. ad_t *has eigenvalues* 64, 16, 2 *and* 0. *The corresponding eigenspaces are of dimension* 1 (*generated by* t *itself*), 4371, 96256 *and* $1 + 96255$ *respectively, with the* 1-*space in the last being generated by* $64 - t$.

Proof. It can easily be seen that ad_t has eigenvalue 64 on t and 0 on $64 - t$ (see case $1A$ above), 16 on $t' - t''$ where $|t'||t''| = |t|$ (see case $2A$), 0 on t' if $|t||t'|$ has class $2B$ (see case $2B$), and 2 on $t' - t''$ where $|t||t'| = |t'||t''|$ is an element of class $3A$ (see case $3A$). When acted on by $C_M(|t|)$, the double cover of the Baby Monster B, the first two vectors are fixed, while the last generates a 96256-space (the unique faithful constituent of the 196884-character restricted to $2B$). The others must, in some order, generate (modulo the space generated by $64 - t$) a 4371-space and a 96255-space; the order can be determined by noting that

$$\mathrm{tr}[\mathrm{ad}_t] = \mathrm{tr}[\mathrm{ad}_1].\frac{(t, 1)}{(1, 1)} = 196884.\frac{4}{3}. \quad \square$$

We introduce the notation (x, y, z) for the triple product $(x * y, z)$, which is totally symmetric in the vectors x, y and z.

We can now deal with the $3C$ case in Table 1. $(t_0, t_1) = 4$ follows from the eigenvalue calculation in the Monster graph. We deduce that the triple products (t_0, t_0, t_1) and $(t_0, t_1, 1)$ are 256 and 4 respectively. Now let $t_0 * t_1 = \alpha t_0 + \beta t_1 + \gamma t_2 + \delta$, the general element fixed by $C_M(<t_0, t_1, t_2>) = Th$. Then $\alpha = \beta$ (by symmetry), $\beta = \gamma + 2$ (because ad(t_0) has eigenvalue 2 on $t_1 - t_2$), and $256\alpha + 4\beta + 4\gamma + 4\delta = 256$ and $4\alpha + 4\beta + 4\gamma + 3\delta = 4$ (from the two triple products above), from which we deduce $\alpha = \beta = 1$, $\gamma = -1$, $\delta = 0$ as required.

The $5A$ case is similar if more elaborate. $C_M(<|t_0|, |t_1|, |t_2|, |t_3|, |t_4|>) = HN$ fixes a 7-space, generated by the t_i plus 1 and a vector w which is fixed by $D_{10} \times HN$ but negated by any element that acts as an outer automorphism of D_{10} and HN. w is orthogonal to 1 and the t_i because they lie in different eigenspaces of such an outer automorphism. As before we have $(t_i, t_j) = 6$ from the eigenvalue calculation on the graph. Calculating with (t_i, t_j, t_k) and $(t_i, t_j, 1)$ gives the formula for $t_0 * t_1$, except for a possible coefficient of w which may be assumed to be 1 by defining w appropriately. Applying an outer element of $[D_{10} \times HN].2$ gives the formula for $t_0 * t_2$. We deduce $t_0 * (t_1 - t_2 - t_3 + t_4) = 2(t_1 - t_2 - t_3 + t_4 + w)$. Known symmetries then give $t_0 * w = \alpha(t_1 - t_2 - t_3 + t_4) + \beta w$, where $\alpha = \beta = 14$ is required to give ad(t_0) the correct eigenvalues on $<t_1 - t_2 - t_3 + t_4, w>$. The triple product (t_0, t_1, w) then gives $(w, w) = 7000$. Again by symmetry, $w * w = \gamma(t_0 + t_1 + t_2 + t_3 + t_4) + \delta$, and the triple products $(w, w, 1)$ and (w, w, t_0) respectively give $20\gamma + 3\delta = 7000$ and $280\gamma + 4\delta = 98000$, whence $\gamma = 350$ and $\delta = 0$, thus completing the proof of the $5A$ case.

3. Algebra closure

The following result was stated in [2].

Theorem 1. *The Griess-Conway algebra product is closed on the lattice generated by products of up to two transposition vectors.*

Proof. We start with

Lemma 2. *A vector of type v is a linear combination of transposition vectors.*

Proof. The square of $|v|$ determines an extraspecial group 2^{1+24} which fixes a 300-space; we work inside such a space. $|v|$ itself will correspond to a Leech Lattice vector of type 3, which we may assume to have shape $(+2^{12}, 0^{12})$. We call the support of this vector, which belongs to the Golay Code, S.

Choose a set T of three coordinate positions, disjoint from S. These determine a splitting of S into four triples such that the union of any two is a special hexad which is completed to an octad by a subset of T. For each such octad, the $(+2^8, 0^{16})$-type vector gives rise to a pair of transpositions. Then the sum of all twelve corresponding transposition vectors, plus or minus other transposition vectors corresponding to lattice vectors of type $(\pm 4^2, 0^{22})$, can be made to yield v. $\square$

Lemma 3. *The triple product of the vectors corresponding to any three transpositions is an integer.*

Proof. There are three cases according to the conjugacy class of the product of the first two.

(a) If the product has class $1A$, $2A$, $2B$, $3C$ or $4B$, the result is obvious; if the class is $4A$, we use Lemma 2.

(b) If the class is $3A$ or $6A$ we need to show that (t, u) is always integral. We use the classification of transpositions under $3Fi_{24}$. Character theory shows that there are 22 orbits and the details are shown in Table 3 (see page 8). It is easy to compute most of the inner products (t, u), together with the $(u, u^{|t|})$'s we shall be needing in (c) below, by expressing $|u|$ as a product of two suitable transpositions $|t_1|$ and $|t_2|$, and using the symmetry of the triple products (t, t_1, t_2) and $(t_1, t_2, u^{|t|})$ and the known values of the trace of ad_u and its square.

(c) For the class $5A$ one uses the following argument. We work inside an A_5 of type $(2A, 3A, 5A)$. As we know, the vector corresponding to an element of class $2A$ (t) or $3A$ (u) is determined by the group it generates, but for a $5A$ the vector (w) is negated if the element is squared. Then the following hold:

The inner product with 1 of a t, u, w or 1 itself is 4, 9, 0, 3 respectively (see Table 1).

The inner product of t's is 256, 32, 13, 6 according as they generate the group 2, V_4, S_3, D_{10} (see Table 1).

The inner product (t, u) is 90, 40, 20 according as $<|t|, |u|> \cong S_3$, A_4, A_5 (see Table 3).

The inner product of u's is 810, 170, 20 according as the corresponding group elements generate the group 3, A_4, A_5 (see Table 3, last column).

The inner product (t, w) is 0, -98, 98 according as tw has class $2A$, $3A$, $5A$. (Express $|w|$ as a product of two transpositions $|t'|$ and $|t''|$ and consider the triple product (t, t', t''). A similar method applies in the following two cases.)

The inner product (u, w) is 350 or -350 if $|u||w|$ has class $3A$ or $2A$. In other cases the inner product can be computed by passing to a power of $|w|$.

The inner product of w's corresponding to elements in the same A_5-conjugacy class and cyclic subgroup is 7000. If just the first, the second, or both conditions are violated then the inner product is respectively -7000, 1624, -1624.

It follows from this that the sum of the 5 cyclic permutations of

$$|(13)(24)| - |(12)(34)| + |(123)| - |(124)| + \frac{1}{6}[|(12345)| - |(13425)|]$$

is zero (as its norm is zero). From this we can deduce that the difference between $|(14)(23)| * |(25)(34)|$ and an even permutation thereof is integrally generated by t's and u's. By connectivity, it then follows that the triple product of any transposition with any pair of transpositions whose group product has class $5A$ is an integer. This completes the proof of Lemma 3. $\square$

We now continue the proof of Theorem 1. First, some notation: with w and t_i $(0 \le i \le 4)$ defined as in Table 1, we write $\tilde{w} = \frac{1}{2}(w + t_0 + t_1 + t_2 + t_3 + t_4)$.

When reduced modulo any prime, the 196883 representation remains irreducible after at most two 1's have been split off. (This can be seen by restricting to the subgroups $2^{1+24}.Co_1$ and $3^{1+12}.2Suz$.) Therefore, if one scales the inner products correctly, the dual quotient of the lattice T generated by the transposition vectors is "small" and acted on trivially by the Monster. The inner product form we are using is clearly integral and non-trivial modulo any prime, which means it is the "correct" form in the above sense. Now we know from Lemma 3 that a vector of type u or $\tilde{w}$ lies in the dual of T; so, by the triviality of the action of the Monster on the dual quotient of T, the difference between two u's or two $\tilde{w}$'s lies inside T.

Let U be the lattice generated by 1, t's, u's and $\tilde{w}$'s. We already know that U lies between T and its dual, and (using Lemma 2) that $T * T$ is in U. Clearly $1 * U$ is in U. As the difference between two u's or $\tilde{w}$'s is in T, we see that if we fix t, whether $t * u$ or $t * \tilde{w}$ lies in U does not depend on which u or $\tilde{w}$ we choose. In particular we may choose u or $\tilde{w}$ to be inverted by t, in which case we already know from Table 1 that $t * u$ or $t * \tilde{w}$ are in U. This proves that $T * U$ is in U. To show that $U * U$ is in U, we apply a similar argument using the values of $u * u$ and $\tilde{w} * \tilde{w}$ (for which see Table 1), and the fact that $u * \tilde{w} = 0$ if $|u|$ commutes with $|w|$ (deducible by the methods of [1, Section 17]. This completes the proof of Theorem 1. □

4. QUILTS AND FOOTBALLS

In [5] Hsu defines the concept of a "quilt" corresponding to sets of triples of involutions. Roughly speaking, a quilt consists of the closure of a triple (a, b, c) under the braiding operations taking this to (b, a^b, c) and (a, c, b^c), which preserve the product abc. We form the equivalence classes of such triples under the closure of conjugation by abc. We may then define the "vertices", "edges" and "faces" of a quilt to consist, respectively, of the closures of such equivalence classes under the operations taking (a, b, c) to (b, c, a^{bc}), (c, b^c, a^{bc}) and (a, c, b^c). This gives rise to a trivalent polyhedron (possibly with some degeneracies, as described in [5]).

If we take transpositions in the Monster, all faces will be at most hexagons because of the Monster's 6-transposition property. For a random triple we expect almost all the faces to be hexagons and pentagons, with the former outnumbering the latter by a factor of about 2 to 1 (this is obtained simply by counting how many transpositions, when multiplied by a given one, belong to a given conjugacy class). As the skeleton of a (European) football, a truncated icosahedron, has approximately this structure, we call the quilts belonging to the Monster *footballs*. There is some hope of classifying triples of transpositions by their footballs, and the following simple theorem is of interest.

Theorem 2. *If $a \leftrightarrow t$, $b \leftrightarrow t'$, and $c \leftrightarrow t''$, then the triple product $(t-2, t'-2, t''-2)$ is a football invariant for (a, b, c).*

Proof. The eigenvalue of ad_t on the 96256-space of $C_M(a)$ is 2. So $x * (t-2) = x^{|t|} * (t-2)$ for any vector x. As the triple product is totally symmetric, the theorem follows immediately. □

5. TRACES AND CHARACTER VALUES

Table 2 (see page 7) shows some formulae. a, b, c, d and e are now vectors; the sums are taken over all cyclic permutations of the relevant subset of $\{a, b, c, d, e\}$. In the fifth formula (only), we assume that all these vectors are orthogonal to 1; this formula can be used to define an alternating form (a, b, c, d, e), unique up to a scalar factor, on the five vectors. $\chi(g)$ denotes the value of the 196884-character on a group element g, and h denotes an element of class $3C$ inverted by the transpositions $|t_0|$, $|t_1|$ and $|t_2|$. We also write $T = t_0 + t_1 + t_2$.

The first five are proved by noting that the left hand side is a multilinear form invariant under the dihedral group on the listed vectors; and the right hand side, with the constant coefficients replaced by parameters, describes all such forms. We can then use known cases to calculate the constants. For the fifth formula, it shortens the argument to consider the cases $c = d = e = 3t - 4$ and $b = c = 3t - 4$, $d = e = 3t' - 4$ where $|t||t'|$ is a transposition, using the formula at the end of Table 1, and also Lemma 1 which enables us to express the cube of $\mathrm{ad}(3t - 4)$ quadratically in terms of t. Before dealing with the last three, we prove:

Lemma 4. *The eigenvalues of ad_u are 90, 30, 18, 3 and 0 with the corresponding eigenspaces having dimensions 1 (generated by u), $783 + 783$, 8671, $64584 + 64584$ and $1 + 57477$ with the 1-space in the last generated by $90 - u$.*

Proof. The 196884-character decomposes as shown over the Fischer group $3.Fi'_{24} = C_M(|u|)$. The eigenvalues for the components of dimension 1, 783 and 57477 can be seen from Table 1 together with details of the action of the Fischer group on transpositions and commuting pairs thereof. The rest can be seen by noting that u is invariant under the full automorphism group $3.Fi_{24}$ and using the formulae we have already proved in Table 2 to compute the traces of various powers of ad_u. $\square$

The last three formulae in Table 2 can now be proved. The last formula in Table 1 shows that the group action of $|t|$ on the 196884-space can be described by a form quadratic in t. Lemma 4 enables us to derive a similar, cubic, formula, describing the group algebra operation taking x to $x^{|u|} + x^{|u|^{-1}}$, which has eigenvalue -1 on the 783- and 64584-spaces and 2 on the other eigenspaces. So there is a formula for the trace of $x \to x^{|t||t'|}$ or $x \to x^{|t||u|} + x^{|t||u|^{-1}}$ which is quadratic in t and t' and cubic in u, i.e. similar to the right hand side of the corresponding formula in Table 2. In this formula, the exact coefficients can now be obtained by substituting elements for t, t' and u where all the character values and inner products are known.

As for the last formula, the 196883-character of the Monster restricts to $S_3 \times Th$ as $(1^+ + 2) \times (1 + 4123) + 2 \times (248 + 61256) + 1^- \times 30628 + 1^+ \times 30875$. We can derive from this a similar formula expressing the group algebra operation taking x to $x^h + x^{h^{-1}}$ cubically in terms of t_0, t_1 and t_2. Again the exact coefficients can be obtained by taking particular cases where all terms on each side are known.

We may also prove

Lemma 5. *The eigenvalues of ad_v are 192, 96, 72, 54, 32, 8, 6 and 0 with corresponding eigenspaces having dimension 1 (generated by v), 23, 552, $2048 + 2048$, 11178, 48600, $24104 + 24104$ and $1 + 275 + 37950$ (with the 1-space generated by $192 - v$).*

Proof. Use the explicit definition of the Griess-Conway algebra in [1]. $\square$

Table 2

$$\text{tr}[\text{ad}_a] = 65628(a,1)$$

$$\text{tr}[\text{ad}_a\text{ad}_b] = 4620(a,b) + 20336(a,1)(b,1)$$

$$\text{tr}[\text{ad}_a\text{ad}_b\text{ad}_c] = 900(a,b,c) + 5952(a,1)(b,1)(c,1) + 1240\sum(a,b)(c,1)$$

$$\text{tr}[\text{ad}_a\text{ad}_b\text{ad}_c\text{ad}_d] = 166[(a*b,c*d)+(a*d,b*c)] - 116(a*c,b*d) +$$
$$52[(a,c)(b,d)+(a,d)(b,c)+(a,b)(c,d)] +$$
$$1664(a,1)(b,1)(c,1)(d,1) +$$
$$320[(a,c)(b,d)(c,1)+(b,d)(a,1)(c,1)] +$$
$$\sum[228(a,b,c)(d,1)+320(a,b)(c,1)(d,1)]$$

$$\text{tr}[\text{ad}_a\text{ad}_b\text{ad}_c\text{ad}_d\text{ad}_e] = (a,b,c,d,e) + \sum 8[(a,b)(c,d,e)+(a,c)(b,d,e)] +$$
$$\sum[30(a*b,c*d,e)+4(a*c,b*d,e)-22(a*d,b*c,e)]$$

$$7\chi(|t||t'|) = 13(t^{|t'|},t) + 20(t,t')^2 + 256(t,t') - 1396$$

$$2\chi(|t||u|) = (u,u^{|t|}) - 2(t,t^{|u|}) + (t,u)^2 - 140$$

$$14\chi(|t|.h) = 856 - 144(t,T) + 7[(t,T)^2+(T^{|t|},T)] -$$
$$14(t,t^h) - 6\sum[(t,t_i)^2+(t_i^{|t|},t_i)]$$

It would be nice to go one stage further and calculate $\text{tr}[\text{ad}_a\text{ad}_b\text{ad}_c\text{ad}_d\text{ad}_e\text{ad}_f]$. If we put $a=b=t$, $c=d=t'$, $e=f=t''$, this would give a formula involving $\chi(|t||t'||t''|)$, which would be of considerable use in football theory. Unfortunately character calculations in the Monster show that there is a totally symmetric sextic form on the 196883-space which cannot be accounted for by combinations of the inner product, algebra product, quinary form and operations derived from the last.

It is, however, worthy of note that the last formula in Table 2 gives results for every football containing a $3C$-type triangle; structure constant calculations show that there are many footballs with this property.

6. THE ACTION OF $3.Fi'_{24}$ ON TRANSPOSITIONS

Table 3 shows the 22 orbits of the $3A$-centralizer in the Monster on transpositions. For the first 18 orbits sample elements can be found inside the A_{12} centralizing an A_5, and are shown as such. In the last four orbits, however, we show sample elements inside the groups S_{12}, $2S_4$ (centralizing a Tits group), $2^{1+24}.Co_1$ and $2^2.M_{21}.2$ (centralizing a $14A$) respectively. In the last three of these, we only show the group elements modulo the covering involutions, and elements of $M_{21}.2$ are shown by their orbit structure inside M_{24} (with the $M_{21}.2$ defined as the subgroup of M_{24} fixing the top left entry and fixing or interchanging the two entries labelled "A"). The involution $|t|$ in $M_{21}.2$ is described by its orbit structure, with the fixed points denoted by o. The various columns show, in order, the elements $|u|$ and $|t|$, the structure of the group $<|t|,|u|>$, the conjugacy class of the product $|t||u|$, and the inner products (t,u), $(t,t^{|u|})$, and $(u,u^{|t|})$.

Table 3

$\|u\|$	$\|t\|$	$<\|t\|,\|u\|>$	$\|t\|\|u\|$	(t,u)	$(t,t^{\|u\|})$	$(u,u^{\|t\|})$
123	12.45	S_3	2A	90	13	810
123	45.67	6	6A	0	256	810
123	12.34	A_4	3A	40	32	170
123	14.56	S_4	4B	10	13	170
123	14.25	A_5	5A	20	6	20
123	12.34.56.78.9x.et	$2 \times A_4$	6C	8	32	170
123	14.25.36.78.9x.et	$3 \times S_3$	6A	18	13	0
123.456	17.24	$L_2(7)$	7A	15	4	25
123.456	17.28	$3 \times A_5$	15A	11	6	47
123.456	17.48	S_4	4A	26	13	42
123.456	17.89	$3 \times S_4$	12C	10	13	80
123.456	12.34.57.68.9x.et	$2 \times L_2(7)$	14A	11	8	57
123.456	12.37.45.68.9x.et	$2 \times A_4$	6A	16	0	42
123.456	12.37.48.59.6x.et	$S_3 \times A_4$	6B	17	5	17
123.456	14.27.38.59.6x.et	$2 \times A_5$	10A	12	6	52
123.456.789.xet	14.28	$3^3.A_4$	9A	13	5	37
123.456.789.xet	15.24.37.6x.89.et	$4^2.S_3$	8B	10	13	90
123.456.789.xet	12.34.57.6x.8e.9t	$L_2(11)$	11A	12	6	42
123.456.789.xet	12.ab	$3 \times S_3$	6D	9	13	81
(ABC)	(CD)	$2S_4$	8C	10	4	48
(234) on rows	$\begin{smallmatrix} -3 & +1 & +1 & +1 & +1 & +1 \\ +1 & +1 & -1 & -1 & -1 & -1 \\ -1 & -1 & +1 & +1 & -1 & -1 \\ -1 & -1 & -1 & -1 & +1 & +1 \end{smallmatrix}$	$4A_4$	12A	12	8	42
(234) on rows	$\begin{smallmatrix} O & O & A & A & B & C \\ D & E & F & B & C & F \\ O & G & O & G & O & E \\ H & O & O & H & O & D \end{smallmatrix}$	$2.4^2.S_3$	8A	14	13	42

We list the centralizers of the groups $<\|t\|,\|u\|>$ in the Monster. In order, they are Fi_{23} (for S_3), $3 \times 2Fi_{22}.2$ (for 6), $^2D_5(2)$ (for A_4), $S_8(2)$ (for S_4), A_{12} (for A_5), $2^{1+12}.(U_4(2) \times S_3)$ (for $2 \times A_4$), $3 \times O_7(3)$ (for $3 \times S_3$), $S_4(4).2$ (for $L_2(7)$), $3 \times A_9$ (for $3 \times A_5$), $2^{11}M_{23}$ (for S_4), $3 \times S_6(2)$ (for $3 \times S_4$), $2^2M_{21}.2$ (for $2 \times L_2(7)$), $2^{11}M_{22}.2$ (for $2 \times A_4$), $U_5(2)$ (for $S_3 \times A_4$), $2M_{22}.2$ (for $2 \times A_5$), $3^5.2^4.A_5$ (for $3^3.A_4$), $2^{10}M_{11}$ (for $4^2.S_3$), M_{12} (for $L_2(11)$), $3^{1+8}.2^{1+6}.3^{1+2}.2A_4$ (for $3 \times S_3$), $2 \times {}^2F_4(2)'$ (for $2S_4$), $4.2^{10}.3^4.5.4$ (for $4A_4$), and $2.2^6.U_3(3)$ (for $2.4^2.S_3$). We may check that the sum of their indices in $3.Fi'_{24}$ is the number of transpositions in the Baby Monster.

7. ATKIN-LEHNER INVOLUTIONS

Proofs are omitted in this section. We start with the following theorem, which leads us into the realm of conjecture.

Theorem 3. *Let S be the subalgebra of the Griess-Conway algebra fixed by an element of class 2A, 3A or 4A. Then S has an outer automorphism.*

This automorphism is the operation that negates the eigenspace of the corresponding vector with dimension 4371, 8671 and 11178 respectively, and fixes all other eigenspaces.

What does this automorphism do ? In the $2A$ case, let our transposition be a. Then it is not too hard to see from Table 1 that the outer automorphism interchanges the vectors corresponding to b and ab when both of these group elements are transpositions, and fixes the vector corresponding to b when ab belongs to class $2B$.

Similarly, if a belongs to class $3A$, the outer automorphism fixes the vector corresponding to b, of class $3A$ and commuting with a, unless just one of ab and $a^{-1}b$ belongs to class $3A$, in which case its vector will be interchanged with that corresponding to b.

Let A be an elementary abelian 2-group or 3-group. Then one may consider the group generated by all the outer automorphisms of the fixed space of A corresponding to the $2A$- or $3A$-elements of A. (It is easy to see that these outer automorphisms will leave this fixed space invariant.) This group need have no relation to the Monster. For example, let $E \cong 2^{1+24}$ be a O_2-subgroup of the $2B$-centralizer. Then there is a natural correspondence between subgroups of E, maximal subject to not containing the centre, and Niemeier lattices (even unimodular lattices in 24 dimensions) including the Leech Lattice. (This can be seen by considering the natural map from the Leech Lattice to $E/Z(E)$; then the inverse image of one of our maximal subgroups will be a rescaled Niemeier lattice.) The outer automorphism group obtained from such a subgroup can be shown to be the reflection group of the Niemeier lattice. Examples of such reflection groups include S_{25} and $S_{13} \times S_{13}$.

For the rest of this section an understanding of [3] is necessary, and we do not redefine terms that were defined in that paper. We move from an elementary abelian group to the cyclic group generated by a Monster element g. In [3] this gives rise to a modular function which turns out to be a Hauptmodul whose fixing group we call G.

Conjecture 1. *If G has form $\Gamma_0(n)+n$, then the subalgebra on the fixed space of g has an involutory outer automorphism, not in the Monster.*

Note that the notation for g used in [3] will be $n+n$ unless n is a prime power, in which case, though $n+n$ will still be a valid notation, $n+$ will be the symbol normally used. Conjecture 1 "nearly" implies:

Conjecture 2. *Each Atkin-Lehner involution in G gives rise to an outer automorphism of the fixed space of g in the Griess-Conway algebra.*

"*Proof*". We take m dividing the order of g, such that g^m gives rise to a modular function with fixing group $\Gamma_0(n)+n$. By Conjecture 1, this gives rise to an outer automorphism of the fixed space of g^m, which can be restricted to the fixed space of g. It follows from [3] that G must contain an Atkin-Lehner involution $+n$, and we consider our automorphism as arising from this Atkin-Lehner involution. We believe it can be shown that this gives rise to a bijection between the products of such outer automorphisms and the Atkin-Lehner involutions of G.

In most cases this is straightforward. For example, if g has class $6A = 6+$, its square and cube have class $3A = 3+$ and $2A = 2+$; their outer automorphisms correspond to the Atkin-Lehner involutions $+3$ and $+2$ of G. Their product therefore corresponds to the Atkin-Lehner involution $+6$.

However, there are more difficult cases. For example, let us take a g of class $30A = 30+6, 10, 15$. Its powers will have class $15+15$, $10+10$, $6+6$. Assuming Conjecture 1, their automorphisms will certainly correspond in a sense to the Atkin-Lehner involutions $+15$, $+10$, $+6$. However, these Atkin-Lehner involutions multiply to the identity; for the bijection to work, this would also have to be true of the corresponding algebra automorphisms. This does not seem to be obvious.

Conjecture 3. *The above automorphisms extend to one of the graded algebra structures associated with the Monster and Moonshine.*

At present we cannot formulate this conjecture more precisely.

In [6] a correspondence is given between elements of the modular group Γ and certain operators on pairs of commuting elements of the Monster. Under this correspondence, operators that fix the pair $(1, g)$ (up to conjugation) tend to give rise to elements of $G \cap \Gamma$, though there are problems with constant factors which prevent this correspondence from being exact. This gives some sort of significance to the elements of $G \cap \Gamma$.

We hope that the arguments of this section can be used to give a similar significance to the Atkin-Lehner involutions of G; together with $G \cap \Gamma$ these generate the whole of G, thus, hopefully, helping to unlock the mysteries of Moonshine.

REFERENCES

1. J. H. Conway, *A simple construction for the Fischer-Griess Monster Group*, Invent. Math. **79** (1985), 513–540.
2. J. H. Conway, R. T. Curtis, S. P. Norton, R. A. Parker and R. A. Wilson, *Atlas of Finite Groups*, Oxford Univ. Press, Oxford, 1985.
3. J. H. Conway and S. P. Norton, *Monstrous Moonshine*, Bull. London Math. Soc. **11** (1979), 308–339.
4. R. L. Griess, *The Friendly Giant*, Invent. Math. **69** (1982), 1–102.
5. T. M. Hsu, *Quilts, T-systems, and the combinatorics of Fuchsian groups*, Ph. D. thesis, Princeton Univ., 1994.
6. S. P. Norton, *Generalized Moonshine*, Proc. Sympos. Pure Math. **47** (1987), 208–209.

DPMMS, 16 MILL LANE, CAMBRIDGE CB2 1SB, ENGLAND
E-mail address: simon@emu.pmms.cam.ac.uk

Contemporary Mathematics
Volume **193**, 1996

MODULAR MOONSHINE?

A. J. E. RYBA

1. INTRODUCTION

In this paper we consider the following sequence of groups (the right hand names use ATLAS notation [6]): ${}^2\mathbf{M} \cong B$, ${}^3\mathbf{M} \cong Fi_{24}{}'$, ${}^5\mathbf{M} \cong HN$, ${}^7\mathbf{M} \cong He$, ${}^{11}\mathbf{M} \cong M_{12}$, ${}^{13}\mathbf{M} \cong L_3(3)$, ${}^{17}\mathbf{M} \cong L_3(2)$, ${}^{19}\mathbf{M} \cong A_5$, ${}^{23}\mathbf{M} \cong S_4$, ${}^{29}\mathbf{M} \cong 3$, ${}^{31}\mathbf{M} \cong S_3$, ${}^{41}\mathbf{M} \cong 1$, ${}^{47}\mathbf{M} \cong 2$, ${}^{59}\mathbf{M} \cong 1$ and ${}^{71}\mathbf{M} \cong 1$. Our notation, ${}^p\mathbf{M}$, indicates that the group arises as a subquotient of the centralizer of a particular $p-$element of the Monster group. The theme of this paper is that much of the monstrous behaviour of the Monster is reflected in the $p-$modular behaviour of the group ${}^p\mathbf{M}$. We remark that the first eight groups of our sequence are finite simple groups, and moreover the first five of these groups are sporadic simple groups. One of the important goals of this paper is to present uniform behaviour of a family that includes several sporadic simple groups.

The particular monstrous properties that we examine concern the Griess algebra [9] and the McKay-Thompson series [7]. These two, originally disparate, aspects of the Monster are now known as basic structural features of the Monster vertex algebra [1]. We speculate that the phenomena of this paper could have a common origin in a sequence of invariant modular vertex algebras for our groups. In Section 6, we suggest a construction for such a sequence of modular vertex algebras.

In Section 2, we give a brief review of some properties of the Griess algebra, the McKay-Thompson series, and the Monster vertex algebra. We also use Section 2 to set up notation and definitions for the later sections.

In Section 3, we construct a sequence of analogues of the Griess algebra. For each group, ${}^p\mathbf{M}$ with $p \not> 2$, our sequence has a ${}^p\mathbf{M}-$invariant $p-$modular nonassociative algebra. The algebras corresponding to ${}^3\mathbf{M} \cong Fi_{24}{}'$, ${}^5\mathbf{M} \cong HN$, ${}^7\mathbf{M} \cong He$ and ${}^{11}\mathbf{M} \cong M_{12}$ are known from [15], [21], [19] and [18]. Theorem 1 places these algebras into a common family. Moreover, in our proof of Theorem 1, we construct new algebras for the groups ${}^p\mathbf{M}$ with $p > 11$. An invariant algebra in characteristic 2 for the group ${}^2\mathbf{M} \cong B$ remains conjectural. In Conjecture 1, we propose that our algebras all arise as subquotient rings of a 196884-dimensional version of the Griess algebra: we verify that this is the case when $p \in \{31, 47, 59, 71\}$. For other values of $p > 2$, we verify that our nonassociative algebras are "locally" images of this version of the Griess algebra.

In Section 4, we refine the main conjecture of Section 3, and in Conjecture 2 we give explicit subquotients of the Griess algebra as likely origins for our modular algebras. We also show how information can be lifted from our algebras to guess structural properties of the Griess algebra itself.

1991 *Mathematics Subject Classification.* 20C34; Secondary 20D08, 20C20, 11F22, 17B67.

In Section 5, we consider a sequence of "head characters" for each of our groups. The head characters are set up in Definition 2 of Section 2. Our choice of head characters differs from earlier choices [16] in that we use modular characters rather than ordinary characters. Conjecture 3 states that for each group our head characters form an increasing sequence. This conjecture provides an algebraic explanation for an increase apparent in the Fourier coefficients of Hauptmoduls for the function fields of the corresponding modular groups.

Our results could be viewed as a series of numerological and algebraic coincidences. However, the analogies between our results are striking and suggestive. We use these analogies to formulate simple general conjectures. Each of our conjectures has a case corresponding to each prime divisor of $|\mathbf{M}|$.

2. NOTATION AND DEFINITIONS

In this section, we establish notation and definitions that will be used throughout the remainder of the paper. Any notation or definition that we introduce later will apply only within its own section or subsection.

Let $\mathbf{M}$ denote the Monster group, and let p represent one of the fifteen prime divisors of $|\mathbf{M}|$. The group $\mathbf{M}$ has a class of $p-$elements that is called $p+$ in [7] and is called pA in [6]: let g_p be an element of this class.

Definition 1. *Let $\widehat{{}^p\mathbf{M}}$ be the group $C_{\mathbf{M}}(g_p)$, and let ${}^p\mathbf{M}$ be the group $C_{\mathbf{M}}(g_p)/\langle g_p\rangle$.*

As p varies, the groups ${}^p\mathbf{M}$ given by this definition form the sequence listed in the introduction.

We use the terms generalized character and generalized $p-$modular character for $\mathbf{Z}-$linear combinations of irreducible characters and of irreducible $p-$modular characters, respectively. We use the following process to associate a generalized $p-$modular character of ${}^p\mathbf{M}$ to each integral valued generalized character of $\mathbf{M}$. A similar procedure is normally used to define generalized decomposition numbers (see, for example, [8, Section 90]).

Construction 2.1. *Let G be a finite group, let q be a prime divisor of $|G|$, and let g be a $q-$element of G. Let χ be an integer valued generalized character of G.*

(i) *For each $q-$regular element, h say, of $C_G(g)$, let ${}^g\widehat{\chi}(h) := \chi(gh)$. Then ${}^g\widehat{\chi}$ is a $\mathbf{Z}-$linear combination of irreducible $q-$modular Brauer characters of $C_G(g)$.*

(ii) *For any $q-$regular element, k say, of $C_G(g)/\langle g\rangle$, there is a unique $q-$regular preimage, h say, in $C_G(g)$. Let ${}^g\chi(k)$ be given by ${}^g\widehat{\chi}(h)$. Then ${}^g\chi$ is a $\mathbf{Z}-$linear combination of irreducible Brauer characters of $C_G(g)/\langle g\rangle$.*

Proof. (i) The function ${}^g\widehat{\chi}$ is a class function on the $q-$regular classes of $C_G(g)$. Therefore ${}^g\widehat{\chi}$ can be written, uniquely, as a linear combination of irreducible $q-$modular characters of $C_G(g)$. We prove our claim of integrality by showing that the coefficients in this linear combination must be simultaneously algebraic integers and rational numbers.

Let χ_i denote a typical irreducible character of $C_G(g)$. Since g is central in $C_G(g)$, we have $\chi_i(g) = \epsilon_i\chi_i(1)$, where ϵ_i is a qth$-$root of unity. We may write $\chi|_{C_G(g)} = \sum_i a_i\chi_i$, where the coefficients a_i are integers. Thus, ${}^g\widehat{\chi}(h) = \sum_i a_i\chi_i(gh) = \sum_i a_i\epsilon_i\chi_i(h) \overset{=}{=} \sum_{i,k} a_i\epsilon_i d_{ik}\phi_k(h)$, where ϕ_k is the kth irreducible Brauer character of $C_G(g)$ in characteristic q, and d_{ik} is a decomposition number. This establishes that the coefficients in the decomposition of $^g\widehat{\chi}$ are algebraic integers.

Let χ_z be the generalized character of $C_G(g)$ given by $\sum a_i \chi_i$, where the sum is taken over those values of i with $\epsilon_i = 1$, and let $\chi_y = \chi|_{C_G(g)} - \chi_z$. For $1 \leq j \leq (p-1)$, we have, $\chi(gh) = \chi(g^j h)$ (since gh and $g^j h$ generate the same cyclic group, and χ is an integral valued generalized character). Thus $^g\widehat{\chi}(h) = \sum_i a_i \epsilon_i^j \chi_i(h)$. Hence $(q-1)\,^g\widehat{\chi}(h) = \sum_j{}^g\widehat{\chi}(h) = \sum_i \sum_j a_i \epsilon_i^j \chi_i(h) = \sum_i a_i \chi_i(h) \sum_j \epsilon_i^j = (q-1)\chi_z(h) - \chi_y(h)$. Therefore, $(q-1)\,^g\widehat{\chi}$ is an integral linear combination of irreducible Brauer characters.

We have now established (i). For (ii), we just need to observe that g is in the kernel of any irreducible q-modular representation of $C_G(g)$. $\square$

Each element of $\mathbf{M}$ has a McKay-Thompson series [7] that gives a Hauptmodul for an empirically determined genus zero subgroup of $PSL_2(\mathbf{R})$. Given $g \in \mathbf{M}$, the McKay-Thompson series, $T(g)$, is a meromorphic function on the upper half plane. The McKay-Thompson series has a Fourier expansion of the form $T(g) = q^{-1} + \sum_{n=1}^{\infty} H_n(g)q^n$ (where $q = e^{2\pi i z}$, and z is in the upper half plane). We regard H_n as a complex valued function on $\mathbf{M}$: it is called the nth "head character" of $\mathbf{M}$ [7]. The coefficients in the q-expansion of $T(g)$ are integers, hence the head characters are integral valued class functions on $\mathbf{M}$. It was conjectured in [7] (and is now known from [2]) that the head characters are all proper characters of $\mathbf{M}$. A weaker version of this theorem (proved by Atkin, Fong and Smith [23]) states that H_n is a generalized character of $\mathbf{M}$: this result allows us to make the following definitions.

Definition 2. *The nth head character of $^p\mathbf{M}$, written ^pH_n, is the generalized p-modular character obtained as $^{g_p}H_n$ by Construction 2.1 (ii).*

Definition 3. *Let g be a p-regular element of $^p\mathbf{M}$. The McKay-Thompson series of g, written $^pT(g)$, is the formal power series $q^{-1} + \sum_{n=1}^{\infty}{}^pH_n(g)$.*

We observe that the McKay-Thompson series of a p-regular element, g in $^p\mathbf{M}$, coincides with the McKay-Thompson series of any p-singular lift of g in $\mathbf{M}$. The origins of Definitions 2 and 3 lie in the work of [17, 16] that defines a McKay-Thompson series for each element of each group $^p\widehat{\mathbf{M}}$. It is noted in [17] that the McKay-Thompson series of a p-regular element of $^p\widehat{\mathbf{M}}$ is identical to the McKay-Thompson series of an appropriate p-singular element in $\mathbf{M}$. However, for any other element of $^p\widehat{\mathbf{M}}$, empirical considerations are needed to define a McKay-Thompson series: these empirically defined functions are not quite so convenient as the McKay-Thompson series of $\mathbf{M}$-elements. (For example, the subgroup of $PSL_2(\mathbf{R})$, that corresponds to a p-singular element of $^p\widehat{\mathbf{M}}$, does not necessarily lie between a group $\Gamma_0(N)$ and its normalizer [17]. Moreover, the head characters of [17] include irrational entries.) The particularly natural behaviour of p-regular elements of $^p\widehat{\mathbf{M}}$ that is apparent in [17, 16] gives the first indication that we should consider modular characters.

The McKay-Thompson series of the groups $^p\widehat{\mathbf{M}}$ [17, 16] give rise to a sequence of head characters; we write the nth as $^p\widehat{H}_n$. These head characters are class functions on $^p\widehat{\mathbf{M}}$ and they are believed to be proper characters, but there is no easy proof that $^p\widehat{H}_n$ is even a generalized character of $^p\widehat{\mathbf{M}}$. Our head character ^pH_n is the $p-$modular reduction of $^p\widehat{H}_n$, and, by settling for a modular reduction, we do have a uniform definition and a uniform proof that the character is at least a generalized modular character.

The first head character, H_1 of $\mathbf{M}$, has degree 196884. We write $(V, *)$ for the $\mathbf{M}-$invariant algebra, with character H_1, that is constructed in [5]. Let $\underline{1}$ be the identity element of $(V, *)$, and let $\langle \ , \ \rangle$ denote the $\mathbf{M}-$invariant bilinear form on $(V, *)$ whose values are exactly twice those of the form that is given in [5]. The bilinear form $\langle \ , \ \rangle$ is associative for $(V, *)$ in the following sense (familiar from Lie algebras).

Definition 4. *Let $(X, *)$ be a nonassociative algebra. A bilinear form $\langle \ , \ \rangle$ on X is called associative for the algebra if $\langle u_1, u_2 * u_3 \rangle = \langle u_1 * u_2, u_3 \rangle$.*

Let U be the orthogonal complement (with respect to $\langle \ , \ \rangle$) of the span of $\underline{1}$ in V. The irreducible $\mathbf{M}-$module U has character $H := H_1 - 1$. It supports an $\mathbf{M}-$invariant product, $*$ say, that is defined by multiplication within $(V, *)$, followed first by orthogonal projection onto U, and then by scalar multiplication by 6. Up to an indeterminate scale factor, this product is identical to the one originally defined in [9]. We equip the algebra $(U, *)$ with an invariant associative bilinear form, $(\ , \ )$, defined by $(u_1, u_2) = 3\langle u_1, u_2 \rangle$. We refer to $(U, *)$ as the Griess algebra. The following general construction recovers the product and the bilinear form on V from our invariants of U.

Construction 2.2. *Let $(X, *)$ be a nonassociative algebra with an associative bilinear form $(\ , \ )$. Let $Y = X \oplus I$ where I is a 1-dimensional space spanned by a vector 1. Let α, β and γ be non-zero scalars. Let $*$ be defined on Y by $(u_1, c_1 1) * (u_2, c_2 1) = (\beta(u_1 * u_2) + c_1 u_2 + c_2 u_1, (c_1 c_2 + \alpha(u_1, u_2))1)$, and let $\langle \ , \ \rangle$ be defined on Y by $\langle (u_1, c_1 1), (u_2, c_2 1) \rangle = \gamma(\alpha(u_1, u_2) + c_1 c_2)$. Then, $(Y, *)$ is a nonassociative algebra with an identity element, and $\langle \ , \ \rangle$ is an associative bilinear form. We call $(Y, *)$ an augmented form of $(X, *)$.*

In the 196884-dimensional algebra, $(V, *)$, we have $\langle \underline{1}, \underline{1} \rangle = 3$ (see [5, Table 3]). Together with our choices for the scale factors in the definitions of products on U, this shows that the algebra $(V, *)$ is an augmented form of $(U, *)$, with $\alpha = 1/9$, $\beta = 1/6$, and $\gamma = 3$.

In Section 3, our main goal is to exhibit analogues of $(V, *)$ and $(U, *)$ for the groups $^p\mathbf{M}$. For each $p > 2$, we produce a $^p\mathbf{M}-$invariant algebra, $(^pV, *)$, with character $^pH_1$. Moreover, in the cases where $41 \geq p \geq 5$, there is a $^p\mathbf{M}-$invariant algebra, $(^pU, *)$, with character ^pH. We regard the algebras on ^pV and ^pU as $p-$modular versions of $(V, *)$ and $(U, *)$, respectively. We conjecture that there is also a $^2\mathbf{M}-$invariant algebra with character $^2H_1$. We guess that there is no natural 3-modular analogue of $(U, *)$.

The following lemma verifies that there are modules that have characters ^pH and $^pH_1$, at least in the cases $p \leq 41$ for ^pH, and in all cases for $^pH_1$. In the lemma, and at later stages of the paper, we describe a (modular) irreducible character by its degree, or in the case of a character of a cyclic group, by its value on a particular generator of that group. Thus ω and $\overline{\omega}$ denote the two faithful characters of a cyclic group of order 3. The zero (generalized) character is written as 0.

Lemma 2.3. *The $p-$modular character ^pH, of $^p\mathbf{M}$, has one of the following decompositions as a sum of irreducible constituents:* $^2H = 1 + 4370$, $^3H = 1 + 781$, $^5H = 133$, $^7H = 50$, $^{11}H = 16$, $^{13}H = 11$, $^{17}H = 6$, $^{19}H = 5$, $^{23}H = 3$, $^{29}H = \omega + \overline{\omega}$, $^{31}H = 2$, $^{41}H = 1$, $^{47}H = 0$, $^{59}H = 0$, $^{71}H = 0$. *The decomposition of $^pH_1$ is obtained by adding a trivial constituent, 1, to the decomposition of ^pH. In particular, $^pH_1$ is a proper $p-$modular character, for any choice of p.*

Proof. We read off the character values of ^pH from the values of the 196883-dimensional, irreducible character of $\mathbf{M}$ [6, pages 220–223]. For $p > 5$, the decomposition follows from the known $p-$modular character table of $^p\mathbf{M}$ [11, 19, 6]. For $p \in \{2, 3, 5\}$ the result follows from the explicitly computed module decompositions of [24], [15] and [21], respectively. $\square$

The final monstrous structure that we use to guide our study of a group $^p\mathbf{M}$ is the Monster vertex algebra (originally described in [1]). It is defined on an $\mathbf{M}-$module, $\mathcal{V}$ say, that can be written as a direct sum $\mathcal{V} = \mathcal{V}_0 \oplus \mathcal{V}_1 \oplus \mathcal{V}_2 \oplus \dots$, where $\mathcal{V}_i$ has character H_{i-1}. In particular, $\mathcal{V}_1 = 0$. The Monster vertex algebra has $\mathbf{M}-$invariant graded products parameterized by the integers, such that the nth product (written $(v, w) \mapsto v_n w$) maps $\mathcal{V}_i \times \mathcal{V}_j$ to $\mathcal{V}_{i+j-n-1}$. The product corresponding to $n = 1$ maps $\mathcal{V}_2 \times \mathcal{V}_2$ to $\mathcal{V}_2$; it is the augmented version of the Griess algebra [1].

There is an essentially unique $\mathbf{M}-$invariant vector, c say, in $\mathcal{V}_2$ [1]. We write L_1, L_0, L_{-1} for the operators c_2, c_1, c_0. The Lie algebra generated by L_1, L_0, L_{-1} is a copy of sl_2, and this algebra commutes with the action of $\mathbf{M}$. A Cartan subalgebra is spanned by the operator L_0 which acts as multiplication by i on $\mathcal{V}_i$. The vertex algebra gives an easy proof of the following lemma that will form the basis for our $p-$modular conjecture in Section 5.

Proposition 2.4. *The head characters of the Monster, $H_1, H_2, H_3, \dots$, form an increasing sequence of characters.*

Proof. Suppose that $v \in ker(L_{-1}) \cap \mathcal{V}_n$. Then, v generates a finite dimensional module, X say, for the Lie algebra spanned by L_1, L_0, L_{-1} (since $L_{\pm 1}$ maps V_i to $V_{i \mp 1}$). Moreover, X contains no eigenvectors of L_0 with negative weight (since L_0 acts as i on $\mathcal{V}_i$). But, the weights of L_0 on any finite dimensional module are closed under sign changes, and therefore $X \leq \mathcal{V}_0$.

It follows that $v \in \mathcal{V}_0$. Hence, $ker(L_{-1}) \leq \mathcal{V}_0$. We deduce that L_{-1} gives an $\mathbf{M}-$invariant injection from $\mathcal{V}_i$ to $\mathcal{V}_{i+1}$, for $i > 0$. $\square$

3. ALGEBRAS

In this section, we explain how, for each $p > 2$, we can regard a $^p\mathbf{M}-$invariant $p-$modular algebra, with character $^pH_1$, as analogous to the augmented Griess algebra. Although it is easy to construct nonassociative algebras with large automorphism groups, it is rare to find an invariant nonassociative product on a small module for a finite simple group. (Indeed, this is why many simple groups of sporadic and exceptional Lie types are distinguished from natural overlying classical groups by means of a nonassociative product.) Thus, the existence of the invariant algebra is fairly surprising, and our evidence that this algebra shares a common local structure with the Griess algebra is very surprising. The local structure that is known in the Griess algebra is determined by the subalgebras generated by axes of transpositions ($2A$–elements of $\mathbf{M}$), it was first investigated in [5].

Definition 5. *Let G be a group, let X be a G–module, and let C be a conjugacy class of G. An axis map is a G–invariant function, $f : C \to X$ (that is a function with $f(c^g) = f(c)g$). If $c \in C$, we call $f(c)$ the axis of c.*

Observe that there is a nontrivial axis map from the class C to the module X if and only if X contains non-zero vectors fixed by the centralizer of an element of C. Axes (in V) of $\mathbf{M}$–elements of classes $2A$, $3A$, $4A$, and $5A$ are given in [5]: these axis maps can be composed with orthogonal projection onto $U = \underline{1}^\perp$ to obtain axes in U. Let $\underline{t}$, $\underline{u}$, $\underline{v}$, and $\underline{w}$ denote the resulting U–axes of $\mathbf{M}$–elements of classes $2A$, $3A$, $4A$, and $5A$, respectively. Let t, u, v, and w denote the corresponding axes in V.

Lemma 3.1. *Axes in U and V are related by the equations: $t = 4/3\underline{1} + \underline{t}$, $u = 3\underline{1} + \underline{u}$, $v = 8\underline{1} + \underline{v}$, and $w = \underline{w}$.*

The lemma is immediate from the inner product information in [5, Table 3].

Definition 6. *Let G be a group, let $(X, *)$ be a G–invariant algebra, let C be a conjugacy class of involutions of G, and let $f : C \to X$ be an axis map. Let D be a dihedral subgroup of G that is generated by a pair of involutions in C. We define $\Delta_X(D)$ to be the subring of $(X, *)$ that is generated by $f(C \cap D)$.*

We call $\Delta_X(D)$ the dihedral subalgebra of $(X, *)$ that corresponds to the dihedral group D.

The group $\mathbf{M}$ has nine orbits on unordered pairs of transpositions [6]. The orbit of a pair of transpositions is completely determined by the $\mathbf{M}$–class of the product of the transpositions. The class of the product is either $1A$, $2A$, $2B$, $3A$, $3C$, $4A$, $4B$, $5A$, or $6A$. We deduce that the group $\mathbf{M}$ has nine orbits on dihedral subalgebras of either $(U, *)$ or $(V, *)$. We write $\Delta_U(1A), \Delta_U(2A), \dots, \Delta_U(6A)$ and $\Delta_V(1A), \Delta_V(2A), \dots, \Delta_V(6A)$ for representatives of the dihedral subalgebras in U and V, respectively. The structures of the dihedral subalgebras of $(V, *)$ are given in [5, Table 3], we use them to obtain the structure constants for the dihedral subalgebras of $(U, *)$, as given in Table 1.

In Table 1, and whenever we need to name the elements of a dihedral group, we adopt the following notational conventions from [5, Table 3]. A generating pair of transpositions is $\{t_0, t_1\}$. We write t_i for $t_0(t_0 t_1)^i$. Any minimal positive power of $t_0 t_1$ that has order 5, 4, 3, or 2 is called w, v, u, or t, respectively.

Table 1 and the invariance of $*$ and $(\ ,\ )$ give enough information to calculate values of the product and the inner product of any pair of vectors in any dihedral subalgebra of $(U, *)$. Table 1 shows that each dihedral subalgebra of $(U, *)$ is additively spanned by a linearly independent set of integral multiples of axis vectors. For each dihedral subalgebra of $(V, *)$, other than $\Delta_V(5A)$, there is a similar spanning set (see [5, Table 3]). For the dihedral subalgebra $\Delta_V(5A)$, the vectors t_0, t_1, t_2, t_3, t_4, and $(t_0 + t_1 + t_2 + t_3 + t_4 + w)/2$ give an integral spanning set.

Table 1. The dihedral subalgebras of $(U, *)$

$\Delta_U(1A)$	$\underline{t_0} * \underline{t_1} = 2^4 \, 23\underline{t_0}, \; (\underline{t_0}, \underline{t_1}) = 2^4 \, 47$
$\Delta_U(2A)$	$\underline{t_0} * \underline{t_1} = 2^3 \, (5\underline{t_0} + 5\underline{t_1} - 6\underline{t}), \; (\underline{t_0}, \underline{t_1}) = 2^4 \, 5$
$\Delta_U(2B)$	$\underline{t_0} * \underline{t_1} = -2^3 \, (\underline{t_0} + \underline{t_1}), \; (\underline{t_0}, \underline{t_1}) = -2^4$
$\Delta_U(3A)$	$\underline{t_0} * \underline{t_1} = 2^4 \, (\underline{t_0} + \underline{t_1}) + 12\underline{t_2} - 18\underline{u}, \; (\underline{t_0}, \underline{t_1}) = 23$
	$\underline{t_0} * \underline{u} = 102\underline{t_0} - 60\underline{t_1} - 60\underline{t_2} + 52\underline{u}, \; (\underline{t_0}, \underline{u}) = 234$
	$\underline{u} * \underline{u} = 504\underline{u}, \; (\underline{u}, \underline{u}) = 3^4 \, 29$
$\Delta_U(3C)$	$\underline{t_0} * \underline{t_1} = -2(\underline{t_0} + \underline{t_1}) - 6\underline{t_2}, \; (\underline{t_0}, \underline{t_1}) = -4$
$\Delta_U(4A)$	$\underline{t_0} * \underline{t_1} = 10(\underline{t_0} + \underline{t_1}) + 6(\underline{t_2} + \underline{t_3} - \underline{v}), \; (\underline{t_0}, \underline{t_1}) = 8$
	$\underline{t_0} * \underline{v} = 312\underline{t_0} - 144\underline{t_1} - 72\underline{t_2} - 144\underline{t_3} + 64\underline{v}, \; (\underline{t_0}, \underline{v}) = 2^8 \, 3$
	$\underline{v} * \underline{v} = 1056\underline{v}, \; (\underline{v}, \underline{v}) = 13248$
$\Delta_U(4B)$	$\underline{t_0} * \underline{t_1} = -2(\underline{t_0} + \underline{t_1}) - 6(\underline{t_2} + \underline{t_3} - \underline{t}), \; (\underline{t_0}, \underline{t_1}) = -4$
$\Delta_U(5A)$	$\underline{t_0} * \underline{t_1} = (\underline{t_0} + \underline{t_1}) - 3(\underline{t_2} + \underline{t_3} + \underline{t_4} - \underline{w}), \; (\underline{t_0}, \underline{t_1}) = 2$
	$\underline{t_0} * \underline{t_2} = (\underline{t_0} + \underline{t_2}) - 3(\underline{t_4} + \underline{t_1} + \underline{t_3} + \underline{w}), \; (\underline{t_0}, \underline{t_2}) = 2$
	$\underline{t_0} * \underline{w} = 84(\underline{t_1} - \underline{t_2} - \underline{t_3} + \underline{t_4}) + 76\underline{w}, \; (\underline{t_0}, \underline{w}) = 0$
	$\underline{w} * \underline{w} = 2100(\underline{t_0} + \underline{t_1} + \underline{t_2} + \underline{t_3} + \underline{t_4}), \; (\underline{w}, \underline{w}) = 21000$
$\Delta_U(6A)$	$\underline{t_0} * \underline{t_1} = -2(\underline{t_0} + \underline{t_1}) - 6(\underline{t_2} + \underline{t_3} + \underline{t_4} + \underline{t_5} - \underline{t} - \underline{u}), \; (\underline{t_0}, \underline{t_1}) = -1$
	$\underline{t} * \underline{u} = -8\underline{u} - 18\underline{t}, \; (\underline{t}, \underline{u}) = -36$

Our main goal is to exhibit a sequence of $^p\mathbf{M}$−invariant nonassociative algebras whose dihedral subalgebras are images of those of $(V, *)$ in the following sense.

Definition 7. *Let $(X, *)$ be a nonassociative $\mathbf{Z}$−algebra with an associative integer valued bilinear form $(\ ,\ )$. Let $(Y, *)$ be a nonassociative $\mathbf{F}_p$−algebra with an associative $\mathbf{F}_p$−valued bilinear form $(\ ,\ )$. We say that Y is an image of X if there is a surjective ring homomorphism $f : X \to Y$ such that $(f(x_1), f(x_2)) = (x_1, x_2)$ (mod p), for all x_1 and x_2 in X.*

In order to check that a given algebra $(\Delta, *)$ is an image of one of the dihedral subalgebras of $(U, *)$, we just need to locate "axis vectors" that span Δ and multiply according to the rules of Table 1. A similar test (based on [5, Table 3]) identifies images of dihedral subalgebras of $(V, *)$.

In each of the cases where $3 \leq p \leq 23$, $p = 31$, or $p = 47$, the group $^p\mathbf{M}$ has even order and it has a uniquely defined class of involutions (whose elements we call transpositions) that have a lift in the class $2A$ of $\mathbf{M}$.

Theorem 1. *Suppose that $p \mid |\mathbf{M}|$, and that $p > 2$. Then there is a non-trivial $^p\mathbf{M}$−invariant $\mathbf{F}_p$−algebra, $(^pV, *)$, with an invariant associative bilinear form such that:*

(i) *The character of the $^p\mathbf{M}$−module ^pV is $^pH_1$.*

Moreover, if $3 \leq p \leq 23$, or $p = 31$, or $p = 47$, then we may choose the algebra $(^pV, *)$, and an axis map from the transpositions of $^p\mathbf{M}$ to ^pV, so that the following additional property holds.

(ii) *Let t_0 and t_1 be $2A$−elements of $\mathbf{M}$ that centralize g_p, and let $^pt_0$, $^pt_1$ be the images of t_0, t_1 in $^p\mathbf{M}$. Then $\Delta_{^pV}(\langle ^pt_0, {}^pt_1 \rangle)$ is an image of $\Delta_V(\langle t_0, t_1 \rangle)$.*

We remark that if $p \in \{29, 41, 59, 71\}$, the group $^p\mathbf{M}$ has odd order, so we can not hope to obtain property (ii). The cases $p = 5$ and $p = 11$ of this theorem are already known from [21] and [18]. The cases corresponding to values of p greater than 23 are all proved in Subsection 3.9, and the cases corresponding to smaller values of p are proved case by case in Subsections 3.1 to 3.8. In each of Subsections 3.1 to 3.8, we concentrate on a single value of the prime p, and we construct a $^p\mathbf{M}$−invariant algebra with an associative $^p\mathbf{M}$−invariant bilinear form and $^p\mathbf{M}$−invariant axis maps. We calculate the dihedral subalgebras, and we verify that they meet the conditions of Theorem 1(ii). (These calculations are carried out by computer when $5 \le p \le 11$, and by hand when $p = 3$ and when $13 \le p$.)

In those cases of Theorem 1 where $p \in \{5, 11, 13, 17, 19, 23, 29, 31, 41\}$, it is convenient to exhibit an algebra, $(^pU, *)$, analogous to $(U, *)$, rather than an analogue of $(V, *)$. The following lemma applies to show that an augmented form of $(^pU, *)$ plays the role of $(^pV, *)$ in Theorem 1.

Lemma 3.2. *Let $p > 3$, and let $(X, *)$ be an $\mathbf{F}_p$−algebra with an associative bilinear form. Suppose that $(Y, *)$ is an augmented form of $(X, *)$ constructed by taking $\alpha = 1/9$, $\beta = 1/6$, and $\gamma = 3$. Suppose that the subring Δ_X, of $(X, *)$, is an image of a dihedral subalgebra $\Delta_U(C)$. Let $\underline{t_0}, \underline{t_1}, \dots, \underline{t_n}$ denote the Δ_X−images of the generating $(2A)$−axes of $\Delta_U(C)$. Let Δ_Y be the subring of $(Y, *)$ that is generated by $4/3\,\underline{1} + \underline{t_0}, 4/3\,\underline{1} + \underline{t_1}, \dots, 4/3\,\underline{1} + \underline{t_n}$. Then Δ_Y is an image of $\Delta_V(C)$.*

Proof. We can completely reconstruct the relations of [5, Table 3] from the relations of our Table 1 and the formulae of Lemma 3.1 for axes in V. □

Our philosophy that the prime divisors of $\mathbf{M}$ should all be treated on an equal footing suggests that property (ii) of Theorem 1 should be a consequence of a more general characterisation of the algebra $(^pV, *)$. The most natural place to look for algebras that contain images of dihedral subalgebras of $(V, *)$ is among the subquotient rings of $(V, *)$. We therefore propose the following uniform approach to Theorem 1.

Conjecture 1. *There is a non-trivial $^p\mathbf{M}$−invariant $\mathbf{F}_p$−algebra, $(^pV, *)$, with an invariant associative bilinear form such that:*

  (i) *The character of the $^p\mathbf{M}$−module ^pV is $^pH_1$.*
  (ii) *The algebra $(^pV, *)$ is an image of a subring of $(V, *)$ that contains the axes of all transpositions of $\mathbf{M}$ that centralize g_p.*

Whereas we are able to verify individual cases of Theorem 1 by working entirely within representations of the "small" groups $^p\mathbf{M}$, verification of Conjecture 1 requires some structural information about the algebra $(V, *)$. In Subsection 3.9, we establish the cases $p = 31$, $p = 47$ and the trivial cases $p \ge 59$ of Conjecture 1 (in these cases, appropriate structural information for $(V, *)$ is given by Table 1 and by [5, Table 3]).

We now give our proof of Theorem 1. An important secondary objective of the following subsections is to provide complete combinatorial definitions of our $^p\mathbf{M}$−invariant algebras and their axis vectors. These definitions make it easy to calculate with the algebras. The reader can get a good feel for the content and the results of the subsections by sampling any one of them.

3.1. The algebra for Fi_{24}'. In this subsection we work with a group $G \cong Aut(^3\mathbf{M}) \cong Fi_{24}$. The group $^3\mathbf{M}$ is isomorphic to Fi_{24}': we identify it with the commutator subgroup of G. Let X be the 783-dimensional, indecomposable, $\mathbf{F}_3G$-module of [6, page 207]. The restriction of X to a $^3\mathbf{M}$−module has character $^3H_1$. Our aim is to show that this restriction of X plays the role of ^3V in Theorem 1.

We note that the transpositions of Theorem 1 are the $2A$−elements of Fi_{24}. In other work on the group Fi_{24}, it is customary to refer to $2C$−elements as (Fischer) transpositions. In order to avoid confusion, we shall refer to all involutions of Fi_{24} by their ATLAS class names.

A spanning set of $24 + 2 \times 759$ vectors for X is given in [6, pages 206–207]. The spanning set includes 24 vectors called e_i, where $1 \leq i \leq 24$. We let 1 denote the vector $\sum_{i=1}^{i=24} e_i$ (this vector spans the 1-dimensional G−invariant subspace in X). For each vector e_i, with $i = 1, 2, \ldots, 24$, there is a unique $2C$−element, t_i say, in G with $Stab_X(C_G(t_i)) = \langle 1, e_i \rangle$. We define e_i to be the axis of t_i. Axes of all other $2C$−elements are determined by the action of G. We shall denote axes by underlining. Thus, for example, e_i is also written as $\underline{t_i}$. Let $\mathcal{A}$ denote the set of axes of all $2C$−elements in G.

There is a G−invariant bilinear form on X with $\langle \underline{t}, \underline{t'} \rangle$ equal to 0 or 2 according as the product tt' has even or odd order (this is obtained as the 3-modular reduction of the sesquilinear form of [6, page 206]). It follows from the construction of [15] that the product $X \times \mathcal{A} \to X$ given by $\underline{v} * \underline{t} = \underline{v} - \langle \underline{v}, \underline{t} \rangle - \underline{v}\,t$ extends to a G−invariant bilinear product $X \times X \to X$. We remark that the algebra product given in [6, page 207] is incorrect — since it is formed by quotienting a non-ideal $\langle 1 \rangle$ from a modified version of our algebra $(X, *)$.

Any $2A$−element of G, t say, determines a unique pair of $2C$−elements, t^+ and t^- say, with the property that $t = t^+t^- = t^-t^+$. We define the axis of t to be the vector $\underline{t} := \underline{t^+} + \underline{t^-}$. Similarly, any $3A$−element of G, u say, belongs to a unique $2C$−generated dihedral group of order 6: we define $\underline{u}$ to be the sum of the axes of the three involutions in this dihedral subgroup. Moreover, any $5A$−element, w say, of G belongs to a unique $2A$−generated dihedral group of order 10 : let t_0 be an involution of this group. Then the w−conjugates of t_0^+ and t_0^- generate a copy of S_5. We may assume that t_0^+ and t_0^- are ordered so that t_0^+, t_0^-, and w correspond to the S_5−elements 12, 35, and 12345. We define the axis of w to be $\underline{w} := \sum_{i=0}^{4}(\underline{t_0}^- - \underline{t_0}^+)w^i$. The elements of classes $3C$ and $4A$ in G should also have axes in X (since they lift to $3A$ and $4A$ elements of $\mathbf{M}$). We identify likely candidates for their axes in the course of our proof of Proposition 3.1.1, but we have not verified that these candidates are stable under the requisite centralizers in G.

Proposition 3.1.1. *Let t_0 and t_1 be $2A$−elements of G, then the dihedral subalgebra of X that corresponds to $\langle t_0, t_1 \rangle$ is an image of any dihedral subalgebra of $(V, *)$ that corresponds to a ($2A$−generated) dihedral lift of $\langle t_0, t_1 \rangle$ in $\mathbf{M}$.*

Proof. The group G has twelve orbits on unordered pairs of $2A$−involutions. The orbit of a pair $\{t_0, t_1\}$ is determined by the G−class of t_0t_1, unless $t_0t_1 \in 2A$, when there are two possibilities for the orbit. The eleven possibilities for the G−class of t_0t_1 are $1A, 2A, 2B, 3A, 3C, 3E, 4A, 4B, 5A, 6A$, and $6F$. We now consider the twelve different dihedral subalgebras of $(X, *)$ and show that each is the image of all corresponding dihedral subalgebras in $(V, *)$.

If $t_0 t_1 \in 1A$, then $(\underline{t_0}, \underline{t_1}) = (\underline{t_0^+} + \underline{t_0^-}, \underline{t_0^+} + \underline{t_0^-}) = 1$, since t_0^+ and t_0^- commute. Moreover, $\underline{t_0} * \underline{t_1} = (\underline{t_0^+} + \underline{t_0^-}) * (\underline{t_0^+} + \underline{t_0^-}) = (\underline{t_0^+} + \underline{t_0^-}) = \underline{t_0}$ (from our formula for multiplication by a $2C$−axis). It follows that $\Delta_X(\langle t_0, t_1 \rangle)$ is an image of both $\Delta_V(1A)$ and $\Delta_V(3A)$.

A very similar analysis covers the cases where $t_0 t_1 \in \{2B, 3A, 3E, 6A\}$, and covers one of the cases with $t_0 t_1 \in 2A$. We characterize representatives of these five orbits by the following relationships between the $2C$−elements t_0^+, t_0^-, t_1^+ and t_1^-: either $|t_0^+ t_1^+| = |t_0^+ t_1^-| = |t_0^- t_1^+| = |t_0^- t_1^-| = 2$ for the case $t_0 t_1 \in 2B$, or $|t_0^+ t_1^+| = 1, |t_0^- t_1^-| = 2$ for the case $t_0 t_1 \in 2A$, or $|t_0^+ t_1^+| = 1, |t_0^- t_1^-| = 3$ for the case $t_0 t_1 \in 3A$, or $|t_0^+ t_1^+| = |t_0^- t_1^-| = 3, |t_0^- t_1^+| = |t_0^+ t_1^-| = 2$ for the case $t_0 t_1 \in 3E$, or $|t_0^+ t_1^+| = |t_0^+ t_1^-| = |t_0^- t_1^+| = 2, |t_0^- t_1^-| = 3$ for the case $t_0 t_1 \in 6A$. These relations allow us to calculate the necessary products and inner products just as in the case where $t_0 t_1 \in 1A$.

For the cases $t_0 t_1 \in \{4B, 5A\}$ and for the second case with $t_0 t_1 \in 2A$, we can exhibit representatives for $t_0^\pm$ and $t_1^\pm$ within a subgroup S_6 of G as follows. To obtain $t_0 t_1 \in 2A$, we take $t_0^+ = 12$, $t_0^- = 34$, $t_1^+ = 23$ and $t_1^- = 14$. To obtain $t_0 t_1 \in 4B$, we take $t_0^+ = 12$, $t_0^- = 34$, $t_1^+ = 23$ and $t_1^- = 56$. To obtain $t_0 t_1 \in 5A$, we take $t_0^+ = 12$, $t_0^- = 35$, $t_1^+ = 45$ and $t_1^- = 13$. In each of these cases a simple calculation within the group S_6 establishes that $\Delta_X(\langle t_0, t_1 \rangle)$ has an appropriate structure. We now illustrate this calculation for the case $t_0 t_1 \in 5A$.

Given $2A$−involutions, t_0 and t_1, with $t_0 t_1 \in 5A$, we may assume that $t_0^\pm, w$ are represented by the S_5−elements given above. We write $[ij]$ for the axis of the S_5−element ij, so that: $\underline{t_0} * \underline{t_1} = ([12] + [35]) * ([13] + [45]) = [12] + [13] - [23] + [35] + [13] - [15] + [35] + [45] - [34] = ([23] + [41]) + ([15] + [24]) + ([34] + [52]) + ([12] + [23] + [34] + [45] + [51] - [35] - [41] - [52] - [13] - [24]) = \underline{t_2} + \underline{t_3} + \underline{t_4} - \underline{w}$. Similar calculations show that $\underline{t_0} * \underline{t_2} = \underline{t_4} + \underline{t_1} + \underline{t_3} + \underline{w}$, $\underline{t_0} * \underline{w} = \underline{t_2} + \underline{t_3} - \underline{t_1} - \underline{t_4} - \underline{w}$, $\underline{w} * \underline{w} = -\underline{w}$, $(\underline{t_0}, \underline{t_1}) = (\underline{t_0}, \underline{t_2}) = (\underline{t_0}, \underline{w}) = 0$, $(\underline{w}, \underline{w}) = 1$. Hence $\Delta_X(\langle t_0, t_1 \rangle)$ is an image of $\Delta_V(5A)$.

For the case $t_0 t_1 \in 4A$, we have $\langle t_0^\pm, t_1^\pm \rangle \cong 2^3{:}S_4$ (where the 2^3−module for S_4 is the space $(1, 1, 1, 1)^\perp$ in the natural permutation module of S_4 over $\mathbf{F}_2$). The group $\langle t_0^\pm, t_1^\pm \rangle$ contains twelve conjugates of t_0^+, including all eight elements $t_i^\pm$, for $0 \leq i \leq 3$. If we set $\underline{v}$ to be the sum of the axes of these twelve $2C$−elements of G, then the vectors $\underline{t_0}$, $\underline{t_1}$, $\underline{t_2}$, $\underline{t_3}$ and $\underline{v}$ satisfy the relations of $\Delta_V(4A)$. This follows from a calculation in $2^3{:}S_4$ similar to our earlier calculation in S_5. Therefore $\Delta_X(\langle t_0, t_1 \rangle)$ is an image of $\Delta_V(4A)$. Note that although we believe that $\underline{v}$ is an axis for the $4A$−element, $t_0 t_1$ of G, we have not yet established this — since $t_0 t_1$ can be written in sixty different ways as a product of a pair of $2A$−involutions.

In the case $t_0 t_1 \in 6F$, we have $\langle t_0^\pm, t_1^\pm \rangle \cong 3^3{:}S_4$ (where the 3^3−module for S_4 is the space $(1, 1, 1, 1)^\perp$ in the natural permutation module of S_4 over $\mathbf{F}_3$). The group $\langle t_0^\pm, t_1^\pm \rangle$ contains eighteen conjugates of t_0^+, including all twelve elements $t_i^\pm$ for $0 \leq i \leq 5$. If we define $\underline{t}$ as $\underline{t_0}\, \underline{t_3^+}$ and $\underline{u}$ so that $-\underline{u} - \sum_{i=0}^{5} \underline{t_i}$ is the sum of the axes of the eighteen conjugates of t_0^+ in $\langle t_0^\pm, t_1^\pm \rangle$, then the vectors $\underline{t_0}$, $\underline{t_1}$, $\underline{t_2}$, $\underline{t_3}$, $\underline{t_4}$, $\underline{t_5}$, $\underline{t}$ and $\underline{u}$ satisfy the relations of $\Delta_V(6A)$. (This follows from a calculation in $3^3{:}S_4$ similar to our earlier calculation in S_5.) Therefore the dihedral subalgebras of $(X, *)$ that correspond to dihedral groups $\langle t_0, t_1 \rangle$ with $t_0 t_1 \in 3C$ and $t_0 t_1 \in 6F$ are images of $\Delta_V(3A)$ and $\Delta_V(6A)$, respectively. Note that although we believe that $\underline{u}$ is an axis for the $3C$−element, $t_0 t_2$ of G, we have not yet established this — since $t_0 t_2$ can be written in 1134 different ways as a product of a pair of $2A$−involutions. □

The $p = 3$ case of Theorem 1 follows immediately from Proposition 3.1.1.

3.2. The algebra for HN. In this subsection we use a group $G \cong Aut({}^5\mathbf{M}) \cong HN{:}2$. Let X be the 133−dimensional $\mathbf{F}_5 G$−module that is constructed explicitly in [22] (the representation can also be obtained from the information in [14]). The restriction of X to a ${}^5\mathbf{M}$−module has character ${}^5 H$. We give a brief summary of the results of [21] that show that the restriction of X to a ${}^5\mathbf{M}$−module plays the role of ${}^5 U$ in Theorem 1.

There is an essentially unique G−invariant bilinear form, (,) say, on X and there is an essentially unique axis map from $2A$−involutions of G to X [21]. Moreover, each $2A$−involution of G has exactly two square roots in the class $4D$ of G [6, pages 164–165]. Given a $2A$−involution, g say, of G, we write $\sqrt{g}$ and $\sqrt{g}^3$ for its $4D$−square roots and we write its axis in X as $\underline{g}$. Let $\mathcal{A}$ denote the set of all axes of $2A$−elements of G.

A generalization of the argument of [15] is used in [21] to show that the map $X \times \mathcal{A} \to X$ given by $x * \underline{g} = 2x + 3x\underline{g} + x\sqrt{g} + x\sqrt{g}^3 + (x, \underline{g})\underline{g}/(\underline{g}, \underline{g})$ extends to a G−invariant bilinear product $X \times X \to X$.

There are nine orbits of G on unordered pairs of $2A$−elements [14]: the orbit of a pair, $\{t_0, t_1\}$ is completely determined by the class of $t_0 t_1$ as one of $\{1A, 2A, 2B, 3A, 4A, 4B, 5A, 5E, 6A\}$.

An augmented version of $(X, *)$, constructed in [21], has the property that for each pair of involutions, $\{t_0, t_1\}$ in G, the corresponding dihedral subalgebra of the augmented algebra is an image of any dihedral subalgebra that corresponds to a ($2A$−generated) dihedral lift of $\langle t_0, t_1 \rangle$ in $\mathbf{M}$. Thus the results of [21] establish the $p = 5$ case of Theorem 1.

3.3. The algebra for He. In this subsection we work with a representation (constructed in [19]) of a group $G \cong Aut({}^7\mathbf{M}) \cong He{:}2$, on a 51-dimensional $\mathbf{F}_7$−module, X say (the module is called V in [19]). The group ${}^7\mathbf{M}$ is isomorphic to He: we identify it with the commutator subgroup of G. The restriction of X to a ${}^7\mathbf{M}$−module has character ${}^7 H_1$. Our aim is to show that this restriction of X plays the role of ${}^7 V$ in Theorem 1. A particularly interesting aspect of this case of Theorem 1 is that each axis is shared by three different transpositions. This observation leads to a simple 7−modular consistency check for many of the entries of Table 1 and of [5, Table 3].

We now give a brief summary of an invariant algebra on X (see [19] for full details). We write the elements of the field $\mathbf{F}_4$ as $\{0, 1, \omega, \overline{\omega}\}$. We reserve the letters i, j, k, l, m and n to denote the different points of the set $\{1, 2, 3, 4, 5, 6\}$, and the symbols χ, χ' to denote distinct short hexacode words. (The hexacode [4] is an isotropic subspace of the standard 6−dimensional unitary space $\mathbf{F}_4{}^6$. The hexacode has size 2^6, it contains 45 "short words" that have exactly four non-zero coordinates.) The cocode of the hexacode is formed by the cosets of the hexacode in $\mathbf{F}_4{}^6$. Since the hexacode is isotropic, the natural unitary form gives an $\mathbf{F}_4$−valued pairing between the hexacode and the cocode. The group $3 \cdot S_6$ acts on both the hexacode and the cocode [4, 19]: it also acts, via its quotient S_6, on $\{1, 2, 3, 4, 5, 6\}$.

Our $\mathbf{F}_7 G$−module, X, has a basis consisting of six vectors, $\underline{x}_i$, and forty-five vectors $\underline{x}_\chi$. The group G contains an elementary abelian subgroup, $T \cong 2^6$, whose elements can be identified with words of the cocode: we name elements of T by representatives of cosets of the hexacode. Each element, $t \in T$, acts on X according to the formulae: $\underline{x}_i t = \underline{x}_i$ and $\underline{x}_\chi t = ||\chi(t)|| \underline{x}_\chi$ (where, $||0|| = ||1|| = 1$ and $||\omega|| = ||\overline{\omega}|| = -1$). We may select a subgroup $S \cong 3 \cdot S_6$, in $N_G(T)$, that acts on our basis of X by subscript permutation. Thus, if $s \in S$, $\underline{x}_i s = \underline{x}_{is}$ and $\underline{x}_\chi s = \underline{x}_{\chi s}$. The group G is generated by adjoining an extra generator to its maximal subgroup $\langle S, T \rangle$. In [19] a suitable generator u is shown to act on X as follows:

$$\underline{x}_i u = 3\underline{x}_i + 2\sum_{j=1}^{6} \underline{x}_j + \sum_\chi (2|\chi(i)| + 1)\underline{x}_\chi, \quad \underline{x}_\chi u = \sum_i (2|\chi(i)| + 1)\underline{x}_i + \sum_{\chi'} a_{\chi\chi'} \underline{x}_{\chi'}$$

where $|0| = 0, |1| = |\omega| = |\overline{\omega}| = 1$, $a_{\chi\chi'} = 1$ if $\chi = \chi'$, $a_{\chi\chi'} = 4$ if $\chi' \in \{\omega\chi, \overline{\omega}\chi\}$, $a_{\chi\chi'} = 3$ if $\chi + \omega^n \chi'$ is short for all values of n, $a_{\chi\chi'} = 5$ if $\chi + \omega^n \chi'$ is short only for $n = 0$ and $a_{\chi\chi'} = 6$ if $\chi + \omega^n \chi'$ is short only for $n \neq 0$.

The module X supports a G−invariant nonassociative product with an invariant associative bilinear form [19]. We use the symmetric G−invariant bilinear form $\langle \, , \, \rangle$ given by $\langle \underline{x}_i, \underline{x}_i \rangle = 4$, $\langle \underline{x}_i, \underline{x}_j \rangle = \langle \underline{x}_i, \underline{x}_\chi \rangle = 0$, $\langle \underline{x}_\chi, \underline{x}_\chi \rangle = 4$ and $\langle \underline{x}_\chi, \underline{x}_{\chi'} \rangle = 0$. (This form is four times the natural orthonormal, invariant bilinear form of [19].) We also use a G−invariant product, $*$ say, obtained by negating the product of [19], so that: $\underline{x}_i * \underline{x}_i = \underline{x}_i$, $\underline{x}_i * \underline{x}_j = 0$, $\underline{x}_i * \underline{x}_\chi = 2|\chi(i)|\underline{x}_\chi$, $\underline{x}_\chi * \underline{x}_\chi = 2\sum |\chi(i)|\underline{x}_i$, $\underline{x}_\chi * \underline{x}_{\chi'} = b_{\chi\chi'} \underline{x}_{\chi+\chi'}$, where $b_{\chi\chi'}$ is 4, 6, 5 or 0 according as $a_{\chi\chi'}$ is 4, 3, 5 or 6. We write $\underline{x}_1 + \underline{x}_2 + \ldots + \underline{x}_6$ as $\underline{1}$, since it is an identity element for $(X, *)$.

The transpositions of $^7\mathbf{M}$ are the $2A$−involutions of G [6]. Let τ be a transposition in $^7\mathbf{M}$. The group $C_{G'}(\tau)$ has structure $2^2.L_3(4).2$—it fixes a 2-dimensional subspace of X, [11]. Hence, we may choose an axis of τ in X. In fact, the group $C_G(\tau)$ has index 3 in a maximal subgroup of G [6], and all of the vectors in X that are fixed by $C_G(\tau)$ are also fixed by this larger group [11]. It follows that any axis of τ in X must be shared by the other two transpositions in $O_2(C_G(\tau))$.

We now choose axes for the transpositions of G. In the next few paragraphs we shall make several assertions about particular vectors in X. These assertions are the results of routine computer calculations made with our explicit representation of G on X. Let τ_1 denote the element of T that corresponds to the cocode word 100000. Then τ_1 is a transposition of G, and its centralizer fixes the vectors in the space spanned by $\underline{x}_1$ and $\underline{1}$. Let the axis of τ_1 be $\underline{\tau}_1 := \underline{x}_1$. The axes of all other $2A$−involutions are determined as G−images of $\underline{x}_1$. In particular, the three transpositions that share the axis $\underline{x}_1$ are the T−elements that correspond to cocode words of the form $a00000$.

The group S has a uniquely determined involution, $\tau_{ij.kl}$ say, that maps to $(ij)(kl)$ in S_6. The involution $\tau_{ij.kl}$ is a transposition of G: its axis is $\underline{\tau}_{ij.kl} := 2(\underline{x}_i + \underline{x}_j + \underline{x}_k + \underline{x}_l) + 4\sum \underline{x}_\chi$, where the sum is taken over the three short hexacode words that are non-zero at positions i, j, k and l. The symmetry of this axis with respect to i, j, k and l shows that $\underline{\tau}_{ij.kl} = \underline{\tau}_{ik.jl} = \underline{\tau}_{il.jk}$. The group S also has 3×15 involutions that map to the fifteen S_6−elements with the cycle type of (ij) [6]. These fortyfive involutions of S are transpositions of G, and they correspond to the short hexacode words. The involution, τ_χ, that corresponds to χ, maps to the S_6−element that interchanges the two zero coordinates of χ. The G−element

τ_χ has axis $\underline{\tau}_\chi := \underline{x}_i + \underline{x}_j + \underline{1} - 2\underline{x}_\chi - \sum \underline{x}_{\chi'}$, where (ij) is the S_6−image of τ_χ, and the sum is taken over the six words χ' with $a_{\chi\chi'} = 5$.

We also need axes of G−elements from classes $3A, 4B$, and $5A$. Let $w \in G$ be an element from the class $5A$. We choose the zero vector of X as the axis of w. (The first hint of this choice can be seen in [5, Table 3], where the inner products and products that depend on a vector w are all zero when reduced modulo 7. Similarly, according to Table 1, the image of $\underline{w}$ spans an isotropic ideal in the 7-modular reduction of $\Delta_U(5A)$.)

There are three elements of S that map to the S_6−element (i, j, k): one of these three elements is $u_{ijk:l} := \tau_{ij.mn}\tau_{ik.mn}$ (the others are $u_{ijk:m} = \tau_{ij.ln}\tau_{ik.ln}$ and $u_{ijk:n} = \tau_{ij.lm}\tau_{ik.lm}$). The element $u_{ijk:l}$ has class $3A$ in G [6], and for its axis we choose $\underline{u}_{ijk:l} := 3\underline{x}_l + 6\underline{x}_m + 6\underline{x}_n + 4\underline{1} + 3\sum_1 \underline{x}_\chi + \sum_2 \underline{x}_{\chi'}$, where the first sum is taken over the six short words that vanish at l and either at m or at n, and the second sum is taken over the nine short words that vanish at l and one of i, j, and k. A computer calculation shows that $\underline{u}_{ijk:l}$ is fixed by $C_G(u_{ijk:l})$. The axes of other $3A$−elements are determined via the action of G.

The G−element $\tau_{ij.kl}\tau_i$ has class $4B$ (since it squares to an element of T that corresponds to a cocode word whose non-zero coordinates are at locations i and j). We choose the axis of this element to be: $2\underline{x}_i + 2\underline{x}_j + \underline{x}_k + \underline{x}_l + \underline{x}_\chi$, where χ is the unique short hexacode word that has $\chi(i) = 1$ and $\chi(m) = \chi(n) = 0$. A computer calculation shows that this vector is fixed by $C_G(\tau_{ij.kl}\tau_i)$. The action of G determines the axis of any $4B$−element.

Let $\underline{t}_1$ and $\underline{t}_2$ be axes of transpositions in G. There are nine ways to select a pair of transpositions in G with these particular axes. We define the type of $\{\underline{t}_1, \underline{t}_2\}$ to be the 9-tuple of G−classes determined by the products of the nine pairs of transpositions.

Lemma 3.3.1. *The group G has six orbits on pairs of axes of transpositions. The types of representative pairs for these orbits are as follows:* $(1A^3, 2A^6)$, $(2B^9)$, $(2A, 3B^4, 4A^4)$, $(3A^3, 5A^6)$, $(4B^9)$, $(6A^9)$.

Proof. The permutation character of He on a maximal subgroup of index 8330 is $1a+51ab+680a+1275a+1920a+4352a$. (Note that this character is given incorrectly in [6, page 104]. The decomposition of [6] would give a negative character value on elements of class $28A$ in He. The correct decomposition can be deduced from the information that the permutation character must be zero on classes $7C$ and $17A$.) It follows that $G \cong He{:}2$ has six orbits on ordered pairs of axes of transpositions.

A routine computer calculation of element orders shows that the types are as stated and that the following pairs of transpositions have inequivalent axis pairs with the respective types: $\{\tau_1, \tau_1\}$, $\{\tau_1, \tau_2\}$, $\{\tau_{12.34}, \tau_{12.56}\}$, $\{\tau_{12.34}, \tau_{12.35}\}$, $\{\tau_{12.34}, \tau_1\}$, $\{\tau_{12.34}, \tau_{\omega\overline{\omega}0101}\}$. $\square$

Proposition 3.3.2. *The group G has ten orbits on its transposition-generated dihedral subgroups. Moreover, if D is such a dihedral subgroup of G, and if $\widehat{D}$ is a ($2A$−generated) dihedral lift of D in $\mathbf{M}$, then the dihedral subalgebra of $(X, *)$ that corresponds to D is an image of $\Delta_V(\widehat{D})$.*

Proof. We first observe that the orbit of an unordered pair, $\{t_0, t_1\}$, of transpositions in G is completely determined by the order of $t_0 t_1$ and the type of $\{\underline{t}_0, \underline{t}_1\}$.

(If $\{t_0, t_1\}$ is one of the pairs of transpositions given in Lemma 3.3.1, and if $\{t'_0, t'_1\}$ and $\{t''_0, t''_1\}$ are pairs of transpositions with $\underline{t_0 = t'_0 = t''_0}$ and $\underline{t_1 = t'_1 = t''_1}$, then $\{t'_0, t'_1\}$ is visibly G−equivalent to $\{t''_0, t''_1\}$ so long as $\overline{|t'_0 t'_1|} = \overline{|t''_0 t''_1|}$.) It follows that G has ten orbits on transposition-generated dihedral subgroups as given in Table 2.

The first two columns of Table 2 give a generating pair of transpositions, t_0 and t_1, for a dihedral group. The next columns give the class of the product of lifting transpositions in $\mathbf{M}$, and the G−class of $t_0 t_1$. We then give the type of the corresponding axis pair, and finally the mutual inner product of these axes (computed from the particular axis vectors that we specified earlier).

Table 2. The dihedral subgroups of He

t_0	t_1	$\mathbf{M}$−Class of $t_0 t_1$	G−Class of $t_0 t_1$	Type of $\{t_0, t_1\}$	$\langle t_0, t_1 \rangle$
τ_1	τ_1	$1A$	$1A$	$(1A^3, 2A^6)$	4
τ_1	τ_1^ω	$2A$	$2A$	$(1A^3, 2A^6)$	4
τ_1	τ_2	$2B$	$2B$	$(2B^9)$	0
$\tau_{12.34}$	$\tau_{12.56}$	$2A$	$2A$	$(2A, 3B^4, 4A^4)$	4
$\tau_{13.24}$	$\tau_{15.26}$	$3C$	$3B$	$(2A, 3B^4, 4A^4)$	4
$\tau_{12.34}$	$\tau_{15.26}$	$4B$	$4A$	$(2A, 3B^4, 4A^4)$	4
$\tau_{12.34}$	$\tau_{12.35}$	$3A$	$3A$	$(3A^3, 5A^6)$	6
$\tau_{12.34}$	$\tau_{13.25}$	$5A$	$5A$	$(3A^3, 5A^6)$	6
$\tau_{12.34}$	τ_1	$4A$	$4B$	$(4B^9)$	1
$\tau_{12.34}$	$\tau_{\omega\bar\omega 0101}$	$6A$	$6A$	$(6A^9)$	5

For each dihedral group $D = \langle t_0, t_1 \rangle$, as given in Table 2, a computer calculation shows that $\Delta_X(D)$ is an image of the corresponding dihedral subalgebra in $(V, *)$. The computation consists of the calculation of eighteen algebra products: this is a tedious but routine application of bilinearity. There are several easy observations that can further reduce the number of computations required. For example, note that: $\underline{\tau}_{12.34} = \underline{\tau}_{13.24}$ and $\underline{\tau}_{12.56} = \underline{\tau}_{15.26}$ and $\underline{\tau}_{34.56} = \underline{\tau}_{53.64}$. Therefore, as soon as we have verified the $2A$-relation that $\underline{\tau}_{12.34} * \underline{\tau}_{12.56} = 8(\underline{\tau}_{12.34} + \underline{\tau}_{12.56} - \underline{\tau}_{34.56})$, we derive the $3C$−relation that $\underline{\tau}_{13.24} * \underline{\tau}_{15.26} = \underline{\tau}_{13.24} + \underline{\tau}_{15.26} - \underline{\tau}_{53.64}$ (since $8 \equiv 1$ (mod 7)). (These short cuts are also 7−modular consistency checks on entries of [5, Table 3].) $\square$

Proposition 3.3.2 is just a restatement of the $p = 7$ case of Theorem 1.

3.4. The algebra for M_{12}. The case of Theorem 1 that corresponds to $p = 11$ is proved in [18]. We now summarize the structure of the $^{11}\mathbf{M}$−invariant algebra that is given in [18]: the particular structure constants that we give can be transformed to those of [18] by a basis change.

The group $^{11}\mathbf{M}$ has structure M_{12}, it acts on a 16−dimensional nonassociative $\mathbf{F}_{11}$-algebra, $(X, *)$, that plays the role of ^{11}U. In our description of this algebra, we shall denote distinct elements of $\{1, 2, 3, 4, 5\}$ by the symbols i, j, k, l and m. We choose a spanning set for X that consists of twenty vectors called $h_{i,j}$ together with twenty vectors called $x_{i,j}$. The linear relationships among these vectors are generated by: $h_{i,j} = h_{j,i}$, $x_{i,j} = -x_{j,i}$ and $x_{i,j} + x_{i,k} + x_{i,l} + x_{i,m} = 0$. A group

$S \cong S_5$ acts naturally on our spanning set of X by subscript permutation. The subspace of X that is spanned by the vectors $h_{i,j}$ gives the 10-point permutation representation of S, while the subspace of X that is spanned by the vectors $x_{i,j}$ gives the skew square of the natural 4-dimensional deleted permutation module for S.

An involution, $t:X \to X$, that maps $h_{i,j} \mapsto h_{i,j}$ and $x_{i,j} \mapsto -x_{i,j}$ commutes with the action of S. The representation of $t \times S$ on X extends to a representation of $M_{12}:2$ on X (matrices are available from [18], and compact formulae to generate the action of M_{12} are given below). We define a bilinear product on X by

$$h_{i,j} * x_{i,j} = 0, \ h_{i,j} * x_{i,k} = 6x_{i,j} - h_{i,k} + 3x_{j,k}, \ h_{i,j} * x_{k,l} = 2x_{k,l}$$
$$h_{i,j} * h_{i,j} = 8h_{i,j}, \ h_{i,j} * h_{i,k} = h_{i,j} + h_{i,k} - h_{j,k} + 8h_{l,m}, \ h_{i,j} * h_{k,l} = 7(h_{i,j} + h_{k,l})$$
$$x_{i,j} * x_{i,j} = 4(h_{i,j} + h_{k,l} + h_{k,m} + h_{l,m}), \ x_{i,j} * x_{k,l} = 3(h_{i,k} + h_{j,l} - h_{i,l} - h_{j,k})$$
$$x_{i,j} * x_{i,k} = 6(h_{i,j} + h_{i,k}) + 8(h_{i,l} + h_{i,m}) + 7h_{l,m}$$

This product is well defined (since multiplication by any relation vector yields a zero product). The algebra is plainly invariant under $t \times S$, and from [18], it is invariant under $M_{12}:2$. (The multiplication structure constants described in [18] are specified with respect to a different basis of an augmented form of the algebra.)

We now describe additional generators for M_{12} in terms of the algebra product. Let $\underline{t}_{i,j,k,l} = -(h_{i,k} + h_{j,l} + (x_{i,j} + x_{j,k} + x_{k,l} + x_{l,i})/2)$, and let $ad(\underline{t}_{i,j,k,l})$ denote the adjoint action of this vector on $(X, *)$. Then the map $1 + 2ad(\underline{t}_{i,j,k,l}) + 7ad^2(\underline{t}_{i,j,k,l}) + 6ad^3(\underline{t}_{i,j,k,l}):X \to X$ is an involution that preserves the algebra $(X, *)$ — it extends $t \times S$ to a group $G \cong M_{12}$. (This is verified by computer calculation, or by a change of basis from the transformation group of [18].)

The G–algebra, $(X, *)$, supports an invariant associative bilinear form whose non-zero values on pairs of vectors from our spanning set are given by $(h_{i,j}, h_{i,j}) = 1, (h_{i,j}, h_{i,k}) = 3, (h_{i,j}, h_{k,l}) = 7, (x_{i,j}, x_{i,j}) = 6, (x_{i,j}, x_{i,k}) = 9$. (This inner product is visibly invariant under $t \times S$, and a computer calculation shows that it is invariant under G.)

Elements of the G–classes $2A$, $3B$, $4A$, $4B$ and $5A$ have axes in X. (We may label the classes of G so that elements of S of type (i,j,k,l) are in class $4A$ of G, and elements of $t \times S$ of type $t \times (i,j,k,l)$ are in class $4B$ of G.) The transformation t has class $2A$ in G, and we define its axis to be $\sum_{i,j} h_{i,j}$. This vector is clearly fixed by $t \times S = C_G(t)$. Axes of all other $2A$–elements are determined via the action of G. For example, the cycle (i,j) of S has class $2A$ in G, and it has axis $2h_{i,j}$. (We note that the transformation $1 + 2ad(2h_{i,j}) + 7ad^2(2h_{i,j}) + 6ad^3(2h_{i,j})$ is identical to the S–element (i,j). It follows that the action of any transposition of M_{12} is determined from its axis by means of a similar formula. This explains the origin of our earlier expression for generators of M_{12}.)

The element (i,j,k) of S has class $3B$ in G, and we define its axis in X to be the vector $4(h_{i,j} + h_{i,k} + h_{j,k} + 2h_{l,m})$. (This vector is visibly fixed by $C_S(i,j,k)$, and a computer calculation shows that it is fixed by the full G–centralizer). The axes of other $3B$–elements are determined by the action of G. A similar computer calculation shows that the following choices generate axes for all elements of orders 4 or 5 in G. For the particular $4A$–element $(1,2,3,4)$, of S, we select the axis $10(h_{1,3} + h_{2,4}) + 2(h_{1,2} + h_{2,3} + h_{3,4} + h_{4,1}) + 8(h_{1,5} + h_{2,5} + h_{3,5} + h_{4,5})$, and for the $4B$–element $t \times (1,2,3,4)$, of $t \times S$, we select the axis $9(h_{1,3} + h_{2,4}) + (h_{1,2} + h_{2,3} + h_{3,4} + h_{4,1}) + 4(h_{1,5} + h_{2,5} + h_{3,5} + h_{4,5})$. For the $5A$–element $(1,2,3,4,5)$, of S, we pick the axis $4(h_{1,2} + h_{2,3} + h_{3,4} + h_{4,5} + h_{5,1}) - 4(h_{1,3} + h_{2,4} + h_{3,5} + h_{4,1} + h_{5,2})$.

Proposition 3.4.1. *The group G has eight orbits on its transposition generated dihedral subgroups: the orbit of a dihedral subgroup is determined by the G-class of the product of its generating transpositions as one of $1A$, $2A$, $2B$, $3B$, $4A$, $4B$, $5A$, or $6A$. Moreover, if D is a dihedral subgroup of $G = {}^{11}\mathbf{M}$, and if $\widehat{D}$ is a $(2A-generated)$ dihedral lift of D in $\mathbf{M}$, then $\Delta_X(D)$ is an image of $\Delta_U(\widehat{D})$.*

Proposition 3.4.1 follows from the analogous result about an augmented form of $(X, *)$ that is proved in [18]. We also verified the proposition by computer. In light of Lemma 3.2, Proposition 3.4.1 establishes the case $p = 11$ of Theorem 1.

3.5. The algebra for $L_3(3)$. In this subsection we use the following notation. Let G denote the group ${}^{13}\mathbf{M} \cong L_3(3)$. Let $P^2(\mathbf{F}_3)$ be the projective plane of order 3. The group G acts naturally on the 13 points and 13 lines of $P^2(\mathbf{F}_3)$. We reserve the letters P, Q and R to represent points of $P^2(\mathbf{F}_3)$. We reserve l and m for lines of the projective plane. We write PQ for the line that joins P and Q, and we write lm for the point where l and m meet.

Let X be the 13-dimensional $\mathbf{F}_{13}G$-module obtained from the permutation action of G on the points of $P^2(\mathbf{F}_3)$. We write $\underline{P}$ for the basis vector of X that corresponds to the point P, and we set $\underline{\Sigma} = \sum \underline{P}$. For each line, l say, we consider a vector $\underline{l} \in X$ with $\underline{l} = 10 \sum_{P \in l} \underline{P}$. Note that $\underline{\Sigma} = \sum \underline{P} = \sum \underline{l}$. Let $(\ ,\ )$ be the G-invariant inner product on X given by:

$$(\underline{P}, \underline{Q}) = \begin{cases} 1 & \text{if } P = Q \\ 0 & \text{otherwise} \end{cases}$$

Bilinearity of $(\ ,\ )$ gives:

Lemma 3.5.1. $(\underline{P}, \underline{\Sigma}) = (\underline{l}, \underline{\Sigma}) = 1, (\underline{\Sigma}, \underline{\Sigma}) = 0,$

$$(\underline{P}, \underline{l}) = \begin{cases} -3 & \text{if } P \in l \\ 0 & \text{otherwise} \end{cases} \qquad (\underline{l}, \underline{m}) = \begin{cases} -3 & \text{if } l = m \\ -4 & \text{otherwise} \end{cases}$$

It follows that $\underline{\Sigma} \in \underline{\Sigma}^\perp$, and therefore $\overline{X} := \underline{\Sigma}^\perp / \langle \underline{\Sigma} \rangle$ is an 11-dimensional $\mathbf{F}_{13}G$-module. Our bilinear form on X induces an invariant bilinear form, that we write as $(\ ,\ )$, on $\overline{X}$. We specify vectors of $\overline{X}$ by naming a preimage in X, so that, for example, $\underline{P} - \underline{l}$ describes an element of $\overline{X}$. We note that the character of $\overline{X}$ is ${}^{13}H$. Our aim is to show that $\overline{X}$ supports an invariant product that makes it play the role of ${}^{13}U$.

Let $* : X \times X \to X$ be the G-invariant commutative bilinear product with

$$\underline{P} * \underline{Q} = \begin{cases} \underline{\Sigma} & \text{if } P = Q; \\ -4\underline{l} + 2\underline{P} + 2\underline{Q} & \text{if } P \neq Q, \text{ and } l = PQ. \end{cases}$$

Bilinearity of $*$ gives:

Lemma 3.5.2.

$$\underline{P} * \underline{\Sigma} = 0, \quad \underline{P} * \underline{l} = \begin{cases} (\underline{P} - \underline{l}) - 3\underline{\Sigma} & \text{if } P \in l \\ -2(\underline{P} - \underline{l}) + 3\underline{\Sigma} & \text{if } P \notin l \end{cases}$$

$$\underline{l} * \underline{m} = \begin{cases} -3\underline{\Sigma} & \text{if } l = m \\ -2\underline{l} - 2\underline{m} + 4\underline{R} - 5\underline{\Sigma} & \text{if } l \neq m, \text{ and } R = lm \end{cases}$$

We deduce that the image of $*$ lies in $\underline{\Sigma}^{\perp}$, and that $\langle \underline{\Sigma} \rangle$ is an ideal of $(X, *)$. We write $(\overline{X}, *)$ for the natural G-invariant subquotient algebra.

Lemma 3.5.3. *The bilinear form* (,) *is associative for* $(X, *)$.

Proof. We check that $(\underline{P} * \underline{Q}, \underline{R}) = (\underline{P}, \underline{Q} * \underline{R})$ for all choices of points P, Q, and R. In cases where $P = R$, this equality is obvious. Otherwise one of the following four cases arises:

(i) $P = Q \neq R$. Here, $(\underline{P} * \underline{Q}, \underline{R}) = 1 = (\underline{Q} * \underline{R}, \underline{P})$.

(ii) $R = Q \neq P$. Here, $(\underline{P} * \underline{Q}, \underline{R}) = 1 = (\underline{Q} * \underline{R}, \underline{P})$.

(iii) P, Q, and R are three collinear points. Here, $(\underline{P} * \underline{Q}, \underline{R}) = -1 = (\underline{Q} * \underline{R}, \underline{P})$.

(iv) P, Q, and R form a triangle. Here, $(\underline{P} * \underline{Q}, \underline{R}) = 0 = (\underline{Q} * \underline{R}, \underline{P})$. $\square$

Corollary 3.5.4. *The bilinear form* (,) *is associative for* $(\overline{X}, *)$.

We now construct dihedral subalgebras of the G-invariant algebra $(\overline{X}, *)$. Let t be an involution of G. Then t fixes four points that form a line, l_t say, and a fifth point P_t. We define the axis of t to be the vector $\underline{t} := \underline{P}_t - \underline{l}_t$. (We remark that we can reconstruct t from the corresponding point-line pair (P_t, l_t) as follows. The involution t fixes P_t and the four points of l_t. Given any other point Q, its image, Qt, is the unique point of $P_t Q$ that is distinct from P_t and Q and does not lie on l_t.) Now let u be a $3A$-element of G, then u fixes exactly four points of $P^2(\mathbf{F}_3)$. These four points form a line l_u. In the action of G on lines of $P^2(\mathbf{F}_3)$, the element u fixes exactly four lines all of which pass through a point P_u (thus $P_u \in l_u$). We write $\underline{u}$ for $8(\underline{P}_u - \underline{l}_u)$. The vector $\underline{u}$ is fixed by $C_G(u)$, and it is our choice for the axis of u.

Proposition 3.5.5. *The group G has seven orbits on its dihedral subgroups. Moreover, if D is a dihedral subgroup of $G = {}^{13}\mathbf{M}$, and if $\widehat{D}$ is a (2A-generated) dihedral lift of D in $\mathbf{M}$, then $\Delta_{\overline{X}}(D)$ is an image of $\Delta_U(\widehat{D})$.*

Proof. A structure constant calculation shows that there are seven orbits on dihedral subgroups of G. The orbit of a dihedral subgroup, generated by transpositions t_0 and t_1, is determined by the G-class of $t_0 t_1$, unless $t_0 t_1 \in 3A$, when there are two possibilities for the orbit. The six possibilities for the G-class of $t_0 t_1$ are $1A$, $2A$, $3A$, $3B$, $4A$, and $6A$. The corresponding possibilities for the $\mathbf{M}$-class of a lift of $t_0 t_1$ (to a 2A-generated dihedral subgroup) are $1A$, $2A$, $3A$, $3C$, $4B$, and $6A$.

In Table 3, we summarize our calculation of the structure constants for the seven dihedral subalgebras of $\overline{X}$. We describe elements of dihedral groups with the notation given before Table 1. For each dihedral subgroup, we give a label to specify the G-class of the element $t_0 t_1$, we then give geometrical information that describes the relationships between the points and lines of $P^2(\mathbf{F}_3)$ that correspond to subgroup elements, and we finish by giving the structure constants for the corresponding subalgebra. The geometrical information, that we provide, allows us to calculate the structure constants directly from our definitions of the product and the inner product on $\overline{X}$.

Table 3. The dihedral subalgebras of $(\overline{X}, *)$

$1A$	$P_{t_0} = P_{t_1}$, $l_{t_0} = l_{t_1}$, $\underline{t_0} * \underline{t_1} = 4(\underline{P}_{t_0} - \underline{l}_{t_0})$, $(\underline{t_0}, \underline{t_1}) = -2$
$2A$	$P_{t_0} \in l_{t_1}$, $P_{t_1} \in l_{t_0}$, $P_t = l_{t_0}l_{t_1}$, $l_t = P_{t_0}P_{t_1}$
	$\underline{t_0} * \underline{t_1} = \underline{P}_{t_0} + \underline{P}_{t_1} + 4\underline{P}_t - \underline{l}_{t_0} - \underline{l}_{t_1} - 4\underline{l}_t$, $(\underline{t_0}, \underline{t_1}) = 2$
$3A$	$P_{t_0} = P_{t_1}$, $l_{t_0} \neq l_{t_1}$, $P_u = P_{t_0} = P_{t_1} = P_{t_2}$, $Q = l_{t_0}l_{t_1}$, $l_u = P_uQ$
	$l_u, l_{t_0}, l_{t_1}, l_{t_2}$ are the lines through Q, $\underline{u} * \underline{u} = 2(\underline{P}_u - \underline{l}_u)$, $(\underline{u}, \underline{u}) = -4$
	$\underline{t_0} * \underline{t_1} = 4\underline{P}_u + \underline{l}_u + \underline{l}_{t_2} - 3\underline{l}_{t_0} - 3\underline{l}_{t_1}$, $(\underline{t_0}, \underline{t_1}) = -3$
	$\underline{t_0} * \underline{u} = 8(\underline{P}_u - 3\underline{l}_{t_0} + \underline{l}_{t_1} + \underline{l}_{t_2})$, $(\underline{t_0}, \underline{u}) = 0$
$3A$	$P_{t_0} \neq P_{t_1}$, $l_{t_0} = l_{t_1}$, $l_u = l_{t_0} = l_{t_1} = l_{t_2}$, $m = P_{t_0}P_{t_1}$, $P_u = l_u m$
	$P_u, P_{t_0}, P_{t_1}, P_{t_2}$ are the points of m, $\underline{u} * \underline{u} = 2(\underline{P}_u - \underline{l}_u)$, $(\underline{u}, \underline{u}) = -4$
	$\underline{t_0} * \underline{t_1} = -4\underline{l}_u - \underline{P}_u - \underline{P}_{t_2} + 3\underline{P}_{t_0} + 3\underline{P}_{t_1}$, $(\underline{t_0}, \underline{t_1}) = -3$
	$\underline{t_0} * \underline{u} = 8(-\underline{l}_u + 3\underline{P}_{t_0} - \underline{P}_{t_1} - \underline{P}_{t_2})$, $(\underline{t_0}, \underline{u}) = 0$
$3B$	$P_{t_0} \notin l_{t_1}$, $P_{t_1} \notin l_{t_0}$, $Q = l_{t_0}l_{t_1} \in m = P_{t_0}P_{t_1}$
	The lines through Q are $m, l_{t_0}, l_{t_1}, l_{t_2}$ and $m = \{Q, P_{t_0}, P_{t_1}, P_{t_2}\}$
	$\underline{t_0} * \underline{t_1} = -2\underline{P}_{t_0} - 2\underline{P}_{t_1} - 6\underline{P}_{t_2} + 2\underline{l}_{t_0} + 2\underline{l}_{t_1} + 6\underline{l}_{t_2}$, $(\underline{t_0}, \underline{t_1}) = -4$
$4A$	$P_{t_0} \notin l_{t_1}$, $P_{t_1} \notin l_{t_0}$, $P_t = l_{t_0}l_{t_1} \notin l_t = P_{t_0}P_{t_1}$
	$P_{t_2} = l_tl_{t_0}$, $l_{t_2} = P_tP_{t_0}$, $P_{t_3} = l_tl_{t_1}$, $l_{t_3} = P_tP_{t_1}$
	$\underline{t_0} * \underline{t_1} = 4\underline{P}_{t_0} + 4\underline{P}_{t_1} + 4\underline{P}_t - 4\underline{l}_{t_0} - 4\underline{l}_{t_1} - 4\underline{l}_{t_2}$, $(\underline{t_0}, \underline{t_1}) = -4$
$6A$	$P_{t_0} \notin l_{t_1}$, $P_{t_1} \in l_{t_0}$, $P_t = l_{t_0}l_{t_1}$, $l_t = P_{t_0}P_{t_1}$, $l_u = l_{t_0}$, $P_u = P_{t_1}$
	The lines through P_t are $l_{t_0}, l_{t_1}, l_{t_3}, l_{t_5}$ and $l_t = \{P_{t_0}, P_{t_1}, P_{t_2}, P_{t_4}\}$
	$l_{t_2} = l_{t_4} = l_{t_0}$, $P_{t_3} = P_{t_5} = P_{t_1}$, $l_{t_3} = P_tP_{t_0}$, $P_{t_4} = l_tl_{t_1}$
	$\underline{t_0} * \underline{t_1} = 4\underline{P}_t + 4\underline{P}_{t_0} + \underline{P}_{t_1} - 4\underline{l}_t - \underline{l}_{t_0} - 4\underline{l}_{t_1}$, $(\underline{t_0}, \underline{t_1}) = -1$
	$\underline{t} * \underline{u} = \underline{P}_u - \underline{l}_u + 8\underline{P}_t - 8\underline{l}_t$, $(\underline{t}, \underline{u}) = 3$

The structure constants given in Table 3, show that the dihedral subalgebras of $(\overline{X}, *)$ are images of the corresponding dihedral subalgebras of the $\mathbf{M}-$invariant algebra $(U, *)$. $\square$

In light of Lemma 3.2, Proposition 3.5.5 verifies the $p = 13$ case of Theorem 1.

3.6. The algebra for $L_3(2)$. In this subsection we use the following notation. Let G denote the group ${}^{17}\mathbf{M} \cong L_3(2)$. Let $P^2(\mathbf{F}_2)$ be the projective plane of order 2. The group G acts naturally on the points of $P^2(\mathbf{F}_2)$. We reserve the letters P, Q, and R to represent points of $P^2(\mathbf{F}_2)$, and we reserve the letters l and m for lines of the projective plane. Let X be the 7-dimensional $\mathbf{F}_{17}G-$module obtained from the permutation action of G on the points of $P^2(\mathbf{F}_2)$. We write $\underline{P}$ for the basis vector of X that corresponds to the point P. For each line, l say, we let $\underline{l} = 3\sum_{P \in l} \underline{P}$. Let $\underline{\Sigma} = \sum \underline{P}$. Let $(\ ,\ )$ be the $G-$invariant inner product on X with

$$(\underline{P}, \underline{Q}) = \begin{cases} 2 & \text{if } P = Q \\ -6 & \text{otherwise} \end{cases}$$

Bilinearity of $(\ ,\ )$ shows that $(\underline{l}, \underline{l}) = 2$ and that $(\underline{l}, \underline{m}) = -6$, when $l \neq m$. We define a $G-$invariant commutative bilinear product, $* : X \times X \to X$, by

$$\underline{P} * \underline{Q} = \begin{cases} -2\underline{P} & \text{if } P = Q \\ 12\underline{l} - 8\underline{P} - 8\underline{Q} & \text{if } P \neq Q \text{ and } l \text{ is the line } PQ \end{cases}$$

Bilinearity of $*$ establishes the following.

Lemma 3.6.1.

$$\underline{P} * \underline{l} = \begin{cases} 4(\underline{P} - \underline{l}) & \text{if } P \in l \\ 8(\underline{P} - \underline{l}) + 6\underline{\Sigma} & \text{if } P \notin l \end{cases}$$

$$\underline{l} * \underline{m} = \begin{cases} 2\underline{l} & \text{if } l = m \\ 8(\underline{l} + \underline{m}) - 12\underline{R} + 2\underline{\Sigma} & \text{if } l \neq m, \text{ and } R \text{ is the meet of } l \text{ and } m. \end{cases}$$

Lemma 3.6.2. *We have* $(\underline{P}, \underline{\Sigma}) = 0$, *and* $\underline{P} * \underline{\Sigma} = 13\underline{\Sigma}$.

Proof. The bilinearity of (,) gives $(\underline{P}, \underline{\Sigma}) = 2 - 6 - 6 - 6 - 6 - 6 - 6 \equiv 0$ (mod 17). Let l_1, l_2 and l_3 denote the three lines through the point P, then $\underline{P} * \underline{\Sigma} = -2\underline{P} + 7\underline{l}_1 + 7\underline{l}_2 + 7\underline{l}_3 - 6\underline{P} - 8\underline{\Sigma} = 34\underline{P} + 13\underline{\Sigma}$ (since $7(\underline{l}_1 + \underline{l}_2 + \underline{l}_3) = 42\underline{P} + 21\underline{\Sigma}$). $\quad\square$

It follows that $\underline{\Sigma}$ spans an ideal of $(X, *)$, and that this ideal is in the radical of (,). We deduce that the 6−dimensional G−module, $\overline{X} = X/\langle \underline{\Sigma} \rangle$, supports an induced G−invariant bilinear form, (,) say, and a G−invariant quotient algebra, with product $*$ say. We specify vectors of $\overline{X}$ by naming a preimage in X. We observe that the character of $\overline{X}$ is ^{17}H. Our aim is to show that the algebra $(\overline{X}, *)$ plays the role of ^{17}U.

Note that $\overline{X}$ has multiplicity 2 in $S^2(\overline{X})$, so there is a 2-dimensional space of G−invariant commutative algebras on the module $\overline{X}$. Of the $18 = (17^2 - 1)/(17 - 1)$, essentially different, nontrivial products on $\overline{X}$, only our choice leads to the "right" dihedral subalgebras. Our algebra, $(\overline{X}, *)$, is also distinguished as one of only two out of the eighteen algebras that is invariant under a larger group isomorphic to $Aut(G) \cong L_3(2).2$.

Lemma 3.6.3. *The bilinear form* (,) *is associative for* $(\overline{X}, *)$.

Proof. We need to check that $(\underline{P} * \underline{Q}, \underline{R}) = (\underline{Q} * \underline{R}, \underline{P})$ for all triples of points, P, Q, and R. The G−invariance of (,) and $*$, covers all cases, except the proof that $(\underline{P} * \underline{P}, \underline{Q}) = (\underline{Q} * \underline{P}, \underline{P})$ when P and Q are distinct points. Here, $(\underline{P} * \underline{P}, \underline{Q}) = 12$, while $(\underline{Q} * \underline{P}, \underline{P}) = -328$, but $12 \equiv -328 \pmod{17}$. $\quad\square$

We now construct dihedral subalgebras of $(\overline{X}, *)$. Any involution of G, t say, fixes three points that form a line l_t, moreover, t fixes three lines that pass though a point P_t. Note that $P_t \in l_t$. We define the axis of t to be the vector $\underline{t} := 4(\underline{P}_t - \underline{l}_t)$. This choice of axis is made from a 2−parameter family of possibilities. Our choice is the only one that leads to the "right" dihedral subalgebras. As in the choice of the product on $\overline{X}$, our choice is guided by the group $Aut(G)$ (since $\underline{t}$ spans the fixed space of $C_{Aut(G)}(t)$).

Let u be a $3A$ element of G. Then u fixes a unique point, P_u say, and a unique line l_u, say. Note that $P_u \notin l_u$. We define the axis of u to be the vector $\underline{u} := 6(\underline{P}_u - \underline{l}_u)$.

Proposition 3.6.4. *The group G has five orbits on its dihedral subgroups. The products of generating transpositions of these subgroups have G−classes $1A$, $2A$, $2A$, $3A$, and $4A$, and the products of the corresponding lifting transpositions in $\boldsymbol{M}$ have $\boldsymbol{M}$−classes $1A$, $2A$, $2A$, $3A$, and $4B$. Representatives of the five classes of dihedral subalgebras of $(\overline{X}, *)$ are images of $\Delta_U(1A)$, $\Delta_U(2A)$, $\Delta_U(2A)$, $\Delta_U(3A)$, and $\Delta_U(4B)$, respectively.*

Proof. A structure constant calculation shows that G has five orbits on unordered pairs of involutions. We now consider the five corresponding dihedral subalgebras of $(\overline{X}, *)$, our notation is as described immediately before Table 1.

Case $t_0 t_1 \in 1A$: Here $P_{t_0} = P_{t_1}$ and $l_{t_0} = l_{t_1}$. Hence, $(\underline{t_0}, \underline{t_1}) = 4 \equiv 2^4 47 \pmod{17}$. We have $\underline{t_0} * \underline{t_0} = 10(\underline{P}_{t_0} - \underline{l}_{t_0}) = 2^4 23 \underline{t_0}$.

Case $t_0 t_1 \in 2A$, and $l_{t_0} = l_{t_1}$, $P_{t_0} \neq P_{t_1}$: We calculate $(\underline{t_0}, \underline{t_1}) = 12 \equiv 2^4 5 \pmod{17}$. Here, $l_t = l_{t_0} = l_{t_1}$, and $3(\underline{t_0} + \underline{t_1} + \underline{t}) = 2\underline{l}_t$. We have $\underline{t_0} * \underline{t_1} = -12\underline{P}_t + 16\underline{l}_t = 8(5\underline{t_0} + 5\underline{t_1} - 6\underline{t})$.

Case $t_0 t_1 \in 2A$, and $P_{t_0} = P_{t_1}$, $l_{t_0} \neq l_{t_1}$: We calculate $(\underline{t_0}, \underline{t_1}) = 12 \equiv 2^4 5 \pmod{17}$. Here, $P_t = P_{t_0} = P_{t_1}$, and $\underline{t_0} + \underline{t_1} + \underline{t} = 5\underline{P}_t$. We have $\underline{t_0} * \underline{t_1} = \underline{P}_t + 12\underline{l}_t = 8(5\underline{t_0} + 5\underline{t_1} - 6\underline{t})$.

Case $t_0 t_1 \in 3A$: Here, P_{t_0}, P_{t_1}, and P_{t_2} give the three points of l_u, and l_{t_0}, l_{t_1}, and l_{t_2} give the three lines through P_u. We calculate $(\underline{t_0}, \underline{t_1}) = 6 \equiv 23 \pmod{17}$, $(\underline{t_0}, \underline{u}) = 13 \equiv 234 \pmod{17}$, and $(\underline{u}, \underline{u}) = 3 \equiv 3^4 29 \pmod{17}$. We have $\underline{t_0} * \underline{t_1} = -(\underline{P}_{t_0} - \underline{l}_{t_0}) - (\underline{P}_{t_1} - \underline{l}_{t_1}) + 12(\underline{P}_u - \underline{l}_u) = 16(\underline{t_0} + \underline{t_1}) + 12\underline{t_2} - 18\underline{u}$. Also, $\underline{t_0} * \underline{u} = 2(\underline{P}_{t_0} - \underline{l}_{t_0}) + (\underline{P}_u - \underline{l}_u) = 102\underline{t_0} - 60(\underline{t_1} + \underline{t_2}) + 52\underline{u}$. And, $\underline{u} * \underline{u} = 15(\underline{P}_u - \underline{l}_u) = 504\underline{u}$.

Case $t_0 t_1 \in 4A$: Here, $P_{t_1} \in l_{t_0}$, $P_{t_0} \notin l_{t_1}$, hence $(\underline{t_0}, \underline{t_1}) = 13 \equiv -4 \pmod{17}$. We have, $P_t = P_{t_1} = P_{t_3}$, $l_t = l_{t_0} = l_{t_2}$, and hence $\underline{t_0} * \underline{t_1} = -\underline{P}_{t_0} + \underline{l}_{t_1} + 7(\underline{P}_t - \underline{l}_t) = -2(\underline{t_0} + \underline{t_1}) - 6(\underline{t_2} + \underline{t_3} - \underline{t})$. $\square$

The $p = 17$ case of Theorem 1 follows from Proposition 3.6.4 and Lemma 3.2.

3.7. The algebra for A_5. In this subsection we use the following notation. Let G denote the group $^{19}\mathbf{M} \cong A_5$. Let $P^1(\mathbf{F}_5)$ be the projective line of order 5. The group G acts naturally on the 6 points, $\{\infty, 0, 1, 2, 3, 4\}$, of $P^1(\mathbf{F}_5)$. We reserve the letters i, j, and k to represent points of the projective line. Let X be the 6-dimensional $\mathbf{F}_{19}G$–module obtained from the permutation action of G on the points of $P^1(\mathbf{F}_5)$. We write $\underline{i}$ for the basis vector of X that corresponds to the point i. Let Σ denote the vector $\sum \underline{i}$. Let $(\ ,\ )$ be the G–invariant inner product on X with

$$(\underline{i}, \underline{j}) = \begin{cases} 11 & \text{if } i = j \\ 13 & \text{otherwise} \end{cases}$$

If i, j, k are distinct points of $P^1(\mathbf{F}_5)$, we define $\epsilon_{i,j,k}$ to be ± 1 according as $\{i, j, k\}$ is, or is not an A_5–image of $\{0, 1, \infty\}$. We set $\epsilon_{i,j,k} = 0$ in cases where the collection (i, j, k) includes a repeated member. (It is easy to calculate $\epsilon_{i,j,k}$ as $\zeta(i - j)\zeta(i - k)\zeta(j - k)$, where $\zeta(1) = \zeta(4) = 1$, $\zeta(2) = \zeta(3) = \zeta(\infty) = -1$, and $\zeta(0) = 0$.) Let $* : X \times X \to X$ be the G–invariant commutative bilinear product defined by $\underline{i} * \underline{j} = 2\sum_k \epsilon_{i,j,k}\underline{k}$.

Lemma 3.7.1. *We have* $(\underline{i}, \Sigma) = 0$, *and* $\underline{i} * \Sigma = 0$.

Proof. The bilinearity of $(\ ,\ )$ gives $(\underline{i}, \Sigma) = 11 + 13 + 13 + 13 + 13 + 13 \equiv 0 \pmod{19}$. The bilinearity of $*$ gives $\underline{\infty} * \Sigma = 0$, and the result follows by letting G act, since Σ is fixed by G. $\square$

It follows that Σ spans an ideal of $(X, *)$, and that this ideal is in the radical of $(\ ,\ )$. We deduce that the 5–dimensional G–module, $\overline{X} := X/\langle \Sigma \rangle$, supports an induced G–invariant bilinear form, $(\ ,\ )$ say, and a G–invariant quotient algebra, with product $*$ say. We specify vectors of $\overline{X}$ by naming a preimage in X. We remark that the character of $\overline{X}$ is ^{19}H. Our aim is to show that the algebra $(\overline{X}, *)$ plays the role of ^{19}U. The symmetry of $\epsilon_{i,j,k}$ shows that the bilinear form (,) is associative for the algebra $(\overline{X}, *)$.

Note that $\overline{X}$ has multiplicity 2 in $S^2(\overline{X})$, so there is a 2-dimensional space of G−invariant commutative algebras on the module $\overline{X}$. Of the $20 = (19^2-1)/(19-1)$, essentially different, nontrivial products on $\overline{X}$, only our choice leads to the "right" dihedral subalgebras. Our algebra, $(\overline{X}, *)$, is also distinguished as one of only two out of the twenty algebras that is invariant under a group isomorphic to $Aut(G) \cong S_5$.

We now examine the dihedral subalgebras of the G−invariant algebra $(\overline{X}, *)$. Each involution, t say, fixes two points, i and j say, in the projective line. We define the axis of t to be the vector $\underline{t} := \sum \epsilon_{i,j,k}\underline{k} = 1/2(\underline{i} * \underline{j})$. Our choice of axis is made from a 2−parameter family of possibilities. We make the only choice of axis that leads to the "right" dihedral subalgebras. As in the choice of the product on $\overline{X}$, our choice is guided by the group $Aut(G)$ (since $\underline{t}$ spans the fixed space of $C_{Aut(G)}(t)$).

The group G has two classes of elements of order 5. We may assume that the classes $5A$ and $5B$ are represented, as permutations of the projective line, by (01234) and its square, respectively. Let w be a $5A$ element of G. Then w fixes a unique point, i say. We define the axis of w to be $-4\underline{i}$, and we define the axis of the $5B$−element w^2 to be $4\underline{i}$.

Lemma 3.7.2. *The group G has five orbits on unordered pairs of involutions. The orbit of a pair, $\{t_0, t_1\}$, is completely determined by the conjugacy class of $t_0 t_1$ in G: this conjugacy class is one of $1A, 2A, 3A, 5A$ or $5B$. The five associated dihedral subalgebras of $(\overline{X}, *)$ are images of $\Delta_U(1A), \Delta_U(2A), \Delta_U(3C), \Delta_U(5A),$ and $\Delta_U(5A)$, respectively.*

Proof. Case $t_0 t_1 \in 1A$: We may assume that $t_0 = t_1$ acts on $P^1(\mathbf{F}_5)$ as $(14)(23)$. Then $\underline{t_0} = \underline{1} + \underline{4} - \underline{2} - \underline{3}$, and $(\underline{t_0}, \underline{t_1}) = -2 - 2 - 2 - 2 \equiv 2^4 47 \pmod{19}$. We have $\underline{t_0} * \underline{t_0} = (\underline{1} + \underline{4} - \underline{2} - \underline{3}) * (\underline{1} + \underline{4} - \underline{2} - \underline{3}) = 7(\underline{1} + \underline{4} - \underline{2} - \underline{3}) = 2^4\, 23\, \underline{t_0}$.

Case $t_0 t_1 \in 2A$: We may assume that t_0 acts on $P^1(\mathbf{F}_5)$ as $(14)(23)$, and that t_1 acts as $(23)(0\infty)$. Hence $t = (14)(0\infty)$. Then, $\underline{t_0} = \underline{1} + \underline{4} - \underline{2} - \underline{3}$, $\underline{t_1} = \underline{2} + \underline{3} - \underline{0} - \underline{\infty}$, and $(\underline{t_0}, \underline{t_1}) = 0 + 0 + 2 + 2 \equiv 2^4 5 \pmod{19}$. We have $\underline{t_0} * \underline{t_1} = (\underline{1} + \underline{4} - \underline{2} - \underline{3}) * (\underline{2} + \underline{3} - \underline{0} - \underline{\infty}) = 12(\underline{1} + \underline{4} - \underline{0} - \underline{\infty}) = 12(\underline{t_0} + \underline{t_1}) = 8(5\underline{t_0} + 5\underline{t_1} - 6\underline{t})$, (since $\underline{t_0} + \underline{t_1} + \underline{t} = 0$, in $\overline{X}$).

Case $t_0 t_1 \in 3A$: We may assume that t_0 acts on $P^1(\mathbf{F}_5)$ as $(14)(23)$, and that t_1 acts as $(04)(2\infty)$. Hence $t_2 = (01)(3\infty)$. Then, $\underline{t_0} = \underline{1} + \underline{4} - \underline{2} - \underline{3}$, $\underline{t_1} = \underline{0} + \underline{4} - \underline{\infty} - \underline{2}$, and $(\underline{t_0}, \underline{t_1}) = 0 - 2 - 2 + 0 = -4$. We have $\underline{t_0} * \underline{t_1} = 11(\underline{0} + \underline{1} - \underline{3} - \underline{\infty}) - 4(\underline{4} - \underline{2}) = -2(\underline{t_0} + \underline{t_1}) - 6\underline{t_2}$.

Case $t_0 t_1 \in 5A$: We may assume that t_0 acts on $P^1(\mathbf{F}_5)$ as $(14)(23)$, and that t_1 acts as $(42)(01)$. Hence $t_2 = (20)(34)$, $t_3 = (03)(12)$, $t_4 = (31)(40)$, and $w = (01234)$. Then, $\underline{t_0} = \underline{1} + \underline{4} - \underline{2} - \underline{3}$, $\underline{t_1} = \underline{4} + \underline{2} - \underline{0} - \underline{1}$. Thus, $(\underline{t_0}, \underline{t_1}) = 2$, and $(\underline{t_0}, \underline{t_2}) = 2$. Also $(\underline{w}, \underline{t_0}) = 0$, and $(\underline{w}, \underline{w}) = 176 \equiv 21000 \pmod{19}$. We have $\underline{w} * \underline{w} = 0 = 2100(\underline{t_0} + \underline{t_1} + \underline{t_2} + \underline{t_3} + \underline{t_4})$ and $\underline{w} * \underline{t_0} = -2\underline{0} - 8\underline{\infty} = 84(\underline{t_1} - \underline{t_2} - \underline{t_3} + \underline{t_4}) + 76\underline{w}$. Also $\underline{t_0} * \underline{t_1} = 2(-\underline{0} + \underline{1} + \underline{2} - \underline{3}) + 10(\underline{4} - \underline{\infty}) = (\underline{t_0} + \underline{t_1}) - 3(\underline{t_2} + \underline{t_3} + \underline{t_4} - \underline{w})$. Similarly, $\underline{t_0} * \underline{t_2} = 2(\underline{0} + \underline{1} - \underline{2} - \underline{4}) + 10(\underline{\infty} - \underline{3}) = (\underline{t_0} + \underline{t_2}) - 3(\underline{t_4} + \underline{t_1} + \underline{t_3} + \underline{w})$.

We observe that in the analysis of the case $t_0 t_1 \in 5A$, we have $t_0 t_2 \in 5B$, and therefore our calculations also cover the case where $t_0 t_1 \in 5B$. $\square$

The $p = 19$ case of Theorem 1 follows from Proposition 3.7.2 and Lemma 3.2.

3.8. The algebra for S_4. In this subsection we use the following notation. Let G denote the group $^{23}\mathbf{M} \cong S_4$. The group G acts naturally on four points that we label as $\{1, 2, 3, 4\}$. We use the letters i, j, and k to represent points of $\{1, 2, 3, 4\}$. Let X be the 4-dimensional $\mathbf{F}_{23}G$–module obtained from the natural permutation action of G. We write $\underline{i}$ for the basis vector of X that corresponds to the point i, and $\underline{\Sigma}$ for $\sum \underline{i}$. Let $(\ ,\ )$ be the G–invariant inner product on X with

$$(\underline{i}, \underline{j}) = \begin{cases} 18 & \text{if } i = j \\ -6 & \text{otherwise} \end{cases}$$

Let $* : X \times X \to X$ be the G–invariant commutative bilinear product with

$$\underline{i} * \underline{j} = \begin{cases} \underline{i} & \text{if } i = j \\ -12\underline{i} - 12\underline{j} & \text{otherwise} \end{cases}$$

Lemma 3.8.1. *We have* $(\underline{i}, \underline{\Sigma}) = 0$ *and* $\underline{i} * \underline{\Sigma} = -12\underline{\Sigma}$.

Proof. The bilinearity of $(\ ,\ )$ gives $(\underline{i}, \underline{\Sigma}) = 18 - 6 - 6 - 6 = 0$. The bilinearity of $*$ gives: $\underline{i} * \underline{\Sigma} = \underline{i} - 24\underline{i} - 12\underline{\Sigma}$. $\square$

It follows that $\underline{\Sigma}$ spans an ideal of $(X, *)$, and that this ideal is in the radical of $(\ ,\ )$. We deduce that the 3–dimensional G–module, $\overline{X} := X/\langle \underline{\Sigma} \rangle$, supports an induced G–invariant bilinear form, $(\ ,\ )$ say, and a G–invariant quotient algebra, with product $*$ say. We specify vectors of $\overline{X}$ by naming a preimage in X. We remark that the character of $\overline{X}$ is ^{23}H. Our aim is to show that the algebra $(\overline{X}, *)$ plays the role of ^{23}U.

Lemma 3.8.2. *The bilinear form* $(\ ,\ )$ *is associative for the algebra* $(\overline{X}, *)$.

Proof. It is enough to show that $(\underline{i} * \underline{i}, \underline{j}) = (\underline{i}, \underline{i} * \underline{j})$, when $i \neq j$. Direct computation shows that both sides of this equation have value -6. $\square$

We now examine the dihedral subalgebras of the G–invariant algebra $(\overline{X}, *)$. Let t be a transposition of G, so that t acts as (ij). We define the axis of t to be the vector $\underline{t} := 4\underline{i} + 4\underline{j}$. Let u be a $3A$ element of G, the element u acts as (ijk), and we define the axis of u to be the vector $\underline{u} := 2(\underline{i} + \underline{j} + \underline{k})$.

Lemma 3.8.3. *The group G has three orbits on unordered pairs of involutions. The orbit of a pair, $\{t_0, t_1\}$, is completely determined by the order of $t_0 t_1$: the order is either 1, 2, or 3. The corresponding dihedral subalgebras of $(\overline{X}, *)$ are images of $\Delta_U(1A)$, $\Delta_U(2B)$, and $\Delta_U(3A)$.*

Proof. Case $|t_0 t_1| = 1$: We may assume that $t_0 = t_1$ acts as (12). Then $(\underline{t_0}, \underline{t_0}) = 16(12 + 12) \equiv 2^4 47 \pmod{23}$. We have $\underline{t_0} * \underline{t_0} = 16(\underline{1} + \underline{2} - 24(\underline{1} - \underline{2})) = 0 = 2^4 23 \underline{t_0}$.
Case $|t_0 t_1| = 2$: We may assume that t_0 acts as (12) and that t_1 acts as (34). Hence $\underline{t_0} + \underline{t_1} = 0$. We have $(\underline{t_0}, \underline{t_1}) = -(\underline{t_0}, \underline{t_0}) \equiv -2^4 \pmod{23}$, and $\underline{t_0} * \underline{t_1} = -\underline{t_0} * \underline{t_0} = 0 = -2^3(\underline{t_0} + \underline{t_1})$.
Case $|t_0 t_1| = 3$: We may assume that $t_0 = (12)$, $t_1 = (13)$, $t_2 = (23)$, and $u = (123)$. We have $(\underline{t_0}, \underline{t_1}) = 16(18 - 6 - 6 - 6) \equiv 23 \pmod{23}$, $(\underline{t_0}, \underline{u}) = 8(18 - 6 - 6 - 6 + 18 - 6) \equiv 234 \pmod{23}$, and $(\underline{u}, \underline{u}) = 4(18 - 6 - 6 - 6 + 18 - 6 - 6 - 6 + 18) \equiv 3^4 29 \pmod{23}$. We calculate $\underline{t_0} * \underline{t_1} = 7(\underline{3} + \underline{2}) = 2^4(\underline{t_0} + \underline{t_1}) + 12\underline{t_2} - 18\underline{u}$. Also, $\underline{t_0} * \underline{u} = 4(\underline{1} + \underline{2} + 2\underline{4}) = 102\underline{t_0} - 60(\underline{t_1} + \underline{t_2}) + 52\underline{u}$. And, $\underline{u} * \underline{u} = 4\underline{4} = 504\underline{u}$. $\square$

The $p = 23$ case of Theorem 1 follows from Proposition 3.8.3 and Lemma 3.2.

3.9. The cases $p \geq 29$. In this subsection, we first consider the cases $p = 31$ and $p = 47$, where the group $^p\mathbf{M}$ has involutions. In these cases we can verify Conjecture 1 as well as Theorem 1. We then consider the cases $p = 29$, $p = 41$, and $p > 47$, where the group $^p\mathbf{M}$ has no involutions. Here, existence of an algebra that meets the conditions of Theorem 1 is an easy matter.

Let $(X, *)$ denote the S_3–invariant dihedral subalgebra, $\Delta_U(3C)$, of $(U, *)$. As an abelian group, X is freely generated by the elements called $\underline{t_0}$, $\underline{t_1}$, and $\underline{t_2}$ in Table 1. Let I be the S_3–invariant additive subgroup of X that is generated by the elements $31\underline{t_1}, 31\underline{t_2}, 31\underline{t_3}$, and $(\underline{t_0} + \underline{t_1} + \underline{t_2})$. Observe that the quotient group, $\overline{X} := X/I$, has the structure of an $\mathbf{F}_{31}$ $^{31}\mathbf{M}$–module with character ^{31}H.

We have: $\langle \underline{t_0} + \underline{t_1} + \underline{t_2}, \underline{t_i} \rangle = 24 \times 31 \equiv 0 \pmod{31}$, and hence there is an induced S_3–invariant bilinear form, $(\ ,\ )$ say, on the module $\overline{X}$. Moreover, $(\underline{t_0} + \underline{t_1} + \underline{t_2}) * \underline{t_i} = 12(31\underline{t_i} - 2/3(\underline{t_0} + \underline{t_1} + \underline{t_2}))$, therefore, I is an ideal of $(X, *)$. Hence, there is a quotient ring $\overline{(X/I, *)}$ that plays the role of ^{31}U. The $p = 31$ cases of Theorem 1 and Conjecture 1 follow.

A similar argument, based on $\Delta_U(1A)$, establishes the cases of Theorem 1 and Conjecture 1 with $p = 47$. The key ingredient is the congruence $\langle \underline{t_0}, \underline{t_0} \rangle = 2^4 \times 47 \equiv 0 \pmod{47}$ (see Table 1). Hence, $(V, *)$ has a subquotient, $\langle \underline{t_0} \rangle / \langle 47\underline{t_0} \rangle$, that plays the role of ^{47}V. (Note that ^{47}H is zero dimensional, so there is no module ^{47}U.)

In the cases $p = 59$ and $p = 71$, the group $^p\mathbf{M}$ is trivial and the character $^pH_1$ is trivial. The following construction meets all of the requirements of Theorem 1 and Conjecture 1. Let $p \in \{59, 71\}$, let ^pV be a 1-dimensional trivial $\mathbf{F}_p$–module for the trivial group. Let $\underline{1}$ be a basis vector of ^pV, and let $\underline{1} * \underline{1} = \underline{1}$, and $(\underline{1}, \underline{1}) = 3/2$.

The group $^{41}\mathbf{M}$ is trivial, and the character ^{41}H has degree 1. Let X be an $\mathbf{F}_{41}$–module, for $^{41}\mathbf{M}$, with character ^{41}H. Any bilinear products and forms on X are $^{41}\mathbf{M}$–invariant. The $p = 41$ case of Theorem 1 follows.

In the case, $p = 29$, the group $^{29}\mathbf{M} \cong 3$. There are many possibilities for a $^{29}\mathbf{M}$–invariant algebra that plays the role of ^{29}U. We now select a particular choice of algebra for its invariance under the larger group $N_{\mathbf{M}}(g_{29})$. Let X be a $^{29}\mathbf{M}$–module (defined over a quadratic extension of $\mathbf{F}_{29}$) with character ^{29}H. Let $\{\underline{x_\omega}, \underline{x_{\overline{\omega}}}\}$ be a basis of X on which $^{29}\mathbf{M}$ acts diagonally. The module X supports an invariant commutative product given by $\underline{x_\omega} * \underline{x_\omega} = \underline{x_{\overline{\omega}}}, \underline{x_\omega} * \underline{x_{\overline{\omega}}} = \underline{x_{\overline{\omega}}} * \underline{x_\omega} = 0, \underline{x_{\overline{\omega}}} * \underline{x_{\overline{\omega}}} = \underline{x_\omega}$. The associative bilinear form on $(X, *)$ defined by $(\underline{x_\omega}, \underline{x_\omega}) = 0 = (\underline{x_{\overline{\omega}}}, \underline{x_{\overline{\omega}}}), (\underline{x_\omega}, \underline{x_{\overline{\omega}}}) = 1 = (\underline{x_{\overline{\omega}}}, \underline{x_\omega})$ is $^{29}\mathbf{M}$–invariant. An augmented form of $(X, *)$ plays the role of $(^{29}V, *)$ in Theorem 1.

4. APPLICATIONS OF THE MODULAR ALGEBRAS

In this section we make use of the explicit descriptions of the invariant algebras that are given in Section 3. Our goals are firstly to obtain a more precise conjecture about the subquotient of the Griess algebra that occurs in Conjecture 1, and secondly to demonstrate how our modular algebras can be used to guess structural information about the Griess algebra by a lifting process. Since the Griess algebra is a large and unexplored structure, it is very useful to possess small models that can illuminate its properties.

Conjecture 1 asserts that for each p, our $^p\mathbf{M}$–invariant algebra $(^pV, *)$ can be written in the form X/I where X is an appropriate subring of the Griess algebra, and I is an ideal of X. Although we can only guess at suitable candidates for X, the following result shows that once X is known, the ideal I is completely determined.

Proposition 4.1. *Suppose that* $(X, *)$ *is a nonassociative* **Z**−*algebra with an associative* **Z**−*valued bilinear form* (,), *and that* $(Y, *)$ *is a nonassociative* F_p−*algebra with a non-singular associative* F_p−*valued bilinear form* (,). *If* $(Y, *)$ *is an image of* $(X, *)$ *then the kernel of the corresponding ring homomorphism is* $\{x \in X \mid (x, x') \equiv 0 \pmod{p}, \text{ for all } x' \in X\}$.

Proof. Suppose that $(x, x') \equiv 0 \pmod{p}$, for all $x' \in X$. Let y be the image of x in Y. Then $(y, y') \equiv 0 \pmod{p}$, for all $y' \in Y$. We deduce $y = 0$, since (,) is non-singular on Y.

Conversely, if $x \in X$ maps to 0 in Y, and $x' \in X$ maps to $y' \in Y$. Then $(x, x') \equiv (0, y') \pmod{p} = 0 \pmod{p}$. □

Note that in all of the modular analogues of the Griess algebra, that we construct in Section 3, the associative bilinear form is non-singular. Hence, Proposition 4.1 applies to show that if $(^pV, *)$ arises as an image of a subring X of the Griess algebra, then ^pV has the form $(X/pX)/((X/pX) \cap (X/pX)^{\perp})$. An obvious, $C_{\mathbf{M}}(g_p)$−invariant, candidate for the subring X is the fixed point subring, L^{g_p}, of the action of g_p on an integral form, L, of the Griess algebra. (An integral form of the Griess algebra, due to Norton, is mentioned in [5].)

Conjecture 2. *There is an integral form of the augmented Griess algebra,* $(L, *)$ *say, that contains the axes of transpositions, such that each quotient algebra of the form* $(L^{g_p}/pL^{g_p})/((L^{g_p}/pL^{g_p}) \cap (L^{g_p}/pL^{g_p})^{\perp})$ *is an* F_p $^p\mathbf{M}$−*invariant algebra with character* $^pH_1$.

We note that Conjecture 1 follows as an immediate corollary of Conjecture 2. We also observe that the algebra described in Conjecture 2 is clearly invariant under the group $N_{\mathbf{M}}(g_p)/\langle g_p \rangle$. (We recall from Subsections 3.6 and 3.7 that invariance under this larger group is another feature of our algebras.)

We now show how our descriptions of the algebras $(^pV, *)$ can be used to guess structural information about the Griess algebra. As a simple illustration of this lifting process, we guess the structure of a subring, R say, of $(V, *)$ that is generated by the axes of the three $2A$−elements and the four $3A$−elements of $\mathbf{M}$ that lie in a subgroup, $H \cong A_4$, that centralizes a group $J \cong O_{10}^-(2)$ in $\mathbf{M}$.

We write elements of H as permutations of the set $\{1, 2, 3, 4\}$, and we use i, j, k, and l to denote distinct elements of this set. We write the axes of A_4−elements $ij.kl$ and ijk as $[ij.kl]$ and $[ijk]$, respectively. From [5, Table 3], we know that $[ij.kl] * [ij.kl] = 64[ij.kl]$, $[ij.kl] * [ik.jl] = 8([ij.kl] + [ik.jl] - [il.jk])$ and $[ijk] * [ijk] = 90[ijk]$, but we have no knowledge of products such as $[ij.kl] * [ijk]$ or $[ijk] * [ijl]$.

We observe that J includes elements of the $\mathbf{M}$−classes $5A$, $7A$, $11A$, and $17A$. Therefore, under the hypothesis that Conjecture 2 does hold, the ring $(R, *)$ has an image in each of the algebras $(^{17}V, *)$, $(^{11}V, *)$, $(^7V, *)$ and $(^5V, *)$. For each of these algebras, we calculate the structure of the subring generated by the axes of elements in an appropriate A_4 group. In each case the resulting subring is additively spanned by its generating axes. This suggests that $(R, *)$ is spanned by its seven generating axes, and under this additional assumption, the structure constants of R are now determined upto integral multiples of $6545 = 5 \times 7 \times 11 \times 17$.

For example, the images of the product $[12.34] * [123]$ in ^5V, ^7V, ^{11}V, and ^{17}V are images of $-2[134] + [124] + [234]$, $3[12.34] - 2[123] + 3[134] + 3[124] + 3[234]$, $-[12.34] + 5[123] + 3[134] - 4[124] - 4[234]$, and $-7[12.34] + 5[123] + 3[134] - 4[124] - 4[234]$, respectively. By lifting these expressions, we guess that in R, $[12.34] * [123] = 10[12.34] + 5[123] + 3[134] - 4[124] - 4[234]$. A similar process suggests that $[123] * [124] = 8[14.23] - 12[12.34] - 12[13.24] + 18[123] + 18[124] - 5[134] - 5[234]$. All of the unknown structure constants in R are now determined via the action of H.

Additional confidence in our guess for the structure of R is gained by the observation that if we have found the correct structure of R, then the eigenvalue spectra of the operators $*[12.34]$ and $*[123]$ on the ring R are $0, 2, 16, 64$ and $0, 3, 18, 30, 90$, respectively. These spectra do coincide with the eigenvalue spectra of the operators $*[12.34]$ and $*[123]$ on V.

We note that at present no identity based definition of the Griess algebra is available. Indeed it is known from [10, Appendix 3] that, apart from consequences of commutativity, no identities of degree less than 6 hold in the algebra. It may be possible to classify low degree identities that hold in our algebras $(^pV, *)$, and this should allow us to study identities of the Griess algebra by a lifting process.

5. CHARACTERS

In this section, our objective is to motivate and test the following conjecture. The conjecture provides a p−modular analogue of Proposition 2.4.

Conjecture 3. *The class functions $^pH_1$, $^pH_2$, $^pH_3$, ... give an increasing sequence of proper p−modular characters of ^pM.*

Our evidence in support of this conjecture is similar to the evidence that [23] gives in support of the (former) conjecture that the head characters of M are proper characters. Specifically, we know from Construction 2.1 that the characters ^pH_n are generalized characters of ^pM (the analogue of Theorem 1.3 in [23]). Moreover, we shall describe a computation that shows that the characters $^pH_n - {}^pH_{n-1}$ are proper modular characters of ^pM so long as $p \geq 7$, and $n < 200$. (This is our analogue of the computational results from [23, Section 2].) For values of p less than 7 we can compute somewhat weaker results.

The tables of [13, 7] suggest that if p is fixed, and n increases, then the values of $H_n(g_p)$ appear to give an increasing sequence of positive integers. We view $H_n(g_p)$ as the degree term, $^pH_n(1)$, so that Conjecture 3 would give a particularly simple algebraic explanation for this numerical observation. We first review the evidence that the degree sequences are increasing. We then present our character theoretic computations that give strong support to Conjecture 3. We finish this section by considering and rejecting an alternative to Conjecture 3 that uses ordinary rather than modular characters.

We now begin our investigation of Conjecture 3 by examining the corresponding degree sequence. The following result is conjectured in [7, page 315]: it is proved in [12, Theorem 2.1]. We continue to use the convention $q = e^{2\pi i z}$ as in Section 2.

Theorem (Kondo and Tasaka). *Let p be a prime divisor of $|M_{24}|$, and let L be the Leech lattice. Let $l \in Aut(L)$ be an element of order p, with trace $t = 24/(p+1)$*

on L. Let ^pL be the lattice of fixed points of the action of l on L, and let $^p\Theta$ be the theta function of ^pL. Then the q–series of $T(g_p)$ is given by the q–expansion of $^p\Theta(q)/\{\eta(z)\eta(pz)\}^t$.

Corollary 5.1. *Suppose that* $p \in \{2,3,5,7,11,23\}$, *then the coefficients*, $^pH_n(1)$, *of the* q–*series of* $^pT(1)$ *are an increasing sequence of positive integers.*

Proof. Recall that $^pT(1) = T(g_p)$. According to the Kondo-Tasaka Theorem, $T(g_p)$ has a q–series given by: $^p\Theta(q)/\{\eta(z)\eta(pz)\}^t$. We expand the eta functions to obtain $^pT(1) = {}^p\Theta(q)\{(\sum_{n=0}^{n=\infty} p(n)q^n)(\sum_{n=0}^{n=\infty} p(n)q^{pn})\}^t/q$. (Here $p(n)$ denotes the number of partitions of n.)

Therefore, $^pT(1)$ has the form $(\sum_{n=-1}^{n=\infty} c_n q^n)(\sum_{n=0}^{n=\infty} p(n)q^n)$, where the coefficients c_n are non-negative integers. Hence, by evaluating coefficients in $^pT(1)$, we deduce that $^pH_n(1) = \sum_{i=-1}^{i=n} c_i\, p(n-i) \le \sum_{i=-1}^{i=n+1} c_i\, p(n+1-i) = {}^pH_{n+1}(1)$. $\square$

The Kondo-Tasaka Theorem gives no information about the cases with $p \in \{13,17,19,29,31,41,47,59,71\}$, but we believe that the coefficients of the q–series of $^pT(1)$ form an increasing sequence in each of these cases too. The first 10 terms of each of the q–expansions are available in [5], the first 50 terms are available in [13], and in our computation of head characters (described below) we find the first 200 terms: in none of these collections of data is there any instance where $^pH_{n+1}(1) < {}^pH_n(1)$. Moreover, the available data suggests that the values of $^pH_n(1)$ actually increase faster than any polynomial in n, and that this "asymptotic behaviour" starts well before the limits of our computation (and before the limits of the tables in [13]). In [3], Borcherds observes that the dominant term in the Rademacher asymptotic expansion for the coefficients of a modular function (coming from the pole at the cusp 0) shows that $\log(^pH_n(1))/\sqrt{n}$ has a positive limit as n tends to infinity. In particular, [3] proves that the degree sequence $^pH_n(1)$ does increase faster than any polynomial in n.

We now report our computations that confirm that the modular character ^pH_n is indeed contained in the modular character $^pH_{n+1}$ so long as either $p \ge 7$ and $200 > n$, or $p = 5$ and $200 > n \ge 76$, or $p = 3$ and $200 > n \ge 112$, or $p = 2$ and $200 > n \ge 136$. For $p \le 5$, the p–modular characters of $^p\mathbf{M}$ are unknown: it is this limitation that prevents us from testing whether $^pH_n \le {}^pH_{n+1}$ for the small values of n.

In practice, for large enough values of n, the character $^pH_n - {}^pH_{n-1}$ is a proper character by virtue of being close to a multiple of the regular representation. (Because, for large n, the degree term is sufficiently dominant over the other character values). The following p–modular adaptation of [23, Lemma 2.1] gives a crude, but effective, test for a p–modular character, χ, to be of this type.

Lemma 5.2. *Let* χ *be a generalized* p–*modular character of a finite group* G, *with* $\chi(1) > 0$ *and* $|\chi(1)/\chi(g)| > |G|$, *for all non-identity* p–*regular elements* $g \in G$. *Then* χ *is a proper* p–*modular character of* G. *Moreover,* χ *involves every* p–*modular irreducible character of* G.

Proof. Observe that the multiplicity of an irreducible constituent in χ is given by the inner product of χ with the corresponding projective indecomposable character. This inner product must be positive (since a projective indecomposable character is a positive linear combination of ordinary irreducibles, and therefore takes its largest absolute value at the identity element).   □

Essentially the same argument establishes the following slightly stronger result, analogous to [23, Lemma 2.2].

Lemma 5.3. *Let χ be a generalized $p-$modular character of a finite group G, such that $\chi(1) \geq \sum |\chi(g)|$, where the sum is taken over all non-identity $p-$regular elements of G. Then χ is a proper $p-$modular character of G.*

In order to apply these lemmas, we need an effective way to compute the head characters. We use the recursive formulae (9.1) of [2] to calculate the head characters of $\mathbf{M}$. Our computations are carried out by a program written in the arbitrary precision character system MOC3. The head characters of $^p\mathbf{M}$ are obtained by extracting appropriate entries from the head characters of $\mathbf{M}$. (We observe that, for $p > 2$, formulae (9.1) of [2] also apply to the head characters of $^p\mathbf{M}$, since the $p-$singular classes of $C_{\mathbf{M}}(g_p)$ are closed under the squaring map. However, we do need the head characters of $\mathbf{M}$ in order to obtain values of ^2H_n, so we make no use of this possible short cut.) After extracting the head characters of $^p\mathbf{M}$, our MOC3 program tests the condition of Lemma 5.3. To describe the output of our program, we introduce a function $N(p)$ given by: $N(2) = 137$, $N(3) = 113$, $N(5) = 77, N(7) = 53, N(11) = 30, N(13) = 22, N(17) = N(23) = 14, N(19) = 12$, $N(29) = 3$, $N(31) = 12$, $N(47) = 9$, and $N(41) = N(59) = N(71) = 2$.

Computation 5.4. *Suppose that $N(p) \leq n \leq 200$, then the character $^pH_n - {}^pH_{n-1}$ meets the conditions of Lemma 5.3.*

We remark that, if $n < N(p)$, then the character $^pH_n - {}^pH_{n-1}$ fails to meet the conditions of Lemma 5.3 unless (p,n) is one of the following pairs: $(47,4)$, $(47,5)$, $(47,6)$, $(47,7)$, $(31,2)$, $(31,10)$, $(23,10)$, $(23,12)$, $(19,4)$, $(19,6)$, $(19,8)$, $(19,10)$, $(17,6)$, $(17,10)$, $(17,12)$, $(13,20)$. This suggests that, for each prime p, there is a rather sharp transition upto values of n for which the hypotheses of Lemma 5.3 are satisfied by $^pH_n - {}^pH_{n-1}$. We take this as evidence that "asymptotic behavior" is already taking place once n exceeds $N(p)$.

In each case where $p > 5$, the $p-$modular character table of $^p\mathbf{M}$ is known [19, 11, 6], and we use it to check that $^pH_n - {}^pH_{n-1}$ is a proper character of $^p\mathbf{M}$ whenever $n < N(p)$. (Note that for $p \geq 17$, the group $^p\mathbf{M}$ has order coprime to p, so that its p-modular character table is identical to its ordinary character table.)

Computation 5.5. *Let $p \geq 7$, let $n < N(p)$, let ψ be the character of a projective indecomposable $\mathbf{F}_p{}^p\mathbf{M}-$module, then the inner product of $^pH_n - {}^pH_{n-1}$ with ψ is non-negative.*

To verify Calculation 5.5, we compute all necessary inner products, for each group $^p\mathbf{M}$, with $p \geq 7$, by using the standard MOC3 function "IPR".

We have now demonstrated that our sequences of head characters are increasing within the stated ranges. We close this section by comparing the head characters ^pH_n of $^p\mathbf{M}$ with the head characters $^p\widehat{H}_n$ of $^p\widehat{\mathbf{M}}$.

It is believed that, for any prime p, the head characters $^p\widehat{H}_n$ are proper characters of $^p\widehat{\mathbf{M}}$ [17]. The truth of this conjecture would immediately show that our characters

^pH_n are proper p–modular characters of $^p\mathbf{M}$. Conversely, we can use our result that ^pH_n is a generalized modular character as follows.

Proposition 5.6. *There is a sequence of generalized characters* $^p\widetilde{H}_1$, $^p\widetilde{H}_2$, ... *of the group* $^p\widehat{\mathbf{M}}$ *such that for each* p'*–element, h say, of* $^p\widehat{\mathbf{M}}$, *we have* $^p\widetilde{H}_n(h) = H_n(g_p h) = {}^p\widehat{H}_n(h)$.

Proof. For any finite group, the **Z**–span of the p–modular Brauer characters is identical to the **Z**–span of the restrictions of the ordinary characters to the p–regular classes. □

Proposition 5.6 does not say anything about the sequence obtained by evaluating a character $^p\widetilde{H}_n$ at a p–singluar element of $^p\widehat{\mathbf{M}}$. One of the remarkable observations of [17] is that we can use the sequence of Fourier coefficients of a Hauptmodul of a genus zero modular group as the sequence of values of $^p\widehat{H}_n$ at any particular p–singular element.

Can the observed increase in the Fourier coefficients of $T(g_p)$ be accounted for by using the ordinary head characters $^p\widehat{H}_n$ in place of our modular head characters ^pH_n? Proposition 5.7 shows that there would be considerable difficulties in such an approach.

Proposition 5.7. *The following are decompositions of virtual characters of* $^p\widehat{\mathbf{M}}$. *In each case the decomposition exhibits a difference* $^p\widehat{H}_n - {}^p\widehat{H}_{n-1}$ *that is not a proper character of the corresponding group* $^p\widehat{\mathbf{M}}$.
$^5\widehat{H}_2 - {}^5\widehat{H}_1 = 760 - 1 - 133b$, $^5\widehat{H}_3 - {}^5\widehat{H}_2 = 1 + 3344 - 760$, $^5\widehat{H}_5 - {}^5\widehat{H}_4 = 760 + 35112 - 8778$, $^7\widehat{H}_2 - {}^7\widehat{H}_1 = 51b + 153a - 51a$, $^7\widehat{H}_3 - {}^7\widehat{H}_2 = 1 + 680 - 51b - 153a$, $^7\widehat{H}_5 - {}^7\widehat{H}_4 = 51a + 51b + 4352 - 1275c$, $^{11}\widehat{H}_2 - {}^{11}\widehat{H}_1 = 45 - 16a$, $^{11}\widehat{H}_4 - {}^{11}\widehat{H}_3 = 16b^2 + 120 - 16a$, $^{11}\widehat{H}_5 - {}^{11}\widehat{H}_4 = 1 + 16a^2 - 16b + 45 + 54 + 66 + 99$,

The decompositions in Proposition 5.7 come from the tables of [17] that give decompositions for the first five head characters of the groups $^5\widehat{\mathbf{M}}$, $^7\widehat{\mathbf{M}}$ and $^{11}\widehat{\mathbf{M}}$.

6. VERTEX ALGEBRAS?

The origins of Conjectures 1 and 3 as analogues of properties of the Monster vertex algebra encourage us to propose that the group $^p\mathbf{M}$ acts on a p–modular vertex algebra whose graded character gives the McKay-Thompson series of $^p\mathbf{M}$. Moreover, Conjecture 2 suggests that such a vertex algebra should be obtained from an integral form, $\mathcal{L}$ say, of the Monster vertex algebra as a subquotient $(\mathcal{L}^{g_p}/p\mathcal{L}^{g_p})/((\mathcal{L}^{g_p}/p\mathcal{L}^{g_p}) \cap (\mathcal{L}^{g_p}/p\mathcal{L}^{g_p})^{\perp})$ —where $\mathcal{L}^{g_p}$ is the fixed space of g_p on $\mathcal{L}$.

We now exhibit one property of the Monster vertex algebra that only seems to have a p–modular analogue for the groups $^p\mathbf{M}$ with $p < 23$. (This failure might indicate that we are not looking at the correct p–modular analogue.)

We obtain a stronger version of Proposition 2.4 by using the Virasoro algebra in place of $sl_2(\mathbf{C})$. This shows that for $n > 0$, the generalized character h_n given by $h_n = H_n + \sum_{l=-\infty}^{l=\infty}(-1)^l(H_{n-(3l^2-l)/2})$ is a proper characters of $\mathbf{M}$. (These combinations are obtained as the q–coefficients of a formal power series $(1-q)(1-q^2)(1-q^3)\dots(\sum_{n=-1}^{n=\infty} H_n q^n)$.) As an attempt to obtain a p–modular version of this property, we let ^ph_n be the character of $^p\mathbf{M}$ defined by $^ph_n =$

$^pH_n + \sum_{l=-\infty}^{l=\infty}(-1)^l\ ^pH_{n-(3l^2-l)/2}$. We use the computational techniques of Section 5 to examine the characters ^ph_n. The following results suggest that the character ^ph_n is a proper p–modular character of $^p\mathbf{M}$ so long as $n > 0$, and $p < 23$. Let $n(p)$ be the function with $n(2) = 118$, $n(3) = 93$, $n(5) = 60$, $n(7) = 42$, $n(11) = 29$, $n(13) = n(17) = 22$ and $n(19) = 21$.

Computation 6.1. *Suppose that $p \leq 19$ and $n(p) \leq n \leq 200$. Then, ^ph_n satisfies the conditions of Lemma 5.3.*

We note that if $p \leq 19$ and $n < N(p)$, then ^ph_n fails to satisfy the conditions of Lemma 5.3 unless the pair (p,n) is one of: $(17,12), (17,20), (19,9), (19,10)$.

For each of the cases $7 \leq p \leq 19$, we use the known p–modular character table of $^p\mathbf{M}$ (as in Section 5) to obtain:

Computation 6.2. *Let $7 \leq p \leq 19$, and let $1 \leq n \leq 200$, then ^ph_n is a proper p–modular character of $^p\mathbf{M}$.*

Further evidence that ^ph_n is a proper character whenever $p < 23$ comes from the Kondo-Tasaka Theorem which shows that if $p < 23$, and $p \mid |M_{24}|$, then $^ph_n(1) > 0$.

If $p = 23$, the Kondo-Tasaka Theorem just shows that $^{23}h_n(1) \geq 0$. Moreover, for any choice of p with $p \geq 29$, there are numerous instances of generalized characters of the form ^ph_n that have negative degree. For example, $^{29}h_6(1) = -1,$ $^{29}h_7(1) = -3$, and $^{31}h_5(1) = -2$. Thus, for values of $p > 23$, the ^ph_n can not all be proper characters.

We remark that a proof that the characters ^ph_n are proper, whenever $p < 23$, would provide very useful data for any computation of the p–modular characters of $^p\mathbf{M}$. (This is because, for small values of n we would expect the character ^ph_n to have few irreducible constituents.) We note that the p–modular character table of $^p\mathbf{M}$ is completely unknown in the cases $p = 2$, $p = 3$, and $p = 5$.

Acknowledgement. I would like to thank Richard Borcherds and Simon Norton for many useful conversations about this paper.

REFERENCES

1. R. E. Borcherds, *Vertex algebras, Kac-Moody algebras, and the Monster*, Proc. Natl. Acad. Sci. USA **83** (1986), 3068–3071.
2. R. E. Borcherds, *Monstrous moonshine and monstrous Lie superalgebras*, Invent. Math. **109** (1992), 405–444.
3. R. E. Borcherds, personal communication.
4. J. H. Conway, *Hexacode and tetracode—MOG and MINIMOG*, Computational Group Theory (M. D. Atkinson, ed.), Academic Press, New York (1984), pp. 359–365.
5. J. H. Conway, *A simple construction for the Fischer-Griess Monster group*, Invent. Math. **79** (1985), 513–540.
6. J. H. Conway, R. T. Curtis, S. P. Norton, R. A. Parker and R. A. Wilson, *An ATLAS of Finite Groups*, Oxford University Press, Oxford, 1985.
7. J. H. Conway and S. P. Norton, *Monstrous moonshine*, Bull. London Math. Soc. **11** (1979), 308–339.
8. C. W. Curtis and I. Reiner, *Representation Theory of Finite Groups and Associative Algebras*, Interscience Publishers, New York, 1965.
9. R. L. Griess, *The friendly giant*, Invent. Math. **69** (1982), 1–102.

10. R. L. Griess, *The Monster and its nonassociative algebra,* Finite Groups — Coming of Age, Proc. 1982 Montreal Conference, (J. McKay, ed.), Contemporary Math. **45** (1985), pp. 121–157.

11. C. Jansen, K. Lux, R. A. Parker and R. A. Wilson, *An ATLAS of Modular Characters,* in preparation.

12. T. Kondo and T. Tasaka, *The theta functions of sublattices of the leech lattice,* Nagoya Math J. **101** (1986), 151–179.

13. J. McKay and H. Strauss, *The q-series of monstrous moonshine and the decomposition of the head characters,* Comm. Alg. **18(1)** (1990), 253–278.

14. S. P. Norton, *F and other simple groups,* Ph. D. thesis, University of Cambridge, 1975.

15. S. P. Norton, *On the group Fi_{24},* Geom. Dedicata **25** (1988), 483–501.

16. L. Queen, *Some relations between finite groups, Lie groups and modular functions,* Ph. D. thesis, University of Cambridge, 1980.

17. L. Queen, *Modular functions arising from some finite groups,* Math. Comput. **37** (1981), 547–580.

18. A. J. E. Ryba, *Algebras related to some sporadic simple groups,* Ph. D. thesis, University of Cambridge, 1985.

19. A. J. E. Ryba, *Matrix generators for the Held group,* Computers in Algebra (M. C. Tangora, ed.), Marcel Dekker, New York, 1988, pp. 135–141.

20. A. J. E. Ryba, *Calculation of the 7-modular characters of the Held group,* J. Alg. **117** (1988), 240–255.

21. A. J. E. Ryba, *A natural invariant algebra for the Harada-Norton group,* Math. Proc. Camb. Phil. Soc. (to appear).

22. A. J. E. Ryba and R. A. Wilson, *Matrix generators for the Harada-Norton group,* Experimental Math. **3** (1994), 137–145.

23. Stephen D. Smith, *On the Head characters of the Monster simple group,* Finite Groups — Coming of Age, Proc. 1982 Montreal Conference, (J. McKay, ed.), Contemporary Math. **45** (1985),pp. 303–313.

24. R. A. Wilson, *A new construction of the Baby Monster, and its applications,* Bull. London Math. Soc. **25** (1993), 431–437.

DEPARTMENT OF MATHEMATICS, STATISTICS AND COMPUTER SCIENCE, MARQUETTE UNIVERSITY, CUDAHY HALL, P.O. BOX 1881, MILWAUKEE, WI 53201-1881

E-mail address: alexr@sylow.mscs.mu.edu

Replicant Powers for Higher Genera

Gene Ward Smith

1   Introduction

In [4], Simon Norton [1] defined the concept of replicant power, for a class
of functions he called replicable. In [1], a more conceptual definition
of replicability was given, and a subclass of functions called completely
replicable was defined. With the exception of some "trivial" rational
functions, the completely replicable functions turned out to be modular
functions invariant under a group containing $\Gamma_0(N)$ for some N, and
such that the fundamental region had genus zero.

The motivation for these definitions came from Moonshine theory,
and about half of the above mentioned functions turned out to be
directly related to the Monster in the manner described in [3].

As it happens, the definition of replicant power can can be generalized to modular functions such that the fundamental region has genus
greater than zero. It also appears to be the case that some kind of
connection with the Monster is still discernible.

2   Definitions

In the case of genus zero, we can define the basic notions "replicant
power" and "extended Hecke operator" for polynomials in the replicable function–which is to say, in the algebra generated by the replicable
function. This suggests giving a more general definition in terms of
more general algebras of functions.

To do this, let us define a notion of what it means for two functions
to be "close" in a region around 0. Suppose f and g are functions
defined in a punctured disk around $q = 0$ by q-series with coefficients
in $\mathbb{C}$. For any ϵ less than the radius of convergence, define

$$|f|_\epsilon = \mathbf{Max}|\epsilon e^{i\theta}|,$$

so that $|f|_\epsilon$ is the maximum modulus of f in a circle of radius ϵ around
0.

---

[1] 1991 Mathematics Subject Classification. 11

Definition 1 *Let us say that f is* dominated *by g, written $f << g$, if there exists a positive δ such that for all $0 < \epsilon < \delta$, we have*

$$|f|_\epsilon < |g|_\epsilon.$$

Definition 2 *Let $\mathbf{A}$ be an algebra of functions meromorphic in a disk D around 0 such that $f \in \mathbf{A}$ has poles in D only at 0, and such that any non-constant element has a pole of some order at 0. We will call any such $\mathbf{A}$* integral.

Lemma 1 *For any function ζ meromorphic on D with its only poles at 0, there exists a unique $\alpha = P_{\mathbf{A}}(\zeta) \in \mathbf{A}$ such that for any $\beta \in \mathbf{A}$ with $\alpha \neq \beta$,*

$$\zeta - \alpha << \zeta - \beta.$$

Proof: From the conditions on ζ, we have

$$\zeta(q) = \sum_{i=n}^{\infty} a_i q^i.$$

Beginning at n (which is zero or negative) and working upwards, there will be a first degree m in q such that a function whose lowest degree is m exists in $\mathbf{A}$. Hence there will be an $\alpha_1 \in \mathbf{A}$ such that

$$\alpha_1 = a_m q^m + \cdots,$$

that is, α_1 has the same m-th coefficient as ζ. Thus $\zeta - \alpha_1$ will be dominated by $\zeta - \beta$ for any $\beta \in \mathbf{A}$ which does not share an m-th coefficient with ζ. Hence our sought for α must be in the subspace of functions which share an m-th coefficient with ζ.

We may now look in this subspace for a function α_2 which matches the corresponding coefficient of ζ for the next possible term. Continuing in this way, we eventually reach an $\alpha_k = \alpha$ for which no further improvement is possible, since $\mathbf{A}$ is integral and hence every non-constant element has a pole of some order; and so any two elements cannot differ by a non-constant function with a zero at $q = 0$.

In cases where it will not lead to confusion, we will abbreviate $P_{\mathbf{A}}$ by P.

Let us now turn to the classic definition of the Hecke operator, and the proposed modification.

Definition 3 *If*

$$f(q) = \sum_{-\infty}^{\infty} a_i q^i,$$

then

$$f|U_n(q) = \sum_{-\infty}^{\infty} a_{ni} q^i.$$

Definition 4 *We define the* Hecke operator T_n *acting on a q-series by*

$$f|T_n(q) = \frac{1}{n} \sum_{ab=n} a f|U_a(q^b),$$

where a, b, and n are positive integers.

While the Hecke operator may be applied to any q-series, the interest in it attaches to the case of modular functions (and forms, under an extended definition.) In this case if f is a modular function of level prime to n, we obtain another modular function of the same level. We can modify the Hecke operator so that it will do something similar in the case of elements of any integral algebra **A**:

Definition 5 *For any $f \in \mathbf{A}$, where $\mathbf{A}$ is an integral algebra, we define*

$$f|\tilde{T}_n = P_{\mathbf{A}}(f|T_n).$$

As stated, the definition depends on the algebra **A**. In the case where f is a modular function, however, we may make it depend on f alone, by introducing an algebra **A** associated to f. To do this, we define in the usual way an action of matrices in $\mathbf{GL}_2^+(\mathbb{R})$ on f.

Definition 6 *Let $f(z)$ be a function defined on the upper half-plane $\mathcal{H}$, and let*

$$M = \begin{bmatrix} a & b \\ c & d \end{bmatrix}$$

be a matrix with values in $\mathbb{R}$ and with positive determinant. Then

$$(f \circ M)(z) = f((az+b)/(cz+d)).$$

Definition 7 *Let G be the subgroup of $\mathbf{PGL}_2^+(\mathbb{R})$ (i.e., of two-by-two real matrices with positive determinant modulo non-zero scalars) defined by those matrices M such that*

$$f \circ M = f.$$

Then G is the automorphism group *of f.*

Suppose that f is a q-series with coefficients in $\mathbb{C}$. If it converges in some disk around $q = 0$, we may continue it analytically. Suppose it has an analytic continuation to $|q| < 1$. Then we may define a corresponding function on the upper half-plane as usual by setting $q = e^{2\pi i z}$. If we do this, we then get a group of automorphisms associated to this function, and hence to the q-series.

If this group G is a Fuchsian group of the first kind then the quotient space of the upper half-plane with cusps appended is a compact Riemann surface [2]. The function field for this Riemann surface is given by the meromorphic modular functions with group G, considered as functions on the quotient space.

Definition 8 *The integral algebra associated with a non-constant modular function f is the algebra of meromorphic modular functions with group G the automorphism group of f, such that the only poles on the associated Riemann surface occur at $q = 0$ (which is to say, at $z = i\infty$.)*

If g is the genus of the Riemann surface, then by the Riemann-Roch theorem, this algebra is generated by a subspace of functions of dimension $g + 2$, corresponding to the divisor $i\infty^{2g+1}$, and so consisting of functions with poles of degree less than $2g + 2$ at $i\infty$ and no other poles.

If f is an element of its associated integral algebra, our definition of $\tilde{T}_n$ is applicable, and we have extended the definition of the Hecke operator to the case where the level of the modular function is not prime to the order of the Hecke operator, and still gives a modular function with the same automorphism group.

Suppose we have f in some integral algebra $\mathbf{A}$, in particular the algebra of meromorphic modular functions with group G and with all poles at $q = 0$ on the associated Riemann surface. Under certain circumstances, we can define the *replicant power* of f.

Definition 9 *Suppose*

$$f|\tilde{T}_n(q) = \frac{1}{n} \sum_{ab=n} a f^{(b)}|U_a(q^b)$$

for certain functions defined by q-series $f^{(b)}$. Then we say that $f^{(b)}$ is the b-th replicant power of f.

In particular, suppose for some prime p and some $f \in \mathbf{A}$ we have that $g = p(\tilde{T}_p - U_p)|f$ is a power series in q^p. Then we define the p-th replicant power $f^{(p)}$ for f to be $g(q^{\frac{1}{p}})$.

If for some f the replicant power map exists for a prime p, we have

$$f|\tilde{T}_p = P(f|T_p) = \frac{f^{(p)}}{p} + U_p|f.$$

If it also exists for $f^{(p)}$ for another prime l, we have

$$(f|\tilde{T}_p)|\tilde{T}_l = (f|\tilde{T}_l)|\tilde{T}_p = P(f|T_{lp})$$

$$= \frac{(f^{(p)})^{(l)}}{lp} + U_p|\frac{f^{(l)}}{l} + U_l|\frac{f^{(p)}}{p} + U_{lp}|f.$$

We see in fact that

$$f^{(lp)} = (f^{(p)})^{(l)} = (f^{(l)})^{(p)},$$

and in general, $f^{(n)}$ is the result of successive replicant power operations via the prime factors of n, and we have a consistent definition of replicant power which is definable in terms of the prime replicant powers.

If the point $i\infty$ of the Riemann surface associated to $\mathbf{A}$ is an ordinary point, we can put our definition into a more concrete form, using the fact that we have generators for $\mathbf{A}$ consisting of functions of degrees $-g-1, \cdots, -2g-1$ in q.

To do so, let

$$f_i = q^{-g-1-i} + b^i_{-g}q^{-g} + \cdots.$$

for i ranging from 0 to g be $g+1$ q-series with rational coefficients which generate $\mathbf{A}$. We can always do this when the algebraic curve associated to $\mathbf{A}$ is defined over $\mathbb{Q}$, as it is in the case of modular function fields. By taking appropriate linear combinations of products of our generators, we may in fact find a unique such f_i for each i, and hence (together with 1) a canonical *basis*, as well as a canonical generating set.

For a q-series e with rational coefficients, we can find $P(e)$ by solving the linear system of equations which matches the coefficients of e for terms of degree less than $-g$ as well as for degree 0. This gives us $P(e)$ expressed as a linear combination of f_i, and hence also as a polynomial in the set of generators, after a finite number of steps using rational arithmetic; in other words we may compute it.

Finally we should note that in the case of an element $f \in \mathbf{A}$ with rational integer coefficients, we have a convenient relationship between f and $f^{(p)}$, namely that $f \equiv f^{(p)} \pmod{p}$. This follows since modulo p we have the following:

$$pf|T_p(q) \equiv f(q^p) \equiv f(q)^p \equiv p\tilde{T}_p|f(q) \equiv f^{(p)}(q^p) \pmod{p},$$

and hence

$$f \equiv f^{(p)} \pmod{p}.$$

3   Examples

Consider a group G containing $\Gamma_0(N)$ for some N, and such that the associated function field is of genus 0. The function field is then $\mathbb{C}(f)$, where f is a single function, the Hauptmodul. If f is chosen to have its poles at $i\infty$, the integral algebra is $\mathbb{C}[f]$. If a is prime to the level N, $f|T_a$ is a polynomial in f, and hence $f|\tilde{T}_a = f|T_a$. From this we can conclude that $f^{(a)} = f$.

Consider for example the function denoted "11+" under the Conway-Norton naming scheme. This can be defined as the function belonging to the normalizer $\Gamma_0(11)+$ of $\Gamma_0(11)$,

$$f(q) = \frac{1}{q} + 5 + 17\,q + 46\,q^2 + 116\,q^3 + 252\,q^4 + \cdots.$$

It can be characterized precisely as the function with a simple pole at 0 satisfying the following polynomial relation:

$$(f^4 + 228\,f^3 + 486\,f^2 - 540\,f + 225)^3 - (11f|\tilde{T}_{11} + f + 1423)j + j^2 = 0,$$

where

$$j(q) = \frac{1}{q} + 744 + 196884q + 21493760q^2 + \cdots$$

is the modular function of level 1. Here we may compute

$$f|\tilde{T}_{11} = \frac{f^{11}}{11} - 5f^{10} + 108f^9 - 1156f^8 + 6330f^7 -$$

$$16128f^6 + 12096f^5 + 12006f^4 - 17037f^3 + 3645f^2 + 2209f - 743$$

to obtain the polynomial explicitly. We then find $f^{(11)} = j - 739$.

This polynomial is of degree two in j, and its conjugate root is $j(q^{11})$, because the fixing group for f contains the Fricke involution $z \mapsto -1/11z$, where $q = e^{2\pi i z}$. On the other hand, for all other primes p application of the Hecke operator $f|T_p$ will give another function of level 11 which is invariant under the Fricke involution, and therefore is a polynomial in f; so that $f^{(p)} = f$.

The definitions we have given allow this phenomenon, studied in [1], [4] to be generalized to the case where f is a modular function of any level with its only poles at $i\infty$. We may start the process of obtaining examples by noting that the polynomial in f and j above is a curve of genus 1.

The discriminant $(j(q^{11}) - j(q))^2$ when computed is seen to be $(hd)^2$, where

$$h = \sqrt{(f+1)(f^3 - 17f^2 + 19f - 7)},$$

and

$$d = f(f-1)(f-3)(f-6)(f-15)(f^2 - 10f + 5)(f^2 - 12f + 9).$$

Here h is a function of degree two at $i\infty$ such that

$$h(q) = \frac{j(q^{11}) - j(q)}{d(q)} =$$

$$\frac{1}{q^2} + \frac{2}{q} - 12 - 116\,q - 597\,q^2 - 2298\,q^3 - \cdots.$$

We therefore have a modular parametrization of a non-singular plane curve of genus one,

$$h^2 = (f+1)\left(f^3 - 17\,f^2 + 19\,f - 7\right).$$

From its construction, we see that the above is a nonsingular affine plane model of the modular curve $X_0(11)$. This model has the advantage that integral values of $j(q)$ and $j(q^{11})$ (representing pairs of 11-isogenous curves both of which have everywhere potentially good reduction) correspond in any ring of integers exactly to integral points on the curve.

It is not, however, in Weierstrauss form. If we set

$$x = (f^2 - 8f + 1 + h)/2,$$

$$y = (f^3 - 12f^2 + f + 2 + (f - 4)h)/2,$$

we can transform the equation into its minimal model, so that

$$y^2 + y = x^3 - x^2 - 10\,x - 20.$$

Since the above equation is in Weierstrauss form, x and y have poles of order 2 and 3 respectively on the $i\infty$-cusp of the fundamental region for $\Gamma_0(11)$, and no other poles. Hence Hecke operators T_n applied to x for values of n prime to 11 will have poles there only as well, and so will be expressible as polynomial functions of x and y. In these cases, $\tilde{T}_n$ and T_n give the same results, and so the replicant powers exist and are equal to the functions themselves.

Now consider $x^{(11)}$. We find from above that

$$x(q) = \frac{1}{q^2} + \frac{2}{q} + 4 + 5\,q + 8\,q^2 + q^3 + \cdots,$$

$$y(q) = \frac{1}{q^3} + \frac{3}{q^2} + \frac{7}{q} + 12 + 17\,q + 26\,q^2 + \cdots.$$

Then $x^{(11)}(q^{11})$ is by definition $(11x|(\tilde{T}_{11} - U_{11})(q)$.

Solving the 23 linear equations in question allows us to find

$$11x|\tilde{T}_{11} = x^{11} + 198\,x^{10} + 1881\,x^9 - 12562\,x^8 -$$

$$34925\,x^7 + 97845\,x^6 + 373087\,x^5 - 42878\,x^4$$

$$-1225114\,x^3 - 1099153\,x^2 + 381975\,x + 540980$$

$$-(22\,x^9 + 913\,x^8 + 110\,x^7 - 24519\,x^6 - 5918$$

$$x^5 + 129921\,x^4 + 160248\,x^3 - 176341\,x^2 - 271766\,x - 42394)y.$$

We now find that, at least for the first few coefficients,

$$x^{(11)}(q^{11}) = j(q^{11})^2 - 1486j(q^{11}) + 158284.$$

However, since both the left and the right side of this equation are modular functions of level 11 whose only poles are at $i\infty$, they must be equal; and so $x^{(11)}$ is a polynomial in the replicable function j.

Similarly, we find that

$$11y|\tilde{T}_{11} = -33x^{16} - 3954x^{15} - 57923x^{14} + 276223x^{13} + 2613548x^{12}$$

$$-5446617x^{11} - 40028169x^{10} + 8466829x^9 + 314519630x^8 + 317278469x^7$$

$$-936204215x^6 - 2004492022x^5 + 33539374x^4 + 2918481030x^3 + 2089725795x^2$$

$$-215650099x-380704104+(x^{15}+479x^{14}+20189x^{13}+68690x^{12}-1143120x^{11}$$

$$-1759416x^{10}+14551289x^9+30061672x^8-66034514x^7-216264127x^6+$$

$$14761616x^5+532734048x^4+414755908x^3-236803935x^2$$

$$-318938344x-47854454)y,$$

and from this that

$$y^{(11)} = j^3 - 2229j^2 + 1065499j - 36392868,$$

so that y also has a replicant power $y^{(n)}$ for every n.

Now consider a function

$$f(q) = \frac{1}{q^2} + 1 + q + 2q^2 + q^3 + 3q^4 - 3q^5 + \cdots,$$

which is algebraically related to x and y by

$$f^2 - (x^2 + x + 2)f + x^3 + 3x^2 + x - 9 + 2xy + 2y = 0.$$

We can take f as defining the x-coordinate for another point on the same elliptic curve

$$g^2 + g = f^3 - f^2 - 10f - 20,$$

so that together with

$$g(q) = \frac{1}{q^3} + \frac{1}{q} + 1 - 2q + 2q^2 - 4q^3 + 6q^5 + \cdots$$

we can define an algebra.

If we denote the function field associated to the modular group $\Gamma_0(N) + i$, where i are certain involutions, by $K_0(N) + i$. Then this defines $K_0(22) + 2$, the subfield of $K_0(22)$ invariant under the Atkin-Lehner involution which exchanges $j(q)$ with $j(q^2)$ and $j(q^{11})$ with $j(q^{22})$.

Calculating as before, we find that

$$g^{(2)} = y - 3x + 1,$$

and

$$g^{(11)} = e^3 - 312e^2 + 19333e - 49671,$$

where e is the replicable function 2+ of level 2,

$$e(q) = \frac{1}{q} + 104 + 4372q + 96256q^2 + 1240002q^3 + \cdots,$$

and where we have the relation

$$(e + 144)^3 - (e^2 - 207e + 3456)j + j^2 = 0$$

between e and j. Finally,

$$g^{(22)} = j^3 - 2232j^2 + 1069957j - 36867719,$$

and since arguments like those already given show that $g^{(p)} = g$ for primes other than 2 and 11, we see that g has replicant powers for all n.

What about f? We can calculate

$$f^{(11)} = e^2 - 208e + 2073$$

as before. However, our definition does not work to give us $f^{(2)}$. We have, however, the following:

$$f^2 + 1 = x(q^2) + 2U_2|f,$$

so that $x + c$ for any constant c acts in some sense like a replicant power of f. This suggests that a better definition of replicant power might be discoverable.

This is not too hard to do if we start from the observation that

$$f^{(p)} = P_{\mathbf{A}^{(p)}}(f),$$

where $\mathbf{A}^{(p)}$ is the algebra to which the replicant powers $f^{(p)}$ belong. We therefore seek to define an extended replicant power map by setting

$$g^{(p)} = P_{\mathbf{A}^{(p)}}(g).$$

We make the following definitions:

Definition 10 *Let G be a group of modular automorphisms, and let $\mathbf{A}$ be the associated integral algebra of modular functions with poles only at $i\infty$. Then $G^{(n)}$ is the group of modular automorphisms common to the set of functions $\{f^{(n)}\}$, for $f \in \mathbf{A}$, and for which the replicant power is defined.*

It should be noted that $G^{(n)}$ is the common group of automorphisms for the replicant powers of those elements of **A** which have replicant powers, by the *narrow* definition used until now. We can then use this to define a broader notion of replicant power; the procedure is convoluted but not circular.

Definition 11 *If* **A** *is the algebra of modular functions with poles only at* $i\infty$ *and automorphism group* G, *then* $\mathbf{A}^{(n)}$ *is the similar such algebra with group* $G^{(n)}$.

Definition 12 *We extend the definition of replicant power by setting*

$$f^{(n)} = P_{\mathbf{A}^{(\mathbf{n})}}(f)$$

in all cases.

We now have a definition which is compatible with our previous one, but which works for every element of **A**.

Definition 13 *We define a new extended Hecke operator* $\hat{T}_n$ *by*

$$f|\hat{T}_n(q) = \frac{1}{n} \sum_{ab=n} a f^{(b)} |U_a(q^b),$$

using the extended definition of the replicant power map.

The definitions of the replicant power map and especially of the extended Hecke operator have, as I've already remarked, become rather convoluted. It might be that a better approach would be to find what $G^{(n)}$ is working by on the level of groups. One might also attempt to fix the definition of the extended Hecke operator more directly.

Moreover, it still only applies to functions and not forms, and only to those functions with poles only at $i\infty$. In some sense, however, it seems to be leading to a better definition of the Hecke operator.

For an example, if f is our f from above, then the projection $P(f)$ to $\mathbb{C}[x, y]$ is $x - 3$; so that $f^{(2)} = x - 3$.

We then find:

$$f|\hat{T}_2 = \frac{f^2}{2} - 1,$$

whereas

$$f|\tilde{T}_2 = \frac{f^2}{2} - f.$$

It is $\hat{T}$ which seems to be the "correct" Hecke operator; and for which we should seek an extension to all modular functions and even modular forms.

The elliptic curve above is parametrized by both x, y and f, g. If we look instead at the curve parametrized by $x^{(11)}$, $y^{(11)}$ we see that it is of genus zero, since this parametrization can be written $j^2 - 1486j + 158284$, $j^3 - 2292j^2 + 1065499j - 36392868$.

Since this curve is congruent to our elliptic curve $X_0(11)$ mod 11, we see that $X_0(11)$ has bad reduction mod 11 and that the replicant power map gives this reduction to a genus zero curve explicitly. On the other hand, for all other primes we simply get the same curve back again; so that the replicant power map reflects the fact that $X_0(11)$ has conductor 11.

If we look at $f^{(11)}$, $g^{(11)}$, we get another parametrization of a genus zero curve, this time $e^2 - 208e + 2073$, $e^3 - 312e^2 + 19333e - 49671$. Modulo 11, this becomes $e^2 + e + 5$, $e^3 - 4e^2 - 5e + 5$; whereas the previous parametrization becomes $j^2 - j + 5$, $j^3 + 4j^2 - 5j + 5$; we are dealing with in essence the same parametrization here. Once again, the replicant power map has taken the curve to a curve of different genus only when the power map was not prime to the conductor.

The above example has the replicant power map taking a curve to the same curve; we can also find higher genus examples where it is taken to a different curve of the same genus.

For instance, let

$$u(q) = \frac{1}{q^2} + \frac{4}{q} + 13 + 33q + \cdots$$

$$v(q) = \frac{1}{q^3} + \frac{6}{q^2} + \frac{25}{q} + 83 + 244q + \cdots,$$

with the relation

$$v^2 + v = u^3 - u^2.$$

The function u can be defined in various ways; for instance it is related to the function g introduced previously by

$$(g-5)^2 u^4 + (-2g^3 - 6g^2 + 56g - 1)u^3 + (g^4 + 2g^3 + 18g^2 - 27g + 20)u^2 -$$

$$(2g^2 + 4g + 7)(g^2 + 8)u + (g^2 + 8)(g^2 + 3g + 4) = 0.$$

The functions u and v together define the subfield $K_0(22)+22$ of $K_0(22)$ invariant under the Fricke involution $z \mapsto -1/22z$.

We find we cannot directly compute either $u^{(2)}$ or $v^{(2)}$, however if we set $w = v - 6u$, we find that

$$w^{(2)} = y - 3x + 5,$$

and

$$w^{(11)} = d^3 + 72d^2 + 901d + 125,$$

where d is the Moonshine function f_{2-}, i.e. the Hauptmodul for $\Gamma_0(2)$, with q-series

$$d(q) = \frac{1}{q} - 24 + 276q - 2048q^2 + \cdots.$$

We can also define d by the fact that

$$j = (d + 256)^3/d^2.$$

We may now conclude that $(\Gamma_0(22) + 22)^{(2)} = \Gamma_0(11)$, and we may then go ahead and compute $u^{(2)} = x + 9, v^{(2)} = y + 3x + 59$.

The replicant power map takes the curve to a curve of lower genus only for powers which divide 11; and in fact the conductor of the above curve is also 11 (it is 5-isogenous to the previous curve,) so again the map takes us to a lower genus only in the case where the power is not prime to the conductor.

The various examples we have considered others as well suggest the following conjecture: $f^{(n)}$ is not computable using the Hecke operators $\tilde{T}_i$ if and only if at least one of the non-zero coefficients of f with negative q-degree has a degree relatively prime to n.

We can also easily give examples where the genus is greater than one.

For instance, consider the functions

$$f_1(q) = \frac{1}{q^3} + \frac{1}{q} + 2q + 2q^2 + 2q^5 + \cdots,$$

$$f_2(q) = \frac{1}{q^4} + \frac{2}{q^2} + \frac{2}{q} + 3q^2 + 4q^3 + 4q^4 + 4q^5 + \cdots,$$

$$f_3(q) = \frac{1}{q^5} - \frac{1}{q} + 2q + 2q^3 + 4q^4 + 4q^5 + \cdots.$$

These are related by the equations

$$f_2^2 - f_3 f_1 - 3f_1^2 - 4f_3 + 3f_2 - 6f_1 - 4 = 0,$$

$$f_3^2 - f_1^2 f_2 + 4 f_3 f_1 + 2 f_2 f_1 + 7 f_1^2 + 2 f_3 - 9 f_2 + 10 f_1 - 12 = 0.$$

These functions and equations define $K_0(22)$.

We now compute

$$f_1 | \tilde{T}_2 = \frac{f_1^2}{2} - f_2 - 2,$$

$$f_1 | \tilde{T}_{11} = \frac{f_1^{11}}{11} - f_1^9 f_2 + 5 f_1^9 + 4 f_1^8 f_3 - 16 f_1^8 - 20 f_1^7 f_2$$

$$+ 16 f_1^7 f_3 + 28 f_1^7 + 60 f_1^6 f_2 + 74 f_1^6 f_3 + 136 f_1^6 + 164 f_1^5 f_2 + 276 f_1^5 f_3$$

$$+ 90 f_1^5 + 396 f_1^4 f_2 + 146 f_1^4 f_3 + 700 f_1^4 + 1389 f_1^3 f_2 + 632 f_1^3 f_3 + 1929 f_1^3$$

$$+ 606 f_1^2 f_2 - 140 f_1^2 f_3 + 56 f_1^2 + 915 f_1 f_2 - 232 f_1 f_3 + 794 f_1 + 84 f_2 - 103 f_3 - 32$$

We may now calculate

$$f_1^{(11)}(q) = \frac{1}{q^3} + \frac{1}{q} + 33882 q - 1845248 q^2 + \cdots,$$

which is

$$d^3 + 72 d^2 + 901 d + 120,$$

where $d = f_{2-}$ is the Hauptmodul for $\Gamma_0(2)$ defined previously.

We also have

$$f_1^{(2)}(q) = \frac{1}{q^3} + \frac{1}{q} + 2q + 2q^2 + \cdots,$$

which is $y - 3x$, where x and y are the functions defining $K_0(11)$ considered previously. We see that $\Gamma_0(22)^{(2)} = \Gamma_0(11)$, and $\Gamma_0(22)^{(11)} = \Gamma_0(2)$, just as we would suspect.

4 Higher Genus Moonshine

Let us consider again the functions x and y defining $K_0(11)$. Since $K_0(11)$ contains the j-invariant $j(q)$, we must be able to express j as a rational function of x and y. Doing this explicitly gives

$$j = ((x^5 + 4518 x^4 + 1304157 x^3 + 65058492 x^2 + 271927184 x - 707351591)$$

$$(x^5 + 192189 x^4 + 3626752 x^3 - 3406817 x^2 - 37789861 x - 37315543) y$$

$$+ 743 x^{11} + 21559874 x^{10} + 19162005343 x^9 + 2536749758583 x^8 +$$

$$821653627660027x^7 + 576036867160006x^6 - 1895608370650736x^5 -$$

$$14545268641576841x^4 + 420015065507429x^3 + 74593328129816300x^2$$

$$+108160113602504237x - 39176677684144739)/(x - 16)^{11}.$$

Now consider the functions f and g defining $K_0(22) + 2$ we considered previously. These parametrize the same elliptic curve. If we substitute f and g for x and y in the expression above, we get

$$\frac{1}{q} + 743 + m_2 q + (m_3 + 2m_2)q^2 + (m_4 + 3m_3 + 6m_2)q^3 + \cdots,$$

where $m_1 = 1, m_2 = 196883, m_3 = 21296876, m_4 = 842609326, \cdots$ are the degrees of the irreducible representations of the Monster.

The terms of this series suggest that it may correspond to an as yet undiscovered moonshine module, and there is no reason to suppose that there are not many more such examples; my attempts to compute further examples have been unsuccessful, but I suspect I am using algorithms which are far from being optimal.

The elliptic curve

$$y^2 + y = x^3 - x^2 - 10x - 20$$

has an involution given by $y \mapsto -y - 1$. If we substitute $-y - 1$ for y into the above expression for j, we obtain another q-series which seems to be related to the Monster. In this case, we seem to be getting virtual characters rather than anything suggestive of an actual moonshine module:

$$-\frac{1}{q} + 742 - (m_2 + m_1)q + (m_3 + 3m_2 - 2m_1)q^2 -$$

$$(m_4 + 5m_3 + 10m_2)q^3 + \cdots.$$

If we substitute in the functions u and v defining $K_0(22) + 22$, we get

$$\frac{1}{q} + 745 + (m_2 + 4m_1)q + (m_3 + 8m_1)q^2 + (m_4 - m_3 - 2m_2 + 20m_1)q^3 + \cdots,$$

whereas in the involution form, we get

$$-\frac{1}{q} + 742 - (m_2 + 4m_1)q - (m_3 + 4m_1 - 9m_1)q^2 + (m_4 + 7m_3 + 14m_2 + 16m_1)q^3 - \cdots;$$

once again we seem to be getting virtual characters.

References

[1] D. Alexander, C. Cummins, J. McKay, and C. Simons, "Completely Replicable Functions", in *Groups and Combinatorics*, Cambridge, 1992.

[2] Toshitsune Miyake, *Modular Forms*, Springer-Verlag, 1989.

[3] John H. Conway and Simon P. Norton, "Monstrous Moonshine", Bull. Lon. Math. Soc., 11:8 (1983), 863-911.

[4] Norton, Simon P., "More on Moonshine", in *Computational Group Theory*, Michael D. Atkinson, ed., Academic Press, 1984.

Department of Mathematics
The University of California at Santa Cruz
Santa Cruz, CA 95064

gsmith@cats.ucsc.edu

Contemporary Mathematics
Volume **193**, 1996

GENERALISED MOONSHINE
AND
ABELIAN ORBIFOLD CONSTRUCTIONS

Michael P. Tuite

0. Introduction. We consider the application of Abelian orbifold constructions in Meromorphic Conformal Field Theory (MCFT) [Go,DGM] towards an understanding of various aspects of Monstrous Moonshine [CN] and Generalised Moonshine [N]. We review some of the basic concepts in MCFT and Abelian orbifold constructions [DHVW] of MCFTs and summarise some of the relevant physics lore surrounding such constructions including aspects of the modular group, the fusion algebra and the notion of a self-dual MCFT. The FLM Moonshine Module, $\mathcal{V}^{\natural}$, [FLM1,FLM2] is historically the first example of such a construction being a $\mathbb{Z}_2$ orbifolding of the Leech lattice MCFT, $\mathcal{V}^{\Lambda}$. We review the usefulness of these ideas in understanding Monstrous Moonshine, the genus zero property for Thompson series [CN] which we have shown is equivalent to the property that the only meromorphic $\mathbb{Z}_n$ orbifoldings of $\mathcal{V}^{\natural}$ are $\mathcal{V}^{\Lambda}$ and $\mathcal{V}^{\natural}$ itself (assuming that $\mathcal{V}^{\natural}$ is uniquely determined by its characteristic function $J(\tau)$) [T1,T2]. We show that these constraints on the possible $\mathbb{Z}_n$ orbifoldings of $\mathcal{V}^{\natural}$ are also sufficient to demonstrate the genus zero property for Generalised Moonshine functions in the simplest non-trivial prime cases by considering $\mathbb{Z}_p \times \mathbb{Z}_p$ orbifoldings of $\mathcal{V}^{\natural}$. Thus Monstrous Moonshine implies Generalised Moonshine in these cases.

1. Meromorphic Conformal Field Theory. In this section, we will review some of the basic properties of Meromorphic Conformal Field Theory (MCFT) (or chiral algebras) as described by Goddard [Go]. This is a physically motivated approach to Vertex Operator Algebras [B1,FLM2,FHL] containing the same essential ideas. Let $\mathcal{H}$ denote some Hilbert space with a dense subspace of states $\{\phi\}$ including a unique 'vacuum state' $|0\rangle$ with properties described below. In a MCFT we define a set of conformal fields or vertex operators $\mathcal{V}$ such that corresponding to each state ϕ there exists an operator $V(\phi, z) \in \mathcal{V}$ acting on $\mathcal{H}$ with

$$\lim_{z \to 0} V(\phi, z)|0\rangle = \phi \qquad (1.1)$$

It is assumed that there exists Virasoro operators L_n which form the modes of $V(\omega, z)$ (see below) for a Virasoro state ω where

$$[L_n, V(\phi, z)] = z^n [z\frac{d}{dz} + (n+1)h_\phi]V(\phi, z) \qquad (1.2a)$$

$$[L_m, L_n] = (m-n)L_{m+n} + \frac{C}{12}m(m^2-1)\delta_{m,-n} \qquad (1.2b)$$

1991 *Mathematics Subject Classification.* 81T40,17B69,20D08,11F11

where h_ϕ is called the conformal weight of ϕ and C is the central charge for the representation of the Virasoro algebra (1.2b). The Virasoro state ω has conformal weight 2. From (1.2a), L_0 defines a discrete grading on $\mathcal{H}$ with $L_0|\phi\rangle = h_\phi|\phi\rangle$. We assume that $\mathcal{V}$ is unitary so that $h_\phi \geq 0$. By a *Meromorphic CFT*, we will mean a CFT for which the conformal weights are integral and where the operators $\mathcal{V}$ obey the (bosonic) Locality Property :

$$V(\phi, z)V(\psi, w) = V(\psi, w)V(\phi, z) \tag{1.3}$$

with $|z| > |w|$ on the LHS analytically continued to $|z| < |w|$ on the RHS. (These assumptions ensure that all correlation functions are meromorphic). These operators can then be shown to satisfy the Duality Property [Go,FHL]

$$V(\phi, z)V(\psi, w) = V(V(\phi, z - w)\psi, w) \tag{1.4}$$

with $|z| > |w|$ and $|z - w| < |w|$ respectively and where $V(\phi, z)$ is extended by linearity to any state in $\mathcal{H}$. These are essentially the defining properties of a vertex (operator) algebra as defined in [B1] and developed in [FLM2]. All the conformal fields in a MCFT also obey the Monodromy condition :

$$V(\phi, e^{2\pi i}z) = V(\phi, z) \tag{1.5}$$

so that the mode expansion for each operator is $V(\phi, z) = \sum_{k \in \mathbb{Z}} \phi_k z^{-k-h_\phi}$ with $\phi_k|0\rangle = 0$ for all $k > -h_\phi$ and $\phi_{-h_\phi}|0\rangle = \phi$ from (1.1) e.g. the modes for the Virasoro (energy-momentum) operator $V(\omega, z)$ are $\{L_n\}$ as above with $\omega = L_{-2}|0\rangle$. Then (1.4) leads to an exact form of the usual operator product expansion of CFT [BPZ]

$$V(\phi, z)V(\psi, w) = \sum_{k=0}^{\infty}(z - w)^{k-h_\phi-h_\psi}V(\chi, w) \tag{1.6}$$

where $\chi = \phi_{-k+h_\psi}(\psi)$ is a state of conformal weight k. We will sometimes schematically write such an expansion as $\mathcal{V}\mathcal{V} \sim \mathcal{V}$.

2. The Modular Group and Self-Dual MCFTs.
Let $\mathcal{V}$ be a MCFT and define the characteristic function (or partition function) for $\mathcal{V}$ by the following trace

$$Z(\tau) = \mathrm{Tr}_{\mathcal{H}}(q^{L_0 - C/24}) \tag{2.1}$$

where $\tau \in H$, the upper half complex plane. In string theory models, $Z(\tau)$ arises when finding the probability for a closed string to form a 2-torus parameterised by τ. The simplest example, is the one-dimensional $C = 1/24$ bosonic string which has characteristic function $1/\eta(\tau)$ where $\eta(\tau) = q^{1/24}\prod_{n>0}(1-q^n)$. For a d dimensional $C = d/24$ string compactified by an even lattice Λ, we obtain a MCFT denoted by $\mathcal{V}^\Lambda$, with $Z(\tau) = \Theta_\Lambda(\tau)/[\eta(\tau)]^d$ where $\Theta_\Lambda = \sum_{\lambda \in \Lambda} q^{\lambda^2/2}$.

The behaviour of $Z(\tau)$ under the action of the modular group $\Gamma = \mathrm{SL}(2, \mathbb{Z})$, generated by $T : \tau \to \tau + 1$ and $S : \tau \to -1/\tau$, is related to the meromorphic properties of $\mathcal{V}$ and to properties of the meromorphic irreducible representations of

$\mathcal{V}$. For a MCFT we clearly have $Z(\tau + 1) = e^{2\pi i C/24} Z(\tau)$ and, in particular, $Z(\tau)$ is T invariant for $C = 24$.

Let us now discuss the meaning of the S transformation. Let $\tilde{\mathcal{V}}$ denote an irreducible meromorphic representation for $\mathcal{V}$ acting on a Hilbert space $\mathcal{H}^K$ and let $\mathcal{V}^K$ be the corresponding set of intertwiners acting on $\mathcal{H}$ that create the states of $\mathcal{H}^K$ from the original vacuum vector $|0\rangle \in \mathcal{H}$ [FHL,DGM]. Then as in (1.3) and (1.4) we have

$$\tilde{V}(\phi, z)\tilde{V}(\psi, w) = \tilde{V}(\psi, w)\tilde{V}(\phi, z) \tag{2.2a}$$

$$\tilde{V}(\phi, z)\tilde{V}(\psi, w) = \tilde{V}(V(\phi, z - w)\psi, w) \tag{2.2b}$$

$$\tilde{V}(\phi, z)V^K(\chi, w) = V^K(\chi, w)V(\phi, z) \tag{2.2c}$$

$$\tilde{V}(\phi, z)V^K(\chi, w) = V^K(\tilde{V}(\phi, z - w)\chi, w) \tag{2.2d}$$

(up to suitable analytic continuations) for $\tilde{V}(\phi, z) \in \tilde{\mathcal{V}}$, $V^K(\chi, z) \in \mathcal{V}^K$ with $\phi, \psi \in \mathcal{H}$ and $\chi \in \mathcal{H}^K$. Given such a representation, we thus naturally extend $\mathcal{V}$ to act on $\mathcal{H} \oplus \mathcal{H}^K$ and henceforth we drop the tilde notation distinguishing the space on which $\mathcal{V}$ acts. We also define the characteristic function $Z^K(\tau) = \mathrm{Tr}_{\mathcal{H}^K}(q^{L_0 - 1})$ for $\mathcal{V}^K$. In general, the conformal weights of $\mathcal{H}^K$ are not integral but are equal mod $\mathbb{Z}$ and hence $Z^K(\tau)$ is T invariant up to a phase.

By a *Rational MCFT*, we will mean a MCFT which has a finite number M of such irreducible representations $\{\mathcal{V}^K\}$, $K = 0, ..., M - 1$ (with $\mathcal{V} \equiv \mathcal{V}^0$) and where every representation of $\mathcal{V}$ is reducible. For a Rational MCFT, Zhu has shown that each characteristic function $Z^K(\tau)$ is holomorphic on the upper half plane H (given a certain growth condition which is conjectured to follow from rationality) and the functions $\{Z^K\}$ transform amongst themselves under the modular group Γ [Z].

These properties can also be understood if $\mathcal{V}$ together with (possibly multiple copies of) its intertwiners form a non-meromorphic CFT which we call the *Dual CFT* to $\mathcal{V}$ and denote by $\mathcal{V}^*$. We can think of $\mathcal{V}^*$ as comprising the maximal (in some sense!) set of vertex operators of central charge C that are local with respect to $\mathcal{V}$. $\mathcal{V}^*$ is expected to satisfy an operator product algebra given by some generalised version of (1.6) where schematically

$$\mathcal{V}^I \mathcal{V}^J \sim \sum_{K=0}^{M-1} N^{IJK} \mathcal{V}^K \tag{2.3}$$

where N^{IJK} are non-negative integers determining the decomposition in terms of irreducible representations of $\mathcal{V}$ of the non-meromorphic algebra - these are the *Fusion Rules* for $\mathcal{V}^*$ [Ve]. The coefficients N^{IJK} satisfy a commutative associative algebra which is diagonalised by $S : Z^I \to S^{IJ} Z^J$ where S^{IJ} is a unitary symmetric matrix [Ve]. In addition, we assume $\mathcal{V}^*$ is a unitary CFT, so that $S^{I0}/S^{00} \geq 1$ with equality iff we have *Abelian Fusion Rules* i.e. for every given $I, J = 0, ...M - 1$, $N^{IJK} = 1$ for some unique K so that every pair of intertwiners fuses to form a unique intertwiner. Assuming Abelian Fusion Rules we then find, since S^I_J is symmetric and unitary, that

$$S : Z(\tau) \to \epsilon_S \frac{1}{\sqrt{M}} \sum_K Z^K(\tau) = \epsilon_S |\mathcal{H}^*/\mathcal{H}|^{-1/2} \mathrm{Tr}_{\mathcal{H}^*}(q^{L_0 - 1}) \tag{2.4}$$

where $|\epsilon_S| = 1$ and $\mathcal{H}^*$ denotes the Hilbert space $\oplus_{K=0}^{M-1} \mathcal{H}^K$ for $\mathcal{V}^* \equiv \oplus_K \mathcal{V}^K$ and $M = |\mathcal{H}^*/\mathcal{H}|$. If furthermore $\{Z^K\}$ is charge conjugation invariant then S^{IJ} is real so that $\epsilon_S = 1$. This formula can be verified for an even lattice Λ MCFT where the irreducible representations for $\mathcal{V}^\Lambda$ are indexed by Λ^*/Λ where Λ^* is the dual lattice [D1]. In this case, $\mathcal{V}^\Lambda$ is naturally embedded in the non-meromorphic CFT $\mathcal{V}^{\Lambda^*}$ so that $(\mathcal{V}^\Lambda)^* = \mathcal{V}^{\Lambda^*}$. Furthermore, the fusion rules are abelian from the abelian structure of Λ^* [DL]. Then, under the action of S, $Z_\Lambda = \Theta_\Lambda/\eta^d \to |\Lambda^*/\Lambda|^{-1/2}\Theta_\Lambda^*/\eta^d$ in the usual way in agreement with (2.4) with $\epsilon_S = 1$. Similarly, (2.4) holds with $\epsilon_S = 1$ for the Abelian orbifold constructions discussed below.

If $Z(\tau)$ is S invariant and hence $\mathcal{V}$ is the unique irreducible representation for itself, we define $\mathcal{V}$ to be a *Self-Dual MCFT*. This is only possible for $C = 0 \bmod 8$ [Go]. For $C = 24$, then $Z(\tau)$ is modular invariant with a unique simple pole at $q = 0$ on H/Γ which is equivalent to the Riemann sphere of genus zero. Hence $Z(\tau)$ is given by $J(\tau)$, the *hauptmodul* for Γ [Se]

$$Z(\tau) = J(\tau) + N_0 \tag{2.5a}$$

$$J(\tau) = \frac{E_2^3}{\eta^{24}} - 744 = \frac{1}{q} + 0 + 196884q + 21493760q^2 + \dots \tag{2.5b}$$

with $E_2(\tau)$ the Eisenstein modular form of weight 4 [Se] and where N_0 is the number of conformal weight 1 operators in $\mathcal{V}$. Examples of such theories are lattice models where Λ is a Niemeier even self-dual lattice. Then $\mathcal{V}^\Lambda$ is meromorphic self-dual because Λ is even self-dual. In particular, for the Leech lattice which contains no roots, $N_0 = 24$. Other examples of self-dual C=24 MCFTs are the Moonshine Module $\mathcal{V}^\natural$ with $Z(\tau) = J(\tau)$ and other orbifold constructions as we now describe. In general, there are thought to be just 71 such independent self-dual MCFTs [Sch].

3. Abelian Orbifolding of a Self-Dual MCFT.

Let $\mathcal{V}$ be a self-dual MCFT and let $Aut(\mathcal{V})$ denote the automorphism group preserving the operator algebra for $\mathcal{V}$ with

$$gV(\phi, z)g^{-1} = V(g\phi, z) \tag{3.1}$$

for each $g \in Aut(\mathcal{V})$. Consider G any finite abelian subgroup of $Aut(\mathcal{V})$ generated by m commuting elements $\{g_1, \dots, g_m\}$ of order $n_1, \dots, n_m$. Let $\mathcal{P}_G \mathcal{V}$ denote the operators invariant under G with projection operator $\mathcal{P}_G = \frac{1}{|G|}\sum_{g \in G} g$. $\mathcal{P}_G \mathcal{V}$ is a MCFT but is not self-dual as can be seen by studying the corresponding characteristic function $\mathrm{Tr}_{\mathcal{P}_G \mathcal{H}}(q^{L_0-1})$ which is not S invariant. In particular, consider the trace for each $g \in G$

$$Z(1, g) \equiv \mathrm{Tr}_\mathcal{H}(gq^{L_0-1}) \tag{3.2}$$

where we introduce standard notation indicating boundary conditions on the 2-torus where the first label 1 refers to the monodromy condition (1.5). Using path integral methods in string theory [DHVW] one can argue that under $S : \tau \to -1/\tau$ the boundary conditions are interchanged for $Z(1, g)$ charge conjugation invariant and $\mathcal{V}$ a self-dual theory so that

$$S : Z(1, g) \to Z(g, 1) = \mathrm{Tr}_{\mathcal{H}_g}(q^{L_0-1}) = D_g q^{E_0^g} + \dots \tag{3.3}$$

the characteristic function for $\mathcal{H}_g$, the 'g-twisted' Hilbert space. We assume that $\mathcal{H}_g$ is uniquely defined (up to isomorphism) for each $g \in G$. The parameters E_0^g and D_g are called the g-twisted vacuum energy and degeneracy. Note also that the remaining coefficients of the powers are necessarily all non-negative integers. We assume that each twisted state $\psi \in \mathcal{H}_g$ of conformal weight h_ψ is created from $|0\rangle$ by the action of a twisted operator with the following g-twisted Monodromy property:

$$V(\psi, e^{2\pi i} z) = g V(\psi, z) g^{-1} = e^{-2\pi i h_\psi} V(\psi, z) \tag{3.4}$$

We denote the set of such operators for each $g \in G$ by $\mathcal{V}_g$. We also assume that $\oplus_{g \in G} \mathcal{V}_g$ satisfies a non-meromorphic version of the Locality property (1.3) and a G-invariant operator product expansion generalising (1.6) (up to suitable analytic continuation) where

$$V(\psi, z) V(\chi, w) = \epsilon_{\psi, \chi} V(\chi, w) V(\psi, z) \tag{3.5a}$$

$$V(\psi, z) V(\chi, w) = \sum_\rho (z - w)^{h_\rho - h_\psi - h_\chi} V(\rho, w) \tag{3.5b}$$

for $\psi \in \mathcal{H}_g$, $\chi \in \mathcal{H}_h$ and $\rho \in \mathcal{H}_{gh}$ for $g, h \in G$. The Locality phase $\epsilon_{\psi, \chi}$ is of order dividing $|G|$ and is unity for $\phi \in \mathcal{P}_G \mathcal{H}$ and any $\psi \in \mathcal{H}_g$. This latter property implies that each twisted sector $\mathcal{V}_g$ is the intertwiner for a meromorphic representation of $\mathcal{P}_G \mathcal{V}$ as in (2.2c,d). This representation can be then further decomposed into $|G|$ irreducible representations $\mathcal{V}_g = \oplus_{j_k} \mathcal{V}_g^{\{j_k\}}$ labelled by the eigenvalues $\{\exp 2\pi i j_k / n_k\}$ of the generators $\{g_k\}$ for G. Thus, in the language of the last section, we have a set of $|G|^2$ irreducible representations for the Rational MCFT $\mathcal{P}_G \mathcal{V}$ which together form the Dual CFT given by $(\mathcal{P}_G \mathcal{V})^* = \oplus_{g, j_k} \mathcal{V}_g^{\{j_k\}}$ with Abelian Fusion Rules : $\mathcal{V}_g^{\{i_k\}} \mathcal{V}_h^{\{j_k\}} \sim \mathcal{V}_{gh}^{\{i_k + j_k\}}$ from (3.5b). Then (2.4) is recovered with $\epsilon_S = 1$ using (3.3) where

$$S : Z(1, \mathcal{P}_G) = \text{Tr}_{\mathcal{P}_G \mathcal{H}}(q^{L_0 - 1}) \to \frac{1}{|G|} \sum_{g \in G} Z(g, 1) = \frac{1}{|G|} \text{Tr}_{(\mathcal{P}_G \mathcal{H})^*}(q^{L_0 - 1}) \tag{3.6}$$

where $(\mathcal{P}_G \mathcal{H})^* \equiv \oplus_{g \in G} \mathcal{H}_g$.

Since the operators of $\mathcal{V}_g$ are eigenvalues of g, the centraliser of $C(g|Aut(\mathcal{V}))$ has a natural extension as the automorphism group, which we denote by C_g, of the non-meromorphic algebra $\mathcal{V} \mathcal{V}_g \sim \mathcal{V}_g$. This extension depends on the g-twisted vacuum degeneracy D_g. Defining $G_n = C(g|Aut(\mathcal{V}))/\langle g \rangle$, in general one finds that $C_g = \hat{L}.G_n$ for some extension $\hat{L} = \langle g \rangle . L$ determined by the automorphism group acting on the twisted vacuum of $\mathcal{H}_g$. (Here $A.B$ denotes a group with normal subgroup A where $B = A.B/A$. If the twisted vacuum is unique ($D_g = 1$), then $C_g = \langle g \rangle \times G_n$. For each $h \in C_g$ we can then generalise (3.2) to define

$$Z(g, h) = \text{Tr}_{\mathcal{H}_g}(h q^{L_0 - 1}) \tag{3.7a}$$

$$T : Z(g, h) \to Z(g, g^{-1} h), \quad S : Z(g, h) \to Z(h, g^{-1}) \tag{3.7b}$$

which transform under T as given in (3.4) and under S by an interchange of g and h boundary conditions assuming (3.3) in general. Then for any $\gamma = \begin{pmatrix} a & b \\ c & d \end{pmatrix}$ in Γ,

$\gamma : Z(g,h) \rightarrow Z(h^{-c}g^a, h^d g^{-b})$. In particular, for all $g, h \in G$, these characters form a basis for the characters of the irreducible representations $\mathcal{V}_g^{\{j_k\}}$ of the Rational MCFT $\mathcal{P}_G \mathcal{V}$ [DVVV]. Thus, each $Z(g,h)$ is expected to be holomorphic on H [Z]. Other important properties of $Z(g,h)$ are that given charge conjugation invariance then $Z(g,h) = Z(g^{-1}, h^{-1})$ so that γ and $-\gamma$ act equally for each $\gamma \in \Gamma$. Finally, given the uniqueness of the twisted sectors, it also clear that under conjugation by any element $x \in Aut(\mathcal{V})$, with $g \rightarrow g^x = xgx^{-1}$, then $x(\mathcal{V}_g)x^{-1}$ is isomorphic to $\mathcal{V}_{g^x}$ so that $Z(g,h) = Z(g^x, h^x)$ for all $x \in Aut(\mathcal{V})$.

The construction of operators obeying (3.4) is only known in string theory-like models [DHVW,L,DGM,DM1] where the automorphism g is lifted from an automorphism of the embedding space of the string, typically a lattice automorphism. The properties of (3.5) are assumed in the physics literature [DFMS,DVVV] and are only so far understood in limited settings for vertex operator algebras [H]. The modular transformation properties (3.7b) for $Z(g,h)$ can be explicitly demonstrated in many cases [Va,DM1].

The G orbifold MCFT is now constructed from the projection $\mathcal{V}_{\text{orb}}^G = \mathcal{P}_G((\mathcal{P}_G \mathcal{V})^*)$ which has characteristic function $Z_{\text{orb}} = \sum_{g,h \in G} Z(g,h)$. In general, g may act projectively on $\mathcal{V}_g$ in (3.4) for a given $g \in G$ of order n so that g^n is a global phase. Then $\mathcal{V}_{\text{orb}}^G$ is not meromorphic and Z_{orb} is not T invariant. Such a 'global phase anomaly' is absent whenever $nE_0^g = 0 \bmod 1$ [Va] so that the operators of $\mathcal{P}_{\langle g \rangle} \mathcal{V}_g$ are of integer conformal weight. Assuming no such anomalies arise then $\mathcal{V}_{\text{orb}}^G$ is a self-dual MCFT and so $Z_{\text{orb}}(\tau) = J(\tau) + N_0^{\text{orb}}$ as in (2.5) where N_0^{orb} is the number of conformal weight 1 operators in $\mathcal{V}_{\text{orb}}^G$.

The OPA (3.5) is also preserved by the action of the dual automorphism group G^*, defined as follows. Recalling that $G = \langle g_1, ..., g_m \rangle$ with g_k of order n_k, we define g_k^* by

$$g_k^* V(\psi, z) g_k^{*-1} = e^{2\pi i r_k / n_k} V(\psi, z) \tag{3.8}$$

for each $\psi \in \mathcal{V}_g$ where $g = g_1^{r_1} ... g_m^{r_m}$. Then $G^* = \langle g_1^*, ..., g_m^* \rangle$ is clearly an automorphism group for (3.5) and is isomorphic to G. We may then consider the orbifolding of $\mathcal{V}_{\text{orb}}^G$ with respect to G^*. The G^* invariant operators of $\mathcal{V}_{\text{orb}}^G$ are $\mathcal{P}_{G^*} \cdot \mathcal{V}_{\text{orb}}^G = \mathcal{P}_G \mathcal{V}$ as before. Therefore the projection of the dual is $P_{G^*}(\mathcal{P}_G \mathcal{V})^* = \mathcal{V}$ i.e. orbifolding $\mathcal{V}_{\text{orb}}^G$ with respect to G^* reproduces $\mathcal{V}$. Thus the two self-dual MCFTs $\mathcal{V}$ and $\mathcal{V}_{\text{orb}}^G$ are placed on an equal footing with each an Abelian orbifolding of the other. Thus we have :

$$
\begin{array}{ccc}
 & (\mathcal{P}_G \mathcal{V})^* & \\
\mathcal{P}_{G^*} \swarrow & & \searrow \mathcal{P}_G \\
 & \xrightarrow{G} & \\
\mathcal{V} & \underset{G^*}{\xleftarrow{\hspace{1cm}}} & \mathcal{V}_{\text{orb}}^G \\
\mathcal{P}_G \searrow & & \swarrow \mathcal{P}_{G^*} \\
 & \mathcal{P}_G \mathcal{V} &
\end{array}
\tag{3.9}
$$

where the horizontal arrows represent an orbifolding with respect to the indicated automorphism group and the diagonal arrows are projections.

4. The Moonshine Module and Monstrous Moonshine. The FLM Moonshine module $\mathcal{V}^{\natural}$ is historically the first example of a self-dual orbifold MCFT [FLM1] and is constructed as a $\mathbb{Z}_2$ orbifolding of $\mathcal{V}^{\Lambda}$, which will denote the Leech lattice MCFT from now on. The $\mathbb{Z}_2$ automorphism r of $\mathcal{V}^{\Lambda}$ chosen is lifted from the lattice reflection $\bar{r}$ so that $\mathcal{P}_{\langle r \rangle} \mathcal{V}^{\Lambda}$ contains no conformal weight 1 operators. The r-twisted space $\mathcal{H}_r$ on the other hand has vacuum energy $E_0^r = 1/2$ (and is hence global phase anomaly free) but likewise contains no conformal weight 1 operators since $E_0^r > 0$. The resulting orbifold MCFT, $\mathcal{V}^{\natural} = \mathcal{P}_{\langle r \rangle}(\mathcal{V}^{\Lambda} \oplus \mathcal{V}_r)$, therefore has characteristic function $J(\tau)$. As shown by FLM, a symmetrisation of the vertex algebra of the 196884 conformal weight 2 operators (including·the Virasoro operator $V(\omega, z)$) forms an affine version of the 196883 dimensional Griess algebra [Gr] whose automorphism group is the Monster $\mathbb{M}$. FLM went on to show that $\mathbb{M} = Aut(\mathcal{V}^{\natural})$ [FLM1,FLM2]. Note that we can identify as in (3.8), the automorphism group $\langle r^* \rangle$ dual to $\langle r \rangle$. By considering $Aut(\mathcal{P}_{\langle r \rangle} \mathcal{V}^{\Lambda})$ and $Aut(\mathcal{P}_{\langle r \rangle} \mathcal{V}_r)$, the centraliser $C(r^*|\mathbb{M})$ can be found to be $C(r^*|\mathbb{M}) = 2_+^{1+24}.Co_1$ where Co_1 denotes the Conway simple group (i.e. the automorphism group Co_0 of Λ modulo $\bar{r}$), 2_+^{1+24} is an extra-special 2-group. Then $\mathbb{M}$ is generated by C and another involution that mixes the twisted and untwisted sectors [Gr,FLM1,FLM2]. Furthermore, $\mathcal{V}^{\natural}$ can be orbifolded with respect to $\langle r^* \rangle$ as in (3.9) to recover $\mathcal{V}^{\Lambda}$.

FLM have conjectured that $\mathcal{V}^{\natural}$ is characterised (up to isomorphism) as follows [FLM2]:

$\mathcal{V}^{\natural}$ **Uniqueness Conjecture.** $\mathcal{V}^{\natural}$ *is the unique CFT with characteristic function* $J(\tau)$.

This is stated in the context of the assumptions of Sections 1 and 2 where $Z(\tau) = J(\tau)$ is modular invariant and hence $\mathcal{V}^{\natural}$ is a self-dual C=24 MCFT. Furthermore, $\mathcal{V}^{\natural}$ forms the unique irreducible representation for itself [D2]. We will now consider briefly some evidence for this conjecture.

We may consider other possible $\mathbb{Z}_n$ orbifoldings of $\mathcal{V}^{\Lambda}$ with characteristic function $J(\tau)$ which should reproduce $\mathcal{V}^{\natural}$ according to this conjecture. In general, we can classify all automorphisms a of $\mathcal{V}^{\Lambda}$ lifted from automorphisms $\bar{a} \in Co_0$, (for which $\mathcal{V}_a$ can be explicitly constructed) so that [T2] :

(i) $\mathcal{P}_{\langle a \rangle} \mathcal{V}^{\Lambda}$ contains no conformal dimension 1 operators i.e. $\bar{a}$ is fixed point free.

(ii) $E_0^a > 0$ i.e. $\mathcal{V}_a$ contains no conformal dimension 1 operators.

(iii) $\mathcal{V}_a$ is global phase anomaly free i.e. $nE_0^a = 0$ mod 1 for $\bar{a}$ of order n.

There are 51 classes of Co_0 obeying (i) and (ii) only and 38 classes satisfying (i), (ii) and (iii). These 38 classes include 5 prime ordered cases for which $(p-1)|24$. These have been considered in much greater detail by Dong and Mason [DM2] who reconstructed $\mathcal{V}^{\natural}$ exactly for $p = 3$ and by Montague who also analysed the $p = 3$ case [M]. For each of these 38 classes, we expect that a self-dual MCFT $\mathcal{V}_{\text{orb}}^{\langle a \rangle}$ with characteristic function $J(\tau)$ exists. Furthermore, orbifolding $\mathcal{V}_{\text{orb}}^{\langle a \rangle}$ with respect to the dual group $\langle a^* \rangle$ defined as in (3.8) reproduces $\mathcal{V}^{\Lambda}$ with $\mathcal{V} = \mathcal{V}^{\Lambda}$, $G = \langle a \rangle$ and $\mathcal{V}_{\text{orb}}^G = \mathcal{V}^{\natural}$ in (3.9). By analysing $Aut(\mathcal{P}_{\langle a \rangle} \mathcal{V}_{a^k})$ for $k = 0, ..., n-1$ we can calculate explicitly the centraliser [T2]

$$C(a^*|Aut(\mathcal{V}_{\text{orb}}^{\langle a \rangle})) = \hat{L}.G_n \tag{4.1}$$

where $G_n = C(\bar{a}|\mathrm{Co_0})/\langle\bar{a}\rangle$ and $\hat{L} = n.L$ is a cyclic extension of $L = \Lambda/(1 - \bar{a})\Lambda$. For the prime ordered cases, this reduces to a well-known centraliser formula for $\mathbb{M}$ [CN]. (4.1) can also be shown to hold for all 51 classes obeying (i) and (ii) once a^* is appropriately defined and is verified for $Aut(\mathcal{V}_{\mathrm{orb}}^{\langle a \rangle}) = \mathbb{M}$ in many cases [T2]. All of this provides evidence that $\mathcal{V}_{\mathrm{orb}}^{\langle a \rangle} = \mathcal{V}^\natural$ in each construction lending weight to the uniqueness conjecture. Further evidence is given below.

Let us now define the Thompson-McKay series $T_g(\tau)$ for each $g \in \mathbb{M}$

$$T_g(\tau) = \mathrm{Tr}_{\mathcal{H}^\natural}(gq^{L_0-1}) = \frac{1}{q} + 0 + [1 + \chi_A(g)]q + ... \qquad (4.2)$$

where $\chi_A(g)$ is the character of the 196883 dimensional adjoint representation for $\mathbb{M}$. This trace is obviously reminscent of (3.2) and this interpretation will be further explored below. The Thompson series for the identity element is $J(\tau)$, which is the hauptmodul for the genus zero modular group $\Gamma = \mathrm{SL}(2, \mathbb{Z})$ as already stated. By calculating the first ten terms of $T_g(\tau)$ for each conjugacy class of $\mathbb{M}$, Conway and Norton [CN] conjectured

Monstrous Moonshine. *For each $g \in \mathbb{M}$, $T_g(\tau)$ is the hauptmodul for a genus zero fixing modular group Γ_g.*

Borcherds has now demonstrated this rigorously although the origin of the genus zero property remains obscure [B2]. In general, for g of order n, $T_g(\tau)$ is found to be invariant under $\Gamma_0(n) = \{\begin{pmatrix} a & b \\ nc & d \end{pmatrix} | \det = 1\}$ up to h^{th} roots of unity where $h|n$ and $h|24$. $T_g(\tau)$ is fixed by Γ_g with $\Gamma_0(N) \subseteq \Gamma_g$ and contained in the normaliser of $\Gamma_0(N)$ in $\mathrm{SL}(2, \mathbb{R})$ where $N = nh$ [CN]. This normaliser always contains the Fricke involution $W_N : \tau \to -1/N\tau$ where $W_N^2 = 1$ mod $\Gamma_0(N)$. We will refer to those classes with $h = 1$ as *Normal* and those with $h \neq 1$ as *Anomalous* i.e. the fixing group of $T_g(\tau)$ is of type $n + e_1, e_2, ...$ for normal classes and of type $n|h + e_1, e_2, ...$ for anomalous classes in the notation of [CN]. This terminology is motivated by whether the corresponding twisted sector $\mathcal{V}_g$ described below has a global phase anomaly or not.

For a normal element $g \in \mathbb{M}$ of prime order p (there is only one anomalous prime class of order 3 with $h = 3$) we find either $\Gamma_g = \Gamma_0(p)$ or $\Gamma_0(p)+ = \langle\Gamma_0(p), W_p\rangle$. $\Gamma_0(p)$ is of genus zero only when $(p - 1)|24$. There is a corresponding class of $\mathbb{M}$, denoted by $p-$, for each such prime with this Thompson series e.g. the involution r^* above belongs to the class $2-$. $\Gamma_0(p)+$ is of genus zero for all the prime divisors of the order of $\mathbb{M}$. There is a class of $\mathbb{M}$, denoted by $p+$, for each such prime with Thompson series fixed by $\Gamma_0(p)+$. In general all the classes of $\mathbb{M}$ can be divided into Fricke and non-Fricke classes according to whether or not $T_g(\tau)$ is invariant under the Fricke involution W_N. It is also observed that the Thompson series for Fricke classes have non-negative integer coefficients whereas the coefficients of non-Fricke Thompson series are integers of mixed sign. There are a total of 51 non-Fricke classes of which 38 are normal and there are a total of 120 Fricke classes of which 82 are normal.

For each of the 38 constructions above based on classes $\{\bar{a}\}$ satisfying the conditions (i)-(iii) we can compute the dual automorphism Thompson series T_{a^*} and this agrees precisely with the genus zero series for the 38 non-Fricke normal classes of the Monster which also obey the centraliser relationship (4.1). Likewise, we can identify the other 13 anomalous non-Fricke classes and find the corresponding correct genus zero Thompson series [T2]. This is further evidence for the assertion that $\mathcal{V}_{\text{orb}}^{\langle a \rangle} = \mathcal{V}^{\natural}$ implied by the uniqueness conjecture for $\mathcal{V}^{\natural}$ which we will now assume to be true from now on.

We now turn to the interpretation of a Thompson series as an orbifold trace with $T_g(\tau) = Z(1, g)$ as in (3.2) where now $\mathcal{V} = \mathcal{V}^{\natural}$. For g in a non-Fricke class, we can construct the twisted sector $\mathcal{V}_g$ by choosing $g = a^*$ as above with characteristic function obeying (3.3). In particular, all the coefficients of $Z(g, 1)$ are non-negative integers and hence $Z(1, g) - T_g(0)$, which is inverted up to a multiplicative constant under the Fricke involution to give $Z(g, 1)(N\tau) - T_g(0)$, has mixed sign coefficients as observed. For the 38 normal non-Fricke classes we may orbifold $\mathcal{V}^{\natural}$ with respect to $\langle a^* \rangle$ to obtain $\mathcal{V}^{\Lambda}$. Then the vacuum energy $E_0^g = 0$ for the twisted sector $\mathcal{V}_g$ so that conformal weight 1 operators are reintroduced. On the other hand, for an anomalous non-Fricke class, a global phase anomaly parameterised by the parameter $h \neq 1$ occurs and we cannot obtain a MCFT from the resulting orbifolding [T1].

Consider next $f \in \mathbb{M}$, a Fricke element of order n. The corresponding twisted sector can be constructed when f is lifted from a lattice automorphism. We will assume that $\mathcal{V}_f$ exists in each case obeying (3.3)-(3.5). For normal elements, no global phase anomaly occurs and we may orbifold $\mathcal{V}^{\natural}$ with respect to $\langle f \rangle$ to obtain a self-dual MCFT $\mathcal{V}_{\text{orb}}^{\langle f \rangle}$. Assuming $T_f(\tau)$ is a hauptmodul we then find that $\mathcal{V}_{\text{orb}}^{\langle f \rangle} = \mathcal{V}^{\natural}$ for each normal Fricke element. The converse is also true, where given that $\mathcal{V}_{\text{orb}}^{\langle f \rangle} = \mathcal{V}^{\natural}$ for some $f \in \mathbb{M}$ then T_f is the hauptmodul for a genus zero modular group containing the Fricke involution [T2]. In general, we find (assuming the uniqueness conjecture for $\mathcal{V}^{\natural}$) that for all normal elements of $\mathbb{M}$

$$\mathcal{V}^{\Lambda} \underset{\langle a^* \rangle}{\overset{\langle a \rangle}{\rightleftarrows}} \mathcal{V}^{\natural} \overset{\langle f \rangle}{\longleftrightarrow} \mathcal{V}^{\natural} \quad \Leftrightarrow \quad T_{a^*}, T_f \text{ are hauptmoduls} \qquad (4.3)$$

(4.3) can be understood briefly for the prime ordered normal Fricke classes as follows. Suppose that f is a $p+$ element with Fricke invariant hautpmodul $T_f(\tau)$. Then $Z(f, 1)(\tau) = Z(1, f)(\tau/p) = q^{-1/p} + 0 + \dots$ so that $\mathcal{V}_f$ has vacuum energy $E_0^f = -1/p$, degeneracy $D_f = 1$, contains no conformal weight 1 operators and has non-negative integer coefficients. Thus $\mathcal{P}_{\langle f \rangle} \mathcal{V}_f$ does not reintroduce conformal weight 1 operators. Similarly $\mathcal{P}_{\langle f \rangle} \mathcal{V}_{f^k}$, $k \neq 0 \bmod p$, contains no such operators (f and f^k are conjugate) so that $\mathcal{V}_{\text{orb}}^{\langle f \rangle} = \mathcal{V}^{\natural}$ since the characteristic function is $J(\tau)$.

Conversely, if $\mathcal{V}_{\text{orb}}^{\langle f \rangle} = \mathcal{V}^{\natural}$ for a prime p ordered element f then since f and f^k are conjugate, $T_f(\tau)$ is automatically $\Gamma_0(p)$ invariant. The fundamental region $H/\Gamma_0(p)$ for $\Gamma_0(p)$ has only two cusp points [Gu] at $\tau = \infty$ where $T_f(\tau)$ has a simple pole of order 1 from (2.2) and at $\tau = 0$ which is singular iff $E_0^f < 0$ with residue D_f from (3.3). We can then argue that since $\mathcal{V}_{\text{orb}}^{\langle f \rangle} = \mathcal{V}^{\natural}$, $E_0^f = -1/p$ with $D_f = 1$. This follows by considering the dual automorphism $f^* \in \mathbb{M}$ to f as in (3.9) and showing that $T_{f^*} = T_f$. Then the corresponding centralisers must be equal which implies that $D_f = 1$, since no extension occurs. Furthermore, $\mathcal{V}_f$ contains no conformal weight 1 operators which implies that either $E_0^f = -1/p$ or $E_0^f > 0$. The latter possibility is ruled out because then $T_f(\tau) = q^{-1} + 0 + O(q)$ would be a hauptmodul for $\Gamma_0(p)$ which implies $E_0^f = 0$ when the constant term of T_f is zero. Thus we must have $E_0^f = -1/p$ with $D_f = 1$. Finally, consider $\phi(\tau) = T_f(\tau) - T_f(W_p(\tau))$ which is $\Gamma_0(p)$ invariant and is holomorphic on the compactification of $H/\Gamma_0(p)$, which is a compact Riemann surface. Hence $\phi(\tau)$ is a constant which is zero since it is odd under W_p. Hence $T_f(\tau)$ is $\Gamma_0(p)+$ invariant and has a unique simple pole on $H/\Gamma_0(p)+$ and is therefore a hauptmodul for $\Gamma_0(p)+$.

This argument can be generalised to any normal Fricke element $f \in \mathbb{M}$ of order n. Then (4.3) is equivalant to the fact that (i) $\mathcal{V}_f$ has vacuum energy $E_0^f = -1/n$ and degeneracy $D_f = 1$ and (ii) if f^r is Fricke then so is f^s with $s = n/(r,n)$ where r and s must be co-prime and where we must also assume that each n contains at most 3 prime divisors (i.e. $n < 2.3.5.7 = 210$) ! These conditions are then sufficient to supply all the poles and residues of $T_f(\tau)$ so that $T_f(\tau)$ is a hauptmodul for some genus zero fixing group [T1,T2]. Finally, the genus zero property for an anomolous class of $\mathbb{M}$, which follows from the Harmonic formula of [CN], is described in [T2].

5. Generalised Moonshine from Abelian Orbifolds.
Let us now consider the more general set of conjectures suggested by Norton [N] concerning Moonshine for centralisers (or extensions thereof) of elements of the Monster $\mathbb{M}$. Specifically, in the notation of (3.7a) we consider:

$$Z(g,h) = \operatorname{Tr}_{\mathcal{H}_g}(h q^{L_0 - 1}) \tag{5.1}$$

for $h \in C_g \equiv Aut(\mathcal{V}_g)$. For all Fricke elements, the twisted Hilbert space vacuum, which we now denote by $\mathcal{H}_g|_0$, is unique and hence $C_g = \langle g \rangle \times G_n$ where $G_n = C(g|\mathbb{M})/\langle g \rangle$ whereas for g non-Fricke $C_g = \hat{K}.G_n$ (for some extension $\hat{K}$). Norton has conjectured :

Generalised Moonshine Conjecture. $Z(g,h)$ is either constant or is a hauptmodul for some genus zero fixing group for every pair of commuting elements $g, h \in \mathbb{M}$.

This conjecture has been explicitly verified for an orbifold construction based on the Mathieu group M_{24} [DM1]. In terms of the orbifold picture reviewed in the earlier sections we can note the following properties for $Z(g,h)$:

(i) $Z(g,h) = Z(g^{-1}, h^{-1})$.

(ii) $Z(g,h) = Z(g^x, h^x)$ for conjugation with respect to any $x \in \mathbb{M}$.

(iii) $S : T_g(\tau) \to Z(g,1) = D_g q^{E_0^g} + \ldots$ is a series with non-negative integer coefficients decomposable into positive sums of the dimensions of the irreducible representations of C_g where E_0^g is twisted vacuum energy and D_g is the vacuum degeneracy, the dimension of $\mathcal{H}_g|_0$, the twisted Hilbert space vacuum.

(iv) From (3.7) we find that $\gamma : Z(g,h) \to Z(h^{-c}g^a, h^d g^{-b})$ for $\gamma = \begin{pmatrix} a & b \\ c & d \end{pmatrix} \in \Gamma$. Note that $g = \exp(-2\pi i E_0^g)$ on $\mathcal{H}_g|_0$. In particular, for a normal Fricke element of order m, $g = \omega = e^{2\pi i/m}$ and each $h \in C_g$ acts as some element of $\langle \omega \rangle$ on $\mathcal{H}_g|_0$.

(v) As a consequence of (iv), $Z(g,h)$ is invariant up to roots of unity under $\Gamma(m,n) = \{\gamma \in \Gamma | a = 1 \bmod m, b = 0 \bmod m, c = 0 \bmod n, d = 1 \bmod n\}$ where

$m = o(g)$, $n = o(h)$. These extra factors appear if $h^{-c}g^{a}$ is anomalous for some co-prime a and c.

(vi) The value of $Z(g,h)$ at any parabolic cusp a/c (a and c co-prime) is determined by the vacuum energy of the $k = g^{a}h^{-c}$ twisted sector from (iv). In particular, only the Fricke classes are responsible for singular cusp points [N]. The residue of these cusps is determined by the action of $h^{d}g^{-b}$ on $\mathcal{H}_{k}|_{0}$. We will assume, as discussed in section 3, that $Z(g,h)$ is holomorphic at all other points on H.

Thus given any commuting pair of elements g, h as above, the location of any singularities for $Z(g,h)$ is known by finding which of the classes $k = g^{a}h^{-c}$ is Fricke for $(a,c) = 1$. The strength of the pole is then determined by the corresponding vacuum energy E_{0}^{k}. However, the residue for each singular cusp still needs to be found. We will argue below that this extra information is also supplied by the constraints of (4.3), at least in the simplest non-trivial prime cases. Once these singularities are known, then $Z(g,h)$ can be shown in each case to be either constant or to be the hauptmodul for a genus zero modular group.

The basic idea is to consider the orbifolding of $V^{\natural}$ with respect to $\langle g, h\rangle$ and to re-express this as the composition of two $\mathbb{Z}_{n}$ orbifoldings. If $\langle g,h\rangle = \mathbb{Z}_{l}$, $l = mn/(m,n)$, then $Z(g,h)$ can always be related to a regular Thompson series via an appropriate modular transformation e.g. for m, n co-prime with $am + bn = 1$ then $\langle g,h\rangle = \mathbb{Z}_{mn}$ and $\begin{pmatrix} 1 & 1 \\ -bn & am \end{pmatrix} : Z(1, gh) \to Z(g,h)$ from (iv). For $\langle g,h\rangle \neq \mathbb{Z}_{l}$ we will consider here the simplest non-trivial case where $\langle g,h\rangle$ contains only normal prime order p elements. Then $Z(g,h)$ is $\Gamma(p) \equiv \Gamma(p,p)$ invariant from (v). We will further assume that $h \overset{C_g}{\sim} h^{a}$ for $a \neq 0 \bmod p$ i.e. conjugate in C_{g}. This is sufficient to ensure that the coefficients in the q expansion of $Z(g,h)$ are rational since all the irreducible characters are rational. Furthermore, this condition restricts the possible conjugacy classes in $\mathbb{M}$ generated by g and h to just three i.e. $g \overset{\mathbb{M}}{\sim} g^{a}$, $h \overset{\mathbb{M}}{\sim} h^{b}$ and $gh \overset{\mathbb{M}}{\sim} g^{a}h^{b}$ for $a, b \neq 0 \bmod p$. From (iv), we have that $Z(g,h)$ is fixed by $\Gamma_{0}^{0}(p) = \{\gamma \in \Gamma | b = c = 0 \bmod p\} \sim \Gamma_{0}(p^{2})$ (under conjugation by $\mathrm{diag}(1,p)$ so that $Z(g,h)(p\tau)$ is $\Gamma_{0}(p^{2})$ invariant). $Z(g,h)$ therefore has parabolic cusps on $H/\Gamma_{0}^{0}(p)$ at $\tau = i\infty, 0, 1, ..., p-1$ [Gu] with behaviour determined, from (vi), by the vacuum energy of the sectors twisted by $g, h, g^{p-1}h, ...g^{2}h, gh$ respectively where the last $p - 1$ classes are conjugate. Within these assumptions we then find that there are 5 possible cases (up to relabelling) that may occur for any p as follows.

Case 1 : $g, h, gh = p-$. None of the cusps are singular and therefore $Z(g,h)$ is holomorphic on $H/\Gamma_{0}^{0}(p)$ and hence is constant.

We may now assume for the remaining 4 cases (without loss of generality by relabelling) that $g = p+$ so that

$$Z(g,h) = q^{-1/p} + 0 + O(q^{1/p}) \tag{5.2}$$

We also note from (iv) that g acts as $\omega = e^{2\pi i/p}$ on $\mathcal{H}_{g}|_{0}$.

Case 2 : $g = p+$, $h, gh = p-$. In this case $Z(g,h)$ has a unique simple pole at $q = 0$ as in (5.2) on $H/\Gamma_{0}^{0}(p)$ and therefore $Z(g,h)$ is a hauptmodul. This is only possible for $p = 2$, 3, 5 (where $Z(g,h)(p\tau)$ is a hauptmodul for $\Gamma_{0}(p^{2})$). For $p = 5$, no such Generalised Moonshine function is actually observed which, interestingly, is also the case for regular Monstrous Moonshine where $25-$ is one of the so-called ghost elements [CN].

Case 3 : $g, h = p+$, $gh = p-$. $Z(g, h)$ has two singularities at $\tau = i\infty$ and 0. Under $S : \tau \to -1/\tau$ we have

$$Z(g, h) \to Z(h, g^{-1}) = \omega^{-k_g} q^{-1/p} + 0 + O(q^{1/p}) \tag{5.3}$$

where $g = \omega^{k_g}$ on $\mathcal{H}_g|_0$, $k_g \in \mathbb{Z}_p$. We may conjugate h to h^{-1} in C_g so that $Z(h, g^{-1}) = Z(h^{-1}, g^{-1}) = Z(h, g)$, from (i), which implies that $2k_g = 0 \bmod 2$. Hence for $p > 2$, $k_g = 0 \bmod p$. For $p = 2$ we will show below that $k_g = 0 \bmod 2$ also. Consider $f(\tau) = Z(g, h)(\tau) - Z(g, h)(S(\tau)) = 0 + O(q^{1/p})$. $f(\tau)$ is $\Gamma_0^0(p)$ invariant without any poles on $H/\Gamma_0^0(p)$ and hence is constant and equal to zero. Therefore, $Z(g, h)$ is $\langle \Gamma_0^0(p), S \rangle \sim \Gamma_0(p^2)+$ invariant with a unique simple pole and is a hauptmodul. This is only possible for $p = 2, 3, 5, 7$. For $p = 7$, no such Generalised Moonshine function is observed which corresponds to the ghost element $49+$ of Monstrous Moonshine !

To understand the $p = 2$ case it is necessary to consider the interpretation of $Z(g, h)$ in terms of a $\langle g, h \rangle = \mathbb{Z}_2 \times \mathbb{Z}_2$ orbifolding of $\mathcal{V}^\natural$. The orbifold so obtained is meromorphic self-dual (since no anomalous Monster elements occur) and is explicitly

$$\mathcal{V}_{\mathrm{orb}}^{\langle g, h \rangle} = \mathcal{P}_{\langle g, h \rangle}(\mathcal{V}^\natural \oplus \mathcal{V}_g \oplus \mathcal{V}_h \oplus \mathcal{V}_{gh}) \tag{5.4}$$

where $\mathcal{P}_{\langle g, h \rangle} = (1 + g + h + gh)/4 = \mathcal{P}_{\langle g \rangle} \mathcal{P}_{\langle h \rangle}$. We can consider this as two successive $\mathbb{Z}_2$ orbifoldings

$$\mathcal{V}_{\mathrm{orb}}^{\langle g, h \rangle} = \mathcal{P}_{\langle g \rangle}(\mathcal{P}_{\langle h \rangle}(\mathcal{V}^\natural \oplus \mathcal{V}_h) \oplus \mathcal{P}_{\langle h \rangle}(\mathcal{V}_g \oplus \mathcal{V}_{gh})) \tag{5.5}$$

i.e. $\mathcal{V}_{\mathrm{orb}}^{\langle g, h \rangle}$ is a Z_2 orbifolding with respect to $\langle g \rangle$ of $\mathcal{V}_{\mathrm{orb}}^{\langle h \rangle} \equiv \mathcal{P}_{\langle h \rangle}(\mathcal{V}^\natural \oplus \mathcal{V}_h)$. Since $h = 2+$, we know that $\mathcal{V}_{\mathrm{orb}}^{\langle h \rangle} = \mathcal{V}^\natural$ and hence $\mathcal{P}_{\langle h \rangle}(\mathcal{V}_g \oplus \mathcal{V}_{gh})$ is a g twisted sector for $\mathcal{V}^\natural$ for g of order two by the assumed uniqueness of the twisted sectors. Thus $g = 2+$ or $2-$ when acting on $\mathcal{V}_{\mathrm{orb}}^{\langle h \rangle}$. However, we can determine from (5.5) that the character for this g twisted sector is $[Z(g, 1) + Z(g, h) + Z(gh, 1) + Z(gh, h)]/2 = q^{-1/2} + ...$ using (5.2). This implies that g is Fricke when acting on $\mathcal{V}_{\mathrm{orb}}^{\langle h \rangle}$ and hence $\mathcal{V}_{\mathrm{orb}}^{\langle g, h \rangle} = \mathcal{V}^\natural$ with $g = 2+$. We can represent this sequence of orbifoldings diagramatically as follows:

$$
\begin{array}{ccccc}
 & & \mathcal{V}^\natural & & \\
 & \overset{\langle h \rangle}{\nearrow} & & \overset{\langle g \rangle}{\searrow} & \\
\mathcal{V}^\natural & & \underset{\langle g, h \rangle}{\longrightarrow} & & \mathcal{V}^\natural
\end{array}
\tag{5.6}
$$

where each copy of $\mathcal{V}^\natural$ is orbifolded with respect to the denoted group.

We can similarly consider the orbifolding of $\mathcal{V}^\natural$ with respect to $g = 2+$ followed by h. The resulting orbifold must be $\mathcal{V}_{\mathrm{orb}}^{\langle g, h \rangle} = \mathcal{V}^\natural$ and hence $\mathcal{P}_{\langle g \rangle}(\mathcal{V}_h \oplus \mathcal{V}_{gh})$ must be a $2+$ twisted sector. This forces $g = 1$ on $\mathcal{H}_h|_0$ as was claimed earlier. In general, for any p, it is straightforward to see that $\mathcal{V}^\natural \overset{\langle g \rangle}{\longrightarrow} \mathcal{V}_{\mathrm{orb}}^{\langle g \rangle} = \mathcal{V}^\natural \overset{\langle h \rangle}{\longrightarrow} \mathcal{V}_{\mathrm{orb}}^{\langle g, h \rangle} = \mathcal{V}^\natural$ in this case.

Case 4 : $g, gh = p+$, $h = p-$. In this case $Z(g,h)$ has singular cusps at $i\infty, 1, ..., p-1$ on $H/\Gamma_0^0(p)$. We can find the residues of these poles by decomposing the orbifolding with respect to $\langle g,h \rangle$ to obtain $\mathcal{V}^\natural \xrightarrow{\langle g \rangle} \mathcal{V}^\natural \xrightarrow{\langle h \rangle} \mathcal{V}_{\text{orb}}^{\langle g,h \rangle}$ since $g = p+$ where we necessarily find that either $\mathcal{V}_{\text{orb}}^{\langle g,h \rangle} = \mathcal{V}^\natural$ or $\mathcal{V}^\Lambda$ from (4.3). If we alternatively orbifold with respect to $h = p-$ first we then obtain $\mathcal{V}^\natural \xrightarrow{\langle h \rangle} \mathcal{V}^\Lambda \xrightarrow{\langle g \rangle} \mathcal{V}_{\text{orb}}^{\langle g,h \rangle}$. In order that $\mathcal{V}_{\text{orb}}^{\langle g,h \rangle} = \mathcal{V}^\natural$, it is necessary that the g twisted sector of $\mathcal{V}^\Lambda$ so obtained, $\mathcal{P}_{\langle h \rangle}(\oplus_{k=0}^{p-1} \mathcal{V}_{gh^k})$, has positive vacuum energy from condition (ii) of section 4. However, from (5.2), this is impossible since $\mathcal{P}_{\langle h \rangle} \mathcal{V}_g$ has character $q^{-1/p} + 0 + O(q^{1/p})$. Hence $\mathcal{V}_{\text{orb}}^{\langle g,h \rangle} = \mathcal{V}^\Lambda$ in this case.

We can similarly decompose $\mathcal{P}_{\langle g,h \rangle} = \mathcal{P}_{\langle gh \rangle}\mathcal{P}_{\langle f \rangle}$ for $f = g^a h$ a $p+$ element with $a \neq 0, 1 \mod p$. Then $\mathcal{V}^\natural \xrightarrow{\langle f \rangle} \mathcal{V}^\natural \xrightarrow{\langle gh \rangle} \mathcal{V}_{\text{orb}}^{\langle g,h \rangle} = \mathcal{V}^\Lambda$. This implies that gh must act as a $p-$ element on $\mathcal{V}_{\text{orb}}^{\langle f \rangle} = \mathcal{V}^\natural$ and hence the corresponding gh twisted sector $\mathcal{P}_f(\oplus_{k=0}^{p-1} \mathcal{V}_{ghf^k})$ has zero vacuum energy from (4.3). In particular, this implies that $\mathcal{P}_f \mathcal{H}_{gh}|_0 = 0$ so that $f = g^a h \neq 1$ on $\mathcal{H}_{gh}|_0$ for any $a \neq 0 \mod p$ (noting that $gh = \omega$ on $\mathcal{H}_{gh}|_0$). Let $h = \omega^r$ be the action on $\mathcal{H}_{gh}|_0$ (from (iv)) so that $g = \omega^{1-r}$. But we can always choose $a \neq 0 \mod p$ such that $g^a h$ acts as unity on $\mathcal{H}_{gh}|_0$ unless $r = 0 \mod p$. Hence the orbifolding is only consistent when $h = 1$ on $\mathcal{H}_{gh}|_0$. In general, by conjugation, we then find that $h = 1$ on $\mathcal{H}_{g^a h^b}|_0$ for all $b \neq 0 \mod p$. Hence, the residue of any of the singular cusps is known. This allows us to find the full fixing modular group.

Let $\gamma_p = \begin{pmatrix} 1 & -p \\ 1 & 1-p \end{pmatrix}$ which is of order p in $\Gamma_0^0(p)$. γ_p permutes the p cusps of $Z(g,h)$ where $\gamma_p : Z(g,h) \to Z(gh, h^{-1}) = q^{-1/p} + 0 + O(q^{1/p})$ and similarly for the other singular twisted sectors. Then $f(\tau) = Z(g,h)(\tau) - Z(g,h)(\gamma_p(\tau))$ is holomorphic on $H/\Gamma(p)$ and is therefore zero. Hence $Z(g,h)$ is $\langle \Gamma_0^0(p), \gamma_p \rangle \sim \Gamma_0(p)$ invariant with a unique simple pole and is a hauptmodul. This is only possible for $p = 2, 3, 5, 7, 13$. Once again, the largest possible case is not observed, $p = 13$, although this does not correspond to a ghost element for regular Moonshine.

Case 5 : $g, h, gh = p+$. In this last case all sectors are Fricke and $Z(g,h)$ has singular cusps at $i\infty, 0, 1, ..., p-1$. With the assumption that all $h \overset{C_g}{\sim} h^a$ for $a \neq 0 \mod p$ we need only in practice consider $p = 2, 3$ and 5 where $C_g = \langle g \rangle \times G_p$ for $G_p = B, Fi'_{24}$ or HN respectively [CCNPW]. Following a general argument as in Case 3, it is easy to see again that $\mathcal{V}_{\text{orb}}^{\langle g,h \rangle} = \mathcal{V}^\natural$ since both g and h are Fricke and (5.2) is obeyed.

For $p = 2$ we again decompose the orbifolding of $\mathcal{V}^\natural$ with respect to $\langle g,h \rangle$. Referring to (5.5), we note that $\mathcal{P}_{\langle h \rangle}(\mathcal{V}_g \oplus \mathcal{V}_{gh})$ has a $2+$ character and that gh is Fricke. Hence $h = -1$ and $g = 1$ on $\mathcal{H}_{gh}|_0$. Similarly, we can orbifold with respect to g first and find that $\mathcal{P}_{\langle g \rangle}(\mathcal{V}_h \oplus \mathcal{V}_{gh})$ also has $2+$ character. Hence $g = -1$ on $\mathcal{H}_h|_0$. Thus all the residues of $Z(g,h)$ are known. In particular, ST of order three permutes the cusps $\{i\infty, 0, 1\}$ with $ST : Z(g,h) \to Z(gh, g) = q^{-1/2} + 0 + O(q^{-1/2})$. Then, by the usual argument, $Z(g,h)$ is a hauptmodul for $\langle \Gamma(2), ST \rangle$ of genus zero, which is of level 2 and index 2 in Γ. In fact, $Z(g,h)$ is invariant under the full modular group Γ up to ± 1 with $Z(g,h) = -Z(g,gh)$ so that $Z(g,h)(2\tau) = E_2(\tau)/\eta^{12}(\tau) - 252$ is the hauptmodul for $2|2$ in the notation of [CN].

For $p = 3, 5$ we may repeat the argument of Case 3 to show that $g = 1$ on $\mathcal{H}_h|_0$ so that $Z(h, g) = q^{-1/p} + 0 + ...O(q^{1/p})$. But $Z(g, h)$ and $Z(h, g)$ have the same singular structure and hence we may interchange g and h. Since $\mathcal{V}_{gh}$ is preserved by this interchange, $g = h$ on $\mathcal{H}_{gh}|_0$ with $gh = \omega$ so that $g = h = \omega^{(p+1)/2}$. Hence by conjugation as in (ii), all the residues of the singular cusps of $Z(g, h)$ are known.

For $p = 3$, let $\gamma_2 = T^{-1}ST$ which is of order 2 and interchanges the cusps $\{\infty, 0\} \leftrightarrow \{2, 1\}$ whereas S interchanges $\{\infty, 1\} \leftrightarrow \{0, 2\}$. Then $\gamma_2 : Z(g, h) \rightarrow Z(gh, h^{-1}g) = q^{-1/3} + 0 + ...$ and $S : Z(g, h) \rightarrow Z(h, g^{-1}) = q^{-1/3} + 0 + ...$ and similarly for the other cusps. By the usual argument, we then find that $Z(g, h)$ is the hauptmodul for the genus zero group $\langle \Gamma(3), S, T^{-1}ST \rangle$ of level 3 and index 3 in Γ. Further analysis shows that $Z(g, h)$ is invariant under Γ up to third roots of unity with $Z(g, h) = \omega^2 Z(g, gh) = \omega Z(g, g^2 h)$, so that $Z(g, h)(3\tau) = E_3(\tau)/\eta^8(\tau) - 368$ is the hauptmodul for $3|3$ in the notation of [CN].

For $p = 5$, let $\gamma_3 = TST^3$ which is of order 3 and cyclically permutes the cusps $\{\infty, 0\} \rightarrow \{1, 4\} \rightarrow \{2, 3\}$ whereas S interchanges the cusps $\{\infty, 1, 2\} \leftrightarrow \{0, 4, 3\}$. Then $\gamma_3 : Z(g, h) \rightarrow Z(gh^{-1}, g^3 h^3) = Z(gh, (gh^{-1})^3) = q^{-1/5} + 0 + ...$ by conjugation and similarly for the other cusps. $Z(g, h)$ is invariant under $\Gamma(5)$ whose normaliser contains Γ and so $Z(g, h)(\tau) - Z(g, h)(\gamma(\tau))$ is holomorphic on $H/\Gamma(5)$ for both $\gamma = S$ and γ_3 and hence is zero. Thus $Z(g, h)$ is the hauptmodul for the genus zero group $\langle \Gamma_0^0(5), S, \gamma_3 \rangle$ which is of level 5 and index 5 in Γ. In the notation of [FMN], the fixing group of $Z(g, h)(5\tau)$ is $5||5$. In this case, there are five independent functions $f_k(\tau) = Z(g, g^k h), k = 0, 1, ..., 4$ which are permuted under Γ.

We summarise Cases 2-5 in the following table where we reproduce the genus zero fixing group for $Z(g, h)(p\tau)$ with $g = p+$ and where only the actual observed values of p are indicated.

	$gh = p-$	$gh = p+$
$h = p-$	$\Gamma_0(p^2),\ p = 2, 3$	$\Gamma_0(p)-,\ p = 2, 3, 5, 7$
$h = p+$	$\Gamma_0(p^2)+,\ p = 2, 3, 5$	$2|2,\ 3|3,\ 5||5$

6. Conclusion. We have shown that the genus zero property for the Generalised Moonshine functions (5.1) follows from the genus zero property for Thompson series in the simplest non-trivial prime cases. It remains a much greater challenge to extend these arguments to all cases. The major difficulties of this method for general commuting elements g, h are (i) the proliferation of possible Fricke classes in $\langle g, h \rangle$ giving the location of poles and (ii) the determination of the corresponding residues. Once this information is known, then any generalised moonshine function should be reconstructible if it is a hauptmodul. Finally, it is interesting that the ghost groups $25-$ and $49+$ are absent from the above table (as indeed is $50 + 50$ from the list of modular groups for the centraliser moonshine of the $5+$ or $10+$ elements of $\mathbb{M}$ where it might be expected to arise). These hauptmoduls are also distinguished by having non-quadratic irrationalities at their non-singular cusps [FMN] suggesting some possibly deeper number theoretic significance for their absence.

REFERENCES

[B1]   R. Borcherds,   *Vertex algebras, Kac-Moody algebras and the monster* ,

Proc.Natl.Acad.Sc.USA **83** (1986), 3068–3071.

[B2]   R. Borcherds, *Monstrous moonshine and monstrous Lie superalgebras*, Invent.Math. **109** (1992), 405–444.

[BPZ]   A. Belavin, A. M. Polyakov and A. B. Zamolodchikov, *Infinite conformal symmetry and two dimensional quantum field theory*, Nucl.Phys. **B241** (1984), 333–380.

[CCNPW]   J. H. Conway, R. T. Curtis, S. P. Norton, R. A. Parker and R. A. Wilson, An atlas of finite groups, Clarendon Press, Oxford, 1985.

[CN]   J. H. Conway and S. P. Norton, *Monstrous Moonshine*, Bull.London.Math.Soc. **11** (1979), 308–339.

[DFMS]   L. Dixon, D. Friedan, E. Martinec and S. Shenker, *The conformal field theory of orbifolds*, Nucl.Phys. **B282** (1987), 13–73.

[DGM]   L. Dolan, P. Goddard and P. Montague, *Conformal field theory of twisted vertex operators*, Nucl.Phys. **B338** (1990), 529–601.

[DHVW]   L. Dixon, J. A. Harvey, C. C. Vafa and E. Witten, *Strings on orbifolds*, Nucl.Phys. **B261** (1985), 678-686; *Strings on orbifoids II*, Nucl.Phys. **B274** (1986), 285–314.

[D1]   C. Dong, *Vertex algebras associated with even lattices*, J.Algebra **161** (1993 pages 245–265).

[D2]   C. Dong, *Representations of the moonshine vertex operator module*, Univ.Cal.Santa Cruz preprint (1992).

[DL]   C. Dong and J. Lepowsky, *Abelian intertwiner algebras - a generalization of vertex operator algebras*, Preprint (1993).

[DM1]   C. Dong and G. Mason, *An orbifold theory of genus zero associated to the sporadic group M_{24}*, Commun.Math.Phys. **164** (1994), 87–104.

[DM2]   C. Dong and G. Mason, *On the construction of the moonshine module as a Z_p orbifold*, U.C.Santa Cruz Preprint (1992).

[DVVV]   R. Dijkgraaf, C. Vafa, E. Verlinde and H. Verlinde, *The operator algebra of orbifold models*, Commun.Math.Phys. **123** (1989), 485-526.

[FHL]   I. Frenkel, Y.-Z. Huang and and J. Lepowsky, *On axiomatic approaches to vertex operator algebras and modules*, Mem.A.M.S. **104** (1993).

[FLM1]   I. Frenkel, J. Lepowsky and A. Meurman, *A natural representation of the Fischer-Griess monster with the modular function J as character*, Proc.Natl.Acad.Sci.USA **81** (1984), 3256–3260.

[FLM2]   _____ , Vertex operator algebras and the monster, Academic Press, New York, 1988.

[FMN]   D. Ford, J. MacKay and S. P. Norton, *More on replicable functions*, Preprint (1993).

[Go]   P. Goddard, *Meromorphic conformal field theory*, Proceedings of the CIRM Luminy conference 1988, World Scientific, Singapore, 1989, p. 556.

[Gr]   R. Griess, *The friendly giant*, Inv.Math. **68** (1982), 1-102.

[Gu]   R. C. Gunning, Lectures on modular forms, Princeton University Press, Princeton, 1962.

[H]   Y.-Z. Huang, *A non-meromorphic extension of the moonshine vertex operator algebra*, Rutgers Univ. preprint and these proceedings (1994).

[L]   J. Lepowsky, *Calculus of twisted vertex operators*, Proc.Natl.Acad.Sci.USA **82** (1985), 8295–8299.

[M]   P. Montague, *Codes lattices and conformal field theories*, Cambridge University Ph.D. dissertation, 1991.

[N]   S. P. Norton, *Generalised moonshine*, Proc.Symp.Pure Math. **47** (1987), 209–210.

[Se]   J.-P. Serre, A course in arithmetic, Springer Verlag, New York, 1970.

[Sch]   A. N. Schellekens, *Meromorphic C=24 conformal field theories*, Commun.Math.Phys. **153** (1993), 159–185.

[T1]   M. P. Tuite, *Monstrous moonshine from orbifolds*, Commun.Math.Phys. **146** (1992), 277–309.

[T2]   M. P. Tuite, *On the relationship between monstrous moonshine and the uniqueness of the moonshine module*, Commun.Math.Phys. **166** (1995), 495–532.

[Va]   C. Vafa, *Modular invariance and discrete torsion on orbifolds*, Nucl.Phys. **B273** (1986), 592–606.

[Ve]   E. Verlinde, *Fusion rules and modular transformations in 2-d conformal field theory*, Nucl.Phys. **B300** (1988), 360–376.

[Z]   Y. Zhu, *Vertex operator algebras elliptic functions and modular forms*, Yale Ph.D. dissertation and these Proceedings, 1990.

DEPARTMENT OF MATHEMATICAL PHYSICS, UNIVERSITY COLLEGE GALWAY, IRELAND

SCHOOL OF THEORETICAL PHYSICS, DUBLIN INSTITUTE FOR ADVANCED STUDIES, 10 BURLINGTON ROAD, DUBLIN 4, IRELAND

E-mail address: michael.tuite@ucg.ie

Other Titles in This Series

(*Continued from the front of this publication*)

(See the AMS catalog for earlier titles)